# SUPERVISORS'
# *Safety* Manual

*10th Edition*

National Safety Council
Itasca, Illinois

Project Editor: Phyllis Crittenden
Interior Design and Composition: Jennifer Villarreal
Cover Design: Jennifer Villarreal

**Library of Congress Cataloging-in-Publication Data**
Supervisors' safety manual.—10th ed.
          p.      cm.
    Includes bibliographical references and index.
    ISBN 978-0-87912-288-1 (alk. paper)
    1. Industrial safety—United States—Management—Handbooks, manuals, etc.
    I. National Safety Council.
    T55.S8156 2009
    658.3'82--dc22
                2009027737

7.5C1012

ISBN: 978-0-87912-288-1
Product Number: 15144-0000

# Contents

# Preface

The roles and responsibilities of first-line supervisors and team leaders have changed. You are still the direct link between management and the workforce. You must still produce quality goods and/or services. You are still responsible for quality job training, employee motivation, development of good safety attitudes, and detection of hazardous conditions and unsafe work practices. And you must do all these things in an environment that is in the midst of the "information explosion."

You need to understand the techniques and psychology of human relations. You should know the fundamentals of loss control, the applicable industry standards, and federal and local safety and health regulations. You must know all of the potential hazards your workers may encounter, how to prevent them, and/or what safeguards and personal protective equipment are needed, how to use them, and how to enforce use of them. You must be able to assist in incident investigations that occur in your area of responsibility. And you should be able to recognize when your workers need to seek help for stress; problems adjusting to work-shift changes, stressful environments, or job changes; personal problems; or drug or alcohol abuse—and how to encourage them to get the help they need.

The 10<sup>th</sup> edition of the *Supervisors' Safety Manual* provides the basic information and resources you will need to meet these responsibilities and duties. It has been revised to reflect changes in organization structure and management philosophies, in particular the emphasis on empowered employees, self-directed teams, and behavior-based safety. All technical information has been reviewed, revised, and updated by experts from a variety of businesses and industries.

The text is written in easy-to-understand language with bulleted lists for quick reference. The objectives at the beginning of each chapter help to focus learning, and the summary of key points, at the end of each chapter, helps aid retention of knowledge. The Case Studies at the end of the book offer readers an opportunity to apply learning and test their knowledge.

Even though much detailed information has been included, the purpose of this book is not to serve as a complete handbook. Rather, it is designed to emphasize the important issues to be considered by the supervisors who are responsible for safety and health in their organizations.

A completely updated Supervisors' Development Program is also available to support training for your responsibilities and duties. It consists of facilitator guides, participant workbooks, videos, and leader guides. *Supervisors' Safety Manual* is the sourcebook for the Supervisors' Development Program.

## Acknowledgments

The National Safety Council gratefully acknowledges the expertise and assistance provided by technical advisors for their contributions to the 10th edition of the *Supervisors' Safety Manual*. The Council would like to thank E. Scott Geller, PhD; Alumni Distinguished Professor; Director, Center for Applied Behavior Systems; Department of Psychology, Virginia Tech, for his contributions on behavior-based safety. In addition, the Council also thanks Philip E. Hagan, Director of Safety & Environmental Management, Department of Human Sciences, Georgetown University, for his work on the Regulatory Issues chapter.

The Council also recognizes with appreciation the technical reviews provided by Council staff, including:

David J. Amos

John Czerniak

JoAnn Dankert

Don Ostrander

Wes Scott

# Introduction

As modern production processes have become more highly sophisticated, so has our understanding of how to create a safe, healthful work environment. In the past, "accident prevention" simply meant freedom from serious injuries and property loss. Today, however, "loss control" covers not only injury but occupational disease and environmental concerns along with fire and property damage control. This new understanding of safety is reflected at all levels of management from the CEO to the line supervisor or team leader. Loss control is considered so essential a function that in most companies safety performance is included as part of a supervisor's or team leader's overall job evaluation.

As a result, today's supervisor or team leader must become skilled in three aspects of loss control: (1) learn to recognize hazards, (2) learn the acceptable level of risk for department operations, and (3) learn how to control hazards to prevent injuries, illnesses, and property damage. The National Safety Council's *Supervisors' Safety Manual* addresses the skills a supervisor or team leader will need to successfully perform these responsibilities.

## Supervisor: Manager of Safety Problems

The supervisor or team leader, as a member of the company's management team, shares responsibility for maintaining a safe, productive workplace. He or she must communicate and enforce rules and procedures, train workers, and represent the interests of both the organization and the employees.

The supervisor or team leader must constantly watch over and inspect both the workplace and work procedures, keeping in mind the three Es of safety: engineering, education, and enforcement. It is the supervisor's job to work with safety and health professionals, designers, engineers, maintenance, and personnel staff to engineer as many hazards out of the workplace as possible, to educate employees in safe work practices and procedures, and to enforce all safety rules and policies. In this role, the supervisor or team leader acts as investigator, safety researcher, and advocate.

For example, if an employee complains of headaches while performing a degreasing procedure, it is the supervisor's responsibility to investigate the situation. It may be that a degreasing chemical is causing the employee's headaches. If an investigation by the safety and health professional shows this is the case, there are three options: eliminate the hazard by substituting a different chemical, isolate the hazard by installing surface or point-source ventilation, or compensate for the hazard by issuing personal protective equipment.

The supervisor, or team leader, may be involved in determining which of the three options is the best solution to the problem. For example, in this case, personal protective equipment is desirable only when all other options have failed. The equipment does not remove the hazard—it only decreases the risk for the worker.

Once the best solution has been chosen, the supervisor may be involved in presenting the plan to upper management for approval of the required action. At this point, the supervisor or team leader becomes an advocate, presenting the benefits of the plan on behalf of the workers as well as the organization. *Supervisors' Safety Manual* is designed to take you step-by-step through the process of learning how to manage and control safety problems.

The supervisor or team leader manages safety problems by managing people, equipment, and environment in the workplace. To assist you in these tasks, this book has been organized as follows:

Chapters 1 through 7 focus on the basic safety knowledge and skills that every supervisor or team leader must acquire: an understanding of loss control and the fundamentals of safety management, communication, the behavior-based safety approach, employee safety and health training, inspection, and incident investigation. These chapters orient the supervisor to his or her management responsibilities on the job.

Chapters 8 through 12 cover what a supervisor must know to meet OSHA and other regulatory standards to maintain a safe, healthful workplace: industrial hygiene, personal protective equipment, ergonomics, hazard communication, and common regulatory issues. Understanding the worker-machine-environment relationship can help to significantly reduce hazards, incidents, and job-related illnesses.

Chapters 13 through 17 focus on the general principles for safeguarding, maintaining, and using tools and equipment safely to prevent incidents, and reducing property damage and loss through an understanding of electrical and fire safety. In these chapters, you will learn how to make the day-to-day production process safer for all concerned and how to work with various safety experts from within and outside the company to solve safety-related problems.

No safety program can eliminate all the risks and hazards from a job, but *Supervisors' Safety Manual* can help you come as close to this ideal as possible. The more you know about the hazards in your workplace and how to control them, the more everyone in the organization will benefit.

# 1

## Safety Management

**After reading this chapter, you will be able to:**

- Recognize significant events in the safety, health, and management fields and determine how these influence your organization's safety management focus
- Recognize fundamental principles of safety management
- Define key safety terms and explain concepts that relate to safety management
- Differentiate between insured and uninsured costs
- Define the five phases of continuous improvement
- Identify the strategies involved for creating large-scale organizational change

Managing safety starts from the premise that most incidents can be prevented. It includes the premise that *incidents* are near accidents and that so-called *accidents* are not random events, but rather preventable events. They can be prevented with proper hazard identification and evaluation, management commitment and support, preventive and corrective procedures, monitoring, evaluation, and training. Team leaders and supervisors are important links in the chain of safety consciousness. They are responsible for watching over the workplace and protecting their employees from faulty equipment, carelessness, and the many other potential hazards on the job.

## TRENDS IN SAFETY MANAGEMENT

Occupational injuries and illnesses have been known to occur in human activity for as long as there have been workplaces, and have been documented in historical records long before the construction of the Egyptian pyramids. However, the concept that work-related injuries and illnesses could be reduced through safety management is relatively new. Only in the past century have work groups paid much attention to organized or systematic safety practices.

### Changing Focus

Until the beginning of the twentieth century, many owners, and society at large, took a fatalistic view of safety. Much of the history indicates that they saw incidents as regrettable but not preventable. As farms failed, farm workers and new immigrants swelled the city labor markets, providing plenty of workers willing to risk life and limb for a decent wage. But tragedies like the Triangle Shirtwaist Company fire on March 25, 1911, in which more than 146 workers—all women and many of whom were minors—died because of locked fire exits, led social reformers and labor unions to push for more humane conditions in the workplace.

**Physical Workplace.** By 1910, the reformers succeeded in getting workers' compensation laws passed. Now employers had a greatly increased financial stake in making the workplace safer, and general support for safe workplaces was increasing in society at large. In 1913, the National Safety Council was established to provide research and information on how to increase safety.

The 1920s were a time characterized by interest in scientific management.

Many companies still treated their employees as interchangeable parts but also recognized that improved safety led to improved productivity. Many employers began inspecting the physical work space, keeping it clean and monitoring working conditions. In the 1930s, H. W. Heinrich wrote the first significant book on modern safety management called *Industrial Accident Prevention*. Companies began to study not just physical workplaces, but also human contributions to incidents. The emphasis was on safety on the factory floor at this point because industrial incidents tended to be far more numerous and more serious than those in office settings.

**Health Issues.** In the 1940s, the focus of safety management was broadened to include prevention of not only incidents but also illnesses. Workers' compensation was updated to cover disability caused by work-related disease and industrial hygiene problems. Management style, however, was still in the "command-and-control" mode, a legacy of the military model.

In the 1950s management adopted the behavioral approach and turned to technology for solutions. Companies began to apply management and motivational principles in an effort to change workers' behavior. At the same time, some came to believe that technology had all the answers. Emphasis was on engineering—prevention of safety and health problems by redesigning or modifying equipment.

**Environmental Issues.** During the 1960s, national concerns about safety broadened to include the office environment, and government began to recognize many serious environmental risks. In the late 1960s, workers' compensation laws were also updated to include compensation for illness and disability caused by noise.

In 1970, the Williams-Steiger Occupational Safety and Health Act was passed, and the federal government established the Occupational Safety and Health Administration (OSHA) in 1971. OSHA was given the power to monitor companies' compliance with newly legislated safety and health standards, and impose heavy fines on those companies not in compliance. Safety and health efforts began to reflect employee participation in management.

For at least two decades, gradual changes in safety and health management occurred until the 1980s, when priorities shifted from hazard identification to hazard prevention. Next came the inception of a general movement toward proactive, rather than reactive, safety and health management. In the 1990s, safety experts worked to create systems to identify and eliminate workplace hazards. The focus is now on integrating safety, health, and environmental management throughout the business strategy. Every employee, from CEO to forklift driver, is now seen as responsible for safety. Most managers today accept that good safety, health, and environmental programs contribute to the health of the company's bottom line.

## Safety Management

The understanding of what causes incidents has changed over the years. At one time, safety experts believed in the *domino theory*. That is, they thought a mishap or loss was caused by one specific unsafe act or condition and that removal of that act or condition would solve the problem.

The overly simplistic domino theory has been replaced by the *theory of multiple causation*. Safety experts now try to look for all of the factors that may have contributed to a given incident. Studying all contributing causes that may have led to the loss helps them to identify the root or key causes.

*Frequency* refers to the number of times a specific incident occurs. *Severity* describes how serious the consequences are (which may be measured for injuries and illnesses by the number of affected employees). Some safety experts once concentrated on reducing only high-frequency incidents, but modern risk assessment considers both variables.

## Changing Management Styles

At one time, the role of managers and supervisors was to dictate what should be done. The role of employees was to follow those orders unquestioningly. Companies have come to recognize that management does not have "all of the answers" and that other employees are capable of contributing far more to the success of the business than was once believed. Just as employees are sharing responsibility for manufacturing and customer service and other "line" decisions, they are also participating more in safety and health decisions. This approach invests all employees in making the workplace safer.

As employees are permitted more participation in making workplace decisions, they tend to take more and more responsibility for their own safety and health. Current management styles of empowerment, total quality management, and continuous improvement recognize that safety is everyone's job, and that the people most likely to spot problems on the job are those who are performing the job. It is becoming more common for decisions to be made by the people who are "closest to the action," which results in empowerment.

Some American corporations have concluded that the use of self-directed work teams is among the most productive ways to organize the workforce. In successful teams, each member buys in to the team's and the company's goals. No matter what product or service the company provides, one of those goals is always to achieve a safe working environment. Each employee's job description should spell out accountabilities for safety and health. In truly effective companies, senior management has committed to the safety and health policy and addressed it in the business plan.

## PRINCIPLES OF SAFETY, HEALTH, AND ENVIRONMENTAL MANAGEMENT

The management of safety, health, and environmental programs has evolved over the years. Some forward-thinking companies now have programs that include written principles that guide safety and health efforts. These principles are called *Key Results Areas*, or KRAs.

### Definitions

To make sense of these principles, it helps to define the terms relating to the topic of safety.

**Acceptable Level of Risk.** A zero-risk environment does not exist. The goal is to reduce risk to an "acceptable" level. Risks are acceptable if judged to be tolerable. For any operation to succeed, risks must be acceptable.

**Control.** A process intended to control or eliminate future hazards by taking preventive action or controlling or eliminating current hazards by taking corrective action. Methods of controls include engineering, administrative, and personal protective equipment.

- **Engineering controls.** Engineer the hazard out by initial design specifications or by applying substitution methods, isolation, enclosure, or ventilation. In the hierarchy of control methods, engineering control methods should be considered first.
- **Administrative controls.** Reduce employee exposures through methods such as education and training, work reduction, job rotation, maintenance/repairs, housekeeping, personal hygiene, and appropriate work practices. Administra-

tive controls depend on constant employee implementation or intervention.

- **Personal protective equipment (PPE).** A device worn by employees to protect them from the environment. PPE includes anything from gloves to full body suits with self-contained breathing apparatus and can be used with engineering and administrative controls.

**Hazard.** Any existing or potential workplace condition that, by itself or by interacting with other variables, can result in death, injury, property damage, or other loss.

**Incident.** Any unplanned event that results in personal injury or damage to property, equipment, or the environment; or an event that has the potential to result in such consequences. Within this context, incident includes all occurrences, regardless of degree, with the purpose of focusing safety control strategies on the elements of causation.

**Incidence Rate.** Incidence rate, as defined by OSHA, is the number of occupational injuries and/or illnesses or lost workdays per 100 full-time employees. The formula for calculating incidence rates is:

$$\frac{\text{number of injuries and illnesses} \times 200{,}000}{\text{total hours worked by all employees during the period covered}}$$

**Injury.** Physical harm or damage to the body, such as a cut, fracture, sprain, amputation, etc., in the work setting. An injury results from an exchange, usually acute, of mechanical, chemical, thermal, or other environmental energy that exceeds the body's tolerance.

**Prevention.** A proactive process to keep something from occurring.

**Risk.** A measure of the probability and severity of adverse effects.

**Risk Assessment.** The ranking of risks by three criteria: exposure, severity, and probability. In addition, it is an effective means of prioritizing actions.

**Safety.** The control and elimination of recognized hazards to attain an acceptable level of risk. Figure 1–1 illustrates the multiplier effect of unrecognized hazards.

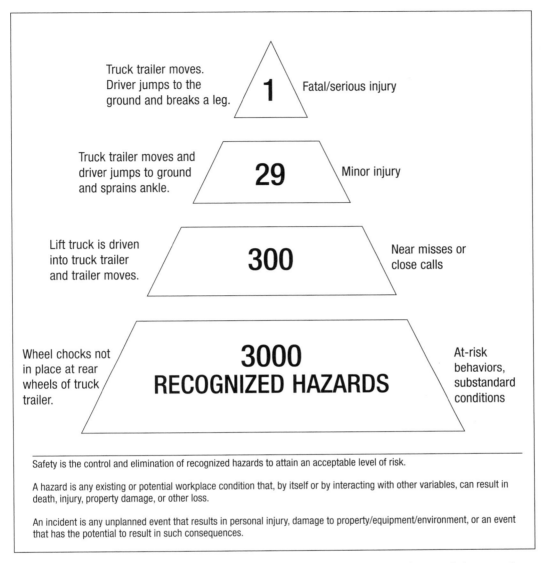

**Figure 1–1.** If left unrecognized and uncontrolled, the hazards that produce today's no-injury or minor-injury incidents will produce tomorrow's disabling injury and/or serious equipment damage. All hazards are important. All should be recognized and evaluated for control action.

Effective safety efforts involve the control and elimination of recognized workplace hazards to attain an acceptable level of risk and to promote the wellness of workers. Optimal safety results from a continuous proactive process of anticipating, identifying, designing, implementing, and evaluating risk-reduction practices.

**Safety Management System.** A safety management system is an organized and structured means of ensuring that an organization (or a defined part of it) is capable of achieving and maintaining high standards of safety performance. The management system should be based on the principles of continuous improvement.

## THE PREVENTION PRINCIPLE

The prevention principle states that the best safety and health management systems don't just eliminate the hazards; they also seek and eliminate the causes of hazards. In doing this, the organization builds a solid foundation for prevention of mishaps.

### Accountability

In order to be effective, employees throughout an organization must be held accountable for their safety and health results. This means that they will be measured and evaluated based on their results in identifying and eliminating workplace hazards and mishaps.

### The Human Factor

Most safety experts agree that most mishaps in the workplace involve human errors of omission or commission. A major step toward improving safety is eliminating those factors within the organizational system that contribute to human error. This includes not only worker errors but also errors by designers, engineers, purchasing agents, supervisors/team leaders, and anyone else.

As industrial designers have become more aware of safety issues, they have begun to invest in and design more provisions for safety: automatic shutoffs for equipment, foolproof operating mechanisms, and ergonomic workstations, for example. In *Turn Signals are the Facial Expressions of Automobiles*, Donald A. Norman points out that a multimillion-dollar jet plane's performance should not depend on the stability of a three-cent polystyrene cup sitting on top of a badly located but crucially important lever in the cockpit (Norman 1992, 167). A custom-designed part in a well-located area may cost more, but will also provide the required control reliably.

Organizations cannot afford to wait until they can build facilities, processes, and equipment to design for safety. They must modify or replace existing ones that are hazardous in any way. Designing for safety has the effect of designing for efficiency, and even productivity, since it reduces employee fatigue, injuries, and illnesses. Safe workplace design reflects physical and psychological compatibility among the employee and the process, methods of operation, equipment, materials, and machinery.

## Business Integration

In the past, safety and health programs and efforts have not been effectively integrated into the business strategy. As a result, they haven't been as successful as they could be. Safety professionals and company chief executives are now seeing the benefit of integrating safety efforts and making them a part of the essential business operations. Like other management processes, safety and health efforts should be planned, budgeted, measured, and evaluated.

## Frequency and Severity

*Frequency* refers to the number of times a particular loss or mishap occurs. *Severity* refers to the seriousness of the loss or mishap. In the past, some safety professionals concentrated their efforts on reducing high-frequency incidents with the belief that overall incident severity would also decrease. This belief has proven to be inaccurate. Modern risk-assessment methods recognize that frequency and severity measure different aspects of incident occurrence and both must be considered when setting loss control priorities.

# INITIATING A SAFETY MANAGEMENT SYSTEM

The term safety management means an organized and systematic approach to managing safety efforts. It identifies, controls, and eliminates the causes of hazards, ultimately supporting good practices and success. Because the scope of safety issues is vast, it requires a carefully planned system, structure, and process to make it successful. Overall, the main route to excellence in safety performance is an effective safety management system.

## Creating a Safety Management System

In creating a safety management system, a company must first clarify and establish its safety and health philosophy, beliefs, and vision or mission. Through these efforts, a culture that promotes safety and health is established.

Once a company has its vision, planners must identify specific elements that support it and design strategies for putting it into action. Then managers must create policy and procedures (steps) to define the who, what, when, where, why and how of achieving it. The process is repeated as the organization builds on successes and learns from less successful efforts.

## Organizing the Safety Management System

A comprehensive safety management system should give equal consideration to the administrative, operational and technical, and cultural issues of safety and health.

The National Safety Council's *Nine Elements of a Successful Safety Management System* provides the structure and organization necessary for a comprehensive and balanced safety management system. It gives equal consideration to the administrative, operational and technical, and behavioral issues necessary for a comprehensive system. Figure 1–2 and Table 1–A show how the elements of NSC's safety management systems are organized according to these basic categories.

## Continuous Improvement of a Safety Management System

The Continuous Improvement Process Model shown in Figure 1–3 illustrates how continuous improvement works. The model is a framework for applying a safety management system on a continuous basis. Applying the process that forms the model requires recognizing causes before designing solutions.

There are five steps in the continuous improvement process for implementing a successful safety management system (see Table 1–B).

**Figure 1–2.** An effective safety management system should give equal consideration to administrative, operational, and cultural safety and health issues.

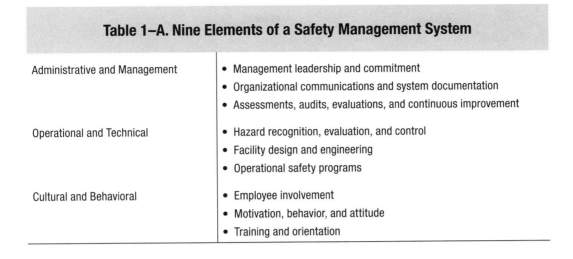

| Table 1–A. Nine Elements of a Safety Management System | |
|---|---|
| Administrative and Management | • Management leadership and commitment<br>• Organizational communications and system documentation<br>• Assessments, audits, evaluations, and continuous improvement |
| Operational and Technical | • Hazard recognition, evaluation, and control<br>• Facility design and engineering<br>• Operational safety programs |
| Cultural and Behavioral | • Employee involvement<br>• Motivation, behavior, and attitude<br>• Training and orientation |

## Budgeting for Safety, Health, and Environmental Management

A budget is a financial plan that shows the funds an organization will need to achieve its goals. Budgets generally describe how much money, time, and material will be spent, and on what. They also show how much money will be earned and from what sources. Budgets are prepared annually, quarterly, monthly, and in other increments, as needed.

There are many expenses involved in running a comprehensive safety, health, and environmental effort. They include everything from incident investigation costs to eyewash fountains, from printing of newsletters to training and incentives. It is important to remember that the effort yields many savings as well. A cost-benefit analysis can prove to management that the program is worth the expense. The question is not whether the program has costs, but whether it is preventive and produces benefits greater than the costs.

In order to convince management to invest in training, equipment, or material, safety professionals detail the reasons why these investments are needed. They can calculate the costs of accidents or injuries, including direct and indirect costs. Direct costs include medical, legal, insurance premium increases, injury compensation, and time and material costs to repair equipment or tools. Indirect costs include hiring and training replacements for injured workers; time lost when other employees help with, watch, discuss, or otherwise react to the incident; machine or equipment downtime; accident investigation costs (interview and administrative time); overtime required to catch up after the incident; and the cost of supervisory time spent handling it.

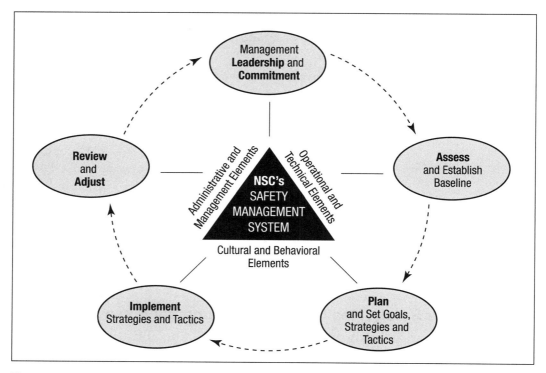

**Figure 1–3.** The continuous improvement model is a framework for applying the safety management system on a continuous basis.

The managers then compare the purchase cost to the cost of one incident, to see how many prevented incidents it takes to pay for the expected costs for training, equipment, or material. They determine the likelihood that the purchase will prevent incidents and the number of potential preventable incidents to calculate the potential dollar cost of loss prevention in the first year. Then they use the company's profit margin to figure how much product the business must sell to make up for the loss. Typically, it takes only a few prevented incidents to pay for the purchase. If the purchase would likely prevent many incidents, it effectively saves the firm money.

Total incident costs can be compared to an iceberg, as shown in Figure 1–4. Experts calculate that for every dollar of direct accident costs, there may be $4 to $10 of insured costs, $5 to $50 of uninsured ledger costs, and $1 to $3 of uninsured miscellaneous costs.

Safety pays in two ways. First, it results in fewer disabling injuries, lower workers' compensation costs, and lower replacement costs. Second, it frees up managers from time spent managing safety crises, so that they can concentrate on quality; productivity; proactive safety, health, and environmental issues; and competitiveness in the market.

## Table 1–B. The five steps in the continuous improvement process

| | |
|---|---|
| Provide Management Commitment and Leadership | • Executive Management must drive the effort<br>• Create a vision, set performance standards, define roles, responsibilities, and accountabilities<br>• Establish systems that measure and evaluate performance<br>• Communicate the vision |
| Assess—Establish a Baseline | • Use NSC's nine elements to assess your current status and determine where to direct future efforts<br>• Communicate the findings |
| Plan—Set Goals, Strategies, and Tactics | • Set measurable improvement goals<br>• Align safety improvement plans with business priorities<br>• Communicate the goals |
| Implement—Strategies and Tactics | • Make the plan happen<br>• Provide training when and where needed<br>• People need to understand that this is as much a part of their jobs as is production<br>• Acknowledge success |
| Review and Adjust | • Learn what worked and what didn't work<br>• Discard what didn't work<br>• Assess again and repeat the process |

# SAFETY, HEALTH, AND ENVIRONMENTAL MANAGEMENT IN TEAM-BASED ORGANIZATIONS

Traditionally, most businesses were organized as hierarchies. Decisions were handed down from managers at the top of the pyramid. Front-line workers had very little say in how they did their jobs, and no say at all in how the company was run. Managers were responsible for their departments and supervisors acted mainly as enforcers for managers. They assumed that if workers were given any autonomy, they would duck out of as much work as possible.

In recent years, however, there has been a major change in the attitudes and theories of American businesses. In many of today's businesses, the managers no longer dictate orders. They share power with their employees, who tend to respond with increased productivity because they are part of the team.

Businesses today are experimenting with several approaches to organizing their workforce. One of the most effective methods is self-directed work teams (SDWTs).

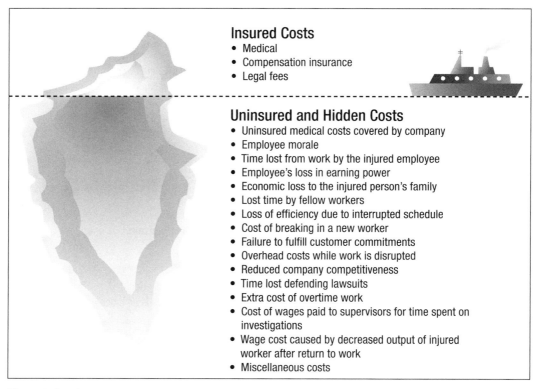

**Insured Costs**
- Medical
- Compensation insurance
- Legal fees

**Uninsured and Hidden Costs**
- Uninsured medical costs covered by company
- Employee morale
- Time lost from work by the injured employee
- Employee's loss in earning power
- Economic loss to the injured person's family
- Lost time by fellow workers
- Loss of efficiency due to interrupted schedule
- Cost of breaking in a new worker
- Failure to fulfill customer commitments
- Overhead costs while work is disrupted
- Reduced company competitiveness
- Time lost defending lawsuits
- Extra cost of overtime work
- Cost of wages paid to supervisors for time spent on investigations
- Wage cost caused by decreased output of injured worker after return to work
- Miscellaneous costs

**Figure 1-4.** Like an iceberg, the hidden costs of incidents are not visible on the surface but are there just the same.

## Characteristics of Team-Based Organizations

A self-directed work team is organized as a group of employees from different functions and departments who meet periodically (perhaps daily) to pool their knowledge and collaborate on creative problem solving. The team is ongoing, so its members may work together for several years. The members are multiskilled and responsible for a whole task or core process. They take ownership of their successes and failures. They constantly cross-train each other, and are always learning and improving. Their range of responsibilities is always increasing, often encompassing duties that were once the province of supervisors. They are empowered to make decisions on scheduling and vacations; to set goals and performance targets; and to interact with customers, vendors, and others. Their rewards are tied to their performance.

Self-directed work teams (also known as high-performance teams) have become popular because of their track records. Many of them have achieved higher quality, lower costs, faster service to customers, higher morale, and greater worker satisfaction than the old hierarchy-based systems. Many companies say that their people are their most important resource, but many of those who have reinvented themselves into self-directed teams can demonstrate that they mean it.

According to Ann and Bob Harper's *Team Barriers*, there are 10 key elements to high-performing teams:

1. shared goals and mission
2. climate of trust and openness
3. open and honest communication
4. sense of belonging
5. respect for differences
6. continuous learning and improvement
7. ability to measure and self-correct
8. interdependence
9. consensus decision making
10. participative leadership

**Cultural Values of Teams.** The self-directed work teams must value productivity above all else. They achieve it by being flexible, imaginative, and motivated. They believe that knowledge is power, and that power is most useful when shared rather than hoarded. This system requires company management to provide feedback to everyone, handing out productivity reports, cash-flow statements, and customer complaints. Every employee has a right to this information; after all, in order to help meet the company's goals, they must know what these goals are.

The SDWT system also values autonomy, respect, and dignity. It is assumed that people can be trusted to do their jobs without external monitoring and that all employees will treat each other as responsible, intelligent adults.

**Leadership Traits.** One thing that the SDWT system has proven is that leadership does not depend on a degree from the right school or from a formal background in corporate management. Anyone can be a leader, given enough training and enough self-confidence. In fact, more and more, leadership duties are shared in rotation among all the members of a team. Each team member is responsible for a designated area (usually operations, training, safety, scheduling, human resources, communication, or quality control) for a specific time period.

For example, one person receives training in safety leadership and, for the next several months, attends all facilty-wide safety meetings as the team's representative. Then a co-worker is trained in safety, and takes over the team's safety leadership role. This approach gives all team members a chance to develop leadership abilities, and also to gain experience in interaction with other teams.

Team-based organizations know that, as W. Edwards Deming said, the job of management is not supervision but leadership. Leading does not mean barking orders, but rather providing guidance, information, encouragement, and inspiration. Leading means motivating employees to share the organization's vision and to strive to achieve it.

Leaders act as role models. By their own behavior, they show employees how to behave and what to value. Role models live up to the highest standards of the company. Leaders are role models every day, on a continuing basis, whether they want to be or not. In addition, they also coach and teach others both practical skills and details of the corporate culture.

## Empowered Employees

*Empowered employees* are those who have the freedom to make many of their own decisions and the training to make good ones. Their jobs are designed to provide ownership of complete task areas and responsibility for their success. Their tasks are defined so that they are responsible for a meaningful process or output, have the power to make decisions and commit organizational resources, and continually measure their own progress. They share authority, resources, and recognition for the work process and results. They receive the skills training they need to do their jobs well, which is tremendously empowering.

Empowerment tends to give workers a reason beyond a paycheck for going to work. It enriches their work life and promotes job satisfaction. Employees who are empowered do more than just stamp out machine parts or answer telephones or blindly follow a limited set of steps. They are in a position to implement the organization's vision and to understand how they personally contribute to achieving it.

For this system to work, employees must be treated honestly and with respect, and management must recognize that front-line employees are the organization. As such, these employees want to know what's really going on. They need information on all aspects of the business, not just their own small segment. They need to know about costs, budgets, quality standards, customers, vendors/suppliers, and competitors. When top management shares all this information freely, the teams are more likely to support its actions.

To be empowered, people need to be trained and educated continuously in three areas:

- technical training (their own skills, new skills, quality, and safety)
- interpersonal/team training (communication, team development, conflict management, leadership skills, decision making, problem solving, etc.)
- administrative skills training (for the tasks that supervisors used to do, like scheduling, budgeting, and interviewing)

It has been demonstrated that employees who have the power to fix problems on their own do so faster and better than when they have to go through layers of management for approval. This leads them to take more pride in their work and counteracts the dehumanizing effects of severely limited, boring, and repetitive tasks. This system is based on the belief that empowerment becomes a way of life and is not just the management "fad of the month."

## Employee Involvement in Safety

Some companies have a safety leader whose job is to get employees involved in the safety, health, and environmental program. In companies where the team system is used, this work is done through management supervisors and team leaders who are responsible for motivating people to care about safety. Employee safety committees, contests, newsletters, and posters are some of the tools they can use. The safety expert may work as an "inside consultant" to provide direction and technical support for others.

The following steps would be helpful for anyone creating a safety, health, and environmental team:

1. Create a character vision for yourself and live it daily.
2. Respect others and accept their differences.
3. Involve others in the process of creating your safety and health vision and goals.
4. Encourage participation by inviting, and listening to, opposing points of view.
5. Involve others in solving, and making decisions related to, safety and health problems.
6. Build an atmosphere of respect, trust, and inquiry.
7. When others need help, the safety, health, and environmental professional provides coaching, resources, support, and assistance.
8. Give recognition and rewards often and consistently (see Chapter 5, Employee Involvement).

## Maintaining Interest and Motivation

Motivation involves moving people to action that will support the company's desired goals. In occupational safety and health, motivation increases employees' awareness, interest, and willingness to act in ways that improve their own safety and that of co-workers. Motivation aims primarily at changing behavior and attitudes. It is generally defined by three factors:

1. direction of behavior
2. intensity of action
3. persistence of effort

Motivation efforts should support the mainline safety, health, and environmental system, not take its place. Chances of success increase when the following factors are present:

- management demonstrates its commitment at every opportunity
- the program is energized through slogans, performance recognition, and discipline
- workplace conditions are safe and healthful
- tools, equipment, and workplace layout are well-designed
- maintenance is effective
- training and supervision are abundant and effective

**Job Safety Analysis.** A job safety analysis can lay the groundwork for behavior change because employees participate in identifying hazards and finding ways to eliminate them. When all workers are involved in formulating safety needs and devising a safety plan, each one has a greater stake in carrying out the plan. Such participation should be reinforced through an open communication system that welcomes employees' safety concerns and responds to them promptly. (For instructions on how to do job safety analyses, see the section "Job Safety Analysis" in Chapter 5, Employee Involvement.)

**Employee Surveys.** Employee surveys help to identify safety priorities and build employee morale. They also provide a baseline for later evaluations of the effectiveness of safety initiatives. Group feedback is important. When a work group accepts safety improvement objectives, its members reinforce each other's behavior through peer pressure.

**Fear.** Many supervisors believe that fear can sometimes be an effective motivator. This strategy attempts to change attitudes about the risks of hazardous behaviors by first instilling fear with statistics, photographs, or other information, and then reducing the fear by providing ways to prevent the danger or lower the risk. Fear messages may work better with new employees than with seasoned workers, who are likely to want to use their own experiences to discount the message. ("For years I've checked to see if circuits were energized by quickly touching them with the back of my hand, and I've never gotten injured!") Fear messages are also more effective on employees who are not under direct supervision because they are on their own and may tend to take the information more seriously.

**Total Quality Management Approach.** One approach to motivating employees that has been effective is the Total Quality Management (TQM) model. TQM works to change employees' attitudes in order to achieve a change in behavior. For example, it strives to train employees to an awareness of the importance of safety. This awareness will lead them to be more careful about always wearing safety equipment and checking the automatic shutoffs on their tools, for example.

TQM emphasizes process, or root cause, improvements. Such process changes flow from workers' involvement and empowerment as both individuals and teams. People are far more likely to support change if both the objectives and the methods used to achieve them are based on their own recommendations, instead of being imposed by management.

Nothing less than a cultural change is called for in order to make TQM work. Management must give up trying to "manage" employees to conform to existing systems. Instead, managers must be willing to completely revamp processes and systems. They must realize that employees are in the best position to know how those processes could work better. The corporate culture must be flexible and responsive to employees' needs. It must engender trust and cooperation between labor and management.

TQM is more difficult and time-consuming than simple reinforcement/feedback motivational techniques. Its effects are, however, more wide-ranging and they last much longer.

## Continuous Improvement Process of the Safety Management System

The continuous improvement process of the safety management system emphasizes the contributions people make to long-range, permanent solutions to problems. Essentially, the process shows how to effectively manage change in a safety management system. As previously discussed, there are five steps in the continuous improvement process:

1. provide management commitment and involvement
2. assess and establish a baseline
3. set goals
4. implement strategies
5. review and adjust

Continuous improvement is the emphasis of total quality management. Because everyone is always working to improve something, the only constant is change. This is also one impediment to a successful TQM program.

Constant change is somewhat uncomfortable because, among other things, it fosters a sense of uncertainty and insecurity. People like to establish a pattern and stick to it. Their natural resistance to change must be overcome if TQM is to be fully successful.

To apply the continuous improvement process, companies must understand causes before they try to design solutions. Improvements, whether dramatic or incremental, occur regularly. The nine components of effective safety, health, and environmental programs listed earlier in the chapter facilitate the continuous improvement process.

## Barriers and Challenges to Team-Based Organizations

Although the team structure has proven its effectiveness again and again, some organizations are still skeptical. There are many barriers that can either keep a company from trying teams or can sabotage the structure if it is undertaken halfheartedly. Among them are the following:

- lack of trust
- a climate of fear
- the uncertainty of constant change
- not involving all of the key stakeholders
- not involving unions (forming a partnership)
- lack of top leadership's demonstrated commitment
- lack of top management responsibility or role
- not allowing enough time to change

- resistance from any of the partners (supervisors, managers, support people, unions, or labor force)
- inadequate training
- systems and structures not designed to support teams
- workers who zero in on their co-workers' imperfections
- failure to redefine the role of leadership
- leaders who won't let go
- too little or too much structure
- no transition plan
- failure to provide work security assistance
- failure to communicate what's happening
- treating change like a program instead of a process
- overwhelming team members with too much responsibility before they have had adequate training
- a history not conducive to employee involvement
- failure to educate everyone in the workplace about self-direction
- failure to understand that TQM self-direction is a total cultural transformation, not just a modification
- amount of time and effort involved

The long-term improvements that teams generate can more than offset these problems. The potential of the self-directed work teams is to boost productivity and competitiveness, enhance people's quality of life, and turn many jobs from drudgery to fulfilling work in which the worker is integrally involved.

## INFLUENCING AND MANAGING CHANGE

The team-based, empowered, continuous-quality workplace is changing all the time. On the one hand, that's a good thing because change is what enables the organization to remain viable and keep up with the competition. On the other hand, people tend to crave stability, and find constant change difficult. They may resist even changes for the better because the known is less threatening than the unknown.

Employees at all levels, especially upper management, may fight change in several ways: logically, emotionally, and through group influence. Even if their real reason is to avoid emotional turmoil, they may be quite good at rationalizing logical reasons why the change is a mistake. When employees offer data that suggest the change may not work, top management must reassure them and explain the benefits the change will bring them.

Sometimes only one or a few people in the group feel strongly opposed to change, and they manage to convince the whole team to resist. The key is to identify those lead resistors and then listen to their reasons. People resist change imposed by someone else, but if the change is their idea, or they have contributed to it in even a small way, they are more likely to welcome it.

Change often precedes perception. Sometimes people cannot understand why they should follow a particular procedure until they try it. Then it begins to make sense to them. Change involves breaking old habits, both individual and organizational, and replacing them with new ones. Such disruption can be wrenching, so people need an incentive to make the effort. They need to be convinced that the change will make their lives better. Without such a conviction, they will just be going through the motions.

Change has a ripple effect. When it occurs in one area of the organization, it will affect other areas. Change reaches critical mass slowly, and then speeds up.

## Strategies for Managing Large-Scale Change

To restructure an entire company, or even a single department, top executives need a change strategy that encompasses a period of months. They should get and use input from as many people as possible before even beginning to design the change. Then they should solicit support from as many people as possible, beginning where they know they have the most support.

Executives need to clearly communicate expected positive results of the change. They will need to plan and communicate the logistics as well. Communications and public relations pieces should be designed to integrate the new, desired culture.

Executives should pace the implementation. A common mistake is to try to make everything happen at once. Executives should study the organization's dynamics and then use them to create change. They need to provide training and education about the change. A checklist for managers will show them how to support and enhance the process.

The change masters should set the example. They should model the behavior they are trying to achieve, whether it be sharing information or wearing protective gear on the factory floor. At the same time, they need to be patient. It is the nature of human behavioral change to start slowly.

## Steps for Influencing Individual Change

Supervisors and team leaders can find themselves in the position of coaxing individuals to embrace change. A meeting to coax an individual to embrace a change should be well-planned, using the following five steps:

Step 1: Plan ahead. Consider beforehand what the person's leadership style is and whether he or she is supportive, neutral, or unsupportive of your goals. How will the change benefit this individual? Practice your meeting with a supportive colleague.

Step 2: Begin the influencing meeting positively. Create a relaxed environment. State your reason for meeting and gain agreement on the goal of the meeting. Establish trust.

Step 3: Discuss the change. Lay out your ideas. Then discuss the other person's concerns. Are they logical, emotional, or group influenced? Deal with those concerns respectfully.

Step 4: Engage in mutual problem solving. List the areas where you agree and those where you disagree. Share important criteria and priorities. Design a solution that addresses both your priorities and the other person's.

Step 5: Gain commitment. If the person is not yet willing to support the change, ask for a commitment to try it out briefly. If the individual won't agree to that, ask for a commitment to just think about what you've said. Think of yourself as a low-key salesperson here.

## Common Mistakes When Influencing Change

Here are common but critical mistakes to avoid when you try to influence change:

- acting without seeking input
- seeking input but ignoring it
- acting without completing all of the proper groundwork
- forgetting to see how the change might be a threat to others
- not tuning into, "What's in it for me?"
- becoming impatient
- ignoring the benefits of small, incremental changes
- trying to change too much too fast

## SUMMARY OF KEY POINTS

Key points covered in this chapter include:

- The focus of safety has changed and expanded over the years to encompass not just the physical workplace but also health and environmental issues. Safety experts know now that hazards have multiple causes, and that both the frequency and the severity of incidents must be controlled.

- All near misses and minor incidents should be investigated. They are an early warning system that can help prevent serious incidents later on. The ultimate goal of any safety program is loss control.

- All employees must be accountable for maintaining safe working conditions and practices. They must remember that safety is everyone's job.

- The safety, health, and environmental policy should be integrated into the company's business plan and supported by top management.

- The National Safety Council has identified nine elements of a safety management system. They include administrative and management, operational and technical, and cultural and behavioral components.

- Indirect costs add significantly to the costs of accidents. Safety experts can run a cost-benefit analysis to demonstrate to management why certain training, equipment, or material is worth the investment.

- Team-based organizations give their employees a great deal of autonomy and a corresponding amount of responsibility. Teams are becoming increasingly prevalent because they can improve quality while reducing costs.

- Empowered employees are those who have the freedom to make their own decisions and the training to make good ones. They receive constantly updated training in technical, interpersonal/team, and administrative skills.

- Companies and team leaders can motivate employees through job safety analyses, in which they participate in devising a safety plan, and employee surveys. Fear can sometimes be an effective motivator, depending on the audience. Total quality management is a good motivator, but requires a cultural change in the organization.

- Continuous improvement is a process-oriented approach that involves five steps for implementing a successful safety management system.

- There are many barriers to the team-based structure, but the payoff is believed to be worth the time and effort it takes to overcome them.

- The team-based, empowered, continuous-quality-improvement workplace is changing all the time. Many people are inclined to resist those changes, but motivational techniques for implementing them can be used both throughout the organization and with individuals.

# 2

# Communication

**After reading this chapter, you will be able to:**

- Identify the components of the communication process
- Analyze advantages and disadvantages of several methods of communication
- Identify ways to improve listening skills
- Identify several interviewing and consulting principles

Communication means sharing information and ideas with others and being understood. Without understanding, there is no communication. An essential element of any safety and health program is good communication. Safety professionals must communicate the program's content to supervisors and team leaders, who must, in turn, effectively communicate to their teams. They must be able to effectively explain their plans for preventing incidents and creating a safer workplace.

It is important to remember that communication is a two-way street. Feedback from employees is critical. Involving employees in all aspects of the safety program will increase employee commitment to carrying it out. If management simply hands down rules, some employees may spend more time and energy trying to outwit those rules than following them. Besides, no one knows the operation or the hazards of a particular job better than the person who accomplishes it.

Giving employees a voice when they disagree with the safety policy already in place is crucial. Employees may feel that safety is not emphasized enough, or they may think a specific procedure doesn't work well. By listening to this feedback, the team leaders and supervisors will gain important insights. Listening is every bit as important a communication skill as speaking. In fact, research shows that most effective supervisors and managers spend 50% more time listening than speaking.

## THE COMMUNICATION PROCESS

Communication is a process that involves a sender, a receiver, a message, and feedback (see Figure 2–1). All components of the process are essential to achieving effective communication. Effective communication occurs when there is a shared understanding between the sender and receiver. It is important to know that these components can create a surprising number of problems. The method of communication that the sender chooses may be unclear, the form of the message may be unappealing, or the receiver may erect barriers that block or filter the message.

Supervisors and team leaders can benefit from using open, two-way communication and from soliciting feedback. They must take into account the receivers' possible filters and try to overcome them.

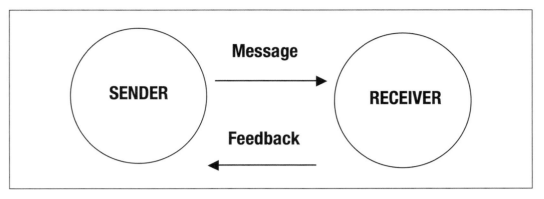

**Figure 2–1.** Effective and successful communication occurs when there is shared understanding between the sender and the receiver.

## Open Communication

An open communication system encourages the free exchange of ideas. It is a bit like the Internet; every member of the department is connected to each other. The supervisor/team leader acts more as a facilitator than a disseminator of information. Employees' opinions are respected in an open communication system. Because of this, it's a good morale builder.

The open system is far more effective (and more realistic) than the old closed or restricted communication system. In a closed communication system, supervisors insisted that all communications pass through them and discouraged employees from talking to each other. The closed system is more about control than it is about communication.

## Feedback

To be effective, a communication system must be two-way; it must provide for feedback. Feedback ensures the sender that the message has been received based on the verbal and visual response to the message. It also provides a way to check on whether the message has been understood.

Communication improves if the receiver has a chance to ask questions and express what he or she perceives the message to be. The sender can then identify any misunderstandings and can adjust the communication to resolve them. Depending on the situation, the best type of communication to use on the job is usually one-on-one and face-to-face.

For example, Marianne Osborne is a supervisor at a fast-food restaurant. She explained to a new employee the steps to take in serving a customer. Then she followed up by saying, "Just to be sure I've covered everything, can you repeat what I said?" When the new employee left out a step, she knew what needed to be repeated or clarified.

Good supervisors also solicit feedback in their written communications. An example is this sentence at the end of an e-mail: "Please contact me by Friday to give me your view of these proposals."

Feedback improves the receiver's understanding. It also helps the sender improve future communications. A supervisor who gets the same questions several times learns what he or she needs to clarify in a message.

Open-ended "you-based" questions can be especially productive because they can build the ego of the person answering the question and may also help to reassure him or her that there are no wrong answers. Examples are, "How do you feel about this approach?" and "What do you like about your job?" Contrast that with, "Is there enough light in this work area?" There's no wrong answer, but there's a very short answer: yes or no. Open-ended questions elicit more information and help to empower the person who answers.

### Communication Needs

The organization needs frequent, informal communication with employees to update them on progress toward safety goals or to ask them to take action. It is important to be brief. Some examples of communication needs include the following:

- Company policy may require a written summary of information.
- Senior management and key staffers in other areas of the company need to be informed about the status of various safety and health programs or told about progress toward previously published goals.
- A team needs an official summary of safety meeting minutes detailing what actions the team has taken and whether they are effective or need adjustments.
- Formal communication is needed with groups or individuals outside the organization.
- Employees need information about safety and health. The results of safety and health programs must be presented to large groups of employees and to management.
- Employees need ongoing reminders of key safety tips and general safety awareness.
- A team, a manager, or a whole department needs to illustrate the results of a safety program or progress toward goals for other employees.

- Employees often need more information than can be presented verbally in a meeting, or the information presented at meetings needs to be reinforced, or an employee who is not able to attend the meeting needs to receive the information.
- Management needs to provide ongoing reinforcement of training and communication of routine safety information.
- Managers must communicate information required by regulatory agencies.
- Companies may make public statements regarding safety information.

## Ways to Communicate

Safety professionals, supervisors, and team leaders need to use various formats to communicate information. Some possible formats include the following:

- An *e-mail* is a good option. For example, a safety professional might send weekly e-mails to people in his or her department to update them on the progress toward a safety goal. An e-mail should be brief, bulleted, and limited to one main topic.
- A *formal report* presents comprehensive and specific information in detail. A typical report includes benefits (such as reductions in workers' compensation costs) of safety and health programs, costs, and relevant statistics.
- *Meeting minutes* provide an ongoing history of events and are an important part of a safety and health program.
- *Letters* that detail relevant information about a safety and health program are often sent to other companies or community groups.
- An *informal talk* is a good way to present such information as safety tips, demonstrations (how to use personal protective equipment), or results of special safety programs. The talk can be supplemented with literature, such as employee booklets.
- A *formal presentation* may be needed. When making a formal presentation, be sure to plan carefully and to use visual aids (charts, graphs, PowerPoints) to summarize key information. A printed handout enhances retention of information.
- *Posters* can provide general safety awareness and remind employees of key safety tips. While they are a valuable addition to a safety and health communication program, posters should never be used as the only method to transmit information.
- *Charts and graphs* are especially helpful in these situations. Consider posting information (especially in chart/graph form) in a cafeteria, break room, and work area.

- *Handouts* such as safety and health booklets are important links in any program. Employees can read them at their own pace. Multilingual versions can help fill the gaps for employees whose first language is not English.
- *Newsletters* can be a good communication tool. They can include safety tips, changes in policies or regulations, results of safety efforts (such as the number of days since a lost-time injury), and messages from management.
- *Legally required communications*, in general, must be made in the format prescribed by the specific laws and regulations.
- *Material Safety Data Sheets* and *HAZMAT* materials can be used for employees responsible for safe handling, transportation, or emergency response to incidents involving hazardous materials.
- Safety information can be included in other public communications, such as *annual* and *quarterly reports* or *press releases*.

## Communication Filters

Communication filters are perceptions—preconceptions, biases for or against various actions, special points of view, and perspectives—through which the receiver interprets messages. Such filtering can change the message in ways the sender never intended. They can even block communication entirely. The more a sender knows about the audience, the better he or she can predict and overcome these filters.

Before supervisors begin to speak or write a message to their employees, they should consider their audience (their receivers) carefully. What are their concerns, problems, knowledge levels, age ranges, biases? Mentally putting oneself in the place of the receivers is a good way to project how they will respond to a message. The supervisor who tailors messages to the audience's needs is more likely to achieve desired understanding.

Some of the most common filters through which people perceive messages are knowledge level, personal biases, moods, and physical comfort levels.

**Knowledge.** Effective communicators adapt their messages to the receiver's level of knowledge. When they are instructing a new employee in his or her first after-school job, they keep the language simple and to the point. By contrast, when they are discussing a new work procedure with an employee who has 20 years of experience, they can use more technical language. They may also acknowledge that experience by asking the employee's opinion about the procedure. He or she may have ideas for streamlining or otherwise improving the procedure.

**Biases.** Biases can have a definite bearing on how people hear and understand communications. Biases are usually based on one's past experiences and on one's attitudes, beliefs, and values. A bias may lead someone to tune out part or even all of the message. For example, consider a 60-year-old union employee listening to a supervisor who is fresh out of school, inexperienced, and perhaps of a different gender or race. This front-line worker may be inclined to assume the supervisor doesn't know anything about the topic of the communication. If the experienced worker feels this way, he or she may ignore the whole message.

These situations can be quite challenging for supervisors. An advantage of the team approach is that the group comes up with solutions instead of having them dictated by management. People often respond better to communications that show respect for their expertise and value of their knowledge and experience.

**Mood.** The receiver's mood can be a serious filter to getting a message through. Listeners who have something else on their minds may not be paying attention. An employee who is worried about the knock in his or her car engine, or the persistent cough of a child at home, may not absorb the message at all.

A partial solution here is to encourage feedback. Suppose the supervisor is warning an employee about the hazards of the task he or she is about to begin. Asking, "At this point, what are some of the problems you will be watching out for?" tests whether the receiver has heard the message. If not, the supervisor repeats it—twice, if necessary. It isn't enough to ask "Do you understand?" because most people simply reply, "yes."

Mood filters are factors in both sides of the two-way communication process. Supervisors must never let their own problems distract them so much that they fail to hear what their employees are saying. People are more inclined to listen to people who listen to them.

**Physical Limitations and Barriers.** Physical limitations, such as poor hearing or vision, can inhibit communication. The limitation does not need to be severe to have an effect. For example, an employee might not realize or want to admit he or she needs glasses and thus may miss information posted on a bulletin board or TV monitor. Or, an employee might have difficulty hearing, but is embarrassed to ask a supervisor to repeat an instruction. Asking the employee to repeat an instruction helps ensure that the message reaches its target.

There can also be physical barriers in the workplace:
- marked out territories into which strangers are not allowed
- closed office doors, barrier screens, separate area for people of different status (management located in offices in the front of a building)
- large working areas or working in one unit that is physically separate from others (geographic location, different floor in a building, remote warehouse)

Research shows that one of the most important factors in building cohesive teams is proximity. As long as people still have a personal space that they can call their own, nearness to others aids in communication because it helps us get to know one another.

**Mental Limitations.** Where people have diminished mental capacity, communication can require more direct, "hands-on" communication and subsequent reinforcement.

**Language.** Language that describes what we want to say in our terms may present barriers to others who are not familiar with our expressions, buzzwords, and jargon. When we express our communication in such language, it can be a way of excluding others. This becomes a critical consideration in the workplace where English is a second language for workers. Communication lapses can also be attributed to cultural differences (ethnic, religious, and social).

## Reinforcement

When a worker does something wrong, most supervisors may be quick to point out the mistake. When a worker does something right, however, far too many of them neglect to comment. Praising employees for a job well done is an effective way to motivate them.

Psychological studies have shown again and again that the carrot is more effective than the stick. Most people like to take pride in their work and make an effort to do it well, especially if they know they're appreciated. As long as praise is sincere, there is no such thing as too much. Supervisors should always follow the rule to praise in public and reprimand in private. Embarrassing workers by pointing out their mistakes in front of others does far more harm than good.

# METHODS OF COMMUNICATION

There are several methods of communication, including verbal, nonverbal, written, and visual images. Different methods are appropriate in different situations. Supervisors may choose to combine the methods. For example, they may explain job procedures face-to-face, pointing out the hazards and showing operators how to do the jobs safely. They may also give workers a completed job safety analysis form that reinforces these points.

## Verbal Communications

Verbal communications have the advantage of being immediate and tailored to the individual listener(s). It is also less formal than written communication.

Verbal communications may take the form of a "tailgate" meeting in which the team leader or supervisor discusses with several people the job they are about to start. The group members have the opportunity to ask questions during the explanation. This immediate feedback can help to ensure that everyone knows what to do. If the workers ask no questions, the group leader should ask questions to make sure they understood the explanation.

Face-to-face communications work best for day-to-day liaison, exchange of information, periodic status reports, and production meetings. Telephone communications are fine for the quick exchange of information, but they lack the warmth of face-to-face discussions. Verbal communications should always be two-way. Team leaders and supervisors can use two-way oral communications to convey information and also to build relationships with workers.

Some people outline on paper what they plan to say. They may not know exactly what they want to say until they think about it enough to write it down. The process of writing clarifies their thinking, even if they do not use their notes in the meeting.

## Written Communications

Messages that are formal, official, or have long-term impact should be written. This method is good for dealing with complicated or technical subjects because it gives the receiver something to refer to. This method allows communication without interruption and provides permanency of a message. Because written communications do not lend themselves to immediate feedback, it's a good idea to follow up with verbal questions to make sure the receiver understood the written message.

Written communications in business are generally in the form of e-mail, letters, reports, faxes, or bulletin board notices. A good example is a work order request that a supervisor sends to the maintenance department, asking it to fix a problem found on a safety inspection. A verbal follow-up to this communication might be a phone call asking when the job can be completed.

### Nonverbal Communications

Actions really do speak louder than words. Nonverbal communications account for 80% of all communication. Supervisors are watched carefully by their people, so the example they set is even more important than what they say. For example, a supervisor who tells employees how important it is to wear personal protective equipment, but then doesn't wear it, is sending conflicting messages. Guess which version employees will pay more attention to.

By the same token, no matter how much pressure supervisors are under, they should strive to not pass on those pressures. Employees who see their boss take shortcuts and work in an unsafe manner are likely to do the same, and serious incidents can result.

Nonverbal communications include body language and facial expressions. A supervisor who ensures workers that he wants to hear their opinions, but then rolls his eyes or drums his fingers while they're talking, is canceling out his spoken message of caring. Employees read and understand nonverbal messages, even unintentional ones, just as clearly as they do verbal and written communications.

Nonverbal communications work both ways. Supervisors and team leaders also notice the body language of their people. It can provide clues to how well they've received messages and to what is really on their minds.

### Visual Images

Visual images are used to convey conscious or unconscious messages. This method of communication includes posters/banners, drawings, photographs, illustrations, and paintings.

## EFFECTIVE LISTENING TECHNIQUES

So far, this chapter has concentrated mainly on the sending portion of communication. The receiving portion is equally important. In every field, effective leaders must be skilled listeners.

Certainly first-line supervisors must know how to listen. Bob Shaffer, a supervisor at a steel mill, complained, "My people just don't listen to me!" But his employees said, "Bob says he has an open-door policy, but when I try to talk to him, he just doesn't listen. Either he's busy doing something else, or he winds up doing all the talking."

Active listening intentionally focuses on who you are listening to, whether in a group or one-on-one, in order to understand what the communicator is saying. As a listener, you should be able to repeat, in your own words, what was said, to the speaker's satisfaction. This does not mean you agree with, but rather understand, what is being said.

The following techniques can help you become an active listener:

- Acknowledging. Send cues to the speaker that allow him to know that you hear him or her.
- Repeating. Say exactly what the sender of the message has said to you.
- Paraphrasing. Repeat what the sender has said to you using your own words.
- Reflecting. Think about what the sender of the message has said, and then respond to some part of his or her message.
- Clarifying. Ask questions to resolve some things that are unclear to you as the listener.
- Probing. Ask questions to get more information.
- Empathizing. Let the sender know that you feel the same way he or she does.
- Giving feedback. By sharing your opinions about the sender's particular ideas, you create a collaborative bond. Focus on affirming his or her insights and ideas, instead of on criticisms.

## The Importance of Effective Listening Skills

How important is listening in a supervisory job? The typical manager spends about 70% of the workday communicating. Studies have classified this communication as follows:

| | |
|---|---|
| Writing | 9% |
| Reading | 16% |
| Speaking | 30% |
| Listening | 45% |

Our educational system does not prepare students well for jobs involving listening, even though nearly every job involves a lot of it. Every student takes classes in reading and writing. Some students take communications and speech classes, but few schools offer much instruction in listening after the elementary grades. Yet this skill accounts for nearly half of a manager's communication time.

How good of a listener are you? Most people rate themselves as average or below-average listeners. Tests conducted at the University of Minnesota showed that immediately after hearing a 10-minute reading, the average listener retained about 50% of the material. A few days later, the retention rate had dropped to between 25% and 30%. In this test, participants were making an effort to listen well. Average listeners retain only a 25% to 33.3% of what they hear, even when they try. Clearly, we all have considerable room for improvement.

Researchers say that like people training for a marathon run, people can learn to improve their listening ability; however, they have to work at it. They must build up their skills gradually by implementing the effective listening techniques presented earlier.

Many incidents in the workplace are caused by poor listening. A worker may be distracted when the supervisor explains how to avoid the hazards of a job. Or perhaps the supervisor doesn't pay attention when the worker mentions a problem that occurred the last time such work was done. The fact is that a better understanding of a problem by both supervisors and employees can reduce incident potential.

While injury-causing incidents are the main concern, there is also the cost of having to redo work or scrapped work caused by poor listening: letters containing errors, damaged customer relations, perhaps even lost business. Improving listening skills could improve companies' bottom lines and workers' paychecks.

## Barriers to Effective Listening

There are many factors that keep people from achieving their potential as good listeners. The barriers may be words, emotions, or distractions. A look at these barriers may reveal some ways to improve in this important area.

**Word Barriers.** Different people have various "turn-off" words or phrases. When one of these is spoken, the listener stops paying attention to what is being said and starts focusing on whatever the turn-off word brings to mind. For example, the word "death" is a turn-off word for some people because it reminds them of a disturbing family tragedy. When they hear it, they may stop listening for 15 seconds—or 15 minutes. Some words that may have the potential to turn people off include layoff, panic, grievance, abortion, taxes, etc., as shown in Figure 2–2. Indeed, this turnoff phenomenon is one reason why people are often advised not to discuss politics or religion.

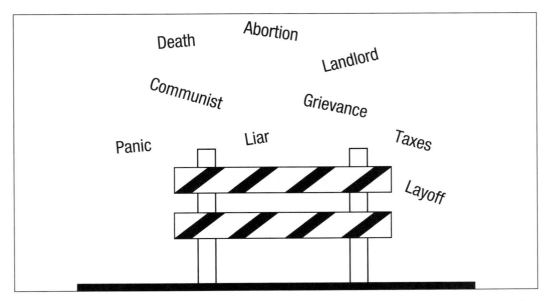

**Figure 2–2.** Turn-off words that recall emotional memories can interrupt your listener's concentration for 15 seconds or even 15 minutes. Avoid using negative words.

What words turn off your listening? To improve your listening ability, figure out which words turn you off and why they affect you, then work to overcome the problem. When you hear one of these words, concentrate on what else the speaker is saying to keep your mind from being distracted by thoughts that the turn-off word evokes. You are the only one who can overcome the effects of turn-off words on your listening ability.

**Emotional Barriers.** Emotions can also block listening. When people become angry, they concentrate on the source of the anger rather than on what is being said. They stop listening and start thinking of things to say that will support their argument and questions that will trip up the speaker.

Anger is just one emotion that can be a barrier to listening. Others include hate, fear, suspicion, jealousy, distrust, and overenthusiasm, as shown in Figure 2–3.

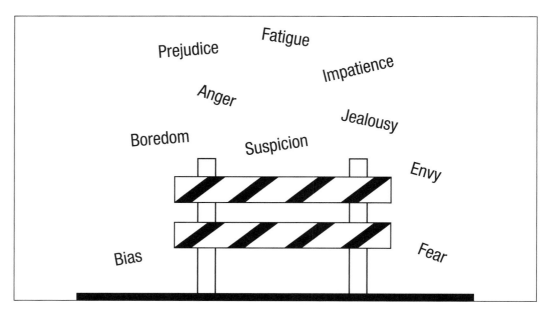

**Figure 2–3.** The listener needs to control emotional barriers in order to hear and understand the entire communication.

Keeping emotions under control can significantly improve listening skills. Even if you are angry at the speaker, or suspicious of his or her motives, try to concentrate on what is being said. Consider how emotions affect your listening during conversations with your employees and your union steward. How about with your spouse and your children? Listening skills are useful in all aspects of life. They will help you communicate better with your family and friends as well.

A good tool for situations in which anger may cloud listening is the heated discussion rule. Participants must agree to its provisions in advance. If the speaker says something you disagree with, signal that you wish to speak. When the speaker is finished speaking, he or she turns the discussion over to you. Before making your statement, you must first state the speaker's position and state why he or she feels that way about the matter. If the speaker agrees with your statement, you can then go on to state your own case. This process forces each person to listen in order to get a turn to speak.

In heated discussions, the points of disagreement are often not as serious as they appear to the combatants. When responding in a sensitive situation, begin by stating the other person's position. A good way to start is, "As I understand what you have said, you feel this way …" Actually listening to the other person's supporting comments instead of thinking of a rebuttal can do a lot to improve understanding. It may turn out that the two viewpoints are not really so far apart.

**Distractions.** Another reason for failure to listen is distractions. External noises, such as machines or equipment operating, can derail one's train of thought. Personal problems can also interfere with effective listening. Supervisors and team leaders should work to overcome the many distractions from within and without so that all members of the team can pay attention. Leaders should also look for ways to reduce distractions for the team, for example, by holding meetings in a quiet spot away from the regular work area.

## Keys to Improved Listening

Most people speak at about 125 to 150 words per minute. Yet most people can comprehend speech at speeds of 400 to 500 or more words per minute. That time differential might seem to ensure good listening. In fact, many people have so much downtime while trying to listen that their minds wander. That is one reason why they hear and remember only 25% to 30% of what is said.

Effective listeners use the time differential to think along with the speaker, mentally outlining the points. They evaluate the speaker's credibility and analyze whether his or her nonverbal messages agree with the verbal ones.

Making good use of the time differential can help people improve their understanding, which is key to communications. Here are tips for becoming an effective listener:

- Stop talking. You can't listen while you are talking.
- Repeat/paraphrase. Once the speaker is finished, repeat what you heard to verify your understanding.
- Clarify/probe. If any part of the message is unclear to you, ask questions until you understand it.
- Maintain eye contact. Meeting the speaker's eyes helps you concentrate on what is being said. It also shows that you are paying attention.
- Empathize. Put yourself in the speaker's place to get a better understanding of why he or she holds the opinions expressed.
- Share responsibility for communication. The receiver is just as responsible as the sender for good communication.
- React to the message. Do not react to the sender or to the delivery of the message.

There are also some don'ts to remember when improving your effective listening skills:

- Don't interrupt.
- Don't criticize the speaker.

- Don't tune out because you think the subject is too dry or boring.
- Don't make assumptions.
- Don't answer too quickly.
- Don't jump to conclusions (think before you respond).
- Don't fake attention.

# INTERVIEWING AND CONSULTING

Safety professionals and supervisors often need to gather information from employees at all levels of the organization. They may need to conduct incident investigations, plant inspections, safety audits, job safety analyses, hiring interviews, and internal consultations. Their communication skills play a large role in the success of all of these missions.

## Principles of Interviewing and Consulting

- *Think of safety and health as a service function.* Instead of seeing your safety and health department as an unwelcomed enforcer, think of the people in your company as your internal customers. Find out what they need and give it to them; be their internal consultant regarding the safety and health needs that they express.
- *Think win-win.* Far from conflicting with other organizational goals, safety and health priorities can enhance them. In the long run, safety does not cost; it pays.
- *Build a relationship on joint problem solving.* Focus on your internal customers' problems and concerns, not just your own. Working together leads to win-win solutions.
- *Build trust and respect over the long haul.* Trust does not occur overnight. It takes time to build a reputation for being fair, honest, and dependable. Expect to earn people's respect and to learn to respect them in turn.
- *Find facts, not fault.* Don't accuse people or look for someone to blame. Instead, show that you want to work with them to resolve the problem.
- *Listen more than you talk.* You can't meet people's needs if you don't know what they are. It is difficult to know what their needs are if you talk more than you listen. Let your internal customers do at least 75 % of the talking.

## Steps for Conducting Interviews and Consultations

Once supervisors and safety professionals master the principles of interviewing and consulting, it is time to think about the actual interviews. Following are five steps that can lead to outstanding interviewing results (see Figure 2–4):

Step 1: Prepare for the interview or consultation. Create an objective, prepare your questions, and have a plan of action. Plan and rehearse your opening, and eliminate distractions. Make sure you allow enough time for the interview. Know your internal customers' leadership style and prepare for it.

Step 2: Open the interview. Be on time and introduce yourself. Use small talk to break the ice. Explain the purpose of the interview and how the information will be used.

Step 3: Conduct the interview. Request a brief overview of the interviewee's job. Use vocabulary that the interviewee can understand. Treat the interviewee as an equal. Move from the general to the specific. Gather detailed information. Summarize the information obtained. Engage in mutual problem solving.

Step 4: Close the interview. Stick to the agreed-upon time limit. Summarize information gathered. End with an agreed-upon plan of action. Close on a positive note. Thank the interviewee for his or her time and cooperation.

Step 5: Document interview results. Record name, title, and job description of person(s) you spoke to. Take notes during the interview. Summarize the outcome.

The following questions may be helpful when interviewing individuals within an organization:

- What are your priorities?
- What are three things our department could do to help you in your safety and health efforts?
- What are your most pressing operational problems right now?
- What does your team think is causing the problems?

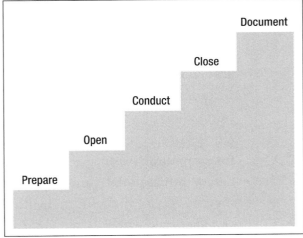

**Figure 2–4.** These five steps help you achieve more effective interviews.

- How can we help? What do you want us to do?
- How would your employees like to be rewarded for their safety and health efforts?

## SUMMARY OF KEY POINTS

Key points covered in this chapter include:

- Effective communication is essential for incident prevention. How well you send, receive, and understand messages will determine your success.
- Communication involves a sender, a receiver, and a message that is understood by both. Two-way, face-to-face communication is the best way to convey messages on the job. It allows the receiver to ask questions. This feedback lets the sender know that the message has been understood. Messages can be filtered by the receiver's level of knowledge, as well as by their biases or moods. It is important to tailor the message to the audience. It is also important to give consideration to the medium you choose to communicate your message.
- Verbal communication is immediate and personal. Written communication can be kept as a reference, so it is good for technical or complex messages. Nonverbal communication (both body language and facial expressions) should reinforce verbal communication. Visual images can be used to convey powerful conscious or unconscious messages.
- Listening is critical to effective communication. Sensing, interpreting, evaluating, and responding to messages are steps in the listening process. People usually hear only 25% to 30% of what is said, but they can double that if they work at it.
- Listeners should overcome word and emotional barriers and avoid distractions. Six rules for better listening are to stop talking, repeat/paraphrase, clarify/probe, maintain eye contact, feel empathy, and share responsibility for communicating.
- Safety professionals draw on all their communication skills in interviewing and consulting with their internal customers. Five steps for conducting a good interview are (1) prepare for the interview, (2) open the interview, (3) conduct the interview, (4) close the interview, and (5) document interview results.

# 3

# Behavior-Based Safety: A Practical Approach

**After reading this chapter, you will be able to:**

- Explain five specific benefits of behavior-based safety (BBS)
- Describe the seven basic principles that define BBS and guide the development of interventions to improve the human dynamics of a work culture and thus prevent workplace injuries
- Conduct an objective and a comprehensive behavioral analysis of an undesirable workplace incident
- Describe the ABC model of BBS and explain its relevance to increasing safe behavior and decreasing at-risk behavior
- Customize a critical behavioral checklist for conducting an injury-preventive interpersonal coaching process
- Define three types of behavior-change interventions and strategies and determine which is most appropriate for particular situations
- Distinguish between "other-directed," "self-directed," and "automatic" behavior and explain the relevance of this categorization to developing self-accountability

This chapter was contributed by E. Scott Geller, PhD; Alumni Distinguished Professor; Director, Center for Applied Behavior Systems, Department of Psychology, Virginia Tech.

The term *behavioral safety*, or *behavior-based safety*, has become quite popular among safety professionals, consultants, and members of safety steering committees. It is commonly used to reflect a proactive approach to injury prevention by focusing attention on at-risk behaviors that can lead to an injury and on safe behaviors that can contribute to injury prevention. However, beyond this general definition there seems to be much confusion or misunderstanding about behavior-based safety (BBS).

The BBS approach is much more than a simple tool or set of procedures. It does more than improve safety-related behavior. It can be the impetus to cultivating a total safety culture. It is not "fuzzy" and does not leave management out of the picture. It is key to helping people transition from a dependent paradigm about occupational safety to an interdependent perspective, and from performing safety-related behaviors because of an external accountability system to a self-directed responsibility mindset. Furthermore, this special behavior-based approach can be effectively applied beyond safety to increase both the quantity and quality of workplace performance.

## THE SPECIAL BENEFITS OF BEHAVIOR-BASED SAFETY

There are specific benefits of a successful BBS process. It goes without saying that injuries will be prevented. Because most injuries are caused in part by at-risk behavior, a reduction in at-risk behavior and an increase in safe behavior will lead to injury prevention. However, it is useful to consider other benefits that result from people contributing interdependently to an effective BBS process. These outcomes are critically important and relate to much more than safety. In fact, they can improve every important function of your organization. The five specific benefits of a successful BBS process are as follows:

1. It focuses evaluation on the right numbers. How is safety performance measured at most facilities? It's measured by final outcomes—injuries that occur during the year. Companies keep score by trying to improve

**Figure 3–1.** Safety prizes based on avoiding injury reward inaction. *Source: Courtesy of Make-A-Difference, LLC.*

their total recordable injury rate (TRIR). Do workers walk around the job thinking about lowering the company's TRIR? Can they relate to that? Of course not. It's too abstract and remote. And it's not really under workers' immediate control.

Figure 3–1 reflects this misguided emphasis on outcomes rather than process. Obviously, the idea of a dead person receiving a safety award is ridiculous. But the type of incentive/reward program illustrated here is quite common. The participants in such a program don't need to do anything in particular to win the safety prize, except perhaps avoid reporting an injury. As such, this type of incentive program can stifle the very thing necessary to lower the TRIR—proactive conversations about situations and actions that need to be changed in order to prevent injuries. Interpersonal learning and coaching is vital to injury prevention, but this can be hindered by a focus on the destination instead of on the journey.

Top management needs to keep worrying about the outcome numbers, but not the people doing the jobs on the floor. Workers need to focus on the process—the day-to-day operations. That is what they can control, and that is the focus of BBS. When line workers concentrate on what they can actually do for safety, they will reach the outcome everyone wants—fewer injuries.

2. It builds positive attitudes. Have you ever noticed how safety-related conversations often resemble an adult-child confrontation? One person holds another accountable for at-risk behavior, as in: "I saw what you did." Then safety becomes a "gotcha game," reinforced by corporate rules and OSHA regulations. This heavy-handed approach only diminishes a worker's feelings of empowerment, importance, belonging, and interpersonal trust. The supervisor in Figure 3–2 might mean well, but the context of the conversation and the attitude of the worker makes the interaction seem confrontational.

BBS focuses on the use of rewards, positive feedback, and interpersonal recognition to motivate and support safe behavior. This encourages workers to get involved in a safety

**Figure 3–2.** Attitude and work culture can be detrimental to constructive feedback. *Source: Courtesy of Make-A-Difference, LLC.*

**Figure 3–3.** External accountability can provoke assertions of personal freedom. *Source: Courtesy of Make-A-Difference, LLC.*

improvement process because they want to, not because they feel threatened or think they have to. Thus, through BBS, people act themselves into a positive attitude.

3. It increases personal responsibility for safety. People often think "accountability" and "responsibility" is the same thing. They are not. When you are held accountable, you are asked to reach a certain objective or goal, often within a designated time period. But that doesn't mean you feel any personal responsibility to meet the deadline. You might feel responsible enough to complete the assignment, but that's all. You will do only what is required and no more. In fact, some work cultures influence some people to try and beat the system, as depicted in Figure 3–3.

Supervisors must understand that when people have tools they can use on a daily basis to prevent injuries, and they have support to use these tools, they have real control over safety. If workers are held accountable for process numbers they can control, and they believe their efforts will prevent workplace injuries, they will feel responsible and do more than what is required. They will feel empowered and want to get involved in an improvement process (Geller 1998a, 2001b). Guidelines for developing a safety accountability system that promotes personal responsibility are discussed in detail, later in this chapter.

4. It facilitates interpersonal coaching and teamwork. Imagine a workplace where everyone coaches each other about the safest way to perform a job. When workers depend on others in this way to improve safety, they understand teamwork. They appreciate how everyone's safe and at-risk behavior influences the safety of everyone. With this interdependent attitude, they are willing to use behavior-based coaching to actively care for their co-workers.

As depicted in Figure 3–4, the five letters of COACH reflect the main ingredients of behavior-based coaching: "C" is for care. Know I care and you'll care what I know. "O" is for observe. I care so much that I'm willing to watch you work, so I can give you behavioral feedback. "A" is for analyze. I'll think about my observations to understand barriers to safe behavior. "C" is for communicate. I'll recognize and support the safe behavior I see, and I'll give corrective feedback for at-risk behaviors in a way that is accepted by the person I observe.

"H" is for help. Behavior-based coaching helps increase safe behavior and decrease at-risk behavior, and this helps to prevent injuries. It also helps to build interpersonal trust and an interdependent mindset (Geller 1998b, 2001b). Much more on interdependency, and on building interpersonal trust and teamwork for safety is addressed later.

5. It teaches and promotes systems thinking. It's easy to get bogged down with handling immediate short-term demands—like production deadlines—and lose sight of the bigger picture. Systems thinkers take a broad and long-term perspective. They look beyond immediate payoffs, like the ease, speed, or comfort they get by taking a risky shortcut. They consider the possibility of a bigger payoff in the distant future. They realize that their safe behavior teaches others by example and protects them from injury.

**C** are

**O** bserve

**A** nalyze

**C** ommunicate

**H** elp

**Figure 3–4.** The letters of COACH reflect a critical BBS process. *Source: Courtesy of Make-A-Difference, LLC.*

The only happy camper in Figure 3–5 illustrates this systems-thinking perspective. Unlike the win/lose individualistic perspective of everyone else in this typical scenario of the hustle and bustle of our everyday lives, our hero sees the bigger picture. His wider view allows for an appreciation of the marvelous interdependent transportation system. He perceives we're all in this together, with winning or losing dependent on our ability to mutually support the system.

Systems thinkers understand the link between behavior and attitude. A small change in behavior can result in a beneficial change in attitude, followed by more behavior change and then more attitude change—eventually resulting

**Figure 3–5.** Systems thinking benefits attitude, behavior, and culture. *Source: Courtesy of Make-A-Difference, LLC.*

in total commitment. So, BBS sets the stage for systems thinking and interdependent teamwork, and this leads ultimately to a total safety culture (TSC).

The successful applications of BBS generally adhere to the seven principles described below. Each principle is broad enough to encompass a wide range of practical operations, but narrow enough to guide the development of cost-effective procedures for managing the human dynamics of safety and health. The research and theoretical foundations of each principle is reviewed, as well as practical ramifications for preventing workplace injuries.

## FOCUS INTERVENTION ON OBSERVABLE BEHAVIOR

The BBS approach is founded on behavioral science as conceptualized and researched by B.F. Skinner (1953). Experimental behavior analysis and applied behavior analysis emerged from Skinner's research and teaching, and laid the groundwork for numerous therapies and interventions to improve the quality of life of individuals, groups, and entire communities (Goldstein and Krasner 1987). Whether working one-on-one in a clinical setting, or with work teams throughout an organization, the intervention procedures always target specific behaviors in order to produce constructive change. In other words, BBS focuses on what people do, analyzes why they do it, and then applies a research-supported intervention strategy to improve what people do.

All interventions to improve safety target behavior. Whether using training, feedback, injury investigation, coaching, or incentives to benefit safety, supervisors must focus on behavior. Why? Well first you can be objective and impersonal about behavior. You can talk about behavior independently from people's opinions, attitudes, and feelings.

Behavior changes according to factors in the external world, including equipment design, the management system, the behaviors demonstrated by others, and various social dynamics. An open discussion about the environmental and interpersonal determinants of safe versus at-risk behavior can lead to practical modifications of the work culture to encourage more safe behavior and less at-risk behavior.

Figure 3–6 illustrates the traditional outcome-based versus behavior-based approach to safety. Tracking the number of injuries or number of days without an injury is not diagnostic with regard to the contributors to an injury. Yet, this is the most common information used to evaluate the impact of an industrial safety program. The BBS approach observes and tracks ongoing work practices in order to learn what behaviors need to be maintained or changed in order to prevent injuries. Also, the environmental barriers to safe work practices are evaluated and changed to make safe behavior more convenient, comfortable, or acceptable. When behaviors are evaluated and tracked, people learn what to do to improve the outcome numbers.

**Figure 3–6.** BBS focuses on upstream work practices rather than downstream outcomes. *Source: Courtesy of Make-A-Difference, LLC.*

## LOOK FOR EXTERNAL FACTORS TO UNDERSTAND AND IMPROVE BEHAVIOR

Skinner did not deny the existence of internal determinants of behavior (such as personality characteristics, perceptions, attitudes, and values). We obviously do what we do because of factors in both our external and internal worlds. However, given the difficulty in objectively defining internal states or traits, it is far more cost effective to identify environmental conditions that influence behavior and to change these factors when behavior change is called for.

Examining external factors to explain and improve behavior is a primary focus of organizational behavior management (Gilbert 1978). In occupational safety, this approach is a BBS analysis, involving a search for answers to the following questions summarized in Figure 3–7.

### Can the Task be Simplified?

Before designing an intervention to improve behavior, it is critical to implement all possible engineering "fixes." Consider the many ways the environment could be changed to reduce physical effort, reach, and repetition. In other words, entertain ways to make the job more user-friendly before deciding what behaviors are needed to prevent injury. This is, of course, the rationale behind ergonomics and

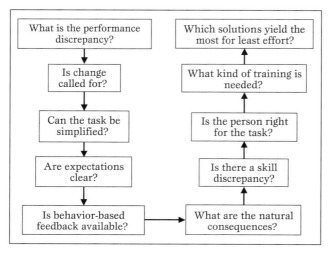

**Figure 3–7.** The 10 sequential questions asked in a BBS incident analysis. *Source: Courtesy of Make-A-Difference, LLC.*

the search for engineering solutions to occupational safety and health.

Sometimes behavior facilitators can be added, such as control designs with different shapes so that they can be discriminated by touch as well as sight, clear instructions placed at the point of application, color codes to aid memory and task differentiation (Norman 1988), or convenient machine lifts or conveyor rollers to help with physical jobs. In addition, complex assignments might be redesigned to involve fewer steps or more people. To reduce boredom or repetition, simple tasks might allow for job swapping. Ask these questions at the start of a BBS analysis:

- Can an engineering intervention make the job more user-friendly?
- Can the task be redesigned to reduce physical demands?
- Can a behavior facilitator be added to improve response differentiation, reduce memory load, or increase reliability?
- Can the challenges of a complex task be shared?
- Can boring, repetitive jobs be cross-trained and swapped?

## Is a Quick Fix Available?

From their more than 60 combined years of analyzing and solving human performance problems, Mager and Pipe (1997) concluded that many discrepancies between real and ideal behavior can be eliminated with relatively little effort. More specifically, behavior might be more at-risk than desired because expectations are unclear, resources are inadequate, or feedback is unavailable. In these cases, behavior-based instruction or demonstration can overcome informal expectations, and behavior-based feedback can enable continuous improvement. A work team should decide what resources are needed to make a safe behavior more convenient, comfortable, or efficient. When conducting this aspect of a BBS analysis, ask these questions:

- Does the individual know what safety precautions are expected?
- Are there obvious barriers to safe work practices?

- Is the equipment as safe as possible under the circumstances?
- Is protective equipment readily available and as comfortable as possible?
- Do employees receive frequent behavior-based feedback related to their safety?

## Is Safe Behavior Punished?

In some work cultures, the interpersonal consequences for reporting an environmental hazard or minor injury are more negative than positive. After all, these situations imply that someone was irresponsible or careless. It is not unusual for people to be ridiculed for wearing protective gear or for using an equipment guard. It might even be considered "cool" or "macho" to work unprotected and take risky short cuts. The hidden agenda might be that "only a 'chicken' would wear that fall protection." Mager and Pipe (1997) refer to these situations as "upside-down consequences," and suggest that they are the cause of many, if not most, of the undesirable behaviors occurring in the workplace. Ask these questions during your BBS analysis:

- What are the consequences for desired behavior?
- Are there more negative than positive consequences for safe behavior?
- What negative consequences for safe behavior can be reduced or removed?

## Is At-Risk Behavior Rewarded?

At-risk behavior is often followed by natural positive consequences, including comfort, convenience, and efficiency. Shortcuts are usually taken to save time and can lead to a faster rate of output. Therefore, taking an at-risk shortcut may be considered "efficient" behavior. Most people perform the way they do because they expect to achieve soon, certain, and positive consequences, or they expect to avoid soon, certain, and negative consequences. People take calculated risks because they expect to gain something positive and/or avoid something negative. Ask these questions:

- What are the soon, certain, and positive consequences for at-risk behavior?
- Does a worker receive more attention, prestige, or status from co-workers for at-risk rather than safe behavior?
- What rewarding consequences for at-risk behavior can be reduced or removed?

## Are Extra Consequences Used Effectively?

Because the natural consequences of comfort, convenience, or efficiency often support at-risk over safe behavior, it's often necessary to add extra consequences. These usually take the form of incentive/reward or disincentive/penalty programs. Unfortunately, many of these programs do more harm than good because they are implemented ineffectively (Geller 1996, 2001b). Disincentives are often ineffective because they are used inconsistently and motivate avoidance behavior rather than achievement. In addition, safety incentive programs based on outcomes can stifle employee participation in the development and administration of an effective BBS incentive program (Geller 2002). Ask these questions when analyzing the impact of using extra consequences to motivate improved safety performance:

- Can the punishment consequences be implemented consistently and fairly?
- Can the safety incentives stifle the reporting of injuries and near misses?
- Do the safety incentives motivate the achievement of safety-process goals?
- Do monetary rewards foster participation only for a financial payoff and conceal the real benefit of safety-related behavior—injury prevention?
- Are workers recognized individually and as teams for completing process activities related to safety improvement?

## Is There a Skill Discrepancy?

What about those times when the individual really doesn't know how to do the prescribed safe behavior? The person is "knowingly at-risk." This situation might call for training, which is a relatively expensive approach to corrective action. Mager and Pipe (1997) claim that most of the time, undesirable work behavior is not caused by lack of knowledge or skill. People can usually perform the recommended safe behavior if the conditions and the consequences are right. So training should be used relatively infrequently for corrective action. Ask these questions to determine whether the behavioral discrepancy is caused by a lack of knowledge or skill:

- Could the person perform the task safely if his or her life depended on it?
- Are the person's current skills adequate for the task?
- Did the person ever know how to perform the job safely?
- Has the person forgotten the safest way to perform the task?

## What Kind of Training Is Needed?

Answers to the last two questions can help pinpoint the kind of intervention needed to eliminate a skill discrepancy. More specifically, a "yes" answer to these questions implies the need for a skill-maintenance program. Skill maintenance might be needed to help a person stay skilled, such as when police officers practice regularly on a pistol range to stay ready to use their guns effectively when necessary. This is the rationale behind periodic emergency training. People need to practice the behaviors that could prevent injury or save a life during an emergency. Fortunately, emergencies don't happen very often; but because they don't, people need to go through the motions just to "stay in practice." Then if the infrequent event does occur, they will be ready to do the right thing.

A very different kind of situation also calls for skill-maintenance training. This is when certain behaviors occur regularly, but discrepancies still exist. Contrary to circumstances requiring emergency training, this problem is not lack of practice. Rather, the person gets plenty of practice doing the behavior ineffectively or unsafely. In this case, practice does not make perfect but rather serves to entrench a bad (or at-risk) habit.

Vehicle driving behavior is perhaps the most common and relevant example of this second kind of situation in need of skill-maintenance training. Most drivers know how to drive a vehicle safely, and at one time performed most of their driving behaviors safely, as prescribed in their driver training classes. For many drivers, however, safe driving practices decrease considerably over time, sometimes dropping out of a driving repertoire completely.

Practice with appropriate behavior-based feedback is critical for solving both types of skill discrepancies. However, if the skill is already used frequently but has deteriorated (as in the driving example), it is often necessary to add extra feedback intervention to overpower the natural consequences that have caused the behavior to drift from the ideal. (BBS techniques for providing this critical component of injury prevention are presented later in this chapter.) Ask these questions to determine whether the cause of the apparent skill discrepancy is due to lack of practice or lack of feedback:

- How often is the desired skill performed?
- Does the performer receive regular feedback relevant to skill maintenance?
- How does the performer find out how well he or she is doing?

### Is the Person Right for the Job?

A skill discrepancy can be handled in one of two ways—change the job or change the behavior. The first approach is exemplified by simplifying the task, while the latter approach is reflected in practice and behavior-based feedback. But what if a person's interests, skills, or prior experiences are incompatible with the job? Before investing in skill training for a particular individual, it's a good idea to assess whether the person is right for the task. If the person does not have the motivation or the physical and mental capabilities for a particular assignment, the cost-effective solution is to replace the performer. Not doing this suboptimizes work output and increases the risk for personal injury. Ask these questions to determine whether the individual has the potential to handle the job safely and effectively:

- Does the person have the physical capability to perform as desired?
- Does the person have the mental capability to handle the complexities of the task?
- Is the person overqualified for the job and thus prone to boredom or dissatisfaction?
- Can the person learn how to do the job as desired?

The bottom line: Before deciding on an intervention approach, conduct a careful analysis of the situation, the behavior, and the individual(s) involved and any observed discrepancy between desired and actual behavior. Avoid the impulse to assume corrective action to improve behavior requires training or "discipline." A behavioral safety analysis, as summarized in this chapter, will likely give priority to a number of alternative intervention approaches.

## DIRECT WITH ACTIVATORS AND MOTIVATE WITH CONSEQUENCES

This principle enables understanding of why behavior occurs, and guides the design of interventions to improve behavior. It runs counter to common sense or "pop psychology." When people are asked why they did something, they make statements like, "Because I wanted to do it," "Because I needed to do it," or "Because I was told to do it." These explanations sound as if the cause of behavior precedes it. This perspective is generally supported by a multitude of "pop psychology" self-help books and audiotapes that claim we motivate our behavior with self-affirmations, positive thinking, optimistic expectations, or enthusiastic intentions.

The fact is, however, we do what we do because of the consequences we expect to get or avoid by doing it. As Dale Carnegie (1936, 62) put it, "Every act you have ever performed since the day you were born was performed because you wanted something." Incidentally, Carnegie cited the research and scholarship of B.F. Skinner as the foundation of this motivation principle.

The important point here is that activators (or signals preceding behavior) are only as powerful as the consequences supporting them. In other words, activators tell us what to do in order to receive a consequence, from the ringing of a telephone or doorbell to the instructions from a training seminar or one-on-one coaching session. We follow through with the particular behavior activated (from answering a telephone or door to following a trainer's instructions) to the extent we expect doing so will give us a pleasant consequence or enable us to avoid an unpleasant consequence.

In other words, activators get their behavioral influence from consequences. Antecedent conditions that tell us what to do are only as influential as the con-

**Figure 3–8.** Consequences increase the behavioral influence of activators. *Source: Courtesy of Make-A-Difference, LLC.*

sequences they reflect. For example, the instructional sign in Figure 3–8 had no effect on behavior because the potential negative consequences were not realized. However, the consequential experiences depicted in Figure 3–8 will make the activator (i.e., the sign) more powerful at directing behavior in the future. This is called "learning the hard way."

This principle is typically referred to as the ABC model, or three-term contingency, with A for activator (or antecedent), B for behavior, and C for consequence. Proponents of the BBS approach use this ABC principle to design interventions for improving behavior at individual, group, and organizational levels. More than 40 years of behavioral science research has demonstrated the success of this general approach to directing and motivating behavior change. The next principle provides more specific direction for intervention design.

## FOCUS ON POSITIVE CONSEQUENCES TO MOTIVATE BEHAVIOR

Skinner's concern for people's feelings and attitudes is reflected in his antipathy toward the use of punishment (or negative consequences) to motivate behavior. "The problem is to free men, not from control, but from certain kinds of control" (Skinner 1971, 41). He goes on to explain why control by negative consequences must be reduced in order to increase perceptions of personal freedom.

To be sure, the same situation can be viewed as control by punishment of unwanted behavior or control by positive reinforcement of desired behavior. Some of the students in a university class, for example, are motivated to avoid failure (e.g., a poor grade), whereas other students are motivated to achieve success (e.g., a good grade or even increased knowledge). Which of these groups of students feel more in control of their class grade and thus have a better attitude toward the class? Of course you know the answer to this question because you can reflect on your own feelings or attitude in similar situations where you perceived your behavior as influenced by positive or negative consequences.

Years ago, John William Atkinson and his associates (Atkinson, 1957; 1964) compared the decision making of individuals with a high need to avoid failure and those with a high need to achieve success, and found dramatic differences. While those motivated to achieve positive consequences set challenging but attainable goals, those participants with a high need to avoid failure were apt to set goals that were either overly easy or overly difficult.

Easy goal setting ensures avoidance of failure, while setting unrealistic goals provides a readily available excuse for failure—termed self-handicapping by later researchers (Rhodewalt 1994; Rhodewalt & Fairfield 1991). Thus, a substantial amount of behavioral research and motivational theory justifies the advocacy of positive reinforcement over punishment contingencies, whether contrived to improve someone else's behavior or imagined to motivate personal rule-following behavior.

Regarding industrial safety, we can often intervene to increase people's perceptions that they are working to achieve success rather than working to avoid failure. Even our verbal behavior directed toward another person, perhaps as a statement of genuine approval or appreciation for a task well done, can influence motivation in ways that increase perceptions of personal freedom and empowerment. Simply put, using positive over negative consequences is critically important. It's relevant to "attitude," and many other internal dimensions of people. Think about it. How does a reward, personal recognition, or a group celebration make you feel compared to a reprimand or criticism? Both consequences can be significant with regard to behavioral impact. The difference is in the accompanying attitude or feeling state.

When positive recognition is delivered correctly, it does more than increase the frequency of the behavior it follows. It also increases the likelihood other safe behaviors will occur and that positive recognition will be used more often to benefit both behavior and attitude. The popular commonsense belief that we learn more from our mistakes than our successes is wrong. We learn more from our successes. So recognizing people's safe behavior will facilitate more learning and positive motivation than will criticizing people's at-risk behavior. Remember that only with positive consequences can you improve both behavior and attitude at the same time.

## APPLY THE SCIENTIFIC METHOD TO IMPROVE INTERVENTION

Systematic and scientific observation enables the kind of objective feedback needed to know what works and what doesn't work to improve behavior. The occurrence of specific behavior can be objectively observed and measured before and after the implementation of an intervention process. This application of the scientific method provides feedback with which improvement can be developed. Supervisors may use the acronym "DO IT" to teach this principle of BBS to employees who are empowered to intervene on behalf of their co-workers' safety and want to continuously improve their intervention skills.

This process puts people in control of improving safety-related behaviors and thereby preventing injuries. It offers a tool for solving the behavioral dimensions of safety problems. It provides objective data for exploring why certain at-risk behaviors are occurring and for evaluating the impact of interventions designed to increase safe behavior or decrease at-risk behavior. If an intervention does not produce a desired effect, it is either refined or replaced with a completely different behavior-change approach. As depicted in Figure 3–9, each letter of "DO IT" reflects one of the steps in this four-stage, continuous improvement process.

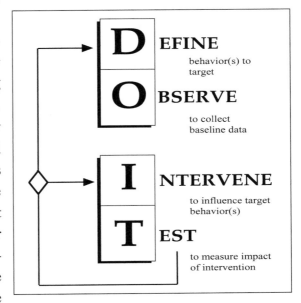

**D**EFINE behavior(s) to target

**O**BSERVE to collect baseline data

**I**NTERVENE to influence target behavior(s)

**T**EST to measure impact of intervention

**Figure 3–9.** The BBS "DO IT" method is a continuous improvement process. *Source: Courtesy of Make-A-Difference, LLC.*

## "D" for Define

The process begins by defining specific behaviors to target. They are at-risk behaviors that need to decrease in frequency or safe behaviors that need to occur more often. Avoiding at-risk behaviors often requires certain safe behaviors, and therefore safe targets might be behaviors to substitute for particular at-risk behaviors. On the other hand, a safe target behavior can be defined independently of an associated at-risk behavior. The definition of a safe target might be as basic as using certain personal protective equipment or "walking within pedestrian walkways." Or, the safe target could be a process requiring a particular sequence of safe behaviors, as when lifting or locking out energy sources.

Deriving a precise definition of a DO IT target is facilitated by considering the various categories of general operating procedures, as exemplified in Figure 3–10, and then deriving the safe versus at-risk method of performing each relevant work practice. Developing such behavioral definitions provides an invaluable learning experience. When people get involved in designing a behavioral checklist, they own a training process that can improve human dynamics on both the outside (behaviors) and the inside (feelings and attitudes) of people.

| Operating Procedures | Safe Method | At-Risk Method |
|---|---|---|
| **Body Positioning/Protecting**<br>Positioning/protecting body parts<br>(e.g., avoiding line of fire, using PPE,<br>equipment guards, barricades, etc.) | | |
| **Visual Focusing**<br>Eyes and attention devoted to ongoing task(s) | | |
| **Communicating**<br>Verbal or nonverbal interaction<br>that affects safety | | |
| **Pacing of Work**<br>Rate of ongoing work<br>(e.g., spacing breaks appropriately,<br>rushing) | | |
| **Moving Objects**<br>Body mechanics while lifting, pushing/pulling | | |
| **Complying with Lockout/Tagout**<br>Following procedures for lockout/tagout | | |
| **Complying with Permits**<br>Obtaining, then complying with permit(s)<br>(e.g., confined space entry, hot work,<br>excavation, open line, hot tap, etc.) | | |

**Figure 3–10.** A tool to facilitate defining target behaviors in a BBS "DO IT" process. *Source: Courtesy of Make-A-Difference, LLC.*

## "O" for Observe

When people observe each other for certain safe or at-risk behaviors, they notice everyone performs at-risk behavior, sometimes without even realizing it. The observation stage is not a fault-finding procedure, but is a fact-finding learning process to facilitate the discovery of behaviors and conditions that need to be changed or continued in order to prevent injuries. Thus, no behavioral observation is made without awareness and explicit permission from the person being observed. The observer should be open to learning as much (if not more) from the post-observation feedback conversation as he or she expects to teach from completing the behavioral checklist.

Regarding an observation process, teams of workers should answer the following critical questions:

- What kind of checklist should be used during one-on-one observations?
- Who will conduct the behavioral observations?
- How often will the observations be conducted?
- How will data from the checklist be summarized and interpreted?
- How will people be informed of the results from an observation process?

There is not one generic observation procedure for every situation, and the customization and refinement of a process for a particular setting never stops. It is often beneficial to begin with a limited number of behaviors and a relatively simple checklist. This reduces the possibility that some people will feel overwhelmed at the start. Starting small also enables the broadest range of voluntary participation, and provides numerous opportunities to successfully improve the process by expanding its coverage of both behaviors and work areas.

Most BBS checklists are specific to a particular task, as is the "Critical Behavior Checklist for Driving" shown in Figure 3–11. This is a checklist used to train a new driver to develop safe driving habits, or at least to demonstrate how to drive safely. The supervisor and the worker can develop such a checklist and behavioral definitions together so that there is a sense of ownership for both participants. This perception of mutual ownership increases when the supervisor allows the worker to use the checklist on him or her.

At the bottom of the checklist is a formula that can be used to summarize a driving trial. Note that it is stated in achievement terms (i.e., percent safe), and it is easy to get a high percentage or "score." This puts the whole process in an achievement mode rather than the typical failure or loss metric used to measure

| Driver: | | Day: | |
|---|---|---|---|
| Observer 1: | | Start Time: | |
| Observer 2: | | End Time: | |
| Weather: | | | |
| Road Conditions: | | | |
| **Behavior** | **Safe** | **At-Risk** | **Comments** |
| **Safety Belt Use:** | | | |
| **Turn Signal Use:** | | | |
| Left turn | | | |
| Right turn | | | |
| Lane change | | | |
| **Intersection Stop:** | | | |
| Stop sign | | | |
| Red light | | | |
| Yellow light | | | |
| No activator | | | |
| **Speed Limits:** | | | |
| 25 mph and under | | | |
| 30 mph-45 mph | | | |
| 55 mph and over | | | |
| **Passing:** | | | |
| **Two-Second Rule:** | | | |
| **Totals:** | | | |

Note: % Safe = $\dfrac{\text{Total Safe Observations}}{\text{Total Safe + At-Risk Observations}} \times 100\% = \underline{\ \ \ }$

**Figure 3–11.** A critical behavior checklist to increase safe driving. *Source: Courtesy of Make-A-Difference, LLC.*

safety performance (e.g., number of injuries or property-damage incidents, and worker' compensation costs).

The BBS checklist shown in Figure 3–12 addresses only one behavior. After defining the safe and at-risk behaviors relevant to each checklist category, a work group can readily use this checklist to hold each other accountable for safe lifting. In this case, the supervisor does not approach workers specifically to observe them, as with the use of the task-specific checklist (Figure 3–11). Instead, members of a work team look for the target behavior to occur. When supervisors see a safe behavior opportunity (SBO), they take out their checklist

| Observer: | Date: | | |
|---|---|---|---|
| Target Behavior | | Safe | At-Risk |
| load appropriate | | | |
| hold close | | | |
| use legs | | | |
| move feet—don't twist | | | |
| smooth motion—no jerks | | | |
| | | | |
| Comments (use back if necessary): | | | |
| | | | |
| | | | |

**% Safe Observations:**

$$\frac{\text{Total Safe Observations}}{\text{Total Safe Observations} + \text{At-Risk Observations}} \times 100 = \underline{\quad\quad} \%$$

**Figure 3–12.** A critical behavior checklist to increase safe lifting. *Source: Courtesy of Make-A-Difference, LLC*

and complete it. If the target behavior is "lifting," for example, observers keep on the lookout for an SBO for lifting. They might observe such an SBO from their workstation or while walking through the facility.

It is critical that everyone observed for safe lifting had previously given permission for their lifting to be observed. However, the only name on the checklist is the observer's. This is not a "gotcha" process. Rather, it is a way to focus people's attention on a certain work behavior and identify how well the group performs the behavior. In the case of behaviors with multiple components (as in lifting), the group learns which aspects of a work practice are relatively safe versus those that are at-risk.

Some checklists target multiple behaviors generic across various jobs. For example, the checklist presented in Figure 3–13 is applicable for any job that requires the use of personal protective equipment (PPE). This checklist was designed to

## Critical Behavior Checklist for Personal Protective Equipment

Observation period (dates): _____

Observer: _____

| TOTAL NUMBER OF EMPLOYEES OBSERVED | TOTAL NUMBER OF EMPLOYEES OBSERVED USING ALL REQUIRED PPE |
|---|---|
|  |  |

| PPE (For Observed Area) | SAFE OBSERVATION (Proper Use of PPE) | AT-RISK OBSERVATION (Improper or No Use of PPE) |
|---|---|---|
| Gloves |  |  |
| Safety Glasses/Shield |  |  |
| Hearing Protection |  |  |
| Safety Shoes |  |  |
| Hard Hat |  |  |
|  |  |  |
|  |  |  |
|  |  |  |
| **TOTAL** |  |  |

% Safe Observations = $\dfrac{\text{Total Safe Observations}}{\text{Total Safe Observation + Total At-Risk Observation}} \times 100 = \_\_\_\_\_ \%$

**Figure 3–13.** A critical behavior checklist to increase the use of PPE. *Source: Courtesy of Make-A-Difference, LLC.*

conduct several one-on-one behavioral audits over a particular time period. Each time observers perform an observation, they place a checkmark in the top left box (for total number of employees observed). If the worker was using all PPE required in the particular work area, a check is placed in the right-hand box. From these entries, the overall percentage of safe employees can be monitored.

The percentage of safe behavior per each PPE category can also be assessed by dividing the total number of safe checks per each PPE category by the total safe and at-risk checks. This allows for the calculation of the percentage of safe behavior for each PPE category. The resultant feedback regarding the relative use of various devices to protect employees might suggest a need to make certain PPE more comfortable or convenient to use. It might also suggest the need for special intervention, which is the next step of the DO IT process.

## "I" for Intervene

During this stage, interventions are designed and implemented in an attempt to increase safe behavior or decrease at-risk behavior. As mentioned previously, intervention means changing external conditions of the system in order to make safe behavior more likely than at-risk behavior. When designing interventions, directing with activators and motivating with consequences is critical. Specifically, the most motivating consequences are soon, certain, and sizable. Positive consequences are preferable to negative consequences because of the undesirable side effects accompanying punitive methods, including negative attitudes and a decrease in self-motivation and ownership, and the probability of discretionary behavior.

The process of observing and recording the frequency of safe and at-risk behavior on a checklist provides an opportunity to give individuals and groups valuable behavior-based feedback. When the results of a behavioral observation are shown to individuals or groups, they receive the kind of information that enables practice to improve performance. Considerable research has shown that providing workers with feedback regarding their ongoing behavior is a very cost-effective intervention approach.

In addition to behavioral feedback, researchers have found a number of other intervention strategies to be effective at increasing safe work practices. These include worker-designed safety slogans, near miss and corrective action reporting, safe behavior promise cards, individual and group goal setting, actively caring thank-you cards, safety coaching, as well as incentive/reward programs for individuals or groups.

## "T" for Test

The test phase of DO IT provides work teams and supervisors with the information they need to refine or replace a behavior-change intervention, and thereby improve the process. If observations indicate significant improvement in the target behavior has not occurred, the supervisor analyzes and discusses the situation, and refines the intervention or chooses another intervention approach. On the other hand, if the target reaches the desired frequency level, the participants can turn their attention to another set of behaviors. They might add new critical behaviors to their checklist, thus expanding the domain of their behavioral observations. Alternatively, they might design a new intervention procedure to focus only on the new behaviors.

Each time the participants evaluate an intervention approach, they learn more about how to improve safety performance. They have essentially become behavioral scientists, using the DO IT process to:

1. diagnose a human factors problem
2. monitor the impact of a behavior-change intervention
3. refine interventions for continuous improvement

The results from such testing provide motivating consequences to support this learning process and keep the participants involved. The systematic evaluation of a number of DO IT processes can lead to a body of knowledge worthy of integration into a theory. This is reflected in the next principle.

## USE THEORY TO INTEGRATE INFORMATION, NOT TO LIMIT POSSIBILITIES

While much, if not most, research is theory driven, Skinner (1950) was critical of designing research projects to test theory. Theory-driven research can narrow the perspective of the investigator and limit the extent of findings from the scientific method. In other words, applying the DO IT process to merely test a theory can be like putting blinders on a horse. It can limit the amount of information gained from systematic observation.

Many important findings in behavioral science have resulted from exploratory investigation. That is, systematic observations of behavior occurred before and after an intervention or treatment procedure to answer the question "I wonder what will happen if...?" rather than "Is my theory correct?" In these situations, the investigators were not expecting a particular result, but were open to finding anything. Subsequently, they modified their research design or observation process according to their behavioral observations, not a particular theory. In other words, their innovative research was data driven rather than theory driven.

This is an important perspective for safety professionals, especially when applying the DO IT process. It is often better to be open to many possibilities for improving safety performance than to be motivated to support a certain process. Numerous intervention procedures are consistent with a BBS approach, and an intervention process that works well in one situation will not necessarily be effective in another setting. Thus, it is recommended that safety leaders make an educated guess about what intervention procedures to use at the start of a BBS process, but be open to results from a DO IT process and refine procedures accordingly.

After many systematic applications of the DO IT process, distinct consistencies are likely. Certain procedures will work better in some situations than others, with some individuals than others, or with some behaviors than others. Summarizing relationships between intervention impact and specific situational or interpersonal characteristics can lead to the development of a research-based theory of what works best under particular circumstances. This implies the use of theory to integrate information gained from systematic behavioral observation. Skinner (1950) approved of this use of theory, but cautioned that premature theory development can lead to premature theory testing and limited profound knowledge.

## DESIGN INTERVENTIONS WITH CONSIDERATION OF INTERNAL FEELINGS AND ATTITUDES

As indicated earlier, Skinner was certainly concerned about unobservable attitudes or feeling states. This is evidenced by his criticism of punishment because of its impact on people's feelings or perceptions. This perspective also reflects a realization that intervention procedures influence feeling states, and these can be pleasant or unpleasant, desirable or undesirable. In other words, internal feelings or attitudes are influenced indirectly by the type of behavior-focused intervention procedure implemented, and such relationships require careful consideration by the developers and managers of a BBS process.

The rationale for using more positive than negative consequences to motivate behavior is based on the different feeling states provoked by positive reinforcement versus punishment procedures. Similarly, the way we implement an intervention process can increase or decrease feelings of empowerment, build or destroy trust, or cultivate or inhibit a sense of teamwork or belonging (Geller

2001b, 2002). Thus, it's important to assess feeling states or perceptions that occur concomitantly with an intervention process. This can be accomplished informally through one-on-one interviews and group discussions, or formally with a perception survey (O'Brien 2000).

Thus, decisions regarding which intervention to implement, and how to refine existing interventions should be based on both objective behavioral observations and subjective evaluations of feeling states. Often, however, it is possible to evaluate the indirect internal impact of an intervention by imagining yourself going through a particular set of intervention procedures and asking the question "How would I feel?" Perhaps in this case, basic common sense is as good as any evidence one could gather from subjective evaluations of other people's feeling states.

## The Flow of Behavior Change

As reviewed above, an effective BBS process requires a careful analysis of the context in which target behaviors occur. Subsequently, a behavior-change intervention needs to be designed, implemented, and evaluated, as reflected in the DO IT process. In fact, each of the BBS principles summarized above manifests guidelines for developing intervention procedures or assessing intervention impact. In addition, the choice of an intervention process to improve safety-related behavior should depend on whether the at-risk behavior is intentional and whether it is other-directed, self-directed, or automatic (Watson and Tharp 2002).

## Three Types of Behavior

Safety-related behavior in an industrial setting usually starts out as other-directed, in the sense that employees follow someone else's instructions. Such direction can come from a training program, an operations manual, or a policy statement. After people learn what to do, essentially by memorizing or internalizing the appropriate instructions, their behavior can become self-directed. They talk to themselves or formulate an image before performing a behavior in order to activate the right response.

Sometimes they talk to themselves after performing a behavior in order to reassure themselves they performed it correctly. Or, they use self-dialogue to figure out ways to do better next time. At this point they are usually open to receiving corrective feedback.

When some behaviors are performed frequently and consistently over a period of time, they become automatic. A habit is formed. Some habits are desirable and some are undesirable, depending on their short- and long-term consequences. If implemented correctly, rewards, recognition, and other positive consequences can facilitate the transfer of behavior from the self-directed state to the habit state.

Of course, self-directed behavior is not always desirable. When workers take a calculated risk, for example, they are choosing intentionally to ignore a safety precaution or take a shortcut in order to perform more efficiently or with more comfort or convenience. In this state, people can be considered "knowingly at-risk." It's often difficult to change self-directed behavior from at-risk to safe, because such a transition usually requires a relevant change in personal motivation.

Before a bad habit can be changed to a good habit, the target behavior must become self-directed. In other words, people need to become aware of their undesirable habit (as in at-risk behavior) before an adjustment is possible. Then, if people are motivated to improve (perhaps as a result of feedback or an incentive/reward program), their new self-directed behavior can become automatic.

## Three Types of Intervention Strategies

The ABC (activator-behavior-consequence) model was reviewed earlier in this chapter as a framework to understand and analyze why behavior occurs, as well as to develop interventions for improving behavior. Consider that activators and consequences are external to the performer (as in the environment), or they can be internal (as in self-instructions or self-recognition). They can be intrinsic or extrinsic to a behavior, meaning they provide direction or motivation naturally as a task is performed (as in a computer game), or they are added to the situation extrinsically in order to benefit performance (as in a pay-for-performance process). An incentive/reward program is external and extrinsic. It adds an activator (an incentive) and a consequence (a reward) to the situation in order to direct and motivate desirable behavior.

**Instructional Intervention.** An instructional intervention uses an activator or antecedent event to get new behavior started or to move behavior from the automatic (habit) stage to the self-directed stage. Or, it is used to improve behavior already in the self-directed stage. The aim is to get the participant's attention and instruct him or her to transition from unknowingly at-risk to knowingly safe. One assumes the person wants to improve, so external motivation is not needed—only external and extrinsic direction.

This type of intervention consists primarily of activators, as exemplified by education sessions, training exercises, and directive feedback. Because the purpose is to instruct, the intervention comes before the target behavior and focuses on helping the performer internalize the instructions. This type of intervention is more effective when the instructions are specific and given one-on-one. Role-playing exercises may provide the supervisor with an opportunity to customize directions specific to individual attempts to improve. They also allow participants to receive rewarding feedback for their achievement.

**Supportive Intervention.** Once a person learns the right way to do something, practice is important for the behavior to become part of a natural routine. Continued practice leads to fluency (i.e., fast and accurate behavior) and in many cases to automatic or habitual behavior. Practice does not come easily and benefits greatly from supportive intervention. Workers need support to reassure them that they are doing the right thing and to encourage them to keep doing it.

While instructional intervention consists primarily of activators, supportive intervention focuses on the application of positive consequences. Thus, when receiving rewarding feedback or recognition for particular safe behavior, workers feel appreciated and are more likely to perform the behavior again (Allen 1990; Daniels 2000; Geller 1997). Each occurrence of the desired behavior facilitates fluency and helps build a good habit. Note that supportive intervention is typically not preceded by a specific activator. In other words, the support of self-directed behavior does not need an instructional antecedent. The person knows what to do.

**Motivational Intervention.** When people know what to do and don't do it, they require some external encouragement or pressure to change. Instruction alone is insufficient because they are knowingly doing the wrong thing. In safety, this is referred to as a "calculated risk." People take calculated risks when they perceive the positive consequences of the at-risk behavior to be more powerful than the negative consequences. The positive consequences of comfort, convenience, and efficiency are immediate and certain, while the negative consequence of at-risk behavior (such as an injury) is improbable and seems remote.

In this situation an incentive/reward program is useful. Such a program attempts to motivate a certain target behavior by promising workers a positive consequence if they perform it. The promise is the incentive, and the consequence is the reward. In safety, this kind of motivational intervention is much less common than a disincentive/penalty program. A disincentive takes the form of a rule, policy, or law that threatens to give people a negative consequence (a penalty) if they fail to comply or if they take a calculated risk.

Often a disincentive/penalty intervention is ineffective because, like an injury, the negative consequence or penalty seems remote and improbable. The behavioral impact of these enforcement programs are enhanced by increasing the severity of the penalty and punishing more people for taking the calculated risk. However, the large-scale implementation of this kind of intervention can seem inconsistent and unfair. In addition, threats of punishment appear to challenge individual freedom and choice (Skinner 1971), and therefore this approach to behavior change can backfire and activate more calculated risk-taking, even sabotage, theft, or interpersonal aggression (Sidman 1989).

Motivational intervention is clearly the most challenging, requiring enough external influence to get the target behavior started without triggering a desire to assert personal freedom (Brehm 1972). Powerful external consequences might improve behavior only temporarily, as long as the behavioral intervention is in place. Hence the individual is "knowingly safe," but the excessive outside control makes the behavior entirely other-directed. Excessive control on the outside of people can limit the amount of control or self-direction they develop on the inside.

## Putting It All Together

Figure 3–14 reviews this intervention information by depicting relationships between four competency states (unknowingly at-risk, knowingly at-risk, knowingly safe, and habitually safe) and three intervention approaches (instructional intervention, motivational intervention, and supportive intervention, leading to self-management). When people are unaware of the safe work practice (i.e., they are unknowingly at-risk), they need repeated instructional intervention until they understand what to do. Then, as depicted at the far left of Figure 3–14, the critical question is whether they perform the desired behavior. If they do, the question of behavioral fluency is relevant. A fluent response becomes a habit or part of a regular routine, and thus the individual is habitually safe.

When workers know how to perform a task safely but don't, they are considered knowingly at-risk. This is when an external motivational intervention can be useful, as discussed above. Then when the desired behavior occurs at least once, supportive intervention is needed to get the behavior to a fluent state.

Figure 3–14 also shows a distinction between knowingly safe/other-directed and knowingly safe/self-directed. If a safe work practice is self-directed,

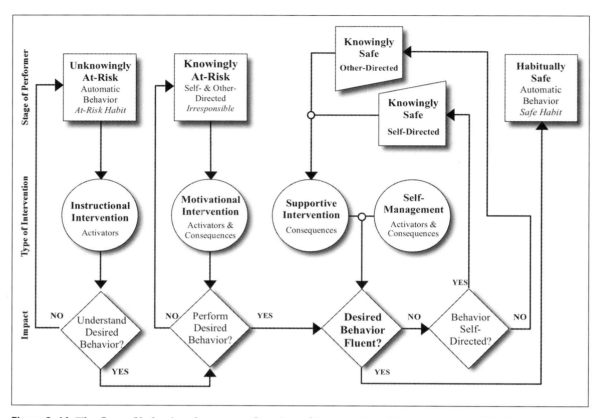

**Figure 3–14.** The flow of behavior change as a function of intervention. *Source: Courtesy of Make-A-Difference, LLC.*

the employee is considered responsible and a self-management intervention is relevant. As detailed elsewhere (Watson and Tharp, 2002), the methods and tools of effective self-management are derived from behavioral science research and are perfectly consistent with the principles of BBS.

In essence, self-management involves the application of the DO IT process to one's own behavior. This means (a) defining one or more target behavior to improve, (b) monitoring these behaviors, (c) manipulating relevant activators and consequences to increase desired behavior and decrease undesired behavior, and (d) tracking continual change in the target behavior in order to determine the impact of the self-management process.

## SUMMARY OF KEY POINTS

Key points covered in this chapter include:
- The BBS approach summarized in this chapter provides tools and methods employees can use to take control of their own safety performance, thereby

enabling a bottom-up empowerment approach to preventing workplace injuries. The five benefits of a behavior-based safety approach are: (1) it focuses evaluation on lowering the total recordable injury rate (TRIR), (2) it builds positive attitudes, (3) it increases personal responsibility for safety, (4) it facilitates interpersonal coaching and teamwork, and (5) it teaches and promotes systems thinking.

- Successful applications of the BBS approach generally adheres to seven principles, each of which is broad enough to encompass a range or practical operations, but narrow enough to guide the development of cost-effective procedures for managing the human dynamics of safety and health and preventing workplace injuries. These seven principles include, (1) focusing intervention on observable behavior, (2) examining external factors to understand and improve behavior, (3) directing with activators and motivating with consequences, (4) focusing on positive consequences to motivate behavior, (5) applying the scientific method to improve intervention, (6) using theory to integrate information, but not limiting possibilities, and (7) designing interventions that take into account internal feelings and attitudes.

- In occupational safety, a behavioral analysis that examines external factors, such as equipment design, the management system, behaviors of others, etc., should be conducted to explain and improve behavior. When conducting a behavioral analysis of risky work practices, the following questions should be asked and answered: Can the task be simplified? Is a quick fix available? Is safe behavior punished? Is at-risk behavior rewarded? Are extra consequences used effectively? Is there a skill discrepancy? What kind of training is needed? Is the person right for the job?

- The ABC, or three-term contingency, model is used to design BBS interventions for individuals, groups, and organizations. The A is for activator (or antecedent), the B for behavior, and the C is for consequence. This model suggests that activators get their behavioral influence from consequences. In other words, antecedent conditions that tell us what to do are only as influential as the consequences they reflect.

- Systematic and scientific observation puts workers in control of improving safety-related behaviors and as a result may prevent injuries in the workplace. Observation should not be used for fault-finding, but for fact-finding to facilitate the discovery of behaviors and conditions that need to be changed or continued in order to prevent injuries. Checklists can be used and it should be determined beforehand who will conduct the observations, how often observations will take place, how the data gathered from observations should be summarized and interpreted, and how workers will be informed of the results of the observation process.

- When choosing an intervention strategy to improve safety-related behavior, one must first understand whether the at-risk behavior is intentional and whether it is other-directed, self-directed, or automatic. Safety-related behavior in most industrial settings starts out as other-directed behavior, meaning direction comes from a training manual, an operations manual, or a policy statement. Self-directed behavior may come after the worker has learned what to do, and has, for the most part, memorized or internalized appropriate instructions. Automatic behaviors are those that are performed frequently and consistently over a period of time, without much thought. Automatic behaviors may be desirable (good habits) or undesirable (bad habits).

- The three major types of intervention strategies for improving safety-related behavior are instructional, motivational, and supportive. An effective BBS process depends on choosing the right behavior-change intervention method and designing, implementing, and evaluating it for your specific workplace.

## REFERENCES

Allen, J. (1990). *I saw what you did and I know who you are: Bloopers, blunders and success stories in giving and receiving recognition.* Tucker, GA: Performance Management Publications.

Atkinson, J. W. (1957). "Motivational determinants of risk-taking behavior." *Psychological Review*, 64, 359-372.

Atkinson, J. W. (1964). *An introduction to motivation.* Princeton, NJ: Van Nostrand.

Brehm, J. W. (1972). *Responses to loss of freedom: A theory of psychological reactance.* New York: General Learning Press.

Carnegie, D. (1936). *How to win friends and influence people.* New York: Simon and Schuster.

Daniels, A. C. (2000). *Bringing out the best in people: How to apply the astonishing power of positive reinforcement* (Second Edition). New York: McGraw-Hill, Inc.

Geller, E. S. (1996). "The truth about safety incentives." *Professional Safety*, 41(10), 34-39.

Geller, E. S. (1997). "Key processes for continuous safety improvement: Behavior-based recognition and celebration." *Professional Safety*, 42(10), 40-44.

Geller, E. S. (1998a). *Applications of behavior analysis to prevent injury from vehicle crashes* (Second Edition). Monograph published by the Cambridge Center for Behavioral Studies, Cambridge, MA.

Geller, E. S. (1998b). *Understanding behavior-based safety: Step-by-step methods to improve your workplace* (Second Edition). Neenah, WI: J. J. Keller & Associates, Inc.

Geller, E. S. (2001a). "Dream – Operationalize – Intervene – Test: If you want to make a difference – Just DO IT." *Journal of Organizational Behavior Management*, 21(1), 109-121.

Geller, E. S. (2001b). *The psychology of safety handbook*. Boca Raton, FL: CRC Press.

Geller, E. S. (2002). *The participation factor: How to get more people involved in occupational safety*. Des Plaines, IL: American Society of Safety Engineers.

Gilbert, T. F. (1978). *Human competence—Engineering worthy performance*. New York: McGraw-Hill.

Goldstein, A. P., and Krasner, L. (1987). *Modern applied psychology*. New York: Pergamon Press.

Mager, R. F., and Pipe, P. (1997). *Analyzing performance problems or you really oughta wanna* (Third Edition). Atlanta, GA: The Center for Effective Performance, Inc.

Norman, D. A. (1988). *The psychology of everyday things*. New York: Basic Books.

O'Brien, D. P. (2000). *Business measurements for safety performance*. New York: Lewis Publishers.

Rhodewalt, F. (1994). "Conceptions of ability achievement goals, and individual differences in self-handicapping behavior: On the application of implicit theories." *Journal of Personality*, 62, 67-85.

Rhodewalt, F., and Fairfield, M. (1991). "Claimed self-handicaps and the self-handicapper: The relations of reduction in intended effort to performance." *Journal of Research in Personality*, 25, 402-417.

Sidman, M. (1989). *Coercion and its fallout*. Boston, MA: Authors Cooperative, Inc., Publishers.

Skinner, B. F. (1950). "Are theories of learning necessary?" *Psychological Review*, 57, 193-216.

Skinner, B. F. (1953). *Science and human behavior.* New York: Macmillan.

Skinner, B. F. (1971). *Beyond freedom and dignity.* New York: Alfred A. Knopf.

# 4

# Safety and Health Training

**After reading this chapter, you will be able to:**

- Describe the goals and benefits of training
- Identify the needs of adult learners
- Develop or select an appropriate training program
- Describe the factors involved in delivering a safety training program
- Explain different methods of training
- Identify key topics for effective new-employee orientation

When a safety or health problem is being caused by a lack of the right behavior, knowledge, or skills, one solution can be appropriate training. Training focuses mainly on behavior or performance change. It can also influence attitude by inspiring confidence and the belief that improvement is possible. In a work setting, the goal of improved training is that learning leads to improved performance on the job. When safety and health are the primary issues, the goal is that employees learn how to keep themselves and others safe and healthy.

A key element in every successful organization is effective job orientation, including safety and health training. The responsibility for employee training rests with top management. For the organization to achieve its objectives, employees must perform at a certain level. To achieve the organization's objectives, those responsible and accountable for safety and health—whether safety and health professionals, front-line supervisors, team leaders, or managers—must:

- be trained in the proper methods of leadership and supervision
- know how to detect and eliminate (or control) hazards, investigate incidents, and handle emergency situations
- understand how to train employees in the safest and most efficient ways to do their jobs
- know and support all company health and safety policies and procedures
- reduce unacceptable risks
- enforce all safety policies and procedures

The safety and health professional must make sure that the training provided meets all regulatory safety training obligations and complements/supports the company's business goals.

Remember, that for any training program to be successful, it must have four components:

1. Training must be performance-based.
2. An implementation plan must be in place before training starts.
3. Measurements of safety performance should be a mix of leading and lagging indicators (proactive and reactive measures).
4. Knowledge and enforcement of safety standards must be accomplished before an injury or illness occurs.

# TRAINING AS PART OF THE SAFETY AND HEALTH PROGRAM

Utilizing effective training programs will help to ensure that employees know how to do their jobs using safe methods. Training provides the "how-to," and usually the "what." It also provides the "why," to the extent that people need to know safe ways to complete their tasks.

Training prepares people for behavior change. In education, the focus is on information that may or may not be used on the job. In training, the focus is specifically on learning how to do the job properly, and how to apply the new information and skills on the job, in order to create and maintain a safer work environment.

Safety and health training is essential for all employees. Both hourly and salaried workers are required to recognize hazards and risks that are unacceptable, and recommend changes in policies and procedures that would increase safety on the job site. Therefore, safety and health training has to be incorporated into two parts. Part one should include all compliance training identified by OSHA, EPA, DOT, and all other regulatory agencies. Part two includes best practices: information and training that the organization needs to improve safety and health performance, but would not necessarily be accomplished by covering strictly compliance topics.

## Roles of the Safety Professional

Safety professionals may play many roles or have a myriad of responsibilities in a comprehensive safety and health process. They may conduct the training, coordinate its development with internal or external instructional designers, personally design and develop the program, purchase materials from vendors, contract with a trainer, deliver the training themselves, or manage the entire training function. No matter what their role, it is important for safety professionals to have a clear idea of the needs and problems of the company or facility and the workers.

The supervisor or team leader will have to interface with the safety professional or human resource function to assist with the completion of employee safety training. In some cases, supervisors may be involved with scheduling their employees to attend training. In other cases, the supervisor may attend a train-the-trainer session so that he or she can deliver the safety training to his or her employees. No matter how training delivery is accomplished, the supervisor or team leader plays a key role in ensuring that it is accomplished.

### Benefits of Training

The benefits of safety and health training include the following:
- a safer work environment with fewer incidents and injuries/illnesses
- reduced costs
- reinforcement of the organization's operational goals, policies, and values
- improved performance
- increased morale
- improved compliance with regulatory, union, and the organization's rules and guidelines
- increased safety and health awareness

### Training and Nontraining Solutions

Before exploring training in more detail, it is important to understand how training fits into an overall plan to address safety and health issues in an organization. In most situations, the success of training is improved if other nontraining solutions for needs and problems are also being implemented in the organization.

Sometimes a training program is not the solution needed in order to affect on-the-job performance. Millions of dollars are wasted every year in training to solve a problem that actually needed a different solution. Nontraining solutions (actions taken to resolve safety and health problems that do not require training) may include actions like the following:
- Workers can be instructed by a sign showing what safety goggles are needed in the work area.
- Supervisors may offer bonuses to employees with good overall safety performance.
- The engineering department may redesign the workflow of a manufacturing site to protect workers who have to walk near heavy machinery.
- Supervisors need to enforce all safety policies and procedures.

Analyzing problems and assessing needs must be done before making decisions about training or nontraining solutions. Management should consider all training and nontraining alternatives to safety and health challenges. For example, suppose a certain task is performed infrequently and workers often fail to remember all steps. Or perhaps it has been discovered that a task is too complex to perform from memory. In situations like these, nontraining solutions such as the ones listed here may be as effective as a training program:

- job aids
    - task procedure instructions (job safety analysis)
    - material safety data sheets (sections on protection information, special precautions, spill/leak procedures, etc.)
    - flowcharts
    - checklists
    - diagrams
    - troubleshooting guides
    - decision tables
- reference manuals
- help desks or hotlines
- reward systems
- improved physical work environments
- improved work processes

Training and nontraining problems or needs and their related solutions can be categorized as: (1) selection and assignment, (2) information and practice, (3) environment, and (4) motivation/incentive.

1. *Selection and assignment* refers to considerations and processes used to hire people and assign them specific responsibilities and on-the-job tasks. Improving the hiring process helps make sure the right (i.e., sufficiently experienced, skilled) worker is assigned to each task.

2. *Information and practice* is the act of providing all necessary information and developing skills needed to use safe practices. This includes communicating tasks, goals, and instructions clearly, as well as providing feedback, training, job aids, guided practice, and experience. This is the area that can often be improved through a training solution, because effective training provides many opportunities for practice.

3. *Environment* refers to all day-to-day influences on performance. These include (but are not limited to) equipment, floor plans, tool access, and working relationships with co-workers, supervisors, and with safety and health professionals. Environmental problems include humidity problems in electrical work areas, slippery walkways, air pollution, etc., which are best solved by nontraining solutions (i.e., engineering controls, writing a maintenance work order, etc.).

4. *Motivation/incentive* solutions refer to any link between incentives for the reduced risk of injury and/or death and safe performance. These include any rewards by management for safe behavior, such as praise and commendations, bonuses, pay increases, and contest prizes. For example, a supervisor may offer a bonus to the employee with the best overall safety record, or 100 days without an injury may be marked with a free lunch for all. Education may be used as a motivator, as well as peer pressure and negative pressures, such as the threat of a loss of bonus, etc. Research shows that negative reinforcement does not produce results as well as positive reinforcement, but unfortunately, it is still frequently used.

The supervisor, manager, or team leader must carefully analyze the needs of his or her group and decide what solutions will best meet those needs. If training and education supplies the best answer, then a safety professional should be consulted to assist in setting up the program.

## Structured Training

When training is selected as the solution to a specific need or problem, the training program should be planned, designed, developed, and delivered in a structured manner. Its potential for solving the problem and changing behavior should be realistically evaluated.

Performance-based training is implemented to solve a specific on-the-job problem or to encourage a specific behavioral change. Management can evaluate performance-based training by analyzing individual workers' performances. This training is directly related to the job the worker is expected to perform and linked to the organization's safety and health goals. It is measured and evaluated by the worker's observable performance and created using a proven systematic approach of five phases:

1. *Assessment.* The safety professional consults with others and conducts research to determine whether training is the right solution and what knowledge, skills, and attitudes people need to develop.
2. *Design.* The designated safety trainer identifies instructional goals, chooses training media, and figures out the sequence in which workers will learn new information and skills. This work is based on needs-assessment results, audience analysis, and specific performance-based objectives.
3. *Materials acquisition and/or development.* The designated safety trainer buys and/or creates training materials.

4. *Delivery.* The designated safety trainer delivers the training.
5. *Evaluation.* The safety professional evaluates the entire design process and how effective it is in achieving the objectives by getting feedback from those involved in the training.

There are three major types of performance-based training.
1. *Instructor-led training.* This training is presented in a classroom-like setting. Everyone follows the pace set by the instructor, trainer, or facilitator. Participant materials and PowerPoint slides are usually included.
2. *Self-paced training.* Employees usually complete this training on their own by working through a textbook and/or workbook. Self-paced training may also be done through computer/Web-based training or computer-assisted instruction. The software provides course content, guides learning activities, and administers tests. If desired, online training can provide a safety professional with a list of who completed the training and individual's scores to monitor performance for record-keeping purposes.
3. *Structured on-the-job training.* Similar to instructor-led training, structured on-the-job training is conducted at the learner's place of work. The instructor is usually a trained supervisor who acts as a coach or guide. This type of training may or may not include participant materials and generally uses job aids.

## DEVELOPING THE SAFETY TRAINING PROGRAM

If the safety professional lacks skills and experience in training development or instructional design, an internal or external instructional designer can be brought in to develop each training program, or the entire program can be outsourced as a complete package. The development process should include a needs assessment, written performance objectives, selection of materials to fit the objectives and the audience, outline of content, timing, testing, and evaluation. The training program should be developed in the five-phase process and should meet the same criteria, whether it is developed internally or selected from outside sources.

### Phase 1—Assessment
To create an effective training program, the safety and health professional needs to assess current employee performance, comparing it to desired employee per-

formance. He or she should also conduct a learner analysis. A needs assessment is important to the organization because it helps to:

- distinguish between training and nontraining needs
- understand the problem before designing a solution
- save time and money by ensuring that solutions effectively address the problems they are intended to solve
- identify what factors will affect the training before it is developed

A needs assessment is the process of determining the who, what, when, where, and why of the training need, after it is determined that a safety or health problem can be solved through training. Identifying the specifics of the training needs is the first phase of developing an effective training program.

The safety and health professional tries to determine who, at a general level, may need training, what they should be trained to do, when they should be trained, where they should be trained, and why they need the training.

- Who does the job? (Which skilled or unskilled employees?)
- What do they do in their jobs? (What are the tasks? How many tasks?)
- When do they perform the task? (All during a shift, in a set sequence, before or after certain other tasks?)
- Where do they perform the task? (Area conditions; proximity to materials, doors, warehouse, docks, open area, etc.)
- Why do they perform the task? (What is the importance of this work?)
- How do they perform the task? (Speed, work practices, safety measures in place or avoided, and any other qualifying characteristics that affect the need for training.)

A needs assessment is conducted primarily to determine the specifics of the problem or need that the training must address. It may show that nontraining solutions are also needed to address the issue completely. When a regulatory agency is not involved, the safety and health professional is responsible for determining whether training is needed at all, before defining the (perceived) training need. In initial stages of assessment, the many solutions that might solve the problem must be considered.

In the needs assessment phase, the safety and health professional should identify a training goal in a statement that describes the general purpose and how the training will satisfy the safety and health need or solve the problem. Training goals are used to provide direction for learner, job, and setting analyses and to develop learning objectives. Training goals are also valuable tools for communicating the purpose of training to others. The training goals should clearly:

- describe how the training will satisfy the safety or health needs, and
- communicate the purpose of the training

While the training goals are being identified, the safety and health professional should also identify the characteristics of the learners who will be involved. Learner characteristics may be observed by study of general demographics and by learners' preferences, attitudes, knowledge, skills, and previous experiences. The results of this learner analysis will affect the training design. Results include specific traits that will affect design and delivery of the training (also called an audience description) and recommendations for how the training design should accommodate learners' specific traits and general adult learning needs. The steps in completing a learner analysis are:

1. Identify general traits of audience members, such as job titles/positions and age.
2. Identify the learners' preferences regarding training.
3. Identify the learners' attitudes or feelings regarding training.
4. Identify the learners' knowledge or skills, especially as they relate to the content of the training.
5. Identify ways to adapt training and delivery to these characteristics.

Job and setting analyses are also conducted in the needs assessment phase. A job analysis is the process used to determine the procedures, decisions, knowledge, and skills required for a worker to perform a job function. A setting analysis is identifying specific characteristics of the training environment (where training will occur) that will affect training. Ideally, it is best to obtain setting information before designing the training. But because this is not always possible, the training course should be adaptable to a variety of settings.

**Adult Learning Needs.** A major factor in the success of training is the extent to which adult learners' needs are considered. Safety and health issues can be technical and difficult to learn. In addition, people may see safety and health training as a disruption in their busy workdays.

To keep the participants motivated and involved, the safety and health professional must allow for their learning needs. There are many adult learning principles. For example, one principle is based on the belief that each person learns best through one or a combination of three senses: hearing, sight, or touch (activity). Training following this model would use media tailored to the learners' best learning modalities.

There are four basic adult learning needs and principles that are common to most adult learners. These principles should help shape any safety and health training developed or contracted.:

- Adults need to know why they are learning a particular topic or skill because they need to apply learning to immediate, real-life changes.
- Adults have experience that they apply to all new learning.
- Adults need to be in control of their learning.
- Adults want to learn things that will make them more effective and successful.

**Need to Know Why.** Adults need to know why they are learning a particular topic or skill because they need to apply learning to immediate, real-life challenges. They learn best when:

- they can see clear demonstrations of how the training applies directly to their jobs
- they can apply the new information or skill to solve problems immediately, during the training
- they have opportunities to think about how they can use the new information or skill

An example of meeting this need is a supervisor demonstrating how to operate a new machine safely, then providing feedback while watching the workers run the machine by themselves.

**Need to Apply Experience.** Adults bring experience with them to all new learning. They learn best when:

- they can share their experiences during training
- they can use comparison and contrast, relationships, and association of ideas
- they have opportunities to think about how the new information or skill relates to their past, present, or future experiences
- their experience and opinions are taken seriously by the trainer

For example, to meet these needs during a training session on personal protective equipment, the supervisor or team leader should give participants time to talk about their past or current PPE practices and those of their co-workers at their facility.

**Need to Control.** Adults need to be in control of their learning. They learn best when:
- they choose (or at least influence) the training they receive
- they are in a flexible learning environment
- they are actively involved in the learning process rather than passive receivers of information (the trainer guides and facilitates versus lectures)
- they have a chance to voice concerns and see them addressed
- they take active part in group discussions, role plays, simulations, and other activities

Meeting this need is illustrated by workers who are part of a safety and health committee. Their responsibilities include assessing workers' needs for training, identifying good training solutions, and evaluating training after it is delivered.

**Need to Succeed.** Adults want to learn things that will make them more effective and successful. They learn best when:
- they know why they are taking the training and how it will benefit them
- they can ask and answer the questions, "What will this do for me? How will it help me get ahead? What hazards will it help me control?"

An example of meeting this need is training for confined spaces that includes an action planning piece for improving the existing program based on gaps discovered during the training session.

**Appropriate Level of Materials.** Another consideration that must be made in meeting adult learning needs is the literacy level of the employee. Some employees read, use math skills, comprehend information, and solve problems at a low- or mid-literacy level. Low-literacy generally refers to workers whose abilities are at or below fourth-grade level. Mid-literacy refers to employees whose abilities are at fifth through eighth-grade levels.

Low- or mid-literacy-level employees may have trouble understanding much of the information that is presented in complex textbooks, manuals, and lectures. The design and delivery of safety and health training programs must take this into account. For instance, training can be written at a fourth- or fifth-grade reading level or can simply rely on less printed material. Use of symbols and illustrations in training and promotion should minimize the risk of the message not being understood. This technique may also work where English is a second language for employees.

Remember to consider the primary language of your work force. English may be their second language. Therefore, you need to design your training programs to address the literacy issues and the language issues of your work force.

**Performance Objectives.** Writing performance objectives and evaluating learning progress are two of the most important steps in developing effective training programs. Performance objectives are specific, measurable, short-term statements that clearly describe what a learner will be able to do (new or different) as a result of training. They provide a structure for developing training.

Performance objectives are important to the employees being trained because they provide a target for performance (or behavior), help learners identify their focus, and tell learners how they will be evaluated. Objectives are especially critical in the safety and health arena because there is too much risk (of potential injury, illness, or loss of life) to employees to allow a hit-or-miss approach. Not determining the training objectives in advance may also waste a vast amount of time and money.

Unlike goals, objectives are measurable and observable. They differ from learning activities in that they describe the results, not how to achieve them. The four parts of an objective (sometimes referred to as the ABCD method of objective writing) are as follows:

- Audience: Describe the intended learner or end user of the instruction.
- Behavior: Identify what learners must do to demonstrate mastery.
- Condition: Identify what learners will or will not be given in order to carry out the behavior (be successful).
- Degree: Specify how well the audience members must perform the behavior in terms of time, accuracy, proportion, quality, etc.

Generally, most people will assume that 100 % is the standard performance if it is not specified otherwise in the objective. Performance can be measured in terms of quantity, quality, or both.

By detailing training objectives, the safety and health professional can develop programs that meet an organization's established safety and health

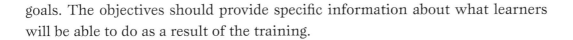

goals. The objectives should provide specific information about what learners will be able to do as a result of the training.

## Phase 2—Design

The training developer organizes and times the training program through outlining the content. To outline, the developer selects an objective, identifies what actions the worker must take to achieve it, and lists related topics that the worker must know to accomplish the actions. The developer then puts the content in order (perhaps with help from the safety and health professional or supervisor) and estimates the amount of time needed for each point. The time available may be predetermined. If it is not enough, the trainer can either ask for more time or narrow the scope of the objectives.

## Phase 3—Materials Acquisition and/or Development

The materials of an effective safety and health training program should provide (1) an introduction, (2) presentation of the information and/or demonstration of actions (3) practice and feedback, and (4) a summary, evaluation, review, or transition. The introduction can be an overview, a rationale, a warm-up activity, questions, or a story or analogy. The information should be presented in a manner that helps workers organize and remember the important facts. Materials that can help achieve this goal include text, graphics, examples, job aids, checklists, charts and graphs, tables, data, reports, relevant articles (such as tools and materials), a glossary, and a table of contents.

Practice and feedback should give workers the opportunity to try out their new knowledge and skills, and to receive a response and feedback about their performance. Materials that help guide the process of practice and feedback include activity directions, activity time allowance, activity worksheets, question-and-answer worksheets, business scenarios, videotapes of performance, simulations, performance checklists, and sample solutions.

The summary, evaluation, review, or transition phase should stress important training points, answer workers' questions, and relate the training experience to the next training topic or to the job. Supporting materials for this phase include text, graphics, question-and-answer sheets, listing of next steps, and goal sheets for implementing new knowledge and skills on the job.

## Phase 4—Delivery

The safety professional or supervisor delivers the training. To be effective, the design must transfer information and skills to the learners so that they effectively understand and complete mastery of the material.

### Phase 5—Evaluation

The safety professional or supervisor evaluates the entire design process and how effective it is in achieving the objectives by getting feedback from those involved in the training. Evaluation can either be formative, to determine inefficiencies before the final product is released, or summative, (done after the final process).

## NEW-EMPLOYEE TRAINING AND ORIENTATION

Safety and health training should begin when a new employee is hired. The person is usually open to ideas and information about how the company operates. From the first day, the new employee formulates opinions about management, supervisors, co-workers, and the organization. Human resource managers sometimes say they have never hired a worker with a bad attitude. While this may be an overstatement, many employees' negative attitudes are developed on the job.

Timeliness of instruction is key to the success of the orientation program. For example, it is generally agreed that new employees are more prone to work-related incidents. This fact is attributed to inexperience, lack of familiarity with procedures and facilities, and zealousness to work. A significant number of employees are also injured after they've been on the job for four to six years. This is generally attributed either to a change in work duties when the employee is transferred or to worker complacency. Covering these issues in an ongoing training program can help to reduce the number of incidents.

The following subjects are suggested as part of the orientation program as it relates to safety and health:

- company orientation: history and goals
- safety, health, and environmental policy statements
- housekeeping standards
- hazard communication
- personal protective equipment
- emergency response procedures
- incident reporting procedures
- incident investigation procedures

- lockout/tagout procedures
- machine guarding (if applicable)
- electrical safety awareness
- ladder use and storage (if applicable)
- confined space entry (if applicable)
- first aid/CPR
- fire prevention and protection
- hand tool safety
- ergonomic principles

These subjects and many others are far too important to overlook or leave to casual learning. A formal safety orientation program should be developed to give employees this information and also to forge a strong link between all employees and the organization's safety and health policy.

Many of these subjects are left to the manager, supervisor, or team leader to cover when new employees arrive at the worksite. Unfortunately, it's hard to find time to cover them, and the importance of doing so is often not recognized until an incident occurs. Supervisors should continually update their information to avoid contradicting the training manual and safety policies. They should also become involved in the development of training programs to make sure the program information is up-to-date and practical in relation to current work demands. If they disregard or contradict the training manual, the entire program and the company's image lose credibility with the workers.

The manager, supervisor, or team leader should reinforce the training program content by demonstrating how it will apply to workers' specific job assignments. For example, the new employee may have been trained on how to read the warning labels on chemical containers. The supervisor can enhance this information by touring the department with the new employee to point out the hazardous materials and identify the protection provided by PPE and control measures.

In addition, all leaders can provide examples by their own conduct. "Lead by example" is a critical concept to the success of the safety and health program, which means that they also should attend the same training as the workers.

The new employee orientation process is not just for employees new to your company. Many times transferred employees who are "new" to a job or an area are left out of the orientation process. Bringing these employees up to speed on specific department or task requirements for safety should also be done. Although not as exhaustive as orientation presented to workers entirely new to your organization, these transferred employees need some measure of department and job safety orientation to make them a successful member of your team. Again, this important task falls to the supervisor or team leader to accomplish.

### Regulatory Obligations

All regulatory agencies publish information about the employer's responsibility to provide training to employees. If the supervisor encounters difficulties with the final interpretation, the best solution is to communicate with the agency involved and with the employer's legal counsel. Information on regulated training requirements and publications for specific industries can be found at the OSHA web site, www.osha.gov.

### Policies and Procedures Manuals

The use of written policies and procedures manuals has long been an efficient way to provide information to new employees. The company should make sure the manual is easy to understand and comprehensive, and that all rules are enforced. The manual should cover such areas as first aid, PPE, electrical safety, and housekeeping, to name just a few. Participants can use the manual as a key reference during training. Manuals should be periodically reviewed and updated, as necessary, to ensure compliance with changes in regulations and other requirements.

## OSHA REQUIREMENTS

The major parts of the OSHA regulations (Title 29—Labor, Code of Federal Regulations) covering training requirements include:
- Part 1910, Safety and Health Training Requirements for General Industry
- Part 1926, Safety and Health Training Requirements for Construction

Consult the individual standards for guidance on specific training requirements.

## TRAINING METHODS

There are various training methods from which designers and trainers can choose when preparing a program. Each method has strengths and weaknesses and serves different needs. The methods selected will depend on the objectives to be met, the type of student participation, time allocated, facility being used, and equipment available. Different people learn at different speeds and feel most comfortable with different methods. Trainers must have the teaching skills to address these elements of human behavior. The following are the most common training techniques used in industry:

- on-the-job training
  - job instruction training
  - coaching/mentoring
- group methods
  - conference
  - brainstorming
  - case studies
  - incident process
  - facilitated discussion
  - role playing
  - lecture
  - question and answer
  - simulation
- individual methods
  - drill
  - demonstration
  - testing
  - video-based training
  - interactive computer-assisted training
  - reading
  - independent study
  - seminars and short courses

## On-the-Job Training

On-the-job-training is widely used because the employee can produce during the training period. However, three considerations must be addressed: (1) The trainer must have good training skills. (2) A training program should make sure all employees are trained in the same way to perform their tasks in the safest and most productive manner. (3) The trainer and trainee must have enough time to cover the subject thoroughly.

A variation of on-the-job training is job instruction training, also referred to as the four-point method. Instruction is broken down into four simple steps: preparation, presentation, observation, and follow-up. The trainer must know the job thoroughly; be a safe worker; and have the patience, skill, and desire to train.

In this four-point method, the trainer-trainee relationship works in the following ways:

1. Preparation. The trainer puts the workers at ease. He or she explains the job and determines what the worker currently knows about the subject. This stage also includes the preparation of the proper learning/working environment.
2. Presentation. The trainer demonstrates the work process. The student watches the performance and asks questions. The trainer should present the job procedures in sequence and stress all key points.
3. Observation. The trainer observes the employee performing the task. The trainer should identify any discrepancies in the work performance and note good performance. The employee should explain the steps being performed. This ensures that the employee not only can perform the task but understands how and why the task is done. This stage continues until the trainer is satisfied with the employee's competence at the job.
4. Follow-up. The trainer and/or the supervisor must monitor the employee's performance to be sure the job is being performed as instructed and to answer any questions the worker might have.

Advantages of the job instruction training method include the following:

1. The personal guidance makes it easier to motivate the employee.
2. The trainer can spot and correct problems as they occur.
3. Results of the training are immediately evaluated as an employee is performing the actual job on actual equipment.
4. Training is practical, realistic, and demonstrated under actual conditions. Employees can ask questions easily.

The most common other method of on-the-job training is coaching or mentoring by a co-worker (the buddy system). It can be quite effective in certain situations, but it has the following drawbacks:

1. Trainers may be selected for their availability rather than for their ability in training. A co-worker who lacks basic skills can completely undermine

a new employee's orientation. An ability to do the task well does not guarantee an ability to teach it.

2. Each trainer may have his or her own individualized methods of doing the tasks being taught. This lack of consistency in training can make it difficult to control hazards in the workplace and can lead to many incidents.

3. Key elements of safety orientation may be overlooked in the training program, and the omissions may not be realized until an incident occurs.

4. Poor techniques or bad habits can spread from the coach to the trainee. Shortcuts or safety violations may be demonstrated as "the way we do it."

5. The training may not emphasize safe performance—job performance should never be separated from safety standards in any training provided to workers.

## Group Methods

Group techniques encourage participation from a selected audience. These methods allow trainees to share ideas, evaluate information, and become actively involved in planning and implementing company policy. All types of group training require skilled facilitators to be successful.

**Conference or Meeting Method.** The conference or meeting method of training is widely used in business and industry for education-sharing purposes and because of the knowledge each participant brings to the group. The trainer/facilitator controls the flow of the session as the participants share their knowledge and experience. The skill of the facilitator can mean the success or failure of these sessions. Facilitators must use various techniques to draw information and opinions from members. The number of people involved should be limited to allow open discussion from all participants. The opinions of each member should be recorded, and a summary of the group conclusions provided to those who were involved, as well as to those who should be kept abreast of the information.

The conference technique is also a valuable method of problem solving. It is important that at the beginning of the conference, members identify their goals and expectations for the session; for example, to make recommendations only, not to establish policy or procedures, etc. When a group is asked to make recommendations, they should be kept informed of the results of those recommendations.

Proper control and guidance of a conference can ensure its success and make it a gratifying experience for the participants.

**Brainstorming.** Brainstorming is a technique of group interaction that encourages each participant to present ideas on a specific issue. The method is normally used to find innovative approaches to issues.

There are four ground rules:

1. Ideas presented are not criticized.
2. "Anything goes" creative thinking and building on ideas are positively reinforced.
3. As many ideas as possible should be presented quickly.
4. Combining several ideas or improving suggestions is encouraged.

A recorder should be selected to write down all the ideas presented. The moderator must control the flow of suggestions, cut off negative comments, and solicit ideas from each member.

Brainstorming allows ideas to be developed quickly, encourages creative thinking, and involves everyone in the process. The group can go beyond old stereotypes and move away from "the way it has always done."

**Case Study.** Case studies are written descriptions of actual or fictitious business decisions or problems that learners will use as a basis for demonstrating predetermined skills and knowledge. The goal of the activity is to develop group members' insight and problem-solving skills. The study normally is presented by defining what happened in a particular incident and the events leading up to the incident. The group is then given the task of determining the actual causes of problems, the significance of each element, and/or the acceptable solutions.

There are two distinct advantages to building case studies: first, case studies provide an opportunity for the learners to use skills and knowledge acquired during the course. Second, case studies can serve as an evaluation tool for trainers to measure the degree of proficiency attained during a course or module. Here are other key benefits:

- improve critical thinking skills
- improve problem-solving skills
- help resolve problems
- encourage reflection, application, and analysis
- reinforce the value of discussion and interaction with others

Planning, thinking about, and adhering to the instructional objectives are paramount in designing an effective case. The key is to start at the end and work backwards toward the beginning. Ask the following questions:

- What questions must be answered?
- What skills or knowledge should be exercised?
- What specific performance objectives should be measured?
- What learning performance objectives are to be addressed by the case study?

**Incident Process.** This is a type of case study in which the group works with a written account of any incident. The group is allowed to ask questions about facts, clues, and details. The trainer provides the answers to the questions, and the group must assemble the information, determine what has happened, and arrive at a decision. The trainer must guide the group to prevent arguments and prevent one or two members from dominating the discussion. This is a useful training method that encourages employee participation in the incident prevention program. Situations can be real cases in the company or developed from potential hazards that exist.

The moderator must be capable of controlling group processes and progress and of preventing the group from missing the true or root causes. For example, an employee was struck in the eye by a foreign body, and an investigation revealed the employee was not wearing safety glasses while operating a bench grinder. The group must seek the root cause of the incident and not settle for the common conclusion, "the worker failed to wear eye protection." They must specify why the supervisor or management failed to enforce the proper procedures (assuming there was an established procedure). Why did management allow this lax supervision?

**Facilitated Discussion.** Facilitated discussion or dialogue is the management of discussion about the course content so that the learning objectives are met, the discussion flows logically from topic to topic, and the applications to the learners' jobs are made clear. Facilitated discussion requires that the trainer/facilitator have the skills to accept all ideas and contributions as valid, show how they relate to the course objectives, and manage the time element and the flow of information to meet the course objectives.

The benefits of facilitated discussion include the following:

- ensures that the learners are involved and challenged
- builds a bond between the trainer and facilitator that encourages the free exchange of ideas and information

**Role Playing.** This training method is effective for evaluating human relations issues. Members attempt to identify the ways people behave under various conditions. Although this technique is not an effective method of problem solving, it can uncover issues not previously considered. This method is particularly helpful in identifying and changing personnel issues such as poor morale or negative attitudes.

**Lecture.** With this method, a single person can share information to a large group in a relatively short time. This method is normally used to communicate facts, give motivational speeches, or summarize events for employees. There is little time or opportunity for interaction by the attendees. Follow-up for these sessions must be well-planned in advance to be successful.

**Question-and-Answer Sessions.** Normally, question-and-answer sessions follow training periods after the trainer has summarized the material presented. Employees can use this method to clarify individual concerns or facts. However, employees will often need time to prepare and organize their thoughts before they can ask questions. In situations where they must absorb a large amount of material, allow them time to reflect on or apply the knowledge and to formulate questions. The trainer can plan to have a follow-up session or allow employees to present their questions personally. Question-and-answer sessions are helpful in clarifying issues of policy or changes in schedules or events.

**Simulation.** When actual materials or machines cannot be used, trainers can use a simulation device. This method is used effectively in aircraft pilot training, railroad engineer training, and other applications. Various methods are employed in management training programs as well, such as the "in-basket technique," "war games," and others. One simulation demonstrates the loss of eyesight to employees to encourage them to wear safety glasses on the job. Safety procedures must be carefully followed to prevent any actual incident.

Simulation is most effective when the employees can participate. Careful planning and attention to detail are required. The initial costs of these sessions can be high because of the equipment required and the time involved in conducting training.

## Individual Methods

With individual methods, employees can work at their own pace. For some learners, this is the most effective approach. The learner does, however, miss out on the give-and-take of ideas. A combination of group and individual methods provides the broadest variety of learning opportunities.

**Drill.** Using the elements of practice and repetition, this method of instruction is valuable for developing worker skill in fundamental tasks and for performing under pressure.

Employees required to perform in crisis situations should be trained under conditions that resemble the crisis as closely as possible. For example, when instructing workers in cardiopulmonary resuscitation (CPR), the trainer must try to instill the tension that workers will experience when they attempt to resuscitate a real victim.

**Demonstration.** As discussed in the section on job instruction training, the method of demonstration allows the trainer to perform the actual task and then have the employee repeat the performance. Trainers must be sure the job is preformed exactly as required to prevent workers from developing poor habits and poor performance standards (and supervisors must see that employees follow the designated procedure). If the conditions used in demonstrations are not similar to the actual workstation or equipment on the job, this method will yield few useful results.

**Testing.** This technique is normally used to determine if employees understand the necessary information and can apply the knowledge when required. Developing good tests is a skill that requires constant review to ensure that training objectives are being met. Poor tests can reduce employees' morale and undermine training objectives.

**Video-Based Training.** The use of videotapes and DVDs does not eliminate the need for professional instruction, but can enhance a classroom presentation. Videos are available from the National Safety Council and numerous private companies. Trainers should screen the materials to make sure they fulfill the needs of the training program.

Some distinct advantages to using a video in a training class include:
- Video offers the learner an opportunity to see examples of tasks and processes being performed correctly.
- The cost of producing training videos has gone up while the difficulty in producing them has gone down.

**Interactive Computer-Assisted Training.** Interactive computer programs allow the workers to receive information by reading or watching a video presentation and then responding to situations and questions. If the correct response is entered, the computer will advance the program to repeat the information and retest the workers. The system is valuable for several reasons:

- Trainees can work at their own pace.
- Records can be automatically kept of all training. The amount and type of records maintained can be modified to meet the company requirements.
- Correct answers are required before a worker can proceed to the next lesson, or remediation methods are built into the program.
- Workers receive training as time is available in their schedule, rather than having to meet training schedules.

This type of training works well for organizations with small workforces or those that cannot remove large groups of workers from their jobs at any one time. Be cautious of off-the-shelf materials. The organization may still need some face time with employees in order to provide the customization of content to their job. Because this method lets trainees work on their own, be sure to set up mechanisms that allow them to ask questions about the material covered.

**Reading.** Companies should provide employees with written safety materials such as monthly newsletters and safety magazines. In addition, organizations may establish a library where employees can research information on subjects such as work procedures, safety, leadership, health care, family or home safety, and other subjects of concern. It is important, however, that management does not assume everyone has the ability to read and comprehend all of the written material provided. Companies cannot replace instruction or training programs simply by handing an employee a training manual. Written material is meant to supplement or to serve as a reference for training.

**Independent Study.** Home study or correspondence courses can help workers advance within the organization or improve their knowledge of their jobs and the industry. A major advantage of this method of training is that the employee does not lose any time from work and can complete the course at his or her own pace. Another advantage is the low cost of implementing and maintaining home-study programs. Normally they are centered around a textbook assignment, followed by self-tests using multiple-choice, true/false, fill-in, or essay questions. Several courses also provide laboratory or performance materials. Some home-study programs come with video presentations for workers to view and follow.

### Seminars and Short Courses

Seminars, short courses, and workshops for safety and leadership information and skills are offered by many colleges and universities, as well as by insurance companies and private organizations. Seminars, short courses, and workshops range from one-hour sessions to several-day sessions. This method of training usually provides employees with specific, relevant and in-depth information on the topics being discussed. Similar to facilitated discussion and lectures, these types of training allow for networking with others—internally or externally to your company—which allows for broader sharing of information and best practices.

## SUMMARY OF KEY POINTS

Key points covered in this chapter include:
- Training focuses mainly on behavior change, showing workers how to perform properly and how to apply their knowledge on the job. In some cases, however, nontraining solutions are more appropriate.
- Training and nontraining problems and solutions are categorized as selection and assignment, information and practice, environment, and motivation/incentive.
- Adult learners have special needs that trainers must recognize in order for the program to be effective. Performance-based instruction generally works well for adult learners.
- To design an effective training program, the safety professional or supervisor must assess learners' (workers') needs, analyze their characteristics, develop specific objectives, create materials and schedules, and design appropriate testing and evaluation methods.
- Training begins with the orientation of each new employee. Using written policies and procedures manuals is one way to help meet new employee training needs and conform to regulatory requirements.
- Training methods include on-the-job training, job instruction training, group methods, and individual methods.

# 5

# Employee Involvement

**After reading this chapter, you will be able to:**

- Establish the importance of involving employees in the safety and health process
- Identify types of employee involvement tools
- Perform a job safety analysis
- Conduct a job safety observation
- Create a safety and health committee
- Conduct a safety and health meeting
- Analyze barriers and challenges to employee involvement
- Recognize steps that can be taken to involve employees in safety and health efforts
- Describe how involvement in safety contacts, safety inspections, perception surveys, employee recognition, and suggestion systems can promote employee investment in safety and health

Employee involvement is an essential building block of any workplace safety and health program. All employees need to understand their responsibility for their own and their co-workers' safety. They have an obvious interest in working with co-workers and employers to improve conditions because their very lives and livelihoods are at stake. Ways to involve employees in promoting safety include safety and health committees, safety and health meetings, involvement in the job safety analysis process and in perception surveys, employee recognition, safety contests and posters, suggestion systems, and safety and health training.

The primary goal of any safety management system is to eliminate or control recognized hazards to attain an acceptable level of risk. For that reason, safety integration is implemented by management's commitment to involve employees from all levels of an organization in proactive (prevention-oriented) activities that focus efforts on recognition, evaluation, and control or elimination of workplace hazards.

Effective and meaningful employee involvement includes these basic requirements:

- management commitment visibly and frequently demonstrated
- individual development and capability (training)
- individual involvement and influence
- constant and varied communication
- recognition, reinforcement, and reward
- employee involvement tools

There are numerous benefits of getting employees involved:

- Employees are often more satisfied and productive when their ideas and suggestions are offered and taken seriously.
- Employees are more likely to support and use processes, polices, and programs in which they have had input.
- Employees who are in contact with potential hazards have a personal interest in recognizing and avoiding those hazards.

## EMPLOYEE INVOLVEMENT TOOLS

The ability to recognize, evaluate, and eliminate or control hazards is developed through several tools or techniques. These tools include job hazard analysis

(JSA), physical hazard inspections, employee safety training, safety meetings, job safety observations, perception surveys, and safety committees.

# JOB SAFETY ANALYSIS PROCESS

Everyone has something to contribute to the improvement of safety and health in the workplace. Management must constantly provide employees with opportunities to participate in the protection of their own well-being.

Job safety analyses formally allow employees to participate in identifying hazards and substandard practices in the workplace and modify work methods to prevent injury, illness, and incidents and improve productivity and product quality. The JSA process encourages safe work practices on a personal level by a process that relies on employees' expertise and knowledge about each job and its related tasks. When used as a tool for continuous improvement, the JSA facilitates better communication between line employees, supervisors, team leaders, and senior managers.

JSA is a procedure used to review job methods and uncover hazards that (1) may have been overlooked in the layout of the facility or building and in the design of the machinery, equipment, tools, workstations, and processes; (2) may have developed after production started; or (3) resulted from changes in work procedures or personnel.

A JSA can be written as shown in Figure 5–1. In the left column, the basic steps of the job are listed in the order in which they occur. The middle column describes all hazards, both those produced by the environment and those connected to the job procedure. The right column gives the safe procedures that should be followed to guard against the hazards and to prevent potential incidents. (See Figure 5–2 for a list of instructions printed on the back of JSA forms published by the National Safety Council.)

For convenience, both the JSA procedure and the written description are commonly referred to as JSA. Health hazards are also considered when creating a JSA.

## Benefits of a JSA

The principal benefits of a JSA include:

- giving individualized training in safe, efficient procedures
- making employee safety contacts

| JOB SAFETY ANALYSIS | JOB TITLE (and number if applicable):<br>Banding Pallets | | PAGE _1_ OF _2_   JSA NO. _105_ | DATE:<br>00/00/00 | ☒ NEW<br>☐ REVISED |
|---|---|---|---|---|---|
| **INSTRUCTIONS ON REVERSE SIDE** | TITLE OF PERSON WHO DOES JOB:<br>Bander | SUPERVISOR:<br>James Smith | | ANALYSIS BY:<br>James Smith | |
| COMPANY/ORGANIZATION:<br>XYZ Company | PLANT/LOCATION:<br>Chicago | DEPARTMENT:<br>Packaging | | REVIEWED BY:<br>Sharon Martin | |
| REQUIRED AND/OR RECOMMENDED<br>PERSONAL PROTECTIVE EQUIPMENT: | Gloves - Eye Protection - Long Sleeves - Safety Shoes | | | APPROVED BY:<br>Joe Bottom | |

| SEQUENCE OF BASIC JOB STEPS | POTENTIAL HAZARDS | RECOMMENDED ACTION OR PROCEDURE |
|---|---|---|
| 1. Position portable banding cart and place strapping guard on top of boxes. | 1. Cart positioned too close to pallet (strike body & legs against cart or pallet, drop strapping gun on foot.) | 1. Leave ample space between cart and pallet to feed strapping - have firm grip on strapping gun. |
| 2. Withdraw strapping and bend end back about 3". | 2. Sharp edges of strapping (cut hands, fingers & arms). Sharp corners on pallet (strike feet against corners). | 2. Wear gloves, eye protection & long sleeves - keep firm grip on strapping - hold end between thumb & forefinger - watch where stepping. |
| 3. Walk around load while holding strapping with one hand. | 3. Projecting sharp corners on pallet (strike feet on corners). | 3. Assure a clear path between pallet and cart - pull smoothly - avoid jerking strapping. |
| 4. Pull and feed strap under pallet. | 4. Splinters on pallet (punctures to hands and fingers) Sharp strap edges (cuts to hands, fingers, and arms). | 4. Wear gloves - eye protection - long sleeves. Point strap in direction of bend - pull strap smoothly to avoid jerks. |
| 5. Walk around load. Stoop down. Bend over, grab strap, pull up to machine, straighten out strap end. | 5. Protruding corners of pallet, splinters (punctures to feet and ankles). | 5. Assure a clear path - watch where walking - face direction in which walking. |
| 6. Insert, position and tighten strap in gun. | 6. Springy and sharp strapping (strike against with hands and fingers). | 6. Keep firm grasp on strap and on gun - make sure clip is positioned properly. |

**Figure 5–1.** This sample completed JSA shows how hazards and safe procedures are identified to help reduce the occurrence of incidents.

- instructing a new person on the job
- preparing for planned safety observations
- giving prejob instructions on irregular jobs
- reviewing job procedures after incidents occur
- studying jobs for possible improvement in job methods

A JSA can be done in three basic steps. However, before initiating this analysis, management must first carefully select the task to be analyzed.

## Selecting the Task

A job is a sequence of separate tasks with steps or activities that together accomplish a work goal. Some jobs can be broadly defined by what is accomplished (for example, making paper, building a facility, and mining iron ore). On the other hand, a job can be narrowly defined in terms of a single action, such as turning a switch, tightening a screw, and pushing a button. Such broadly or narrowly defined jobs are unsuitable for JSA.

# Instructions For Completing the Job Safety Analysis Form

Job Safety Analysis (JSA) is an important analyzing tool that works by finding hazards and eliminating or minimizing them *before* the job is performed, and *before* they have a chance to become injuries or damage. Use JSA for job clarification and hazard awareness, as a guide in new employee training, for periodic contacts and for retraining of senior employees, as a refresher on jobs that run infrequently, and for informing employees of specific job hazards and protective measures. It can also be used as an incident investigation tool.

Set priorities for doing JSAs: jobs that have a history of causing injury or damage, jobs that have produced disabling injuries, jobs with high potential for disabling injury or death, and new jobs.

Select a job to be analyzed. Before filling out this form, consider the following: The purpose of the job—What has to be done? Who has to do it? The activities involved—How is it done? When is it done? Where is it done?

In summary, to complete this form you should consider the purpose of the job, the activities it involves, and the hazards it presents. If you are not familiar with a particular job or operation, interview an employee who is. In addition, observing an employee performing the job, or "walking through" the operation step by step may give additional insight into potential hazards. You may also wish to videotape the job and analyze it.

Here's how to do each of the three parts of a Job Safety Analysis:

## Sequence of Basic Job Steps

Examining a specific job by breaking it down into a series of steps or tasks will enable you to discover potential hazards employees may encounter.

Each job or operation will consist of a set of steps or tasks. For example, the job might be to move a box from a conveyor in the receiving area to a shelf in the storage area. To determine where a step begins or ends, look for a change of activity, change in direction or movement.

Picking up the box from the conveyor and placing it on a handtruck is one step. The next step might be to push the loaded handtruck to the storage area (a change in activity). Moving the boxes from the truck and placing them on the shelf is another step. The final step might be returning the handtruck to the receiving area.

Be sure to list *all* the steps needed to perform the job. Some steps may not be performed each time; an example could be checking the casters on the handtruck. However, if that step is generally part of the job it should be listed.

**Figure 5–2.** Instructions for completing the JSA. ©2009 National Safety Council.

## Potential Hazards

A hazard is a potential danger. The purpose of the JSA is to identify all hazards—both those produced by the environment or conditions and those connected with the job procedure.

Examine each step carefully to find and identify hazards—the actions, conditions, and possibilities that could lead to injury, illness, or damage. Consider the following hazard types:

### Chemical Hazards

___ Inhalation
___ Skin contact
___ Absorption
___ Injection
___ Ingestion

### Biological Hazards

___ Bloodborne Pathogens
___ Brucellosis
___ Building-Related Illness (BRI)
___ Legionnaires' Disease
___ Mold
___ Plant and Insect Poisons
___ Tuberculosis (TB)
___ Water and Wastewater

### Physical Hazards

___ Electrical
___ Fire/Explosion
___ Noise
___ Radiation
___ Thermal Stress
___ Caught In/On/Between
___ Pinch Points
___ Slips/Falls
___ Striking Against
___ Struck By

### Ergonomic Hazards

___ Repetition
___ Forceful Exertions
___ Awkward Postures
___ Contact Stress
___ Vibration
___ Work Area Design

## Recommended Action or Procedure

Using the first two columns as a guide, decide what actions or procedures are necessary to eliminate or minimize the hazards that could lead to an injury, illness, or damage. Begin by trying to: (1) engineer the hazard out; (2) provide guards, safety devices, etc.; (3) provide personal protective equipment; (4) provide job instruction training; (5) maintain good housekeeping; and (6) ensure good ergonomics (positioning the person in relation to the machine or other elements in such a way as to improve safety).

List the recommended safe operating procedures. Begin with an action word. Say exactly what needs to be done to correct the hazard, such as "lift using your leg muscles." Avoid general statements such as "be careful."

List the required or recommended personal protective equipment necessary to perform each step of the job.

Give a recommended action or procedure for each hazard.

Serious hazards should be corrected immediately. The JSA should then be changed to reflect the new conditions.

Finally, review your input on all three columns for accuracy and completeness. Determine if the recommended actions or procedures have been put in place. Reevaluate the JSA as necessary.

Tasks suitable for JSA are those assignments that a line supervisor may make. Operating a machine, tapping a furnace, and piling lumber are good subjects for JSAs because they are neither too broad nor too narrow.

Selection of jobs to be analyzed and establishment of the order of analysis should be guided by the following factors:

- *Frequency of incident.* The greater the number of incidents (injuries, illnesses, equipment damage, chemical spill, etc.) associated with the job, the greater its priority claim for JSA.
- *Rate of disabling injuries.* Every job that has had disabling injuries should be given a JSA, particularly if the injuries prove that prior preventive action was not successful.
- *Severity potential.* Some jobs may not have a history of incidents but may have the potential for producing severe injury.
- *New tasks.* Changes in equipment or in processes obviously have no history of incidents, but their incident potential may not be fully appreciated. A JSA of every new task should be made as soon as the task has been created. Analysis should not be delayed until incidents or near misses occur.

After the task has been selected, the three basic steps in making a JSA are:
1. Break the task down into successive steps or activities and observe how these actions are performed.
2. Identify the hazards and potential incidents. This is the critical step because only an identified problem can be eliminated.
3. Develop safe job procedures to eliminate the hazards and prevent the potential incidents.

## Breaking the Task Down into Steps

Before the search for hazards begins, a task should be broken down into a sequence of steps, each describing what is being done. Avoid two common errors: (1) making the breakdown so detailed that an unnecessarily large number of steps results, or (2) making the job breakdown so general that basic steps are not recorded.

To do a task breakdown, select the right worker to observe—an experienced, capable, and cooperative person who is willing to share ideas. If the employee has never helped out on a JSA, explain the purpose—to make a job safe by identifying and eliminating or controlling hazards—and show him or her a completed JSA. Reassure the employee that he or she was selected because of experience and capability.

Observe the employee perform the task and write down the basic steps. Consider videotaping the job as it is performed for later study. To determine the basic task steps, ask "What step starts the task?" Then, "What is the next basic step?" and so on.

Completely describe each step. Any possible deviation from the regular procedure should be recorded because it may be this irregular activity that leads to an incident.

To record the breakdown, number the job steps consecutively as illustrated in the first column of the JSA training guide, illustrated in Figure 5–3. Each step tells what is done, not how.

The wording for each step should begin with an action word like "remove," "open," or "weld." The action is completed by naming the item to which the action applies, for example, "remove extinguisher," "carry to fire."

Check the breakdown with the person observed and obtain agreement of what is done and the order of the steps. Thank the employee for cooperating.

---

**JOB SAFETY ANALYSIS WORK SHEET**
**JOB: Using a Pressurized Water Fire Extinguisher**

| WHAT TO DO<br>(Steps in sequence) | HOW TO DO IT<br>(Instructions)<br>(Reverse hands for left-handed operator.) | KEY POINTS<br>(Items to be emphasized.<br>Safety is always a key point.) |
|---|---|---|
| 1. Remove extinguisher from wall bracket. | 1. Left hand on bottom lip, fingers curled around lip, palm up. Right hand on carrying handle palm down, fingers around carrying handle only. | 1. Check air pressure to make certain extinguisher is charged. Stand close to extinguisher, pull straight out. Have firm grip to prevent dropping on feet. Lower, and as you do, remove left hand from lip. |
| 2. Carry to fire. | 2. Carry in right hand, upright position. | 2. Extinguisher should hang down alongside leg. (This makes it easy to carry and reduces possibility of strain.) |
| 3. Remove pin. | 3. Set extinguisher down in upright position. Place left hand on top of extinguisher, pull out pin with right hand. | 3. Hold extinguisher steady with left hand. Do not exert pressure on discharge lever as you remove pin. |
| 4. Squeeze discharge lever. | 4. Place right hand over carrying handle with fingers curled around operating lever handle while grasping discharge hose near nozzle with left hand. | 4. Have firm grip on handle to steady extinguisher. |
| 5. Apply water stream to fire. | 5. Direct water stream at base of fire. | 5. Work from side to side or around fire. After extinguishing flames, pour water on smouldering or glowing surfaces. |
| 6. Return Extinguisher. Report Use. | | |

**Figure 5–3.** This JSA worksheet shows how to break down a job by illustrating what to do, rather than how to do it.

### Identifying Hazards and Potential Incident Causes

Before filling in the next two columns of the JSA—Potential Hazards and Recommended Action or Procedure—begin the search for hazards. The purpose is to identify all hazards—both those produced by the environment and those connected with the task. Each step, and thus the entire task, must be made safer and more efficient. Look for chemical, physical, biological, and ergonomic hazards. To do this, ask these questions about each step:

- Is there a danger of striking against, being struck by, or otherwise making harmful contact with an object?
- Can the employee be caught in, by, or between objects?
- Is there a potential for a slip or trip? Can the employee fall on the same level or to another?
- Can strain be caused by pushing, pulling, lifting, bending, or twisting?
- Is the environment hazardous to safety or health? For example, are there concentrations of toxic gas, vapor, mist, fume, dust, heat, or radiation? (See discussion in the National Safety Council's *Fundamentals of Industrial Hygiene*, 5th edition.)

Close observation and knowledge of the particular task are required if the JSA is to be effective. The job observation should be repeated as often as necessary until all hazards and potential causes for an incident or injury have been identified.

When inspecting a particular machine or operation, ask the question, "Can an incident occur here?" More specific questions include:

- Is it possible for a person to come in contact with any moving piece of machine equipment?
- Are rotating equipment, set screws, projecting keys, bolt heads, burrs, or other projections exposed where they can strike or snag a worker's clothing?
- Is it possible to be drawn into the in-running nip point between two moving parts, such as a belt and sheave, chain and sprocket, pressure rolls, rack and gear, or gear train?
- Do machines or equipment have reciprocating movement or any motion where workers can be caught on or between a moving part and a fixed object?
- Is it possible for a worker's hands or arms to make contact with moving parts at the point of operation where milling, shaping, punching, shearing, bending, grinding, or other work is being done?
- Is it possible for material (including chips or dust) to be kicked back or ejected from the point of operation, injuring someone nearby?

- Are machine controls safeguarded to prevent unintended or inadvertent operation?
- Are machine controls located to provide immediate access in the event of an emergency?
- Do machines vibrate, move, or walk while in operation?
- Is it possible for parts to become loose during operation, injuring operators or others?
- Are guards positioned or adjusted to correspond with the permissible openings?
- Is it possible for workers to bypass the guard, thereby making it ineffective?
- Do machines, equipment, and their accessories receive regular maintenance?
- Are machines placed so operators have sufficient room to safely work with no exposure to aisle traffic?
- Is there sufficient room for maintenance and repair?
- Is there sufficient room to accommodate incoming and finished work as well as scrap that may be generated?
- Are the materials-handling methods adequate for the work process and the tooling associated with it?
- If tools, jigs, and other work fixtures are required, are they stored conveniently, where they will not interfere with the work?
- Is the work area well illuminated with specific point-of-operation lighting where necessary?
- Is ventilation adequate, particularly for those operations that create dusts, mists, vapors, and gases?
- Is the operator using personal protective equipment?
- Is housekeeping satisfactory with no debris or tripping hazards or spills on the floor?
- Are there places where employees have access to machines (for example, the backside)?
- Are energy sources heat controlled for protection?
- Are energy sources controlled for maintenance?

All of these questions will be of most value if they are incorporated into an inspection form that can be filled out at regular intervals. Even though a question at first may not seem to apply to a specific operation, on closer scrutiny it may be found to apply. Using a checklist is a good way to make sure nothing is overlooked.

To complete column two of the JSA, the analyzer should list all potential causes for incidents or injuries and all hazards yielded by a survey of the machine or operation. Record the type of potential cause of incident and the agents involved. For example, to note that the employee might injure a foot by dropping a fire extinguisher, write down "struck by extinguisher."

Again, check with the observed employee after the hazards and potential causes of incidents have been recorded. The experienced employee will probably offer additional suggestions. You should also check with others experienced with the job. Through observation and discussion, the analyzer can develop a reliable list of hazards and potential incidents.

## Developing Solutions

The final step in a JSA is to develop a recommended safe job procedure to prevent the occurrence of incidents and injuries. The principal solutions are as follows:

1. Find a new way to do the task.
2. Change the physical conditions that create the hazards.
3. Change the work procedure.
4. Reduce the frequency (particularly helpful in maintenance and materials handling).

To find an entirely new way to do a task, determine the work goal of the task, and then analyze the various ways of reaching this goal to see which way is safest. Consider work-saving tools and equipment.

If a new way cannot be found, then ask this question about each hazard and potential cause listed: "What change in physical condition (such as change in tools, materials, equipment, layout, or location) will eliminate the hazard or prevent the incident?"

When a change is found, study it carefully to find other benefits (such as greater production or time saving) that will accrue. These benefits are good selling points and should be pointed out when proposing the change to management.

To investigate changes in the job procedure, ask the following questions about each hazard and potential incident cause listed: "What should the employee do—or not do—to eliminate this particular hazard or prevent this potential incident?" "How should it be done?" Because of his or her experience, in most cases the supervisor can answer these questions.

Answers must be specific and concrete if new procedures are to be any good. General precautions—"be alert," "use caution," or "be careful"—are of little use. Answers should precisely state what to do and how to do it. This recommendation—"Make certain the wrench does not slip or cause loss of balance"—is incomplete. It does not tell how to prevent the wrench from slipping. Here, in contrast, is an example of a recommended safe procedure that tells both what and how: "Set wrench properly and securely. Test its grip by exerting a light pressure on it. Brace yourself against something immovable, or take a solid stance with feet wide apart, before exerting full pressure. This prevents loss of balance if the wrench slips."

Some repair or service jobs have to be repeated often because a condition needs repeated correction. To reduce the need for such repetition, ask "What can be done to eliminate the cause of the condition that makes excessive repairs or service necessary?" If the cause cannot be eliminated, then ask "Can anything be done to minimize the effects of the condition?"

Machine parts, for example, may wear out quickly and require frequent replacement. Study of the problem may reveal excessive vibration is the culprit. After reducing or eliminating the vibration, the machine parts last longer and require less maintenance.

However, reducing the frequency of a job contributes to safety only in that it limits the exposure. Every effort should still be made to eliminate hazards and to prevent incidents through changing physical conditions or revising job procedures or both.

A task that has been redesigned may affect other jobs and even the entire work process. Therefore, the redesign should be discussed not only with the worker involved, but also co-workers, the supervisor, the facility engineer, and others who are concerned. In all cases, however, check or test the proposed changes with those who do the job. Their ideas about the hazards and proposed solutions can be of considerable value. They can judge the practicality of proposed changes and perhaps suggest improvements. Actually, these discussions are more than just a way to check a JSA. They are safety contacts that promote awareness of job hazards and safe procedures.

## SAFETY OBSERVATIONS

Also known as job behavior observations, a job safety observation (JSO) is a process of observing an employee as he or she completes his or her work. This observation is a means of determining if job training has been effective. Job training is based on work instructions that include JSA information. Supervisors can use the JSA or other work instructions/procedures as a tool to develop the JSO forms and checklists.

Besides determining strengths and weaknesses in the employee's work knowledge, this process allows for the correction of inappropriate or at-risk behavior. It also allows the reinforcement of correct work-completion actions.

After effective JSO training, a first-line supervisor/team leader, manager or peer worker may complete JSOs using checklists. The most effective programs use all three as observers at one time or another.

The data from these observations is collected and analyzed to determine workplace strengths and weaknesses. These areas of identified weaknesses can be used to improve the job instruction training. Ultimately, this process improves the knowledge of the employee and encourages the proper completion of future tasks.

As an additional benefit, JSOs effectively communicate management's commitment to safety.

## SAFETY CONTACTS

Safety contacts are commonly initiated after JSOs or as part of a JSA process. This type of contact is highly individual in nature, involving a one-on-one safety-related meeting or discussion between supervisor and employee. After observing the employee at work, the supervisor takes the opportunity to both recognize and reinforce safe job performance, and to offer constructive criticism and further instruction as necessary. The employee, on the other hand, is given the opportunity to voice opinions regarding his or her own performance and company safety and health rules, procedures, and programs in a forum where he or she is ensured of being heard.

## PHYSICAL HAZARD INSPECTIONS

More commonly known as a safety inspection, a physical hazard inspection is another employee involvement tool. Safety inspections are regularly completed at most organizations. Applying this proactive, hazard-recognition tool to employee involvement means having workers directly involved in the inspections.

Over a period of time, all employees should be involved with physical hazard inspections. In this way, each worker is learning how to recognize hazards in their workplace.

Developing inspection checklists also includes employees within the various departments. In this way, employees have a built-in "buy-in" to the process and know the reason for completing this activity. See Chapter 6, Safety and Health Inspections, for additional information.

# SAFETY AND HEALTH COMMITTEES

Employee involvement can be required, but a cooperative spirit cannot. One of the best ways to create not only awareness, but acceptance, of safety and health goals is to get employees involved in running a safety and health committee or team. Involving employees in the decision-making process can earn their respect and lead to increasing their commitment to the company's safety and health goals.

Two styles of committees are used: employee and joint (both employee and management). Either type gives workers a chance to contribute to the overall safety and health of the organization, but the joint committee is usually more effective in a team-based organization. The existence, composition, and purposes of safety and health committees in many workplaces are mandated by law or by labor union contracts.

## Best-Practice Approach for a Joint Committee

The best-practice approach to developing a safety and health committee involves gathering equal representation by leaders and team members. The leaders or supervisors may be appointed by senior management, but cooperation is maximized when worker and management committee members volunteer or are elected by their peers.

The committee should have balanced representation from the various company departments or operating units. The size of the committee should be workable, with enough members to make the workload comfortable, but not so many that they become disorganized. Time commitment to the committee should be defined. Members can have staggered one- to two-year terms, with a quarter of the members being replaced every six months. Officers should also be rotated and should be nonmanagement members at least half the time. The committee should use the company's safety and health professional as its advisor, but all members should have extra training on safety and health issues as needed.

Committee guidelines should be established and followed. Annual goals should be identified and progress measured. The committee should develop a written policy and define member responsibilities and duties. Meetings should be held at least monthly, and should be short and structured. They should be coordinated with other business priorities and schedules. The safety and health professional should evaluate the committee's effectiveness periodically. The committee must be given time to do its work and should be recognized for its successes. Its success can be greatly enhanced with visible support and constant commitment from top management. The committee structure should facilitate communication with all of the employee population.

## Functions of Safety and Health Committees

The safety and health committee usually has some combination of the following functions and roles:

1. policy, practice, and process
   - recommends safety and health policy changes to management
   - communicates safety and health issues and solutions
   - performs JSAs
   - conducts reviews of data from audits and other sources to decide company safety and health approach
   - investigates and acts on substandard conditions, practices and/or procedures for occupational injuries or illnesses
   - participates in or oversees motivational programs
   - oversees emergency response preparedness and drills

2. inspection and investigation
   - conducts facility inspections to correct hazards
   - recommends safety changes to production units
   - uses root-cause analyses to investigate injuries or illnesses
   - recommends changes to policy and procedures

3. promotion, education, and motivation
   - promotes interest in the safety effort
   - ensures compliance with safety procedures
   - develops good safety attitudes in employees

   ○ develops and evaluates safety and health training programs

4. problem solving and analysis
   ○ uses problem solving to focus on solutions to safety and health problems
   ○ recommends solutions to appropriate persons

## SAFETY AND HEALTH MEETINGS

No one structure is ideal for employee involvement in safety and health. To be effective, the system must fit comfortably into the organization it serves. In some companies, formal safety and health committees are not essential. Regular safety and health meetings with all employees are enough to create team spirit and to keep the issue of safety foremost in workers' minds. Every meeting should have the goal of motivating workers to take an interest in and responsibility for safety and health.

Safety meetings should be used to:

- orient individuals on safety and health techniques and philosophy
- empower employees as problem solvers and decision makers
- identify and correct hazards through review of JSAs, inspections, and observations
- develop leadership skills as facilitators

The first question the organizer must address is "What is the purpose?" The safety coordinator or department should be able to provide a multitude of ideas, based either on company experience or on National Safety Council suggestions. (Many models are provided in the materials listed in the bibliography at the end of this book.)

If meetings are to be held monthly, for example, themes can be assigned for a whole year of meetings in advance. In this way, they can tie into each other and build on knowledge shared at the previous meeting. The organizer could also choose to alternate on-the-job safety with off-the-job safety in assigning program topics. For example, the April meeting might discuss ways to protect against violence in the workplace, and the May meeting might address domestic violence and provide information on stress management and local assistance resource groups. The focus of safety meetings should be for problem solving and decision making. Safety training should be for skill and knowledge improvement.

## Running Effective Meetings

Some practical steps can be taken to make meetings go more smoothly. First, preparation should include at least the following items. A written agenda ensures that essential points are covered and keeps speakers from straying from the subject. Speakers should rehearse and time their remarks. The meeting organizer should preview visual aids and check equipment and exhibits to make sure everything is operating properly.

A comfortable location will help attendees get the most out of the meeting. People become irritated and unresponsive if they must stand too long, are too warm or too cold, or feel ill at ease. Background noises or glaring lights can also distract people from the message. The organizer should appoint a committee member to ensure the location is in good order.

Second, the meeting itself should feature no more than three main ideas. If a meeting organizer tries to cover too much for the time available and the nature of the topics, participants simply won't be able to absorb everything. If there are too many topics to address, then coverage can be spread out over several meetings.

Keeping the session short and simple is key to keeping attention. An employee safety and health meeting should not run longer than 30 minutes without a compelling reason.

Tips for successful meetings:
1. Feature no more than three ideas.
2. Meet in a comfortable place.
3. Plan the agenda and visual aids.
4. Keep it short and simple.

Only the most exceptionally appealing material justifies a longer session. (These guidelines apply only to meetings, not to instructional courses.)

Plenty of time should be set aside at the end of the meeting for feedback and questions. This lets the meetings fill two bills at once: (1) disseminating useful information, and (2) soliciting input from employees. Circulating the meeting agenda in advance gives attendees time to consider what points they want to raise.

## Variations on Meetings

Often, each department or area coordinates its own meetings for its own members. When company policies require discussion or the topic is general causes of incidents or injuries, large meetings of several departments are also valuable. Some meetings may be purely motivational; some are held to present safety awards.

Production huddles, tailgate conferences, or five-minute safety talks are instructional sessions about a specific job. When these sessions incorporate the safety message into the job routine, they can be among the most effective safety meetings of all. For example, public utility crews often start a job with a tailgate conference. The crew members gather around their trucks and discuss the work, laying out the tools and materials required and choosing which part of the job each person should handle. Such sessions can be indispensable, especially in highly hazardous situations. Placing the safety discussion at the start of such meetings accomplishes two goals: (1) it places safety as a primary concern from the start, and (2) it ensures that safety is not overlooked due to time constraints.

## PERCEPTION SURVEYS

Perception surveys are a means of giving all employees a voice in the direction of the safety and health program. The corporate safety and health program cannot be truly effective unless it is perceived as having value, and acted on accordingly, by employees. These surveys quantify the attitudes that influence acceptance and safe job performance. A companywide or facilitywide perception survey can pinpoint program inadequacies and credibility gaps in management's commitment to safety. Moreover, because the survey is anonymous and administered to all levels, all employees have an equal voice—even those who might be unable or unwilling to voice dissatisfaction in other circumstances. Results of the survey must be made known to all participants in a timely manner.

## SUGGESTION SYSTEMS

Every employee has some contribution to make to improve safety and health. Employee suggestion systems should encourage all workers to contribute their ideas or concerns on an ongoing basis. The system should incorporate a process for identifying unacceptable hazards and risks (anonymously if necessary) with suggestions for solutions. This gives all employees a chance to buy in to the company's procedures by inventing or modifying them. It also provides the company with an important source of knowledge and creativity.

A well-organized suggestion system encourages employees to contribute their ideas or concerns to help the company improve work methods or reduce unacceptable work hazards and risks. The system can also stimulate participants' thinking about problem solving. Employees may be rewarded with a percentage of the savings resulting from increase in production or decrease in incidents, injuries, and work-related illnesses. Solutions to safety problems can be rewarded in proportion to the usefulness of the idea.

The company should provide special forms for submitting suggestions and set out suggestion boxes in convenient locations. Suggestions should be gathered frequently and acknowledged promptly. Management must accept or reject each suggestion as soon as possible. When a suggestion is not accepted, the employee should be told why.

To be successful, a suggestion system must have a clearly explained operating plan. Employees must know that the plan is fair, impartial, and potentially profitable. The results of useful suggestions should be publicized, along with the names of the employees who made them. However, employees should be able to make anonymous suggestions if they wish. Anyone who suggests that attention be paid to dangerous or unethical activities should receive protection from retaliation.

## EMPLOYEE SAFETY AND HEALTH TRAINING

Employee safety training provides the skills and knowledge required for an employee to complete his or her job tasks without injury or property damage. Safety training is enhanced through the completion of JSAs. JSA information is integrated into the standard operating procedures that in turn are used to develop operator job-instruction training programs. Employee safety training also includes regulatory training such as hazard communication, confined space, and lockout/tagout. The need for testing is important to determine if the audience has comprehended the presented material.

At minimum, all employees should receive training in emergency procedures, including first aid and disaster plans. Companies with good emergency response training plans have found that the benefits are twofold. This training not only helps to save lives and reduce the severity of injuries, but also makes employees more safety-conscious and less likely to have incidents in the first place. The value of emergency training is most obvious in operations that are far

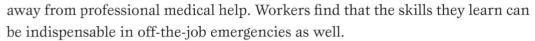

away from professional medical help. Workers find that the skills they learn can be indispensable in off-the-job emergencies as well.

Beyond emergency preparedness training, employees may need specific job-related hazard training such as chemical safety or lockout/tagout training. General safety awareness education and training sessions may be necessary to raise the level of understanding and to identify common objectives and goals. Additional training may be necessary for safety committee members and also for those who organize or coordinate safety meetings.

## BARRIERS AND CHALLENGES TO EMPLOYEE INVOLVEMENT

Many potential barriers and challenges may exist that can prevent an organization from realizing the full potential of meaningful employee involvement. If attempts at involving employees in a meaningful way in safety efforts have not been successful, or if employees have largely resisted efforts to participate in the safety process, consider if any of these potential barriers may be the reason:

- lack of trust
- a climate of fear
- amount of time and effort involved
- uncertainty brought on by constant change
- not involving key stakeholders
- not involving unions (forming a partnership)
- lack of demonstrated commitment from top leadership
- lack of responsibility from top management
- not allowing enough time to change
- resistance from any of the partners (supervisors, managers, support people, unions, labor force)
- inadequate training
- systems and structures not designed to support teams
- workers who zero in on their co-workers' imperfections
- failure to redefine the role of leadership
- leaders who won't let go
- too little or too much structure
- no transition plan
- failure to communicate what's happening

- treating change like a program instead of a process
- overwhelming team members with too much responsibility before they have had adequate training
- team members assuming too much responsibility before they have had adequate training
- a history not conducive to employee involvement
- failure to educate everyone in the workplace about self-direction

## OVERCOMING BARRIERS AND CHALLENGES TO EMPLOYEE INVOLVEMENT

In overcoming barriers or challenges to employee involvement, we have to engage three groups of people: our supporters, the fence-sitters, and those who oppose employee involvement.

It may seem natural to assume that those who are supporting and are involved in our safety efforts are on autopilot. But in reality we have to look for ways to continually keep these employee engaged. Brainstorm ways that you can recognize and reward those who continually participate in moving the safety and health effort forward.

The natural expectation is to work on getting the opposition not just neutralized, but fully participating and supporting the effort. Sometimes it is enough to get those who are opposed neutralized—perhaps not fully supporting the safety and health effort, but not sabotaging the effort either. Look for ways to reward and support the fence-sitters as well. Moving them over to become supporters may not be easy but is worth the effort when building and strengthening your safety and health process.

## SUMMARY OF KEY POINTS

Key points covered in this chapter include:
- Employee involvement is an essential building block of any workplace safety and health program. Ways to involve employees include JSAs, JSOs, safety and health committees, inspection programs, meetings, perception surveys, employee recognition, and suggestion systems.
- JSAs are a great way to proactively analyze tasks for existing and potential

hazards. Using employees to help perform JSAs may lead to acceptance, as well as to practical ideas on how to control identified hazards.

- JSOs offer the ability to praise workers for tasks performed safely and coach or counsel employees who may be exhibiting at-risk behaviors.

- An inspection program not only helps attain OSHA compliance, but can be a proactive method for hazard identification. Setting up a period inspection program staffed jointly by management and employees can also build morale.

- In a joint safety and health committee, about half the members should be workers and half management. The workers should be either elected by their peers or volunteers. Members should have staggered (one- and two-year) terms.

- Safety and health committees are usually responsible for (1) policy, practice, and process, (2) inspection and investigation, (3) promotion, education, and motivation, and (4) problem analysis and solving.

- Safety and health meetings should motivate workers to take an interest in and responsibility for safety and health. Meeting topics should be designed to be productive and of value to the employees.

- Each meeting should feature no more than three ideas and should be held at a comfortable location. There should be a written agenda, circulated to attendees beforehand. The meeting should be short and simple. Production huddles should incorporate the safety message into the job routine.

- Recognizing employees' contributions to safety and health shows them they are valued and encourages them to contribute more. Recognition may be as simple as publishing names, pictures, and safety suggestions of outstanding workers in the company newsletter. Or it may involve plaques issued at annual ceremonies or participation in National Safety Council or OSHA award programs. Recognition should be in proportion to the significance of the contribution.

# 6
# Safety and Health Inspections

**After reading this chapter, you will be able to:**

- Describe the purpose of safety and health inspections
- Distinguish between inspections, audits, and assessments
- Describe how to prepare for an inspection
- Describe how often and what to inspect
- Describe all the items to be observed during an inspection
- Write, and do the follow-up on, the inspection report

Knowing where and how specific safety policies and programs are succeeding or failing is crucial to continuous improvement. To measure how an organization is doing, formal, nonroutine audits or informal, routine inspections are conducted. At periodic intervals, some organizations will also conduct safety management system assessments.

An assessment is a detailed examination of an organization's safety management system. The purpose of the assessment is to identify any gaps in the safety management system that may be preventing the organization from eliminating or minimizing employee injuries and illnesses. During an assessment, it is common for the assessors to interview first-line supervisors to determine the degree of acceptance and completeness of safe operating procedures and best safety practices.

An audit is a methodical examination of a facility's existing procedures and practices to verify whether they comply with legal requirements, internal policies, and good practices. Safety and health audits normally examine operating policies and procedures relating to occupational safety, employee involvement activities, industrial hygiene, occupational health and medicine, and emergency response.

Audits are usually performed by objective parties from outside the department or outside the company. External auditors may be hired by a company or they may be government regulators. It is common practice for auditors to meet with supervisors to gain more detailed explanations of how policies and practices are implemented in the supervisor's work area.

Inspections are examinations of operating conditions and behaviors that exist within a defined work area. It is common practice to expect first-line supervisors to perform work area inspections.

Inspections will be examined in more detail in this chapter.

## SAFETY AND HEALTH INSPECTIONS

The basic purpose of safety and health inspections is to ensure compliance with a facility's safe operating procedures as well as to evaluate a supervisor's safety and health performance activities. Inspections are often a regular element of standard operating procedures.

In the process of conducting a safety and health inspection, the supervisor may detect potential hazards that require immediate correction to prevent employee injuries or illnesses. Just as inspections of the manufacturing process are important to quality control, safety and health inspections are vital to maintaining a high level of safety performance. Inspections in the workplace have been expanded in scope in recent years. In addition to the conditions that have always been observed during an inspection, the inspector is encouraged to observe employee work practices. Numerous studies show that most workplace injuries and illnesses are caused by "at risk" employee behaviors as opposed to unsafe conditions. Observing that work is being performed in compliance with safe job procedures and company policies will further protect the safety and health of all employees.

Dependant on structure, an inspection can be viewed either:
- negatively as fault finding, with the emphasis on criticism
- positively as fact finding, with the emphasis on controlling and eliminating recognized hazards that may affect the safety and health of all employees

The second viewpoint is more effective. This preferred viewpoint is more common in organizations where the management group is committed to continuous improvement and sustainability of operations. The management group is critical in seeing that checklists or other yardsticks, which are effective in measuring a particular situation, are developed. The management group also ensures that inspectors are trained how to compare existing conditions or practices (what is) to desired conditions or practices (what ought to be) (Firenze 1978). The view of using safety inspections as part of an operation's continuous improvement process also tends to facilitate developing a process for resolving noted hazards as well as determining the root causes of unsafe conditions and at-risk behaviors. Failure to analyze inspection reports for the root causes of system defects ultimately means a failure of the continuous improvement process. Without root cause analysis, corrective action may fix a specific hazard noted in an inspection but fail to fix the system defect or deficiency that allowed the hazard to occur in the first place.

Prompt correction of substandard or hazardous conditions detected in an inspection shows employees that the company is serious about incident prevention. A supervisor or team leader who finds that workers are not following safe operating procedures should determine the reasons for noncompliance. If the procedures are outdated or unworkable, the supervisor should work with the employees and safety manager to develop realistic safe operating procedures. If an employee does not have the skills or knowledge to perform a task safely, then he or she should be trained. Do not train employees who understand the safe operating procedures but are not complying with the rules for other reasons. Determine what is motivating the worker to be noncompliant and then address those factors that are motivating the employee to engage in at-risk behaviors.

## Major Elements of an Inspection Program

A safety and health inspection program should answer the following questions:

- What items need to be inspected?
- What aspects of each item need to be examined?
- What conditions need to be inspected?
- How often must items be inspected?
- Who will conduct the inspection?

## Frequency of Inspection

When should a safety and health inspection be conducted? Some supervisors might answer, "The third Friday of every month." A better answer would be, "Every time I go through my department." The goal is continuous inspection; every member of the department should constantly be on the lookout for hazards on the job. Safety and health inspections should be part of every phase of production, service, etc. Required safety inspections are defined in the OSHA standards (for example, monthly inspections of area fire extinguishers).

In addition to continuous inspections, formal inspections should be made routinely at least once a month or as required by standards. The supervisor or team leader who conducts the inspection should have a checklist like the one in Figure 6–1.

Checklists or documentation should be completed, dated, and signed by the inspectors. Action items should be corrected and recorded as completed.

There are three types of scheduled inspections:

1. *Periodic inspections* are inspections of specific items conducted weekly, monthly, semiannually, or at other given intervals.
2. *Intermittent (special) inspections* or assessments are performed at irregular intervals. An injury or near miss in another department that involves machinery similar to one's own would lead to an intermittent inspection of the equipment.
3. *General inspections* are designed to include all areas that do not receive periodic inspections—for example, parking lots, sidewalks, and fences.

## Supervisor's Facility and Administrative Inspection Checklist

**Building/Department** _____

This checklist is intended only as a guide in reviewing general facility and administrative items. Only unsatisfactory items and their location need to be identified by a check. Those items identified as unsatisfactory should be targeted for corrective action.

| FACILITY AND OPERATIONS<br>**Machinery and Equipment** | Check If<br>Action<br>Required | Location/Comments/Action Required |
|---|---|---|
| General safeguarding provided and in place | _____ | _____ |
| Operators properly attired (no loose clothing, jewelry) | _____ | _____ |
| Point of operation safeguarding provided and functioning properly | _____ | _____ |
| Proper tools provided for cleanup and adjustments | _____ | _____ |
| Other: | _____ | _____ |
| **Materials Handling and Storage** | | |
| Manual materials-handling equipment in good condition | _____ | _____ |
| Powered materials-handling equipment in good condition | _____ | _____ |
| Hazardous and toxic materials handled, stored, and transported in accordance with regulatory requirements | _____ | _____ |
| Storage areas properly illuminated | _____ | _____ |
| Cylinders transported and stored in upright position; properly secured | _____ | _____ |
| Shipping/receiving areas in good condition | _____ | _____ |
| Racking and other storage procedures followed | _____ | _____ |
| Wheel chocks and restraining devices available/ functioning properly | _____ | _____ |
| Other: | _____ | _____ |
| **Hand and Portable Power Tools** | | |
| Correct tools provided | _____ | _____ |
| Hand tools and power equipment in good condition | _____ | _____ |
| Guards are in place, adjusted properly | _____ | _____ |
| Grinding wheel tool rest is within 1/8 in. of wheel | _____ | _____ |
| Stored tools are locked and/or secured | _____ | _____ |
| Electrical tools GFCI protected | _____ | _____ |
| Electrical tools and receptacles grounded | _____ | _____ |
| Other: | _____ | _____ |

**Figure 6–1.** A supervisor's facility and administrative inspection checklist, such as this one, can be used as a guide for reviewing general facility items.

**Fire Protection**

Provided as required

Inspected as marked

Location identified

Locations are readily accessible

Alarm system tested (as required)

Fire doors in good operating condition

Exits marked and accessible

Fire detectors working

Other:

**Electrical**

Outlet boxes covered

Electric cords properly placed

Outlet circuits properly grounded

Portable electric tools:

- GFCI protected

- Double insulated

- Grounded as required

Switches in clean, closed boxes

Switches properly identified

Circuit fuses, circuit breakers identified

Motors are clean, free of oil, grease, and dust

Approved extension cords in good condition

Other:

**Housekeeping/Maintenance**

Work areas maintained in clean and orderly condition

Floors, aisles, work areas free of obstruction, slipping and tripping hazards

Washrooms and change facilities clean and well maintained

Tools, equipment, and materials properly stored when not in use

Waste materials stored in appropriate containers and disposed of in a safe manner

Scheduled maintenance

General ventilation systems

Local exhaust systems (paint booths, welding areas, etc.)

Machinery (lubrication, belts, servicing, etc.)

Other:

**Figure 6–1.** (Continued)

**Personal Protective Equipment**

Equipment (determined by exposure):
- Head protection _____ _____
- Eye protection _____ _____
- Ear protection _____ _____
- Foot protection _____ _____
- Clothes _____ _____
- Hand protection _____ _____
- Respiratory protection _____ _____

Personal protective equipment procedure in place _____ _____

Other: _____ _____

**Administrative**

Training records:
- Safety and health orientation _____ _____
- Hazard communication (right to know) _____ _____
- Safe operating procedures _____ _____
- Confined space entry procedures _____ _____
- Lockout/tagout _____ _____
- Evacuation emergency response _____ _____
- Equipment/vehicle operation _____ _____
- Fire protection equipment use _____ _____
- Other _____ _____

Plans:
- Disaster preparedness _____ _____
- Chemical emergencies/spills _____ _____
- Fire/evacuation _____ _____
- Emergency medical _____ _____
- Equipment maintenance _____ _____
- Other _____ _____

Records/reports:
- Injury/illness _____ _____
- Accidents/incidents _____ _____
- MSDSs _____ _____
- Inspection summaries _____ _____
- Noise surveys _____ _____
- Equipment service logs _____ _____
- Other: _____ _____

Other:
- OSHA-required postings _____ _____
- Emergency phone listings _____ _____
- Required labeling _____ _____
- Defective equipment procedure in place _____ _____

Completed by: _____

Date: _____

Route to:
○ Maintenance    ○ Engineering    ○ Other

**Figure 6–1.** (Concluded)

The frequency of inspections can be determined by the answers to these five questions:

1. *Are there required inspections?* Regular inspections of certain equipment may be mandated by regulation, by a manufacturer's recommendation, or by OSHA or other government agencies. Inspections of such equipment should be documented.

2. *What is the loss severity potential of the problem (human, equipment, facilities, products, services)?* Example: A frayed wire rope on an overhead crane block has the potential to cause a much greater loss than a defective wheel on a wheelbarrow. The rope needs to be inspected, therefore, more frequently than the wheelbarrow.

3. *What is the potential for injury to employees?* If the item or a critical part failed, how many employees would be endangered and how frequently? For example, a stairway that is used all the time needs to be inspected more frequently than one that is used rarely.

4. *How quickly can the item or part become hazardous?* Equipment and tools that get heavy use become damaged or defective, or wear out faster, than those that are not often used. In addition, an item in one location may be exposed to greater potential damage than an identical item in a different location.

5. *What is the past record of failures?* Maintenance and production records and incident investigation reports can provide valuable information both about how often items have failed, and what results have occurred, in terms of injuries, damage, delays, and shutdowns.

## What to Inspect

Many different types of inspection checklists are available. They vary in length from a few items to hundreds. Compare the checklist in Figure 6–1 to the one that follows:

1. environmental factors (illumination, dust, gases sprays, vapors, fumes, noise)
2. hazardous chemicals and materials (explosives, flammables, acids, caustics, toxic materials, or by-products), including labeling, signage, placards, and MSDSs
3. production and related equipment (mills, shapers, presses, borders, lathes)
4. power equipment (steam and gas engines, electric motors)
5. electrical equipment (switches, fuses, breakers, outlets, cables, extension and fixture cords, grounds, connectors, connections)

6. hand tools (wrenches, screwdrivers, hammers, power tools)

7. personal protective equipment (hard hats, safety glasses, safety shoes, respirators)

8. personal service and first-aid facilities (drinking fountains, washbasins, soap dispensers, eyewash fountains, first-aid supplies, stretchers)

9. fire protection and extinguishing equipment (alarms, water tanks, sprinklers, fire doors, standpipes, extinguishers, hydrants, hoses)

10. walkways and roadways (ramps, docks, sidewalks, aisles, vehicle ways)

11. elevators, electric stairways, and lifts (controls, wire ropes, safety devices)

12. working surfaces (floors, ladders, scaffolds, catwalks, platforms, sling chairs)

13. materials-handling equipment (cranes, dollies, conveyors, hoists, forklifts, chains, ropes, slings)

14. transportation equipment (cars and trucks, railroad cars, front-end loaders, helicopters, motorized carts and buggies)

15. warning and signaling devices (sirens, crossing and blinker lights, klaxons®, warning signals)

16. containers (scrap bins, disposal receptacles, carboys, barrels, drums, gas cylinders, solvent cans)

17. storage facilities and areas, both indoor and outdoor (bins, racks, lockers, cabinets, shelves, tanks, closets), including labels, signs, placards, and MSDSs

18. structural openings (windows, doors, stairways, sumps, shafts, pits, floor openings)

19. buildings and structures (floors, roofs, walls, fencing)

20. grounds (parking lots, roadways, sidewalks)

21. loading and shipping platforms

22. outside structures (small, isolated buildings)

23. miscellaneous (any items that do not fit in the preceding categories)

24. emergency equipment (fire extinguishers, deluge showers, spill kits, telephones, etc.)

25. environmental controls (pollution control devices, etc.)

Generally, more detailed checklists are keyed to OSHA standards. These checklists are useful in determining how the safety standards or regulations apply to individual situations. Once the relevant standards are identified, a checklist can be customized and computerized for easy follow-up.

General checklists serve as reminders of what to look for, and to document what has been covered in past inspections. The checklists provide direction, and permit easy, on-the-spot recording of all findings and comments. If an inspection is interrupted, checklists provide a record of what has already been covered. An inspector who does not have a printed checklist should at least carry a notebook to jot down items during continuous inspections.

An effective method for developing area-specific checklists and increasing employee sense of influence over the safety management system is relatively simple to execute. A supervisor can hold a meeting and ask each employee to complete the following statement: "I think an item that should always be checked during a safety inspection is ..." During this meeting, the supervisor needs to go from crew member to crew member requiring each to complete the statement. The crew member should either complete the sentence or say, "pass" if he or she cannot think of anything. One crew member may say "that the fire extinguisher is in place" and a second may say "that the floors are clean." A person should be writing down each crew member's answers on a flip chart. The team leader should move from meeting participant to meeting participant in order until everyone has passed. The ideas collected on the flipchart will form the basis of a department-specific checklist.

If this procedure is used in different areas of a facility, quite different checklists would be developed. The concerns of the employees working on a shipping dock would be markedly different than employees working in a chemical mixing area. The summation of the checklists created by the different departments within a facility would constitute that site's hazard control inventory.

It is important to remember that checklists are merely an aid to the inspection process, not an end in and of themselves. A safety and health inspection requires much more than simply checking items off a list. Any hazard observed during an inspection (even if it is not on the list) must be recorded and corrected. Any hazard posing imminent danger to employees must be corrected on the spot or the area must be closed down until the hazard is corrected. Delay could lead to an incident.

**High-Risk Areas.** Some items—such as floors, stairways, housekeeping procedures, fire hazards, electrical installations, and chains/ropes/slings—are especially high-risk areas in a department or facility. These items should be inspected often.

Floors should be inspected carefully, no matter what material they are made of, especially those floors that are slippery or in areas with heavy traffic. Critical factors include:

- Is the surface damaged or wearing out too rapidly?
- Is the floor material shrinking?
- Are there slippery areas?
- Are there holes or unguarded areas?
- Are there indications of cracking, sagging, or warping?
- Are replacements necessary due to deterioration?
- Is shoring up, strengthening, resurfacing, or reattaching necessary?
- Is structural investigation indicated?

Stairs must be clear, sturdy, slip-resistant, and must have adequate safety rails, treads, and enclosures. They should never be used for storage. Stairways should be checked for the following:

- Are treads and risers in good condition and of uniform height and width?
- Are handrails secure and in good condition?
- Is lighting on stairs bright enough?
- Are access areas at the top and bottom open and free of clutter?
- Is the overall structure sturdy and strong enough for its purpose?

General housekeeping throughout the facility should be checked regularly. Aisles should be marked off with painted lines and kept clear; burned-out lights should be replaced promptly; locks, handles, door frames, appliances, and furnishings should be kept in good repair; tools should be stored in organized storage; and general cleanliness and good order should be maintained.

Fire is one of the greatest hazards, especially to industrial facilities. All fire protection equipment (including sprinkler systems, alarms, extinguishers, standpipes, and hoses) should be inspected often and kept in working order. All exits from the building and emergency lighting systems should be routinely inspected and tested. Fire protection equipment must be in the right place and not blocked or obscured.

Electrical installations should be in compliance with the National Electrical Code, ANSI/NFPA 70 1996, published by the National Fire Protection Association. In this book, refer also to Chapter 16, Electrical Safety.

A qualified expert should inspect chains, wire and fiber ropes, slings, and other equipment that is subject to severe strain in handling heavy equipment and materials. Careful records of each inspection should be maintained.

**Work Practices.** Another important purpose of a safety and health inspection is to observe work practices. Are employees following specific safety and health procedures as they do their jobs? The best guide for determining how a job should be done is a job safety analysis.

The following questions highlight some work practice problems that should be inspected:

- Are machines or tools being used without authorization?
- Is equipment being operated at unsafe speeds?
- Are guards or other safety devices being removed or rendered ineffective?
- Are defective tools or equipment being used, or are tools or equipment being used unsafely?
- Are employees using their hands or bodies instead of tools or push sticks to move items in hazardous situations?
- Is overloading or crowding occurring? Are workers failing to store materials properly?
- Are materials being handled in unsafe ways? For example, are employees lifting loads improperly?
- Are employees repairing or adjusting equipment while it is in motion, under pressure, or electrically charged?
- Are employees not using (or not using properly) PPE or other safety devices?
- Are unsafe, unsanitary, or unhealthy conditions being created by workers' poor personal hygiene—for example, smoking in unauthorized areas or using compressed air to clean their clothes?
- Are employees standing or working under suspended loads, scaffolds, shafts, or open hatches?

## Preparation

Inspections should be scheduled when there is a maximum opportunity to view operations and work practices with a minimum of interruptions. Although the areas and routes should be planned in advance, the day and time on which formal inspections are done should be varied, in order to check the greatest possible variety of conditions.

It is a good idea to review all incidents that have occurred in the area in the past. The inspector should review copies of previous inspection reports. This review will reveal whether earlier recommendations to remove or correct hazards were followed.

## Results and Reports

The supervisor or team leader should discuss the results of inspections with the employees. If poor work habits have been observed, the inspector should point them out to employees immediately, determine the root cause of the at-risk behavior, and eliminate those root causes.

Communicating the good news as well as the bad is essential. Many supervisors forget to mention the positive actions and practices that take place in their departments. People who follow good work practices should be encouraged, and their success applauded, rather than ignored. Comments like, "I'm glad to see that you always check the condition of your tools before using them," can give workers positive reinforcement. An inspector who finds it necessary to correct someone's actions should be sure to acknowledge any resulting improvement in the person's work patterns.

A clearly written report must follow each formal inspection. The report should specify the department or areas inspected (identifying the boundaries or location, as needed) and the date and time of the inspection.

One way to begin the report is to copy hazards identified on the last report that have not been completely corrected, numbering each item consecutively. The inspector should specify the recommended corrective action, the date by which the hazard should be corrected, and the person responsible for fixing it.

The report should describe in detail each hazard uncovered during the current inspection. Machines and specific operations should be identified, and locations should be precisely described. Instead of merely noting "poor housekeeping," for example, the report should give clear details: "Empty pallets left in aisles, slippery spots on floor from oil leaks, a ladder lying across empty boxes, scrap piled on floor around machines." Instead of noting "guard missing," the report should read, "guard missing on shear blade of No. 3 machine, SW corner of Bldg. D."

## Follow-Up

The supervisor or team leader should follow up on inspection reports to correct any problems that were discovered. If a problem can be resolved internally, workers should be assigned to fix it promptly. For example, if materials are not properly stacked, employees should be asked to restack them safely. If the condition cannot be corrected by team members, the team leader should write a maintenance work-order request.

Resolving underlying causes of hazards is as important as correcting them immediately. For example, wiping up an oil spill removes the immediate problem, but it does not eliminate the cause. If the oil leak came from a forklift truck, the truck must be identified and repaired to prevent further oil spills.

If permanent correction of the problem will take time, the supervisor should consider temporary measures that will help to prevent an incident. Roping off the area, tagging or locking out equipment, and posting warning signs are examples of intermediate actions. They are not ideal, but they are steps in the right direction. They can prevent injuries or damage, while team members seek permanent solutions to the problem.

If the inspector discovers other dangerous conditions during follow-up work, they should be reported immediately to the appropriate person or management. The report should include recommendations for removing or correcting the conditions. If, for any reason, it is impossible to recommend the optimal best-practice, long-term solutions, then intermediate or short-term solutions must be recommended and identified as such. If identified, the best-practice solution should also be listed, and its advantage described, along with recommendation for it being applied as soon as possible, when conditions permit. Key inspection results should be shared so that others can learn and avoid similar problems.

# CONTROL OF WORKPLACE HAZARDS

When worker exposure to health or physical hazards is found to pose a risk to employee safety, a control program must be implemented. The control method chosen depends on the:

- specific hazard
- magnitude of exposure
- effectiveness
- available technology

Three categories of methods are used to control hazards: engineering, administrative, and personal protective equipment.

## Engineering Controls

The first and most reliable strategy is to control a hazard at its source—design it out of the process. Engineering controls do this, unlike other controls that generally focus on an employee exposed to the hazard. The basic premise behind these controls is that, to the extent possible, the work environment and job must be designed to eliminate hazards or reduce exposure to hazards. If hazards cannot be eliminated, engineering controls often reduce them.

Today, companies are more likely than in the past to examine all possible ways to engineer out hazards. Cost of toxic materials disposal, rising health care costs and workers' compensation, and other protective devices are high. It is more cost-effective to eliminate hazards and engineer safety into design than to wait until a hazard arises.

Examples include:

- Substitution or elimination. Replace or eliminate toxic chemicals with less hazardous or nontoxic substitutes.
- Engineering. Change the process.
- Isolation. Enclose the process to reduce the number of exposed workers.
- Barriers. Install barriers or machine safeguards.

## Administrative Controls

Administrative controls minimize worker exposure by using policies, procedures, and rules such as standard operating procedures. This type of control is normally used with other controls that more directly prevent or control exposure to the hazard.

Examples include:

- Housekeeping. Practice proper chemical storage procedures as well as daily cleaning and organization.
- Education/training. Educate and train employees on safe job practices, hazard recognition, and control methods.
- Relief workers/rotation of workers. Rotate employees in and out of jobs for relief in challenging work environments.
- Breaks. Set up break schedules that complement challenges faced in a job or work setting.

## Personal Protective Equipment Controls

When exposure to hazards cannot be engineered completely out of normal operations or maintenance work, and when administrative controls cannot provide sufficient additional protection, a supplementary method of control (the last line of defense) is the use of personal protective clothing or equipment.

PPE, such as respirators, gloves, and safety glasses, are articles of clothing or equipment worn by workers for protection. These do not reduce hazards, but provide a barrier between the worker and the hazard. Use PPE only after exhausting other controls or while engineering controls are being implemented. For specific OSHA requirements on PPE, see OSHA's standard, 1910 Subpart I.

### Hierarchy of Controls

Figure 6–2 shows the hierarchy of control methods described above. For example, in a very hot work area, the best approach to prevent heat stress is to implement engineering controls such as shields, cooling fans, and air conditioning. Administrative controls may include helping employees adjust to heat by longer (gradual) periods of exposure, increased fluids, and education. A final approach is the use of PPE such as coolant suits or reflective clothing.

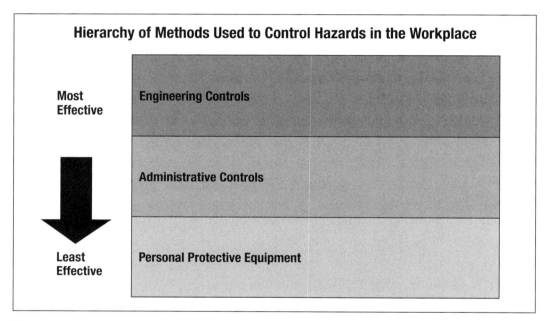

**Figure 6-2.** The most effective way to minimize hazards in the workplace is to incorporate engineering controls in the design process before an incident can occur.

## SUMMARY OF KEY POINTS

Key points covered in this chapter include:
- Safety and health assessments and audits are keys to continuous improvement. They usually cover everything from occupational safety and industrial hygiene to process and product safety, occupational health and medicine, and the overall functioning of the safety management system. Safety and health

inspections are one of the safety management system elements to be audited.

- The safety and health inspection program should address the issues of how often inspections should occur, what items should be inspected, what aspects of each item need to be examined, what conditions should be inspected, and who will conduct the inspection. Supervisors or team leaders should be trained  to conduct both continuous and periodic safety and health inspections, as well as intermittent and general inspections. The timing of inspections should be staggered to check the greatest possible variety of conditions.

- High-risk items should be inspected often. They include floors; stairways; exits and entrances; housekeeping procedures; fire hazards; electrical installations; and chains ropes, and slings.

- The supervisor should discuss the results of inspections with employees, making sure to give positive reinforcement for good work practices. The written report should describe in detail each hazard uncovered during the inspection. Response to reports should include immediate follow-up to correct problems and resolve their underlying causes.

# 7

# Incident Investigation

**After reading this chapter, you will be able to:**

- Describe the purpose and objectives of incident reporting
- List and use the proper incident investigation steps
- Select and interview incident witnesses
- Fill out an incident report
- Determine root causes using a causal factors analysis

The word "accident" often brings to mind circumstances that are both uncontrollable and unpredictable. Many times, the use of the word "accident" may discourage a supervisor from doing an in-depth investigation. Why spend a lot of time investigating an "accident," if it is just going to happen again?

Many organizations that excel in safety management use the term "incident" instead of the term accident. The word incident is defined to mean injuries, illnesses, near misses, and other situations that have the potential to cause harm. By conducting appropriate incident investigations, the root cause of the incident can be discovered and corrected, preventing any recurrence of the incident.

Effective incident investigation is an essential step toward making changes that can prevent recurrence of injuries or illnesses from similar causes. If anything positive results from an employee injury, illness, or near miss, it is the opportunity to determine the root causes and ways to eliminate them. Thorough incident investigation can identify hazardous conditions and at-risk behaviors within an organization. Effective incident investigations conducted on near misses have the added bonus of preventing injuries or illnesses before they occur.

Effective incident investigation involves more than merely filling out forms. Well-managed organizations insist on quality incident investigations, just as they insist on efficient, quality production. Part of the performance evaluation for a supervisor or team leader may be based on how well an incident investigation is handled.

## INCIDENT REPORTING

A good incident investigation program begins with an effective, thorough incident notification plan. Many organizations make the mistake of requiring notification and investigation of "serious" injuries or illnesses only. All incidents, including near misses, should be reported as soon as possible—whether the incident did or could have resulted in personal injuries, illnesses, or property damage. For example, a tool falling off a platform that is missing toeboards might not hurt anyone. However, it just as easily could have fallen on someone's head, inflicting a serious injury. Regardless of the outcome, this kind of incident should be reported, the causes investigated, and corrective action taken to prevent a recurrence. When the causes are removed, injuries and illnesses can be prevented. To help establish an effective incident reporting system, make sure that the incidents are not just events that cause injuries or property damage, but can be near misses that result in neither injuries nor damage. A supervisor or team leader has a personal interest in every incident that occurs in his or her department. In the interest of prevention, team leaders and supervisors must learn the contributing causes for every incident so they can be eliminated.

## FINDING CAUSES

The purpose of incident investigation is to determine root causes and recommend corrective actions to eliminate or control these hazards. Incident investigation should be aimed at fact-finding rather than fault-finding; otherwise, the investigation can do more harm than good. The emphasis should be on identifying all possible causes, not on who should be blamed for the incident. An attempt to place blame damages the investigator's credibility and generally reduces the amount and accuracy of information received from workers. This does not mean oversights or mistakes should be ignored, or that personal responsibility should not be determined, when appropriate. It means that the investigation should be concerned only with the facts. To do a quality job of incident investigation, the investigator must be objective and analytical.

There are many factors that combine to cause incidents. The theory of multiple causation states that it is the random combination of various factors that results in injuries and illnesses.

A report must do more than just identify at-risk acts or hazardous conditions surrounding an incident. These may be only contributing causes and not the root causes of a specific incident. For example, an incident report might state that the incident was caused by an oily patch on the floor. Instead of merely noting there was oil on the floor, the supervisor must determine how and why the oil got there. If there was an equipment lubrication leak, had the condition been previously reported? If poor maintenance was the cause, had that been reported? If so, when? If not, why not? These are the kinds of questions that must be asked and answered (i.e., who, what, where, when, how, and why).

Generally, the procedures followed during an investigation are designed to elicit clues to other problems. Inadequate maintenance, poorly designed equipment, untrained or ill-trained employees, and lack of policy enforcement or standard procedures (management control) are all causative factors (Figure 7–1).

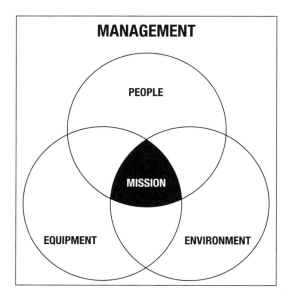

**Figure 7–1.** This diagram shows a basic system in which people, equipment, and the environment are managed to accomplish a mission. Sometimes they work together in unplanned ways to produce incident tips.

## Taking Immediate Action

The safety and health of employees and visitors must be the primary concern when an incident occurs. If an injury or illness results, make sure the affected persons get immediate medical attention and take steps to provide for emergency rescue or evacuation, as necessary. In addition, take any precautions that will prevent or minimize the probability of further injuries or illnesses occurring as a result of the initial incident.

## Securing the Incident Site

After rescue and damage control are completed, the incident site must be secured for the duration of the investigation. It is essential to barricade or isolate the scene with ropes, barrier tapes, cones, and/or flashing lights or other devices to warn people or otherwise restrict access to the area. In extreme cases, guards may be posted to make sure no unauthorized persons enter the scene.

Nothing should be removed from the site without approval of the person in charge of the investigation. The site should be maintained, as much as possible, just as it was at the time of the incident to help investigators identify and examine evidence.

## Preserving Evidence

Time is critical when investigating incidents. The faster the investigator gets to the incident scene, the less chance there is that details will be lost. A prompt and careful investigation will help to answer the questions of who, what, where, when, how, and why. For example, witnesses will remember more facts, chemical spills will not have had time to evaporate, and dust and debris will remain as it was.

Preserving evidence at the incident scene ultimately makes the investigative process much less frustrating. Observing and recording evidence—such as instrument readings, control panel settings, and details of weather and other environmental conditions—can greatly improve investigation results. Evidence can be preserved on film, recorded on tape, diagramed, or sketched. Detailed notes must accompany any pictures or drawings.

### Recording Visual Evidence

Visual evidence should be recorded using cameras, video camcorders, drawings, and notes. The camera is one of the most valuable tools for studying incidents. A digital camera may be preferred by those with no photography experience. Photographs can be studied in detail and at length to look for things initially overlooked. General and specific scenes should be photographed in order to make comprehensive visual records. No one can predict in advance which data will be most useful, so take photographs from many different angles. An old saying for accident photographers is "overshoot and underprint." This means take every possible photo that might be needed, and make necessary enlargements after the proofs have been studied. In addition, accurate, complete, and annotated sketches and diagrams of the scene should be made.

Photographs of objects involved in the incident should be identified and measured to show proper scale and perspective. A ruler or coin can be placed next to objects photographed up close to demonstrate the object's size. Accurate measurements of the area, equipment, and materials involved can be vital to an investigation, with or without photographs.

## EFFECTIVE USE OF WITNESSES

### Identifying Witnesses

Avoid restricting the search for witnesses to those who saw the incident happen. Anyone who heard or knows something about the event may offer useful information. Witnesses who smelled something may be valuable to the investigation. The smell of electrical insulation or forklift brakes are smells that remain with a person long after the initial exposure. Individuals who operated the equipment on the prior shift might have valuable information as to the operation of the equipment that will help in the investigation. Ask witnesses to identify and document the names of others who were in the area, so that everyone can be contacted.

### Interviewing Process

Witnesses should be interviewed one at a time in private as soon after the incident as possible. The accuracy of what people recall is highest immediately following an event. A prompt interview minimizes the possibility of a witness subconsciously adjusting his or her story. If too much time elapses, many things can cloud a person's memory. For example, hearing other opinions or reading stories about the incident can influence a witness. A person with a vivid imagination can "remember" situations that did not actually occur. Make sure to set aside time at the beginning of the investigation to interview as many witnesses as you can.

Whenever possible, interviews should be conducted at the incident site. This gives witnesses opportunity to describe and point out what happened. Also, being at the scene can spark a person's memory. Tactful, skilled investigations will usually elicit greater cooperation from employees. Remind witnesses that you are interested in the facts of the incident, not in placing blame. Such reassurance can dispel people's fear that they may incriminate themselves or others.

**How to Interview.** When conducting a witness interview, establish a relaxed atmosphere. Be a good listener. Witnesses should be allowed to tell their stories without interruption or prompting. More detail can be sought after the full story has been told. Interruptions can derail a person's train of thought, influence answers, and inhibit responses. To elicit the most information, open-ended questions without leading words are best. For example, asking "Where was he standing?" is preferable to "Was he standing there?" Remember that the purpose is to determine the facts and to gain as much information from witnesses as possible. Questions should cover the specifics of who, what, where, when, how, and why.

Take notes and record employee statements, as unobtrusively as possible, for later review. In some cases, it may be best to wait until the employee has finished explaining what happened before making notes or recording details. If the employee's statement is recorded, get his or her permission and have it typed for the employee's review, corrections, and signature. After the witness is through, always repeat the information as it is understood. This feedback technique reduces misunderstandings and often leads to further clarifications.

Employees who have been directly involved with the incident should be contacted first, followed by eyewitnesses, and those who were nearby. Always use discretion in situations where employees have been injured. Fellow workers might be in shock, unable to speak coherently, or have no memory of the incident. Sometimes it is better to postpone interviews until the employee or witness who has been involved in the incident is in a more stable physical or emotional condition. A part of the emergency response should include care for traumatized co-workers.

It is helpful to solicit ideas from the interviewed employees regarding ways to prevent a recurrence of the incident. Often they will have the best suggestions. This will also help them to feel involved in the investigation and provide an opportunity to recognize their willingness to participate. Be sure to record these as suggestions made before causes have been determined.

The interviews should end on a positive note. Thank each individual for his or her time, information supplied, and ideas offered.

**Reenacting the Incident.** The investigator may want to ask employees to show "what they mean" or "how it happened." In some cases, reenacting an incident can provide valuable information about how and why it occurred. Expert investigators have learned to use this technique with caution, however, so the reenactment doesn't cause another injury or illness.

Before asking people to reenact the scene, follow these steps:

1. Ask employees to explain what happened first. This preliminary explanation should give additional insight about the incident.
2. Make sure that the witnesses thoroughly understand that they are to go through only the general motions of the event under close supervision. They should not repeat the exact actions that produced the original event.

**Selecting a Location.** When it is not possible to interview people at the incident site, choose another location free from distractions and away from other witnesses. Privacy is essential. When people are discussing what they saw, they can influence other witnesses. You may want to avoid having people revise their stories. Do not pressure or influence witnesses. If possible, do not use your office to conduct interviews; employees may find the supervisor's office intimidating, inhibiting, or distracting.

## INCIDENT INVESTIGATION REPORTS

The purpose of incident reporting is to alert and inform management and other concerned people about the circumstances surrounding an incident. The report should record in clear, concise language all appropriate facts of the incident. A thorough causal factors (root cause) analysis is a critical part of the investigation. Enter all causal factors that might have led to the event. These factors will be assigned to one or more of the following causal categories: equipment, environment, personnel, and management.

Responsibility for carrying out recommended corrective actions and a timetable for their completion should be established. Completion reports should be required as a means for ensuring the recommended actions have been carried out.

## Filling Out a Report

The following instructions apply to using the Incident Investigation Report shown in Figure 7–2.

This report is designed primarily for investigation of incidents involving injuries or near misses. However, it also can be used to investigate occupational illnesses arising from a single exposure (for example, dermatitis caused by splashed solvent, or a respiratory condition caused by the release of a toxic gas). In property damage accidents, simply write "N.A." (not apply) across Items 1 through 17 and fill out the remainder of the form.

All questions on this form should be answered. If no answer is available, or the question does not apply, the investigator should indicate this on the form, such as placing N.A. in the box. Answers should be complete and specific. Supplementary sheets can be used for other information, such as drawings and sketches, and should be attached to the report. A separate form should be completed for each employee who is injured in a multiple-injury incident.

The report form meets the record-keeping requirements specified in OSHA Form 301. The individual entries are explained below.

**Department.** Enter the department or other local identification of the work area to which the injured is assigned (for example, maintenance shop or shipping room). In some cases, this may not be the area in which the incident occurred.

**Location.** Enter the location where the incident occurred if different from the employer's mailing address.
1. Name of injured. Record the last name, first name, and middle initial.
2. Social Security Number.
3. Sex.
4. Age.
5. Date of accident or date of initial diagnosis of illness.
6. Home address.
7. Employee's usual occupation. Give the occupation to which the employee is normally assigned (for example, assembler, lathe operator, clerk, etc.).
8. Occupation at time of incident. Indicate occupation in which the injured is

working at the time of incident, if other than his or her usual occupation.

9. Length of employment. Check appropriate box to indicate how long the employee has worked for the organization.

10. Time in occupation at time of incident. Record the total time employee has worked in occupation indicated in Item 8.

11. Employment category. Indicate injured worker's employment category at the time of incident (for example, regular, part-time, temporary, or seasonal).

12. Case numbers and names of others injured in same incident. For reference purposes, names and case numbers of all others injured in the same incident should be recorded here.

13. Nature of injury and part of body. Describe exactly the kind of injury, or injuries, resulting from the incident, and the part, or parts, of the body affected. For an occupational illness, give the diagnosis and the body part, or parts, affected.

14. Name and address of physician.

15. Name and address of hospital.

16. Time of injury. In part B, indicate as closely as possible the hour and minute that the injury-producing incident occurred.

17. Severity of injury. Check the highest degree of severity of injury. The options are listed in decreasing order of severity.

18. Specific location of incident. Indicate whether incident or exposure occurred on employer's premises. Record the exact location of the incident (for example, at the feed end of No. 2 assembly line, at support column No. 32 in the warehouse, or in the employees' lunchroom). Attach a diagram or map if it would help to identify the exact location.

19. Phase of employee's workday at time of injury. Indicate what phase of the workday the employee was in when the incident occurred, in which hour of the shift the injury occurred (for example, first hour). In part C, record the type of shift (for example, rotating or straight day). If "other," be specific.

20. Describe what happened. What physical things and procedures were involved? Provide a complete, specific description of what happened. Tell what the injured and others involved were doing prior to the incident; what relevant events preceded the injury; what objects or substances, operations, equipment, tools, etc., were involved; how injury occurred and the specific object or substance that inflicted the injury; and what, if anything, happened after the incident. Include only facts obtained in the investigation. Do not record opinions or place blame.

# Incident Investigation Report

Case Number _____

| Company | Address |
|---|---|
| Department | Location |

| 1. Name of injured | 2. Social Security Number | 3. Sex | 4. Age | 5. Date of accident ___ / ___ / ___ |
|---|---|---|---|---|

| 6. Home Address | 7. Employee's usual occupation | 8. Occupation at the time of the accident |
|---|---|---|

**9. Employment category**
☐ Regular, full-time  ☐ Regular, part-time  ☐ Temporary
☐ Seasonal  ☐ Non-Employee

**10. Length of employment**
☐ Less than 1 mo.  ☐ 1-5 mos.
☐ 6 mos. - 5 yrs.  ☐ More than 5 yrs.

**11. Time in occup. at time of the accident**
☐ Less than 1 mo.  ☐ 1-5 mos.
☐ 6 mos. - 5 yrs.  ☐ More than 5 yrs.

**12. Nature of injury and part of the body**

**13. Case numbers and names of others injured in same accident**

**14. Name and address of physician**

**15. Name and address of hospital**

**16. Time of injury**
A.        a.m.
           p.m.

B. Time within shift _____

C. Type of shift _____

**17. Severity of injury**
☐ Fatality
☐ Lost  workdays - days away from work
☐ Lost  workdays - days of restricted activity
☐ Medical treatment
☐ First Aid
☐ Other, specify _____

**18. Specific location of incident**

On employer's premises?  ☐ Yes  ☐ No

**19. Phase of employee's workday at time of injury**
☐ During rest period          ☐ Entering or leaving plant
☐ During meal period          ☐ Performing work duties
☐ Working overtime            ☐ Other

**20. Describe how the incident occurred**

**21. Incident sequence.** Describe in reverse order the occurence of events preceding the injury and accident.
Start with the injury and moving backward in time, reconstruct the sequence of events that led to the injury.

A. Injury event _____

B. Incident event _____

C. Preceding event #1 _____

D. Preceding event #2, 3, etc. _____

**22. Task and activity at time of incident**
General type of task _____
Specific activity _____
Employee was working:
☐ Alone
☐ With crew or fellow worker
☐ Other (specify)

**23. Posture of employee**

**24. Supervision at time of incident**
☐ Directly supervised ☐ Indirectly supervised
☐ Not supervised     ☐ Supervision not feasible

**Figure 7–2.** Complete an Incident Investigation Report after you have gathered all possible evidence. See the text for information on how to complete this form.

**25. Cost estimates**

☐ Actual estimates      ☐ Formula-generated average      ☐ Both

**A. Actual cost estimate**

| | | | |
|---|---|---|---|
| Property and equipment damage | $ _____ | Product spoilage | $ _____ |
| Production interruption | $ _____ | Lost work time (other than injured) | $ _____ |
| Retraining | $ _____ | Administrative time | $ _____ |
| Workers' compensation | $ _____ | | |
| (medical + indemnity paid + indemnity reserve) | | Total | $ _____ |

Total cost $ _____      ☐ estimated      ☐ actual

**B. Formula-generated average cost**

Average hourly wage plus fringe benefits for this job $ _____

Injured body part:  ☐ Head, face   ☐ Eye   ☐ Neck, shoulder   ☐ Hearing loss   ☐ Chest, lower trunk   ☐ Heart attack
                    ☐ Back   ☐ Rib   ☐ Hernia, rupture   ☐ Arm, elbow   ☐ Thumb, finger   ☐ Wrist, hand
                    ☐ Hip   ☐ Leg, knee   ☐ Foot, ankle   ☐ Toe   ☐ Other (estimate work hours lost: ____ hours)

Injury type:   ☐ Burn   ☐ Amputation   ☐ Strain, sprain, crush or mash   ☐ Death
               ☐ Fracture   ☐ Bruise, abrasion   ☐ Cut, puncture or laceration   ☐ Other _____

Formula-generated cost estimate (from computer calculation) $ _____

☐ Yes   ☐ No    Do you use ANSI Z16.2 coding? If no, proceed to Causal factors.

| | | | |
|---|---|---|---|
| Nature of injury code | _____ | Body part affected code | _____ |
| Injury source code | _____ | Incident type code | _____ |
| Hazardous source condition | _____ | Agency of incident code | _____ |
| Agency of incident part code | _____ | Nature of illness code (BLS) | _____ |
| Other code | _____ | | |

**26. Causal factors.** Events and conditions that contributed to the incident.
Include those identified by use of the **Causal Factors & Corrective Actions** form.

_____
_____
_____
_____
_____
_____
_____
_____
_____
_____

**27. Corrective Actions.** Those that have been, or will be, taken to prevent recurrence.
Include those identified by use of the **Causal Factors & Corrective Actions** form.

_____
_____
_____
_____
_____
_____
_____
_____
_____
_____
_____

**Prepared by** _____        **Approved** _____
**Title** _____        **Title** _____ **Date** _____
**Department** _____ **Date** _____        **Approved** _____
        **Title** _____ **Date** _____

**Figure 7–2.** Concluded.

21. Incident sequence. Provide a breakdown of sequence of events leading to the injury. This breakdown enables the investigator to identify additional areas where corrective action may be taken.

In most incidents, the incident event and the injury event are different. For example, suppose a bursting steam line causes burns to an employee's hands or a chip of metal strikes an employee's face during a grinding operation. In these cases, the incident event—the steam line bursting or the metal chip flying up— is separate from the injury event—the steam burning the employee's hands or the chip cutting the employee's face. The question is designed to draw out this distinction and to record other events that led to the incident event.

There also may be events preceding the incident that, although not incident events themselves, contributed to the incident. These preceding events can take one of two forms. They can be something that happened that should not have happened, or something that did not happen that should have happened. The steam line, for example, may have burst because of excess pressure in the line (preceding event #1). The pressure relief valve may have been corroded shut, preventing the safe release of excess pressure (preceding event #2). The corrosion may not have been discovered and corrected because a regular valve inspection/test was not carried out (preceding event #3).

To determine whether a preceding event should be included in the incident investigation sequence, the investigator should ask whether its occurrence (if it should not have happened) or nonoccurrence (if it should have happened) affected the event sequence that produced the incident and brought about injury.

Take enough time to carefully think through the sequence of events leading to the injury and record them separately in the report. The information found in the Finding Causes section earlier in this chapter can be used to help identify management system defects that may have contributed to events in the accident sequence. By identifying such defects, management may help prevent other types of incidents in addition to the one under investigation (see Item 25).

For example, identifying failure to detect a faulty pressure relief valve would lead to a management review of all equipment inspection procedures. This review could prevent other incidents that might have resulted from failure to detect faulty equipment in the inspection process.

Additional sheets may be needed to list all of the events involved in the incident sequence.

## Incident Investigation

22. Task and activity at time of incident. First, record the general type of task the employee was performing when the incident occurred (for example, pipe fitting, lathe maintenance, or operating a punch press). Then record the specific activity in which the employee was engaged when the incident occurred (for example, oiling shaft, bolting pipe flanges, or removing material from the press). Check the appropriate box to indicate whether the injured employee was working alone, with a co-worker, or with a crew.

23. Posture of employee. Record the injured worker's posture in relation to the surroundings at the time of the incident (for example, standing on a ladder, squatting under a conveyor, or standing at a machine).

24. Supervision at time of accident. Indicate whether, at the time of incident, the injured employee was directly supervised, indirectly supervised, or not supervised. If appropriate, indicate whether supervision was not feasible at the time.

## Determining Causal Factors (Root Causes)

An important part of any incident investigation is the causal factors analysis (root causes), or determining those causes that directly contributed to the incident. Identifying all factors that came into play to cause an injury or near miss and correcting those items is the only way to ensure proper steps have been taken to prevent a recurrence. In using a causal factors analysis, the opportunity to identify all of the facts, and not focus on blame is established. Causal factors to any incident are grouped into four main areas: equipment, environment, personnel, and management.

The causal factors analysis approach to identifying those areas that have a causative relationship to the incident greatly improves the organization's chances of preventing future similar incidents. In addition, this approach to incident investigation allows for more employee involvement as well as objectivity in determining the root cause of the incident (Figure 7–3).

25. Causal factors. Record causal factors (events and conditions that contributed to the incident) that were identified by use of the causal factors analysis checklist.

26. Corrective actions. Describe recommended corrective actions to be taken after the incident to prevent recurrence. These include immediate, temporary, or interim actions (for example, removed oil from floor) and permanent actions (for example, repaired leaking oil line).

## GUIDE FOR IDENTIFYING CAUSAL FACTORS & CORRECTIVE ACTIONS

**Case Number**

Answer questions by placing an X in the "**Y**" circle or box for yes or in the "**N**" circle or box for no.

### PART 1 EQUIPMENT

**○ □**
**Y   N**

**1.0  WAS A HAZARDOUS CONDITION[S] A CONTRIBUTING FACTOR?**
If yes, answer the following. If no, proceed to Part 2.

| | Causal Factors | Comment | Possible Corrective Actions | Recommended Corrective Actions |
|---|---|---|---|---|
| ○ □ <br> Y  N | 1.1  Did any defect(s) in equipment/tool(s)/material contribute to hazardous condition(s)? | | Review procedure for inspecting, reporting, maintaining, repairing, replacing, or recalling defective equipment/tool(s)/material used. | |
| □ ○ <br> Y  N | 1.2  Was the hazardous condition(s) recognized? <br><br> If yes, answer A and B. If no, proceed to 1.3. | | Perform job safety analysis. Improve employee ability to recognize existing or potential hazardous conditions. Provide test equipment, as required, to detect hazard. Review any change or modification of equipment/tool(s)/material. | |
| □ ○ <br> Y  N | A. Was the hazardous conditions(s) reported? | | Train employees in reporting procedures. Stress individual acceptance of responsibility. | |
| □ ○ <br> Y  N | B. Was employee(s) informed of the hazardous condition(s) and the job procedures for dealing with it as an interim measure? | | Review job procedures for hazard avoidance. Review supervisory responsibility. Improve supervisor/employee communications. Take action to remove or minimize hazard. | |
| □ ○ <br> Y  N | 1.3  Was there an equipment inspection procedure(s) to detect the hazardous condition(s)? | | Develop and adopt procedures (for example, an inspection system) to detect hazardous conditions. Conduct test. | |
| □ ○ <br> Y  N | 1.4  Did the existing equipment inspection procedure(s) detect the hazardous condition(s)? | | Review procedures. Change frequency or comprehensiveness. Provide test equipment as required. Improve employee ability to detect defects and hazardous conditions. Change job procedures as required. | |
| □ ○ <br> Y  N | 1.5  Was the correct equipment/tool(s)/material used? | | Specify correct equipment/tool(s)/material in job procedures. | |
| □ ○ <br> Y  N | 1.6  Was the correct equipment/tool(s)/material readily available? | | Provide correct equipment/tool(s)/material. Review purchasing specifications and procedures. Anticipate future requirements. | |

**Figure 7–3.** A causal factors analysis breaks down each incident into areas that contribute to an injury: equipment, environment, personnel, and management. By answering each question and placing an X in either a circle or a box will determine that item's relationship to the injury as a causal factor.

| | Causal Factors | Comment | Possible Corrective Actions | Recommended Corrective Actions |
|---|---|---|---|---|
| ☐ ○ <br> Y N | 1.7 Did employee(s) know where to obtain equipment/tool(s)/material required for the job? | | Review procedures for storage, access, delivery, or distribution. Review job procedures for obtaining equipment/tool(s)/material. | |
| ○ ☐ <br> Y N | 1.8 Was substitute equipment/tool(s)/material used in place of correct one? | | Provide correct equipment/tool(s)/material. Warn against use of substitutes in job procedures and in job instruction. | |
| ○ ☐ <br> Y N | 1.9 Did the design of the equipment/tool(s) create operator stress or encourage operator error? | | Review human factors engineering principles. Alter equipment/tool(s) to make it more compatible with human capability and limitations. Review purchasing procedures and specifications. Check out new equipment and job procedures involving new equipment before putting into service. Encourage employees to report potential hazardous conditions created by equipment design. | |
| ○ ☐ <br> Y N | 1.10 Did the general design or quality of the equipment/tool(s) contribute to a hazardous condition? | | Review criteria in codes, standards, specifications, and regulations. Establish new criteria as required. | |
| ○ | 1.11 List other causal factors in "Comment" column. | | | |

**PART 2 ENVIRONMENT**

| | | | | |
|---|---|---|---|---|
| ○ ☐ <br> Y N | **2.0 WAS THE LOCATION OF EQUIPMENT/MATERIALS/EMPLOYEE(S) A CONTRIBUTING FACTOR?** <br> If yes, answer the following. If no, proceed to Part 3. | | | |

| | Causal Factors | Comment | Possible Corrective Actions | Recommended Corrective Actions |
|---|---|---|---|---|
| ○ ☐ <br> Y N | 2.1 Did the location/position of equipment/material/employee(s) contribute to a hazardous condition? | | Perform job safety analysis. Review job procedures. Change the location, position, or layout of the equipment. Change position of employee(s). Provide guardrails, barricades, barriers, warning lights, signs, or signals. | |
| ☐ ○ <br> Y N | 2.2 Was the hazardous condition recognized? <br><br> If yes, answer A and B. If no, proceed to 2.3. | | Perform job safety analysis. Improve employee ability to recognize existing or potential hazardous conditions. Provide test equipment, as required, to detect hazard. Review any change or modification of equipment/tools/materials. | |
| ☐ ○ <br> Y N | A. Was the hazardous condition reported? | | Train employees in reporting procedures. Stress individual acceptance of responsibility. | |

**Figure 7–3.** Continued.

| | Causal Factors | Comment | Possible Corrective Actions | Recommended Corrective Actions |
|---|---|---|---|---|
| ☐ Y ○ N | B. Was employee(s) informed of the job procedures for dealing with the hazardous condition as an interim action? | | Review job procedures for hazard avoidance. Review supervisory responsibility. Improve employee/supervisor communications. Take action to remove or minimize hazard. | |
| ☐ Y ○ N | 2.3 Was employee(s) supposed to be in the vicinity of the equipment/material? | | Review job procedures and instruction. Provide guardrails, barricades, barriers, warning lights, signs, or signals. | |
| ☐ Y ○ N | 2.4 Was the hazardous condition created by the location/ position of equipment/ material visible to employee(s)? | | Change lighting or layout to increase visibility of equipment. Provide guardrails, barricades, barriers, warning lights, signs or signals, floor stripes, etc. | |
| ☐ Y ○ N | 2.5 Was there sufficient workspace? | | Review workspace requirements and modify as required. | |
| ○ Y ☐ N | 2.6 Were environmental conditions a contributing factor (for example, illumination, noise levels, air contaminant, temperature extremes, ventilation, vibration, radiation)? | | Monitor, or periodically check, environmental conditions as required. Check results against acceptable levels. Initiate action for those found unacceptable. | |
| ○ | 2.7 List other causal factors in "Comment" column. | | | |

**PART 3 PEOPLE**

○ Y ☐ N  **3.0 WAS THE JOB PROCEDURE(S) USED A CONTRIBUTING FACTOR?**
If yes, answer the following. If no, proceed to Part 3.6.

| | Causal Factors | Comment | Possible Corrective Actions | Recommended Corrective Actions |
|---|---|---|---|---|
| ☐ Y ○ N | 3.1 Was there a written or known procedure (rules) for this job? <br><br> If yes, answer A, B, and C. If no, proceed to 3.2. | | Perform job safety analysis and develop safe job procedures. | |
| ☐ Y ○ N | A. Did job procedures anticipate the factors that contributed to the accident? | | Perform job safety analysis and change job procedures. | |
| ☐ Y ○ N | B. Did employee(s) know the job procedure? | | Improve job instruction. Train employees in correct job procedures. | |
| ○ Y ☐ N | C. Did employee(s) deviate from the known job procedure? | | Determine why. Encourage all employees to report problems with an established procedure to supervisor. Review job procedure and modify if necessary. Counsel or discipline employee. Provide closer supervision. | |

**Figure 7–3.** Continued.

| | Causal Factors | Comment | Possible Corrective Actions | Recommended Corrective Actions |
|---|---|---|---|---|
| ☐ ◯ <br> Y  N | 3.2 Was employee(s) mentally and physically capable of performing the job? | | Review employee requirements for the job. Improve employee selection. Remove or transfer employees who are temporarily, either mentally or physically, incapable of performing the job. | |
| ◯ ☐ <br> Y  N | 3.3 Were any tasks in the job procedure too difficult to perform (for example, excessive concentration or physical demands)? | | Change job design and procedures. | |
| ◯ ☐ <br> Y  N | 3.4 Is the job structured to encourage or require deviation from job procedures (for example, incentive, piecework, work pace)? | | Change job design and procedures. | |
| ◯ | 3.5 List other causal factors in "Comment" column. | | | |

◯ ☐
Y  N  **3.6  WAS LACK OF PERSONAL PROTECTIVE EQUIPMENT OR EMERGENCY EQUIPMENT A CONTRIBUTING FACTOR IN THE INJURY?**
If yes, answer the following. If no, proceed to Part 4.
Note: The following causal factors relate to the *injury.*

| | Causal Factors | Comment | Possible Corrective Actions | Recommended Corrective Actions |
|---|---|---|---|---|
| ☐ ◯ <br> Y  N | 3.7 Was appropriate personal protective equipment (PPE) specified for the task or job? <br><br> If yes, answer A, B, and C. If no, proceed to 3.8. | | Review methods to specify PPE requirements. | |
| ☐ ◯ <br> Y  N | A. Was appropriate PPE available? | | Provide appropriate PPE. Review purchasing and distribution procedures. | |
| ☐ ◯ <br> Y  N | B. Did employee(s) know that wearing specified PPE was required? | | Review job procedures. Improve job instruction. | |
| ☐ ◯ <br> Y  N | C. Did employee(s) know how to use and maintain the PPE? | | Improve job instruction. | |
| ☐ ◯ <br> Y  N | 3.8 Was the PPE used properly when the injury occurred? | | Determine why and take appropriate action. Implement procedures to monitor and enforce use of PPE. | |
| ☐ ◯ <br> Y  N | 3.9 Was the PPE adequate? | | Review PPE requirements. Check standards, specifications, and certification of the PPE. | |

**Figure 7–3.** Continued.

| | Causal Factors | Comment | Possible Corrective Actions | Recommended Corrective Actions |
|---|---|---|---|---|
| ☐ ○ <br> Y  N | 3.10 Was emergency equipment specified for this job (for example. emergency showers, eyewash fountains)? <br><br> If yes. answer the following. If no, proceed to Part 4. | | Provide emergency equipment as required. | |
| ☐ ○ <br> Y  N | A. Was emergency equipment readily available? | | Install emergency equipment at appropriate locations. | |
| ☐ ○ <br> Y  N | B. Was emergency equipment properly used? | | Incorporate use of emergency equipment in job procedures. | |
| ☐ ○ <br> Y  N | C. Did emergency equipment function properly? | | Establish inspection/monitoring system for emergency equipment. Provide for immediate repair of defects. | |
| ○ | 3.11 List other causal factors in "Comment" column. | | | |

## PART 4 MANAGEMENT

| ○ ☐ <br> Y  N | **4.0 WAS A MANAGEMENT SYSTEM DEFECT A CONTRIBUTING FACTOR?** <br> If yes, answer the following. If no, STOP. Your causal factor identification exercise is complete. |

| | Causal Factors | Comment | Possible Corrective Actions | Recommended Corrective Actions |
|---|---|---|---|---|
| ○ ☐ <br> Y  N | 4.1 Was there a failure by supervision to detect, anticipate, or report a hazardous condition? | | Improve supervisor capability in hazard recognition and reporting procedures. | |
| ○ ☐ <br> Y  N | 4.2 Was there a failure by supervision to detect or correct deviations from job procedure? | | Review job safety analysis and job procedures. Increase supervisor monitoring. Correct deviations. | |
| ☐ ○ <br> Y  N | 4.3 Was there a supervisor/ employee review of hazards and job procedures for tasks performed infrequently? (Not applicable to all incidents.) | | Establish a procedure that requires a review of hazards and job procedures (preventive actions) for tasks performed infrequently. | |
| ☐ ○ <br> Y  N | 4.4 Was supervisor responsibility and accountability adequately defined and understood? | | Define and communicate supervisor responsibility and accountability. Test for understandability and acceptance. | |
| ☐ ○ <br> Y  N | 4.5 Was supervisor adequately trained to fulfill assigned responsibility in accident prevention? | | Train supervisors in accident prevention fundamentals. | |
| ○ ☐ <br> Y  N | 4.6 Was there a failure to initiate corrective action for a known hazardous condition that contributed to this incident? | | Review management safety policy and level of risk acceptance. Establish priorities based on potential severity and probability of recurrence. Review procedure and responsibility to initiate and carry out corrective actions. Monitor progress. | |
| ○ | 4.7 List other causal factors in "Comment" column. | | | |

**Figure 7–3.** Concluded.

It is critical that an organization periodically review existing incident investigation reports to verify that all corrective actions have been completed. Each report should document the date that the corrective actions were completed and the names of the person or persons who completed that corrective actions as well as the name of the person(s) who verified that the corrective actions were adequately completed.

Users may add other data to the form to fulfill local or corporate requirements. Types of data that might be added include information typical to a specific industry or organization. For example, an establishment regulated by the Mine Safety and Health Administration (MSHA) may wish to add an MSHA identification number, required training, and training received.

## SUMMARY OF KEY POINTS

Key points covered in this chapter include:

- The purpose of an incident investigation is to determine causes and recommend corrective actions to eliminate or control hazards and prevent recurrence of similar incidents. All incidents should be investigated, not only those that cause serious injury or property damage. The investigation should emphasize fact-finding, not fault-finding.

- An organization should set up a site-specific plan for responding to incidents. The plan should specify what should be done in case of an incident, list names and phone numbers of people to call, and assign specific responsibilities to all persons involved in the emergency response plan. All supervisors expected to respond to an incident should be trained on the specific incident investigation steps as well as causal factor analyses.

- When an incident occurs, supervisors should first see that injured persons get immediate medical attention and then secure the incident site for the investigation duration. Evidence can be preserved on film, recorded on tape, diagrammed, or sketched.

- Witnesses are often the best source of information about an incident. Supervisors should speak with anyone who was in the incident area in addition to those who actually witnessed the event.

- Witnesses should be interviewed one at a time as soon after the incident as possible to ensure accurate recall. The supervisor may ask witnesses to reenact, with appropriate safeguards, how the incident happened. Interviews should be conducted in a sensitive manner and in a comfortable location to avoid intimidating witnesses or influencing their stories.

- Incident reports are designed to inform management and other concerned people about the circumstances surrounding an incident. Causal factors leading to an incident generally fall into one or more of the following categories: equipment, environment, personnel, and management. These reports must be filled out as completely and thoroughly as possible.

- A causal factors analysis should be conducted for all incidents to obtain an objective determination of what actually influenced the incident. Based on the determination of these factors, controls or corrective actions should be decided and tracked.

- Periodic reviews of existing incident reports should occur to verify that all corrective actions have been completed and are documented as closed.

# 8

# Industrial Hygiene

**After reading this chapter, you will be able to:**

- Describe various chemical, physical, ergonomic, or biological health hazards commonly encountered on the job
- Recognize environmental health hazards in one or more of these four categories in your work area
- List basic methods of controlling various harmful environmental hazards and stresses
- Describe the concept and purpose of exposure limits
- Ensure that standard operating procedures are established to help provide a safe, healthy work environment

In addition to safety responsibilities, supervisors—together with management and safety personnel—must make sure that the work area is free from conditions that could be detrimental to health. Understanding the basics of industrial hygiene can assist the supervisor in providing a safe and healthy work environment. This chapter provides information to help you recognize environmental health hazards in your workplace. You can then request assistance from industrial hygienists, who work with medical, safety, and engineering personnel to eliminate or safeguard against such hazards.

Industrial hygienists define their work as "the anticipation, recognition, evaluation, and control of environmental conditions that may have adverse effects on health, that may be uncomfortable or irritating, or that may have some undesired effect upon the ability of individuals to perform their normal work." It is possible to group these environmental conditions or hazards into four general categories: chemical, physical, ergonomic, and biological. Their effects are often systemic and latent. Each category is discussed in some detail in this chapter.

## CHEMICAL HAZARDS

Chemical compounds in the form of solids, liquids, and gases may cause health problems by inhalation (breathing), skin absorption or contact, ingestion (eating or drinking), or injection.

### Inhalation

A major route of entry for employee exposure to chemical compounds is through the inhalation of airborne contaminants. Contaminants inhaled into the lungs can be in the form of gases, vapors, and solid or particulate matter. Particulate matter includes dust, mists, fumes, fibers, smoke, or aerosols.

### Absorption

Absorption through the skin can occur quite rapidly if the skin is punctured, cut, or abraded. Unfortunately, many compounds that exist either in liquid or gaseous form, or both, can be absorbed through intact skin. Some are absorbed through the hair follicles while others penetrate by dissolving into the fats and oils of the skin.

Examples of chemical compounds that can be hazardous by skin absorption are alkaloids; phenols; lead acetate; salts of lead, arsenic, and mercury;

nitrobenzene; nitrotoluene; aniline; and nitroglycerine. Other harmful compounds include triorthocresyl phosphate, tetraethyl lead, and parathion and related organic phosphates. Compounds such as toluene and xylene that are good solvents for fats may also be absorbed through the skin.

## Ingestion

Ordinarily, people do not knowingly eat or drink harmful materials. However, toxic compounds capable of being absorbed from the gastrointestinal tract into the blood—for example, lead oxide—can create serious exposure problems if people working with these substances are allowed to eat, drink, apply cosmetics, or smoke in their work areas. Careful and thorough washups should be required before eating or smoking, and at the end of every shift. When necessary, workers should change their clothes before leaving work to avoid contaminating their home environment or other clean work areas.

## Physical Classification of Airborne Materials

Because inhalation of airborne compounds or materials is a common problem, the supervisor should know the physical classifications of these substances.

**Dusts.** These are solid particles generated by handling, crushing, grinding, rapid impact, detonation, and decrepitation (breaking apart by heating) of organic or inorganic materials, such as rock, ore, metal, coal, wood, and grain. Dust is a term used in industry to describe airborne solid particles that range in size from 0.1 to 25 μm (micrometers or microns) in diameter (μm = 1/10,000 cm = 1/25,000 in.

A person with normal eyesight can detect individual dust particles as small as 50 μm in diameter. Dust particles below 10 μm in diameter cannot be seen without a microscope. High concentrations of suspended small particles may look like haze or smoke.

Dusts settle to the ground under the influence of gravity. The larger the particle, the more quickly it settles. Particles larger than 10 μm in diameter settle quickly while those under 10 μm remain suspended in air for much longer. Particles less than 5 μm, called "respirable dusts," can penetrate into the inner recesses of the lungs. Nearly all larger particles are trapped in the nose, throat, trachea, or bronchi from which they are either expectorated or swallowed.

Some larger-sized particles can also cause difficulty, however. Ragweed pollen, which ranges from 18 to 25 μm in diameter, can trigger an allergic reaction known as hay fever when particles enter the upper respiratory tract. Other allergenic dusts, as well as some bacterial and irritant dusts, can also cause respiratory problems in workers exposed to them.

Dust may enter the air from various sources. It may be dispersed when a dusty material is handled—for example, when lead oxide is dumped into a mixer, when talc is dusted on a product, or where asbestos-containing acoustical and/or fireproofing materials are being removed.

When solid materials are reduced to small sizes in such processes as grinding, crushing, blasting, shaking, and drilling, the mechanical action of the grinding or shaking device can disperse the dust formed. When dusty materials are handled or transported, dust may be dispersed to other facility areas, unless good controls are used.

**Fumes.** A fume is formed when volatilized solids, such as metals, condense in cool air. The solid particles that make up a fume are extremely fine, usually less that 1 µm in diameter. In most cases, hot material reacts with air to form an oxide. Examples are lead oxide fume from smelting and iron oxide fume from arc welding.

A fume also can be formed when a material such as magnesium metal is burned or when welding or gas cutting is done on galvanized metal. Gases and vapors are not fumes, though they are often incorrectly called that.

**Smoke.** Smoke is made up partly of carbon or soot particles less than 0.1 µm in size that result from incomplete combustion of such carbonaceous materials as coal or oil. Smoke generally contains liquid droplets as well as dry particles. Tobacco, for instance, produces a wet smoke composed of minute tarry droplets. The particle size in tobacco smoke is about 0.25 µm in diameter.

**Aerosols.** Liquid droplets or solid particles fine enough to be dispersed and to remain airborne for a prolonged period of time are called aerosols. If inhaled, some of these may irritate or injure workers' mucus membranes, eyes, noses, throats, and lungs.

**Mists.** Mists are suspended liquid droplets generated by chemicals condensing from the gaseous to the liquid state, or by a liquid breaking into a dispersed state by splashing, foaming, or atomizing. Mist is formed when a finely divided liquid is suspended in the atmosphere. Examples are the oil mist produced during cutting and grinding operations, acid mists from electroplating, acid or alkali mists from pickling operations, spray-paint mist from spraying operations, and condensation of water vapor into fog or rain.

**Gases.** Usually gases are formless fluids that occupy the space or enclosure in which they are confined. Gases can be changed to the liquid or solid state only by the combined effect of increased pressure and decreased temperature. Gases spread out, or diffuse, into the surrounding atmosphere easily and readily. Examples are welding gases, internal combustion engine exhaust gases, and air.

**Vapors.** Gaseous forms of substances that are normally in solid or liquid state (at room temperature and pressure) are called "vapors." Vapors can be changed back to the solid or liquid state either by increasing the pressure or by decreasing the temperature. Evaporation is the process by which a liquid is changed into the vapor state and mixed with the surrounding atmosphere. Solvents with low boiling points—for example, acetone—will volatilize (evaporate) readily at room temperature.

**Hazards Involved.** The hazard associated with breathing a gas, vapor, or mist usually depends on the solubility of the substance. For example, if the compound is very soluble—such as ammonia, formaldehyde, sulfuric acid, or hydrochloric acid—it is rapidly absorbed in the upper respiratory tract and does not penetrate deeply into the lungs. Consequently, the nose and throat become so irritated that a person is driven out of the exposure area before he or she is in much danger of toxicity from the chemical. (Toxicity is the ability of a substance to produce disease or physical harm.) Nevertheless, exposures, even for brief periods, to high concentrations of these compounds can produce serious health effects.

Compounds that are not soluble in body fluids can penetrate deeply into the lungs before the person can sense exposure. Thus, a serious hazard can be present but not be immediately recognized. Examples of such gases are nitrogen dioxide and phosgene. The immediate danger from these compounds in high concentrations is acute edema (abnormal accumulation of serous fluid in body tissues or serous cavities) or, possibly, pneumonia or circulatory impairment.

However, numerous chemical compounds do not follow the general solubility rule. They are not especially soluble in water and yet are irritating to the eyes and respiratory tract. They can cause lung damage and even death under the right conditions. One example is acrolein, which belongs to the chemical family called aldehydes.

## Particulates

To evaluate particulate exposures properly, you must know the chemical composition, particle size, dust concentration in air, method of dispersion, and many other factors described in this section. With the exception of certain fibrous materials, dust particles must usually be smaller than 5 µm to enter the alveoli or inner recesses of the lungs. Although a few particles up to 10 µm in size may enter the lungs, nearly all larger particles are trapped in the nasal passages, throat, larynx, trachea, and bronchi, from which they are expectorated, or swallowed, entering the digestive tract.

**Dusts.** Most industrial dusts consist of particles that vary widely in size, with small particles greatly outnumbering large ones. Consequently, with few exceptions, when dust can be seen in the air around an operation, there are probably more invisible than visible particles present. The main hazard to personnel occurs when dust becomes airborne. Also, airborne dusts can be flammable and potentially explosive.

A number of occupational diseases result from exposures to such nonmetallic dusts as asbestos. Although specialized knowledge and instruments are needed to determine the severity of a hazard, you should be trained to recognize a hazard and to ask for expert help in controlling it.

A process that produces dust fine enough to remain suspended in the air should be regarded as hazardous until proven safe. An air-monitoring survey of airborne chemicals present will determine employee exposure levels and the overall relative safety. Processes that can generate excessive exposure include:
- abrasive blasting or machining
- bagging and handling of dry materials (i.e., grains, cement)
- ceramic coating
- dry mixing
- metal forming
- grinding

Consider, for example, free silica, which can cause a damaging lung disease, silicosis. Silica dust is produced in hard-rock mines and by quarrying and dressing granite. Grinding castings that contain mold sand also represent a silica hazard.

Methods of drilling rock with power machinery produce more dust than old-fashioned hand methods. This dust, however, is controlled by applying water to the drill bit so that the dust forms a slurry instead of being suspended in air.

Processes in which materials are crushed, ground, or transported are also potential sources of dust. They should either be controlled by use of wet methods or should be enclosed and ventilated by local exhaust. Points where conveyors are loaded or discharged, transfer points along the conveying system, and heads or boots of elevators should be enclosed and, usually, exhaust-ventilated.

Built-in dust controls must be checked periodically to see that no one cancels their effectiveness by tampering with or improperly using them. Required respiratory equipment must be worn by trained workers who need supplementary protection.

**Fumes.** Welding and other hot operations produce fumes, which may be harmful under certain conditions. For example, arc welding volatizes metal that then condenses—as the metal or its oxide—in the air around the arc. In addition, the rod coating is in part volatized. Because they are extremely fine, these fumes are readily inhaled.

Highly toxic fumes, such as those formed when welding galvanized metal or structures painted with red lead, may rapidly produce severe symptoms of toxicity. Fumes should be controlled by using good local exhaust ventilation or by protecting the welder with approved respiratory protection equipment.

When brass is poured, zinc volatizes from the molten mass and oxidizes in air to produce zinc fume, which (in high concentrations) may produce the disease known as "metal fume fever." If lead is present, it will also become airborne. As a result, brass foundry workers may incur either chronic lead poisoning if lead is present in minute amounts, or acute poisoning if it is present in substantial amounts.

Most soldering operations, fortunately, do not require temperatures high enough to volatize an appreciable amount of lead. However, some lead in the molten solder is oxidized by contact with air at the surface. If this oxide, often called dross, is mechanically dispersed into the air, it may produce a severe lead-poisoning hazard.

In operations where this condition may happen—for example, soldering or lead battery making—preventing occupational poisoning depends on scrupulous housekeeping to prevent the lead oxide from dispersing into the air. You can enclose melting pots, dross boxes, and similar operations, and provide exhaust ventilation to capture airborne contaminants.

**Gases.** Gases, which are used or generated in many industrial processes, may be toxic. For example, welding in the presence of chlorinated solvent vapors can produce phosgene, a very toxic gas that causes respiratory distress and damage. Propane-fueled forklift trucks or any process or equipment that burns fuel or other organic materials have the potential to generate carbon monoxide. Carbon monoxide prevents the body from absorbing oxygen, causing headache, nausea, confusion, dizziness, and in severe cases, coma and death. In addition, carbon monoxide and other toxic gases like hydrogen sulfide and methane may be found in confined spaces where organic substances have deteriorated.

### Assessing Respiratory Hazards

When a respiratory hazard exists or is suspected, the actual airborne concentration of the contaminant(s) must be measured by an industrial hygienist. While conditions are sometimes similar in different areas within one facility or in different facilities within an industry, the degree of respiratory hazard must be assessed in each case by scientifically valid methods, such as air sampling. Some typical industrial processes and the types of air contamination generated are listed in Table 8–A.

You must also be aware of the hazards of oxygen deficiency. Oxygen deficiency results when work area air contains less than the normal amount of oxygen found in the atmosphere, about 21 %. OSHA regulations define immediately dangerous to life or health as starting at 19.5 % for confined space.

You should sample the air for toxic chemicals and test for oxygen and flammable gases before allowing workers to enter any confined space. Many gases are heavier than air and thus may remain present even in confined spaces that are open to the atmosphere. Many gases are odorless and colorless, which makes detection unlikely unless appropriate air-sampling equipment is used.

Air sampling may be necessary around workers and process equipment to make certain that hazard controls are working. For example, you may need to sample for ozone and other hazardous gases to document a welder's exposure. Industrial hygiene, safety, or engineering personnel should provide you with information on gases that might be used or generated in your work areas.

The degree of hazard involved is an important factor when analyzing respiratory conditions. Some hazards, such as gases and vapors, can produce an immediate threat to life and health when present in high concentrations; and oxygen deficiency, by its very nature, is automatically dangerous to life and health.

## Table 8–A. Potentially Hazardous Operations and Air Contaminants

| Process Types | Contaminant Types | Contaminant Examples |
|---|---|---|
| **Hot Operations** | | |
| Welding | Gases | Chromates |
| Chemical reactions | Particulates (p) | Zinc and compounds (p) |
| Soldering | (dusts, fumes, mists) | Manganese and compounds (p) |
| Melting | | Melting oxides (p) |
| Molding | | Carbon monoxide (g) |
| Burning | | Ozone (g) |
| | | Cadmium oxide (p) |
| | | Fluorides (p) |
| | | Lead (p) |
| | | Vinyl chloride (g) |
| **Liquid Operations** | | |
| Painting | Vapors (v) | Benzene (v) |
| Degreasing | Gases (g) | Trichloroethylene (v) |
| Dipping | Mists (m) | Methylene chloride (v) |
| Spraying | | Hydrochloric acid (m) |
| Coating | | Sulfuric acid (m) |
| Etching | | Hydrogen chloride (g) |
| Cleaning | | Cyanide salts (m) |
| Dry cleaning | | Chromic acid (m) |
| Pickling | | Hydrogen cyanide (g) |
| Plating | | TDI, MDI (v) |
| Mixing | | Hydrogen sulfide (g) |
| Galvanizing | | Sulfur dioxide (g) |
| Chemical reactions | | Carbon tetrachloride (v) |
| **Solid Operations** | | |
| Pouring | Dusts (d) | Cement |
| Mixing | | Fibrous glass |
| Separations | | |
| Extraction | | |
| Crushing | | |
| Conveying | | |
| Loading | | |
| Bagging | | |
| **Pressurized Spraying** | | |
| Cleaning parts | Vapors (v) | Organic solvents (v) |
| Applying pesticides | Dusts (d) | Chlordane (m) |
| Degreasing | Mists (m) | Parathion (m) |
| Sand blasting | | 1,1,1-trichloroethylene (v) |
| Painting | | Methylene chloride(v) |
| | | Quartz (free silica) |
| **Shaping Operations** | | |
| Cutting | Dusts (d) | Asbestos |
| Grinding | | Beryllium |
| Filing | | Uranium |
| Milling | | Zinc |
| Molding | | Lead |
| Sawing | | |
| Drilling | | |

(Reprinted with permission from *Occupational Exposure Sampling Strategy Manual*, NIOSH Pub. No. 77–173b.)

Abbreviations: d = dust, g = gases, m = mists, p = particulates, v = vapors.

## Respiratory Protection

It is important to understand the basic concepts of respiratory protection to adequately protect your workers. A competent industrial hygienist or safety professional should decide what kind of respiratory protection is needed. However, the more you know about respiratory hazards, and the types and selection of respirators, the better prepared you will be to make sure that employees are protected. Selection of the proper respirator is discussed in Chapter 9, Personal Protective Equipment.

## Liquid Chemicals: Solvents

Commonly used liquid chemicals include fuel or fuel additives, pesticides, lubricants, detergents and cleaning agents, or degreasing or processing solvents. Solvents are perhaps the most widespread class of chemicals in manufacturing. Many of these solvents evaporate readily into the air; therefore, their use can pose immediate exposure problems.

Solvents can be further categorized as aqueous or organic. Aqueous solvents are those that readily dissolve in water. Many acids, alkalis, or detergents, when mixed with water, form aqueous solvent systems.

The term solvent, however, is commonly used to mean organic solvents. Many of these chemicals do not mix easily with water but dissolve other organic materials such as greases, oils, and fats. Important types and specific examples of organic solvents include aliphatic, cyclic, aromatic, halogenated, esters, ketones, alcohols, ethers, glycols, aldehydes, freons; and hexane, gasoline, turpentine, benzene, trichloroethylene, ethyl acetate, acetone, formaldehyde, methanol, ethyl ether, and ethylene glycol.

Organic solvents generally have some effect on the central nervous system. With short-term (acute) exposures, they may cause nervous system depression, in which the victim experiences dizziness, feelings of intoxication and nausea, and a decrease in muscular coordination. Higher levels or more prolonged exposure may cause loss of consciousness, coma, and, in some cases, death.

Other effects of solvents vary. Some can cause long-term damage to the liver or other organs, or affect the worker's reproductive ability. A few have been found to cause cancer. Some research suggests that long-term (chronic) exposure to low levels of certain solvents may cause long-lasting effects, including chronic headaches, inability to concentrate, memory loss, and reduced nervous system and muscular functioning.

As the supervisor, you can assist the industrial hygienist and safety professional in controlling exposure to such chemicals. Substantial exposures, fortunately, can be controlled. Spray-painting booths can be ventilated and degreasing tanks can be exhausted. These will not ordinarily present serious hazards as long as the equipment works. It is the job of the supervisor, engineering, human resources, and safety personnel to make sure that controls and personal protective equipment (PPE) are properly maintained and used at all times. Even small exposures on jobs that come up infrequently or that involve small amounts of solvents not covered by standard operating procedures can produce significant exposure hazards.

What is most important is not how much solvent is used at the job site, but the actual degree of exposure by inhalation or skin absorption. A close check must be kept on all minor, as well as major, uses of solvents. These chemicals should be issued only after determining that they can be used properly and safely. Industrial hygienists or safety professionals can provide information about which solvents to use and for what purposes. The supervisor or team leader is responsible for seeing that this information is used and that employees follow the recommendations.

**Selection and Handling.** Getting the job done without hazard to employees or property is dependent on the proper selection, application, handling, and control of solvents and an understanding of their properties.

One problem encountered with chemical selection is the similarity of names for the various kinds of solvents in the chlorinated hydrocarbon family. Perchloroethylene, trichloroethylene, trichloromethane, and dichloromethane are a few examples of such solvents that have similar-sounding names, yet possess characteristic hazards.

**Hazard Communication.** Many state regulations and the Federal Hazard Communication Standard (29 CFR 1910.1200) require that management provide information about chemical hazards to the workforce.

The role of a supervisor includes making sure this information is provided to employees. Many of these regulations require:

- an inventory and assessment of chemical hazards in the workplace
- development and use of labels that describe chemical hazards and protective measures to use
- Material Safety Data Sheets (MSDSs) that detail chemical hazard and precaution information

- training on identifying hazards, including specific chemicals or groups of chemicals with which employees work, reading labels, and understanding MSDSs
- a written program that describes how the company intends to accomplish these tasks and provides documentation that workers have been trained

Many companies have developed programs using in-house or outside experts to inform employees about general classes of chemicals and chemical hazards. Supervisors and other departmental personnel are often expected to provide specific information on the chemical hazards found in the department, and measures that should be used to control them. Such measures could include engineering controls (ventilation or isolation of a hazardous substance), work-practice controls such as housekeeping programs, or PPE programs. The company should train supervisors to recognize the hazards and to tell their workers how they can best protect themselves.

If the supervisor finds a new chemical for which training has not been provided, or which lacks labels, the supervisor will need to discuss it with the industrial hygienist, safety professional, or chemist, as well as with the administrator of the hazard communication program. They may also need to contact purchasing to request notifying the supplier that an appropriate label and MSDS are needed.

**Degree of Hazard Severity.** The severity of hazard in the use of organic solvents depends on the following facts:
- how the solvent is used
- how much solvent is used
- type of job operation (determines how the workers are exposed)
- work pattern
- duration of exposure
- operating temperature
- exposed liquid surface
- ventilation efficiency
- evaporation rate of solvent
- pattern of air flow
- concentration of vapor in workroom air
- housekeeping

The solvent hazard, therefore, is determined not only by the toxicity of the solvent itself, but by the conditions of its use—who, what, when, where, and how long. Precautionary labeling and/or the MSDS should indicate the major hazards and safeguards.

For convenience, job operations employing solvents may be divided into three categories:

1. Direct contact. This occurs when skin, usually the hands, contacts the solvent directly. Emergency repair of equipment, spraying or packaging volatile materials without ventilation, cleanup of spills, and manual cleaning using cloths or brushes wetted with solvent are examples of jobs in which employees may have direct contact with the solvent.

   Remember that many solvents can penetrate the skin and be absorbed into the body. Solvents are a leading cause of industrial dermatitis or skin disease. It also is important to note that many solvents can penetrate various glove materials. Thus, properly selected chemical-resistant gloves and clothing must be used to prevent solvents from soaking through and contacting the skin.

2. Intermittent or infrequent contact. This occurs when the solvent is contained in a semi-closed system where exposure can be controlled. Examples are: paint spraying in an exhaust-ventilated spray booth, vapor degreasing in a tank with local lateral slot exhaust ventilation, and charging reactors or kettles in a batch-type operation in which the worker is exposed only at infrequent intervals.

3. Minimal contact. This is characterized by remote operation of equipment totally isolated from the work area. This type of operation includes directing chemical facility operations from a control room, mechanical handling of bulk-packaged materials, and other operations where the solvent is contained in a closed system and is not discharged to the atmosphere in the work area.

It is unfortunate that the term "safety" solvent has been applied to some proprietary cold degreasers because the term is not precise and is subject to various interpretations. For example, a "safety" solvent may be considered by some users as nondamaging to surfaces being cleaned. Other users may consider it to be free from fire or toxicity hazards. Depending on conditions of use, neither of these criteria can be met by a so-called "safety" solvent.

These solvents are prepared as mixtures of halogenated hydrocarbons and petroleum hydrocarbons to be used for cold degreasing. The halogenated hydrocarbons are effective grease and oil solvents and generally are not flammable. They also are relatively expensive and may be toxic under adverse conditions. The petroleum hydrocarbons are effective solvents and are inexpensive, but they are flammable (flash points below 100 F) and can also be toxic.

To combine the best qualities of each solvent, manufacturers mix them in an attempt to produce a cold degreaser that has a flash point higher than that of the flammable petroleum hydrocarbon alone. Such a mixture, however, can present both fire and toxicity hazards, depending on the evaporation rates of the solvents used. If the flammable liquid is more volatile than the nonflammable solvent, the vapors from the mixture can be highly flammable. Conversely, if the nonflammable solvent is more volatile, it can evaporate to leave a flammable liquid.

Supervisors must keep these considerations in mind when using such solvents. Never forget that, under the right conditions, the vapor from a "safety" solvent can be just as deadly and flammable as the vapor from an extremely toxic solvent. If there is a chance that workers might be exposed to these hazardous chemicals in the work area, have the air monitored for contaminant levels.

## PHYSICAL HAZARDS

The second category of environmental factors or hazards involves physical agents. This category includes such hazards as noise, ionizing and nonionizing radiation (including visible radiation and lasers), temperature extremes, and pressure extremes. You should be alert to these hazards because they can have immediate or cumulative effects on employee health.

### Noise

Noise (unwanted sound) is a form of vibration that can be conducted through solids, liquids, or gases. The effects of noise on people include:

- Psychological effects. Noise can startle, annoy, and disrupt concentration, sleep, or relaxation.
- Interference with verbal communication, and as a consequence, interference with job performance and safety.
- Physiological effects. Noise-induced hearing loss, aural pain, or even nausea

(when exposure is severe). Some research links long-term overexposure to noise to circulatory problems and heart attack.

## Noise Measurement

A source that emits sound waves produces minute changes in air pressure. We usually measure these pressure changes with sound-level meters and noise dosimeters. The human ear can hear sound over a wide range of pressures, with the ratio of highest to lowest pressures about 10 million to 1.

To deal with the problem of this huge pressure range, scientists have developed the decibel (dB) scale, which is logarithmic. Decibels are not linear units like miles or pounds. Rather, they are points on a sharply rising curve. Thus, 10 decibels is 10 times greater than one decibel; 20 decibels is 100 times greater (10 × 10); 30 decibels is 1,000 times greater (10 × 10 × 10); and so on. The rustle of leaves is rated at 20 dB. Typical office noise has a level of about 50 dB. A vacuum cleaner runs at about 70 dB, while a typical milling machine from 4 ft (1.2 m) away is rated at 85 dB. The sound of a newspaper press is about 95 dB, a textile loom is 105 dB, a rock band is about 110 dB, a large chipping hammer is 120 dB, and a jet engine registers about 160 dB.

The pitch or frequency of sound is also important in noise measurement because the ear hears some frequencies better than others. Our ears tend not to hear very low or very high frequencies as well as those in the middle range. Thus, noise-measuring instruments are designed to mimic the response of the human ear on what is called the "A" weighting scale. As a result, standards for noise are written for the dBA scale. The OSHA Noise Standard requires that noise monitoring instruments be capable of measuring noise between 80 and 130 dBA. Noise above 90 dBA is to be avoided to prevent injury and unprotected exposure should not be permitted. Workers exposed to noise above 90 dBA should use hearing protection or have their exposure time reduced, as discussed below. Typically, if you have to shout to be heard, you are in a noise zone.

Measurement of workplace noise can be quite complex and usually requires the services of a trained safety professional or industrial hygienist. They can make decisions regarding engineering controls or other methods to deal with the noise hazards. They will also furnish data to the company's hearing conservation program administrator, who can decide what types of hearing protection are needed. Details and definitions of noise, and its measurement and control, are given in the National Safety Council's *Fundamentals of Industrial Hygiene* and *Noise Control: A Guide for Employees and Employers.*

*Factors in Hearing Loss.* If the ear is subjected to high levels of noise for sufficient time, hearing loss may occur. A number of factors can influence noise exposure effects. Among these are:

- variation in individual susceptibility
- total energy (loudness) of the sound
- frequency distribution of the sound
- other characteristics of noise exposure, such as whether it is continuous, intermittent, or made up of a series of impacts
- total daily duration of exposure
- length of exposure in a noisy environment

Criteria have been developed to protect workers against hearing loss. OSHA 29 CFR 1910.95(a) and (b), Occupational Noise Exposure, sets allowable noise levels based on the number of hours of exposure. This standard limits unprotected exposure to eight hours at 90 dBA, four hours at 95 dBA, and so on (Table 8–B). Because noise may affect some workers' hearing more than others, it is good practice to require hearing protection when workers are exposed to levels above 85 dBA, regardless of length of exposure.

The OSHA standard requires employers to reduce noise exposures with administrative and engineering controls, where feasible. The standard also requires audiometric testing, training, initial monitoring, regular remonitoring, and remonitoring whenever changes in production, processes, or controls increase noise exposure.

Some individuals may experience hearing loss below 90 dBA. To further protect worker hearing, OSHA enforces its Hearing Conservation Amendment to the noise standard, which requires employers to:

- provide annual audiometric tests to all employees exposed to noise over 85 dBA (for 8 hours)
- offer optional hearing protection to workers exposed above 85 dBA and to make protection mandatory where noise exposures exceed 90 dBA (for 8 hours)
- ensure that workers with existing hearing loss wear protection when exposed to noise levels above 85 dBA (for 8 hours)

### Table 8–B. Permissible Exposures

| Duration per Day (Hours) | Sound Level dB(A)* |
|---|---|
| 8 | 90 |
| 6 | 92 |
| 4 | 95 |
| 3 | 97 |
| 2 | 100 |
| 1½ | 102 |
| 1 | 105 |
| ¾ | 107 |
| ½ | 110 |
| ¼ | 115 |

*Sound level in decibels as measured on a standard level meter operating on the A-weighting network with slow meter response.

- provide additional training in:
  - effects of noise on hearing
  - proper selection, fitting, use, and care of hearing protection, and advantages and disadvantages of each type
  - explanation of the purpose and methods of the hearing test

The regulation is very specific concerning the ways a hearing test is conducted. To be successful, the test requires the close cooperation of the supervisor. If possible, you should schedule employees so that their hearing can be tested before they are exposed to noise on the job or at least 14 hours after exposure. If this is not possible and employees must be called off the job, make certain that workers selected for hearing tests wear earplugs or muffs properly before their test. This will help to prevent confusing a temporary detrimental effect on hearing with a more serious long-term effect. Tests must be conducted in a noise-free environment, such as a testing booth.

**Audiograms.** The first audiogram is called the baseline audiogram because it is the basis of comparison for subsequent yearly audiograms required to assess hearing. Employees must be notified if a significant shift in hearing is discovered.

As a supervisor, you may be asked to ensure that employees wear their hearing protection, both before audiometric testing and on a regular basis. One of the best ways to achieve employee cooperation is to set an example by wearing protection yourself wherever it is required. This is true also for managers, visitors, and other employees who visit noisy areas.

**Hearing Protection.** The most commonly used hearing protection equipment includes earplugs, canal caps, and earmuffs. Earplugs are available in preformed and disposable types. Many styles (in plastic, foam, or other material) are available. Earplugs can also be custom-made to fit an individual employee's ear.

Canal caps are typically cone-shaped caps attached to a headband. They are most often used when the employee goes into noisy areas for brief periods, as they allow for easy insertion and removal.

Earmuffs fit over the entire outer ear. How effective they are depends on the tightness of the seal between the ear and muff. The earpiece of eyeglasses, for example, can create a significant noise "leak" if it passes between the muff and the wearer's skin.

Each type and brand of protection has been given a *noise reduction rating* (NRR), which describes how much an employee's exposure can be reduced by wearing the device. Not all types offer the same protection. In general, custom-molded plugs offer the greatest NRR. Earplugs offer the next best NRR, with muffs and caps offering somewhat less. It is important to note that not all brands of one type of protector offer the same noise reduction.

Noise reduction ratings were developed in a laboratory setting. In the actual work situation, earplugs, muffs, or caps may not provide the expected degree of protection. Therefore, many health and safety personnel select hearing protection devices that provide an extra margin of safety. They try to provide hearing protection that will reduce an employee's exposure below 85 dBA. (However, hearing protection often is not the sole answer to noise exposure. At sound levels over 90–100 dBA, engineering controls must be considered to reduce exposure.)

In selecting protection, the company and employee should also consider the advantages and disadvantages of the different types of hearing protection for their workplace. For example, it might not be appropriate for a mechanic or someone whose hands are routinely soiled by work to wear foam or other insert plugs. The workers will transfer grit and grime to the plug that can contribute to ear canal irritation and possible infection.

Hearing protectors must be comfortable when worn properly or employees will likely find ways to avoid wearing them. Even earplugs, which seem like simple devices, need to be seated properly each time they are worn. They must be inserted fully into the ear canal, not just "propped" at the opening. The administrator of the hearing conservation program should be responsible for ensuring that employees receive proper training and are comfortable with the hearing protection selected. It is up to the supervisor to ensure that employees wear hearing protection when needed. The supervisor must check for use of hearing protection when doing routine inspections, and also be a good role model by wearing the protection regularly. Another way to get employees to use their hearing protection is to get them involved in the selection process and get their "buy-in" on the protection chosen for their jobs.

## Ionizing Radiation

Although ionizing radiation is a complex subject, a brief discussion is included here because it is one of the more dangerous physical agents. Radiation is a form of energy. Familiar forms of radiant energy include light, which enables us to see; and infrared, which we feel as heat. The relationship among the different forms of electromagnetic radiation is shown in Figure 8–1.

Ionizing radiation refers only to those types that involve ionization (a change from the normal electrical balance in an atom). Only the forms of radiation discussed below possess sufficient energy to change this electrical balance, which can produce damage in living cells. Other forms of radiation that are not ionizing may also cause damage, but usually to a lesser extent and by a different mechanism.

Certain chemicals, when exposed to radiation, may form hazardous compounds or more hazardous forms of an already-toxic substance. For example, when oxygen is electrified, ozone is formed. Ozone is highly irritating to nose, throat, and lung tissues. When working with chemicals and ionizing radiation, it is important to know what reactions between the two can take place.

**Gamma Radiation.** This type of radiation, produced from radioactive materials and x radiation, is highly penetrating and can damage body tissues. The amount of damage depends, in part, on the energy of the radiation and on the relative sensitivity of the tissue.

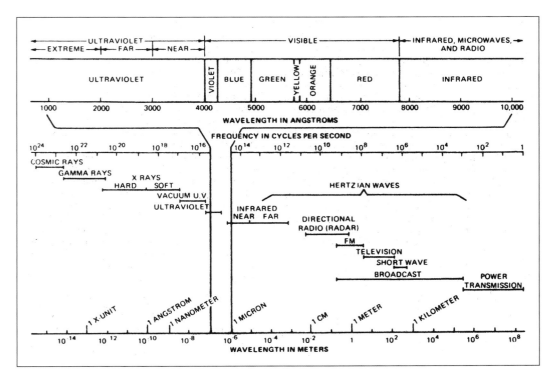

**Figure 8–1.** The electromagnetic spectrum, encompassing the ionizing radiations and the nonionizing radiations (expanded portion and right). Top portion expands spectrum between $10^{-7}$ and $10^{-8}$ m. Note: cycles per second (cps) = hertz (Hz).

**Alpha and Beta Radiation.** These types of ionizing radiation take the form of particles. Alpha particles are heavy and do not travel far in air. Although alpha particles are usually stopped by the skin, they can do considerable internal damage if they are inhaled or ingested. Beta particles are lighter and thus travel farther. If a beta source is left on the skin, it can damage the tissue. Proper shielding material is necessary to protect employees from beta radiation.

**High-Energy Protons and Neutrons.** In handling of some types of radioactive materials and in operation of high-powered particle accelerators, high-speed protons and high-speed and thermal neutrons can be created. A harmful dose is dependent on both the number of these particles and their energy distribution. Exposure potential must be evaluated by a health physicist (radiation specialist).

**X Radiation.** This hazard is produced by high-potential electrical discharge in a vacuum. It should be anticipated in evacuated electrical apparatus operating at a potential of 10,000 volts or more. Although x radiation is certain to be produced in such apparatus, it is dangerous only if it penetrates the apparatus jacket and enters the inhabited part of the workroom.

If a worker is overexposed to radiation of a comparatively low-voltage x-ray tube, dermatitis of the hand is generally the first result. It is characterized by rough, dry skin; a wart-like growth; and dry, brittle nails. Continued exposure and more penetrating x rays can cause bone destruction.

Beta sources and x and gamma radiation are found in some gages to measure the thickness of various materials, or in beta-source thickness gages. Sealed gamma sources and x-ray generators are used as various types of level gages, for example, to measure the liquid level inside a tank. They are also used to x ray welds at joints. Gamma radiation is also used to sterilize some health care products.

These types of radiation cause injury by ionizing tissue in which they are absorbed. Such injuries differ widely in location and extent, but the determining factors are the ability of the particular radiation to penetrate through the tissues, and the amount of ionization produced by a given amount of radiation. Alpha radiation, for instance, is classed as 20 times more biologically active than x rays because alpha particles are heavy, and can greatly disrupt and ionize tissue molecules with which they come in contact.

**Radiation Protection.** There are three general methods for preventing injuries from any penetrating radiation. First, workers can be separated from the hazard by distance sufficient to reduce radiation received to below the maximum permissible dose. Second, the time of exposure can be limited so workers will not receive a harmful dose. Third, specific shielding materials can be utilized to protect workers from exposure.

Rules for working with ionizing radiation are set by industrial hygienists or health physicists. They must be strictly followed at all times to avoid the chance of radiation injuries and illnesses.

## Nonionizing Radiation

Electromagnetic radiation has varying effects on the body, depending largely on the particular wavelength of the radiation involved. Following, in approximate order of decreasing wavelength and increasing frequency, are some hazards associated with different regions of the nonionizing electromagnetic radiation spectrum.

**Low Frequency.** The longer wavelengths—including power frequencies, broadcast, radio, and short-wave radio—can produce general heating of the body. The health hazards of low-frequency radiation were previously thought to be minimal. Now data suggest that exposure to radiation at electric power frequencies may be associated with adverse health effects, including cancer. Scientists have not yet discovered how or why these effects occur. Protection from electric field exposure can be provided by shielding and by limiting access to affected areas.

**Microwaves.** These are wavelengths of 3 m to 3 μm (100 to 100,000 megahertz). They are found in radar, communications, and diathermy applications. Microwave intensities may be sufficient to cause significant heating of tissues.

The effect is related to wavelength, power intensity, and time of exposure. Generally, longer wavelengths will produce a greater temperature increase in deeper tissues than shorter wavelengths. However, for a given power intensity, workers are less aware of the heat from longer wavelengths because it is absorbed beneath the skin.

An intolerable rise in body temperature, as well as localized damage, can result from exposure to wavelengths of sufficient intensity and time. In addition, flammable gases and vapors may ignite when they are inside metallic objects bombarded by a microwave beam.

Power intensities from microwaves are given in units of watts per square centimeter ($w/cm^2$). Areas with a power intensity greater than 0.01 $w/cm^2$

should be avoided. In such areas, dummy loads should be used to absorb energy output while equipment is being operated or tested. If a dummy load cannot be used, adjacent populated areas should be protected by adequate shielding.

**Infrared Radiation.** This type of radiation does not penetrate much below the superficial layer of the skin, so its major effect is to heat the skin tissues immediately below this layer. Longer wave infrared is completely absorbed in the surface layer of skin. Midrange infrared can cause skin burns and increased skin pigmentation. Short-wavelength IR radiation from furnaces can cause eye problems known as "glassblower's cataract" or "heat cataract." Exposed workers need to wear appropriate eye protection with darkened filtered lenses (see Chapter 9, Personal Protective Equipment).

**Ultraviolet and Visible Radiation.** These two types of radiation do not penetrate appreciably below the skin. Their effects are essentially to heat the surface. However, high-intensity visible radiation can severely damage the retina of the eye, even causing blindness in a relatively short exposure time. This is the reason behind insistent warnings to the public not to stare directly at the sun during periods of partial solar eclipses. The automatic defense mechanism of closing the eyes or looking away usually prevents damage. It should be emphasized that this defense mechanism is not effective against visible light emitted by lasers.

The most common exposure to ultraviolet radiation is from direct sunlight. UV radiation is responsible for "suntan" and "sunburn." People may develop tumors on exposed skin areas. These tumors occasionally become malignant.

The effects of UV light are generally more of a problem because UV can produce a severe burn with no warning, and excessive exposure can result in significant damage to the lens of the eye. Because UV radiation in industry is found around electrical arcs, they should be shielded by materials opaque to UV light.

Electrical welding arcs and germicidal lamps are the most common strong producers of UV radiation in industry. The ordinary fluorescent lamp generates a good deal of UV light inside the bulb, but the UV is essentially absorbed by the glass bulb and its fluorescent coating.

Some industrial materials also influence effects of UV radiation from the sun on human skin. After exposure to compounds such as cresols, for example, the skin is exceptionally sensitive to the sun. Even short exposure in late afternoon when the sun is low is likely to produce severe sunburn. Other compounds minimize the effects of UV rays. Some are used in certain protective creams (for example, sunblock found in various suntan lotions).

**Visible Radiation or Lighting.** Visible radiation, which falls about midway in the electromagnetic spectrum, should concern the supervisor because it can affect both the quality and accuracy of employees' work. Good lighting invariably results in increased product quality with less spoilage and increased production. Good lighting contributes to sanitary, clean, and tidy operations.

**Proper Lighting.** Good lighting is the result of several factors: the amount and color of the light, direction and diffusion, and the nature of illuminated surfaces. A dark-gray, dirty surface may reflect only 10 % or 12 % of a light source while a light-colored, clean surface may reflect more than 90 %.

Lighting should be bright enough for easy viewing of work tasks and directed so that it does not create glare. The level should be high enough to permit efficient sight. The National Safety Council's *Accident Prevention Manual for Business & Industry* and *Fundamentals of Industrial Hygiene* list the illumination levels and brightness ratios recommended for various occupational tasks by the Illuminating Engineering Society.

A major problem associated with lighting is glare—brightness within the field of vision that causes discomfort or interferes with proper vision. The brightness can be caused either by direct or reflected light. To prevent glare, keep the light source well above the line of vision, or shield it with opaque or translucent material. You can dull reflections by having such surfaces matte-finished (dull) rather than polished or smooth.

The video display terminals of computers are especially likely to be affected by glare, which can be reduced by dimming overhead lights and using a task light instead, by placing the terminal parallel to windows, or by installing an antiglare screen.

Another visual problem occurs when an object is too bright for its surrounding field. A highly reflecting white paper in the center of a dark, nonreflecting surface, or a brightly illuminated control handle or dial on a dark or dirty machine are two examples. The contrast can irritate the eyes. To prevent this condition, keep surfaces light or dark with little differences in surface reflectivity. Instead, use color contrasts to set objects apart from their backgrounds.

Although it is generally best to provide even, shadow-free light, some jobs require contrast lighting. In these cases, keep the general or background light well diffused and free of glare, and add a supplementary source of light directed to cast shadows as desired. Too much contrast in brightness between the work area and its surroundings can be as distracting and harmful as direct or reflected glare.

**Lasers.** The word laser is an acronym for "light amplification by stimulated emission of radiation." Lasers emit beams of coherent light of a single color (or wavelength). Contrast this with conventional light sources that produce random, disordered light-wave mixtures of various wavelengths. A laser is made up of light waves that are nearly parallel to each other (collimated), and that are all traveling in the same direction. Electrons of atoms capable of emitting excess energy in the form of visible light are "pumped" full of enough energy to force them into higher-energy-level orbits.

These electrons quickly return to their normal-energy-level orbits, but, in so doing, they emit a photon of visible light. The light waves given off are directed to produce the coherent laser beam.

Lasers are extremely versatile, and new applications for this invention are continually being developed. They are used for welding microscopic parts, for welding heat-resistant exotic metals, for communications instruments, for many kinds of guidance systems, for various surgical procedures, in chemistry laboratories, in bar-code readers, and in numerous other applications.

**Biological Hazards.** The eye is the organ most vulnerable to injury induced by laser energy. This is because the cornea and lens focus the parallel laser beam onto a small spot on the retina. Supervisors must make sure workers are instructed never to observe a laser beam or its reflection directly, or to look at it with an optical aid, such as a binocular or a microscope.

The work area should contain no reflective surfaces (such as mirrors or highly polished furniture) as even a reflected laser beam can be hazardous. Suitable shielding to contain the laser beam should be provided. Other safety measures may also be necessary, such as safety interlocks and lockout procedures, storing combustible solvents and other materials in proper containers, shielding them from laser beams or induced electrical sparks, and safeguarding all potential electrical hazards. The fact that infrared radiation of certain lasers may not be visible to the naked eye contributes to the potential hazard. Eyes must be protected. Lasers generating in the ultraviolet range of the electromagnetic spectrum produce corneal burns rather than retinal damage because of the way the eye handles ultraviolet light.

Other factors that influence the degree of eye injury induced by laser light include the following:

• pupil size—the smaller the pupil diameter, the smaller the amount of laser energy permitted to the retina

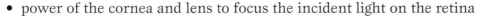

- power of the cornea and lens to focus the incident light on the retina
- distance from the energy source to the retina
- energy and wavelength of the laser
- pigmentation of the subject
- focal point of light on the retina
- divergence of the laser light
- presence of scattering media in the light path

Control of exposure to laser radiation should include use of protective devices built into the laser equipment, training employees in eye hazards, and use of protective eyewear as necessary. To develop a program of controls for the more hazardous lasers may require the help of a laser specialist.

## Temperature Extremes

General experience shows that extremes of temperature affect the amount of work people can do and the manner in which they do it. In industry, people are more often exposed to hazards associated with high temperatures than with low temperatures.

The body is continuously producing heat through its metabolic processes. Because these processes are designed to operate efficiently within only a narrow temperature range, the body must dissipate excess heat as rapidly as it is produced. Body temperature is regulated by a complex set of thermostatic controls that react quickly to significant changes in internal and external temperatures.

**Sweating.** Sweating is the most important of the temperature-regulating and heat-dissipating processes. Almost all parts of the skin are provided with sweat glands that excrete a liquid (mostly water and a little salt) to the body surface. This process goes on continuously, even under conditions of rest. In an individual who is resting and not under stress, the sweating rate is approximately 1 liter per day, which is evaporated in the air as rapidly as it is excreted. Under the stress of heavy work or high temperature, the sweating rate may increase to as much as 4 liters (approximately one gallon) in 4 hours. Up to 10 to 12 gm (150 to 190 grains) of salt per day will be lost with the water. Both water and salt loss must be replaced promptly to ensure good health. Such stress is especially dangerous for those with heart problems because loss of salt and water can affect the heart's ability to function.

If sweat evaporates as rapidly as it forms, and if heat stress does not exceed the maximum sweating rate of which the body is capable, an individual can maintain the necessary constant temperature. Rate of evaporation depends on the moisture content and temperature of the surrounding air (heat and humidity) and the rate of air movement. The industrial hygienist must consider all these factors when determining the effect of the thermal environment on humans.

**Radiant Heat.** Radiant heat is electromagnetic energy that does not heat the air it passes through. It affects the body's ability to remain in equilibrium with its surroundings. For instance, an object, such as a glass window, may be far below the body temperature of a nearby worker. As a result, a large amount of heat will be radiated from the person near the window, who may feel chilled even if the air in the immediate environment is fairly warm. Conversely, if an object in the surroundings, such as a furnace wall, is above body temperature, people can receive considerable heat by radiation. They will find it difficult to keep cool even if the room is air conditioned.

Thus, when radiant heat strikes an object (such as a person), it is absorbed and produces the sensation of heat. Merely blowing air around offers no relief from the source. The only protection is to set up a barrier. Placing radiant-reflective shielding between the heat source and the worker is usually the most cost-effective way to reduce employee exposure. Use of heat-reflective personal protective clothing may be considered when exposures are short and intermittent.

**Heat Conduction and Convection.** If heat can be conducted through clothing and dissipated into the air, this will cool the body to some degree. Conduction usually is not an important means of cooling, however, because conductivity of clothing and heat capacity of air are usually low.

Conduction and convection become an important means of heat loss when the body is in contact with a good cooling agent, such as water. For this reason, when people are exposed to cold water, they become chilled much more rapidly than when exposed to air at the same temperature.

Air movement cools the body by conduction and convection. More importantly, moving air removes the saturated air layer around the body (which is formed rapidly by evaporation of sweat) and replaces it with a fresh layer, capable of accepting more moisture.

**Effects of High Temperature.** Effects of high temperature are counteracted by the body's attempt to keep the internal temperature down through increasing the heart rate. The capillaries in the skin dilate to bring more blood to the surface so that the cooling rate increases and, gradually, the body temperature stabilizes.

If the thermal environment is tolerable, these measures will soon restore the body's equilibrium to the point where the heart rate and the body temperature remain constant. If this equilibrium is not reached by the time the body temperature is about 102 F (38.9 C), corresponding to the sweating rate of about 2 liters per hour, there is imminent danger of heatstroke. Intermittent rest periods for people exposed to extreme heat reduces this danger.

***Heatstroke or Sunstroke.*** Heatstroke is not necessarily the result of exposure to the sun. It is a condition caused by exposure to an environment in which the body is unable to cool itself sufficiently. Sweating stops and the body can no longer rid itself of excess heat. As a result, the body soon reaches a point where the heat-regulating mechanism breaks down completely and internal temperature rises rapidly.

The symptoms are hot, dry skin (which may be red, mottled, or bluish); severe headache; visual disturbances; rapid temperature rise; and confusion, delirium, and loss of consciousness. The condition is recognizable by a flushed face and high body temperature. The victim should be removed from the heat immediately and cooled as rapidly as possible, usually by being wrapped in cool, wet sheets. This is a medical emergency, and the supervisor must get assistance. Studies show that for people admitted to hospital emergency rooms for heatstroke, the higher the body temperature upon admission, the higher the mortality rate. Because the condition can be fatal, medical help should be obtained immediately.

Heatstroke is a much more serious condition than heat cramps or heat exhaustion, which are discussed next. An important factor is excessive physical exertion. The only method of control is to reduce the surrounding temperature or to increase the ability of the body to cool itself so that body temperature does not rise.

***Heat Cramps.*** These may result from exposure to high temperature for a relatively long time, particularly if accompanied by heavy exertion and excessive loss of salt and moisture from the body. Even if the moisture is replaced by drinking plenty of water, an excessive loss of salt may provoke heat cramps or heat exhaustion.

Heat cramps are characterized by the cramping of the muscles of either the skeletal system or the intestines. In either case, the condition may be relieved in a few hours under proper treatment, although soreness may persist for several days.

***Heat Exhaustion.*** This may result from sustained physical exertion in a hot environment, especially if a worker has not been acclimatized to heat and has not replaced water lost in sweating. Symptoms include relatively low temperature, pallor, weak pulse, dizziness, profuse sweating, and cool, moist skin.

**Preventing Heat Stress.** Most heat-related health problems can be prevented, or, at least, the risk can be reduced. The following basic precautions help to reduce heat stress:

- Acclimatization. Acclimatize workers to heat by giving them short exposures, followed by gradually longer periods of work in the hot environment. New employees and workers returning from an absence of two weeks or more should have a five-day period of acclimatization. This period might begin with 50% of the normal workload and time exposures the first day and gradually build up to 100% on the fifth day.
- Mechanical cooling. A variety of engineering controls, including general ventilation and spot cooling by local exhaust ventilation at points of high-heat production, may be helpful. Evaporative cooling, mechanical refrigeration, and cooling fans are other ways to reduce heat, along with eliminating steam leaks. Making equipment modifications, using power tools to reduce manual labor, and using personal cooling devices or protective clothing are other ways to reduce heat exposure for workers.
- Procedures. Work practices can also be tailored to prevent heat disorders. Companies can provide a period of acclimatization for new workers and those returning from vacation. They can make plenty of cool drinking water available, as much as a quart per worker per hour. Companies should train first-aid workers to recognize and treat heat stress and should post the names of trained staff in a location known to all workers. Always consider individual workers' physical conditions when determining their fitness for working in hot environments. Older workers, obese workers, and workers on certain types of medication are at greater risk for heat-stress illnesses.

  If you alternate work and rest periods, with longer rest periods in a cool area, you can help workers avoid heat strain. If possible, schedule heavy work during cooler parts of the day and provide appropriate protective clothing. Allow workers to stop for rest if they become extremely uncomfortable in the heat.

- Rehydration. Employee education is vital so that workers are aware of the need to replace fluids and salt lost by sweating, and so they can recognize dehydration, exhaustion, fainting, heat cramps, heat exhaustion, and heatstroke as heat disorders. Workers should be advised to drink water beyond the point of thirst. Adding salt to the diet is not usually recommended or needed for a worker who is acclimatized to heat. Workers should also be informed of the importance of weighing themselves daily before and after work to avoid dehydration. If the worker has health conditions aggravated by this stressor, the worker's physician must be consulted.

Companies may want to supply a cooled carbohydrate electrolyte beverage to encourage rehydration. Such a beverage should have 6–8% carbohydrate to provide entry and stimulate fluid absorption, as well as some sodium to help maintain the desire to drink and offset loss of sodium from sweating. A number of carbohydrate electrolyte beverages are on the market.

Workers should be educated about the negative effects of high-carbohydrate (above 10%) diet and caffeinated beverages. The former will tend to decrease fluid absorption, and the latter will increase fluid loss through urination. Alcoholic beverages should also be avoided.

## Dermatitis

Skin problems in industry account for the single largest cost of all hygiene problems. Individual skin disorders usually can be traced to exposure to some kind of chemical compound, or to some form of physical abrasion or irritation of the skin. Although rarely a direct cause of death, skin disorders cause much discomfort and are often hard to cure. Even normally harmless substances can cause irritations of varying severity in sensitive skin.

**Sources of Hazards.** Causes of occupational skin disorders are classified in these ways:

1. mechanical agents—friction, pressure, trauma
2. physical agents—heat, cold, radiation
3. chemical agents—organic and inorganic subdivided according to action on the skin as primary irritants
4. biological agents—bacteria, fungi, parasites
5. plant poisons—several hundred plants and woods, such as poison ivy

There are two general types of skin reactions: primary irritation dermatitis and sensitization dermatitis. Nearly all people suffer primary dermatitis from mechanical, physical, or chemical agents. Brief contact with a high concentration of a primary irritant or prolonged exposure to a low concentration will result in inflammation. Allergic reaction is not a factor in these conditions. Sensitization dermatitis, on the other hand, is the result of an allergic reaction to a given substance. Once sensitization develops, exposure even to small amounts of the material may cause symptoms. Some substances can produce both types of dermatitis. Examples are organic solvents, formaldehyde, and chromic acid.

**Occupational Acne.** Organic-based cutting fluids are frequently responsible for occupational acne. When the skin becomes infected, oil pimples or boils form that resemble adolescent acne. Dirty oil, poor materials handling, and addition of germicides in the fluid increase the possibility of occupational acne.

To reduce the possibility of this condition, keep machines and areas around them clean. Substitute an aqueous-based cutting fluid for the organic cutting fluid. Change cutting fluids at regular intervals. Have employees use barrier creams or gloves, aprons, or face shields. Encourage employees who work in these areas to shower or thoroughly wash at the end of each shift. Work clothes should be changed before going home. Abrasive cleaners or brushes should be avoided. Instead, workers should use warm water, mild soap, and a soft brush. They should be discouraged from wearing oil-soaked clothing. Some sensitive skins may require special cleaners.

Certain factors help account for some skin disorders. Any investigation of occupational skin disease should take into consideration:
1. degree of perspiration
2. personal cleanliness
3. preexistence of skin disorders
4. allergic conditions
5. diet

## Atmospheric Pressures

From the beginning of caisson work, dating back to about 1850, it has been recognized that those who work under greater-than-normal pressures are subject to various disorders. The main effect, decompression sickness (commonly known as the bends), results from the release of nitrogen bubbles into the bloodstream and surrounding tissues during decompression. The bubbles lodge at joints and under muscles, causing severe cramps. To prevent this condition, decompression is carried out slowly and by stages so the nitrogen can be eliminated without forming bubbles.

Deep-sea divers are supplied with a mixture of helium and oxygen for breathing. Because helium is an inert diluent and is less soluble in blood and tissue than nitrogen, it presents a less formidable decompression problem.

Common troubles encountered by workers operating under compressed air include pain and congestion in the ears caused by inability to ventilate the middle ear properly during compression and decompression. As a result, many workers under compressed air suffer from temporary hearing loss. The cause is considered to be obstruction of the eustachian tubes that prevents proper equalization of pressure on the middle ear. Occasionally, permanent hearing loss may result.

The effects of low pressure are much the same as the effects of decompression from high pressure. If pressure is reduced too rapidly, decompression sickness and ear disturbances similar to, if not identical with, the divers' conditions may result.

People working at reduced pressure are also subject to oxygen deprivation, which can have serious and insidious effects on the senses and judgment. There is considerable evidence showing that exposure for three to four hours at an altitude of 9,000 ft (2.7 km) above sea level, without breathing an atmosphere enriched with oxygen, can result in severely impaired judgment. Even if pure oxygen is provided, the altitude should be limited to that giving the same partial pressure of oxygen as air at 8,000 ft (2.4 km).

Reduced pressure is not the only condition under which oxygen deprivation may occur. Deficiency of oxygen in the atmosphere of confined spaces is commonly experienced in industry. (Normal air contains about 21% oxygen by volume.) The oxygen content of any tank or other confined space should be monitored before workers are allowed to enter. Instruments, such as an oxygen analyzer, are commercially available for this purpose.

The first physiological signs of a deficiency of oxygen (anoxia) are increased rate and depth of breathing. Oxygen concentrations of less than 19% by volume cause dizziness, rapid heartbeat, and headache. One should never enter or remain in areas where tests have indicated such low concentrations unless wearing self-contained, supplied-air respiratory equipment. Oxygen concentrations less than 16% can cause loss of consciousness and death.

In an oxygen-deficient atmosphere, workers may find it hard to move and have a hazy lack of concern about the imminence of death. In cases of sudden entry into areas containing little or no oxygen, the individual usually has no warning symptoms, but immediately loses consciousness and has no recollection of the incident, if rescued and revived.

Atmospheric contaminants are discussed in this chapter in the sections Threshold Limit Values and Standard Operating Procedures. Entry into confined spaces also requires compliance with lockout/tagout procedures that should be used to reduce hazards from chemical, mechanical, electrical, and other physical hazards. These are discussed in Chapter 13, Machine Safeguarding.

## ERGONOMIC STRESSES

The third category of environmental factors is ergonomic. Involved are human reactions to posture, force, frequency duration, vibration, mechanical stresses, and temperature. This subject is discussed in greater detail in Chapter 10, Ergonomics.

The term *ergonomics* literally means the customs, habits, and laws of work. According to the International Labor Office, ergonomics is:

...the application of human biological science in conjunction with the engineering sciences to achieve the optimum mutual adjustment of man and his work, their benefits being measured in terms of human efficiency and well-being.

The ergonomics approach goes beyond productivity, health, and safety. It includes consideration of the total physiological and psychological demands of the job on the workers.

The human body can endure considerable discomfort and stress and can perform many awkward and unnatural movements, but only for a limited time. When unnatural conditions or motions continue for prolonged periods, they may exceed workers' physiological limitations. To ensure a continuously high level of performance, work systems must be tailored to human capacities and limitations.

*Biomechanics* is that phase of engineering devoted to the improvement of the worker-machine-task relationship in an effort to reduce operator discomfort and fatigue. *Biotechnology*, a broader term, encompasses biomechanics, human factors engineering, and engineering psychology. To arrive at biotechnological solutions to work-stress problems, the sciences of anatomy, physiology, psychology, anthropometry, and kinesiology need to be brought into play. Areas of concern are:

- strictly biomechanical aspects—the consideration of stress on muscles, bone, nerves, and joints

- sensory aspects—the consideration of eye fatigue, odor, audio signals, tactile surfaces, and the like
- external environment aspects—the consideration of lighting, glare, temperature, humidity, noise, atmospheric contaminants, and vibration
- the psychological and social aspects of the working environment

Information obtained from studying these factors can be translated into tangible changes in work environments. Reducing fatigue and stress with redesigned hand tools, adjustable chairs and workbenches, better lighting, control of heat and humidity, and noise reduction are definitely rewarding improvements.

In the broad sense, the benefits that can be expected from designing work systems to minimize physical stress on workers are:

- more efficient operation
- fewer accidents
- lower cost of operation
- reduced training time
- more efficient use of personnel

See Chapter 10, Ergonomics, for more detailed information on ergonomic stresses.

# BIOLOGICAL HAZARDS

Biological hazards include any viruses, bacteria, fungi, or any other living organisms capable of causing disease in humans.

## Tuberculosis

Tuberculosis and other bacterial infections are classified as occupational diseases when they are contracted by nurses, doctors, or health care workers caring for tuberculosis patients or during autopsy or laboratory work where the bacilli may be present.

## Fungal Infections

A number of occupational infections are common to workers in agriculture and in closely related industrial jobs. Grain handlers who inhale grain dust are likely to come in contact with some of the fungi that may contaminate grain. Some of these fungi can and do flourish in the human lung, causing a condition called "farmer's lung." Fine, easily spread fungi, such as rust and smut, can be contacted once the covering of infected grain is broken and may produce sensitization in the exposed individuals.

One example, coccidioidomycosis or Valley Fever, is a disease caused by a fungus present in the soils of southwestern states and the San Joaquin Valley in California. Valley Fever may be either mild or severe. Because the spores are found in the soil, Valley Fever outbreaks sometimes occur during excavation and construction work.

## Byssinosis

Byssinosis occurs in individuals who have experienced prolonged exposure to heavy air concentrations of cotton dust. Flax dust also has been incriminated.

The exact mode of action of the cotton dust is unknown, but one or more of these factors may be important: (1) toxic action of microorganisms adhering to the inhaled fibers, (2) mechanical irritation from the fibers, and (3) allergic stimulation by the inhaled cotton fibers or adherent materials. It takes several years of exposure before manifestations are noticed.

## Anthrax

Anthrax is a highly virulent bacterial infection. In spite of considerable effort in quarantining infected animals and in sterilizing imported animal products, this disease remains a problem. With prompt detection and modern methods of treatment, it is less likely to be fatal than it was a few years ago. Vaccination of high-risk workers and use of personal protective equipment can prevent infection.

## Q Fever

There have been reports of infection (Q fever) with the rickettsial organism among meat and livestock handlers. It is similar to, but apparently not identical with, tick fever, which has been known for many years. Q fever may come from contacting freshly killed carcasses or droppings of infected cattle. Probably in the latter case, transmission is by inhalation of the infectious dust. Prevention undoubtedly depends on recognition and elimination of the disease in the animal host. Workers should use personal and respiratory protection when working in dusty and contaminated environments, and should receive immunization.

## Brucellosis

Brucellosis (undulant fever) has long been known as an infection produced by drinking unpasteurized milk from cows suffering from Bang's disease (also called infectious abortion). Because it can also be contracted by handling the animals or their flesh and is transmitted by swine and goats, it is an occupational ailment of slaughterhouse workers, as well as of farmers, although more common among the latter. Here also, preventive measures are primarily proper animal testing and control to eradicate the disease. Personal protection when handling infected materials is important.

## Leptospirosis

Leptospirosis can infect livestock and wild animals. It causes a flu-like illness and, in severe cases, liver damage. Hunters and people working in rice paddies may come in contact with water containing liquid animal waste. Control includes immunization of animals and appropriate personal protection for workers.

## Psittacosis

Psittacosis is a risk for poultry workers and poultry-processing workers. This disease causes a mild flu-like illness and conjunctivitis. Sanitation and personal hygiene are important in prevention.

## Lyme Disease

Lyme disease is caused by bacteria that live in the gut of deer ticks. It causes flu-like symptoms and may cause a bulls-eye-shaped rash. If left untreated, it can lead to rheumatoid arthritis. Employees who work outdoors, especially in woods or brush, or who have potential contact with wildlife, or who handle stock or domestic animals, are most vulnerable. Prevention requires awareness of the signs and symptoms, often difficult to identify at the time of the bite. Precautions should include covering exposed skin with clothing, tucking pant legs into boots, and surveying the body for ticks when leaving the area. Any ticks found should be removed carefully, using protective gloves. The tick should not be squeezed or crushed, as this could release the bacteria. First aid should be administered to the wound and contact made to the worker's physician for follow-up treatment.

## HIV

Acquired immune deficiency syndrome (AIDS), caused by the human immunodeficiency virus, (HIV) is usually contracted during sexual contact with an infected person, during intravenous drug use involving shared contaminated needles, before or during birth from an infected mother, or by contact with infected body fluids. It cannot be transmitted by casual contact with an infected person in the workplace. The virus is present in the blood and other body fluids of infected persons, and can be a hazard for workers who might have contact with these fluids: medical and emergency rescue personnel, and hospital and laboratory workers who may have contact through cuts or breaks in the skin.

The Centers for Disease Control and Prevention have developed guidelines for workers in health care and hospital settings, and OSHA has a regulation for exposure to bloodborne pathogens covering all potentially affected employees. The CDC guidelines specify that all patients should be assumed to be infectious for HIV and other bloodborne illnesses (such as hepatitis B). Precautions to avoid the spread of bloodborne disease include:

- Proper handling and disposal of all needles in impervious containers labeled as biohazardous.
- Prohibiting recapping of needles and other measures to avoid needlestick injuries.
- Using protective personal resuscitation clothing, including impervious gloves, aprons, boots, face shields, resuscitation masks, and goggles in some situations.
- Using resuscitation devices instead of mouth-to-mouth techniques.
- Adopting appropriate sterilization, disinfection, and housekeeping methods where potential for exposure to blood or body fluids is involved. In addition, the OSHA standard has requirements for a written exposure control plan, employee training, and medical surveillance.

## Upper Respiratory Tract Infections

Workers exposed to dusts from such vegetable fibers as cotton, bagasse (sugar cane residues), hemp, flax, or grain may develop upper respiratory tract infections ranging from chronic irritation of the nose and throat to bronchitis, complicated by asthma, emphysema, or pneumonia, or a combination of these. Dust from any of these fibers may also be a source of allergens, histamine, and toxic metabolic products of microorganisms.

# PERMISSIBLE EXPOSURE LEVELS

The Occupational Safety and Health Act of 1970 (OSHAct) is administered by (OSHA). The agency has the regulatory authority to protect workers from physical hazards, hazardous substances, and forms of energy (such as noise and radiation) in the work environment. OSHA monitors health and safety in the workplace, setting standards for worker exposure to specific chemicals, for permissible exposure levels (PELs), and for monitoring procedures. It also provides research, information, education, and training in occupational safety and health. By establishing the National Institute for Occupational Safety and Health (NIOSH), the act provided for studies to be conducted so that regulatory decisions can be based on the best available information.

OSHA has adopted PELs for more than 400 substances. Many are based on Threshold Limit Values (TLVs) suggested for industrial chemical exposures by the American Conference of Governmantal Industrial Hygienists.

The exposures listed in 29CFR 1910 subpart Z, Toxic and Hazardous Substances, are enforceable by OSHA and exhibit the minimum levels that every workplace should comply with.

# THRESHOLD LIMIT VALUES

Guidelines called Threshold Limit Values (TLVs) have been established for airborne concentrations of many chemical compounds. Supervisors should understand the meaning of TLVs and the terminology in which concentrations are expressed.

## Concept and Purpose

The basic idea of the TLV is fairly simple. TLVs represent exposure levels under which it is believed that nearly all workers may be repeatedly exposed without adverse effect. However, because individual susceptibility varies widely, exposure of some individuals at (or even below) the threshold limit may not prevent discomfort, aggravation of a preexisting condition, or occupational illness. In addition to the TLVs set for chemical compounds, there are limits for physical agents, such as noise, microwave radiation, and heat stress.

The TLV may be a time-weighted average value that would be acceptable for an eight-hour exposure. For some substances, such as an extremely irritating one, an eight-hour time-weighted average concentration would not be acceptable, so a ceiling value is established. The ceiling limit means that at no time during the eight-hour work period should the airborne concentration exceed that limit. It is important to understand that TLVs are not intended to be fine lines between safe and unsafe conditions. The supervisor should use considerable judgment when deciding on an appropriate exposure level in a particular work situation. Many companies choose to set an in-house "action level" of one-half or less of the TLV. When the concentration of the chemical exceeds the "action level," further air monitoring is often conducted, and measures are taken to reduce employee exposure.

As more toxicological information has become available, the wisdom of this practice is clear. Many of the TLVs have been reduced as additional epidemiological and laboratory studies have become available that show health effects at lower exposure levels. NIOSH also has set Recommended Exposure Levels (RELs) for a number of substances; many of these RELs are below the TLVs.

### Establishing Values

Threshold Limit Values have been established for more than 600 substances, as well as for heat and cold stress, vibration, radiation, and biological exposure indexes. The TLVs are developed and reviewed by the American Conference of Governmental Industrial Hygienists, a voluntary organization composed of industrial hygienists, toxicologists, and medical personnel who work in government, educational institutions, and private industry.

The data for establishing a TLV comes from animal studies, human studies, and industrial experience. The limit may be selected for one or several reasons. A substance may be very irritating to the majority of people exposed to concentrations above a given level. Or the substance may be an asphyxiant. Other reasons for establishing a limit might be that the chemical compound is an anesthetic, or fibrogenic, or can cause allergic reactions or malignancy. Some TLVs are established because, above a certain airborne concentration, a substance poses a nuisance to workers.

The concentrations of airborne materials capable of causing problems are quite small. Consequently, industrial hygienists use special terminology to define these concentrations. They often talk in terms of parts (of contaminant) per million (parts of air) when describing the airborne concentration of a gas or vapor. If measuring airborne particulate matter, such as dust or fume, the term milligrams (of dust) per cubic meter (of air), or $mg/m^3$, is used to define concentrations. For materials composed of fibers, such as asbestos or fibrous glass, industrial hygienists may use the number of fibers present in one cubic centimeter of air being tested. In this case, the descriptive term becomes the number of fibers per cubic centimeter, or fibers/cc.

As an example of the small concentrations involved, the industrial hygienist commonly samples and measures substances in the air of the working environment in concentrations ranging from 1 to 100 ppm. Some idea of the magnitude of these concentrations can be appreciated when one realizes that 1 in of 16 miles, one cent in $10,000, 1 oz of salt in 62,500 lb of sugar, 1 oz of oil in 7,812.5 gal of water—all represent one ppm.

# STANDARD OPERATING PROCEDURES

One of your primary functions as supervisor is to assist in developing standard operating procedures for jobs that require industrial hygiene controls, and to enforce these procedures once they have been established. Common jobs involving maintenance and repair of systems for storing and transporting fluids or entering confined spaces for cleaning and repairs are controlled almost entirely by the immediate supervisor.

## Industrial Hygiene Controls

Finally, the supervisor should be aware of the various kinds of controls used to maintain a healthy work environment. The type and extent of these controls depend on the physical, chemical, and toxic properties of the air contaminant, the evaluation made of the exposure, and the operation that disperses the contaminant. The extensive controls needed for lead oxide dust, for example, would not be needed for limestone dust because limestone is less toxic.

General methods of controlling workplace hazards include the following:

1. Substitution of a less harmful material for one that is dangerous to health. Make sure a professional carries out this task so that a more hazardous material is not mistakenly substituted.
2. Change or alteration in a process to minimize worker contact.
3. Isolation or enclosure of a process or work operation to reduce the number of persons exposed, or isolation of the workers in remote locations.

4. Wetting down dust during working operations and cleanup to reduce generation of dust.

5. Local exhaust ventilation at the point of contaminant generation and dispersion.

6. General or dilution ventilation with clean air to provide a less contaminated atmosphere. (Do not use dilution ventilation improperly for toxic chemicals; local exhaust ventilation is the way to control these.)

7. Personal protective equipment, such as special clothing, and eye and respiratory protection. (These are discussed in Chapter 9, Personal Protective Equipment.)

8. Good housekeeping, including cleanliness of the workplace, proper waste disposal, adequate washing and eating facilities, potable drinking water, and control of insects and rodents.

9. Special control methods for specific hazards, such as reduction of exposure time, film badges and similar monitoring devices, and continuous sampling with preset alarms to detect intake of toxic materials.

10. Training and education to supplement engineering controls. (Training in work methods and in emergency response procedures—medical, gas release, etc.—to teach who should and should not respond in specific emergencies.)

11. Medical surveillance programs to monitor employee health before, during, and after exposure.

The supervisor plays an important role in making sure that controls are effective. You must observe and study process changes, shielding or enclosures, use of wet methods, local exhaust ventilation, and general ventilation carefully and systematically to make sure that they are functioning properly, or to determine if maintenance or repair work is needed. Even more important is the part you must play in enforcing the use of personal protective equipment, such as hearing protection and respirators, in areas where they should be worn. If this is not done, your company can be cited and penalized by regulating agencies.

Good housekeeping, special monitoring devices, and safety and hygiene training and education are dependent on your full cooperation. In fact, as a supervisor, you are a key person where safety and industrial hygiene controls are concerned. Without your help and guidance, these controls are of questionable value in ensuring a safe, healthy workplace.

## SUMMARY OF KEY POINTS

Key points covered in this chapter include:

- The more supervisors know about industrial hygiene, the better they can fulfill the objectives of their job: to ensure safe production. Supervisors should know how to identify health hazards and take steps to correct them in four key areas of industrial hygiene: chemical, physical, ergonomic, and biological hazards.

- Chemical hazards exist in the form of dusts, fumes, smoke, aerosols, mists, gases, vapors, and liquids. They can cause health problems when workers inhale, absorb (through the hair or skin), contact, inject, or ingest one or more of these substances.

- The major exposure hazard to chemical compounds is through inhalation of airborne contaminants in gases, vapors, and particulate matter. Supervisors should know the physical classifications of these substances.

- The hazard associated with breathing an airborne contaminant depends on the nature of the chemical. Symptoms can range from mild irritation of the nose and throat to serious injuries deep within the lungs. Concentration of airborne contaminants must be measured by an industrial hygienist. Supervisors should also be alert to the hazards of an oxygen-deficient environment. Workers can guard against respiratory hazards through use of respiratory protective equipment and reducing their exposure to the hazards.

- Absorption of chemicals can occur through breaks in the skin or through direct contact with intact skin. Solvents, classified as aqueous or organic, are among the most widely used liquid chemicals in industry and present the greatest hazard to workers. A solvent hazard is determined not only by toxicity of the chemical but also by the conditions of its use.

- Adverse effects of solvents vary from mild skin irritation, to more serious disorders of the central nervous system and major organs, to the formation of cancerous tumors. Even low-level exposure over time may cause long-lasting effects on workers' health. Preventive and protective measures include the proper selection, application, handling, and control of solvents; a working knowledge of each chemical's risks to the workers; a sound training program to educate workers in chemical hazards; and an effective program of engineering controls, work-practice controls, and personal protective equipment.

- Toxic chemicals can be absorbed from the gastrointestinal tract into the blood. Employees working with toxic materials should not eat, drink, smoke, or apply cosmetics in the work area; should wash their hands after every job; and should maintain a clean environment.

- Physical hazards include noise, ionizing and nonionizing radiation, temperature, and pressure extremes.
- Noise can have harmful psychological and physiological effects on workers as well as interfere with communication on the job. Effects vary, based on individual differences among workers and conditions on the job. Supervisors should try to reduce or eliminate noise hazards as well as take protective and preventive measures to ensure workers' health.
- According to OSHA's Noise Standard, noise above 90 dBA is harmful to workers. When exposed to these (or higher) noise levels, employees must either be given hearing protection equipment, have exposure time reduced, or both, to ensure proper working conditions. Companies are required to administer annual audiometric tests to all workers exposed to noise over 85 dBA. Hearing protection equipment includes earplugs, canal caps, and earmuffs. Supervisors should make sure these devices fit workers properly and are correctly used.
- Ionizing radiation involves a change in the normal electrical balance in an atom. The most common forms, which may be encountered on the job, include gamma radiation, alpha and beta radiation, high-energy protons and neutrons, and x-radiation. Chemicals also can interact with ionizing radiation to form highly toxic substances. This type of radiation causes injury by ionizing the tissues in which they are absorbed. Injuries can range form mild skin irritations to radiation poisoning, to the formation of various types of cancers. Workers can be protected from ionizing radiation by separating or shielding from the hazard or by reducing time of exposure.
- Nonionizing radiation includes low-frequency, microwaves, infrared, and ultraviolet and visible radiation. These types of radiation heat the tissues in which they are absorbed. Injuries range from mild irritation to severe burns and cancerous tumors as well as possible damage to the eyes. Protective measures include shielding workers from the radiation source and providing adequate protective gear.
- Visible radiation or light can affect both quantity and accuracy of employees' work. Proper lighting in the work area is the result of the amount and color of light, direction and diffusion, and nature of illuminated surfaces. Lighting should be bright enough to produce easy viewing and directed so that it does not create glare. Supervisors should work to eliminate or reduce glare off all reflective surfaces and visual display terminals. Visible light tools, such as

lasers, can injure the retina of the eye. Protective measures include shielding laser sources, training workers in eye hazards, and use of protective eyewear.

- Temperature extremes present health hazards when the body cannot heat or cool itself adequately to overcome the extreme. Most work hazards are associated with high temperatures. The body adapts to high temperatures by sweating, radiating heat, and through heat conduction or convection. If the body cannot overcome the extreme, a worker may suffer heatstroke, heat cramps, or heat exhaustion. Supervisors and workers must know the appropriate emergency medical steps to counteract these potentially life-threatening conditions.

- Workers can be protected against high temperatures by allowing them to acclimatize to the hot environment, providing plenty of fluids, allowing intermittent rest periods in a cooler environment, and providing proper work clothing and protective gear.

- Occupational dermatitis, or skin disorders, are generally caused by mechanical, physical, chemical, or biological agents, and plant poisons. Skin reactions are classified as primary irritation dermatitis and sensitization dermatitis. Mechanical, physical, and chemical agents produce primary irritation dermatitis. Sensitization dermatitis, on the other hand, is the result of an allergic reaction to a given substance. Occupational acne is caused by certain cutting fluids. Protective and preventive measures include scrupulous housekeeping, showering or washing at the end of each shift, use of protective clothing, and changing work clothes before leaving the job.

- Atmospheric pressures related to high altitude or underwater work can cause decompression sickness, hearing problems, and oxygen deprivation. An oxygen-deficient environment can impair workers' judgment or cause sudden loss of consciousness and even death in a short time. Workers should be allowed sufficient time to decompress from underwater or high-altitude work and be equipped with proper independent air supplies when entering oxygen-deficient environments.

- Ergonomics involves creating the best "fit" between worker and job, including such factors as work environment (seating, desks, lighting, noise, etc.) body postures and movements and psychological and social aspects. The disciplines of biomechanics and biotechnology are used to solve work-stress problems on the job. Minimizing ergonomic stress on workers can raise efficiency and reduce costs, number of accidents, and training time.

- Biological hazards include any virus, bacterium, fungus, or other living organism capable of causing disease in humans. Biological hazards in workplaces include tuberculosis, fungal infections, byssinosis, anthrax, Q fever, brucellosis, leptospirosis, psittacosis, Lyme disease, HIV, and upper respiratory tract infections. Supervisors must know how to prevent organisms from entering the workplace, what steps to take if an infection occurs, and how to protect workers.

- Threshold Limit Values (TLVs) have been established for airborne concentrations of many chemical compounds. TLVs represent conditions under which it is believed that nearly all workers may be repeatedly exposed without adverse effect. These values have been established for more than 600 substances and may be used to calculate exposure times for employees who work in areas where airborne contamination exists.

- Supervisors are key personnel in creating and supporting effective industrial hygiene controls. They must make use of substitution, engineering controls, good housekeeping measures, special monitoring devices, and safety/hygiene training and education to help ensure a safe, healthy workplace.

# 9

# Personal Protective Equipment

**After reading this chapter, you will be able to:**

- Recognize the requirements of a personal protective equipment hazard assessment
- Develop an effective personal protective equipment program
- Overcome employees' objections to using personal protective equipment and educate them about its importance
- Know the types of protective equipment available for the head, face, eyes, ears, respiratory tract, torso, and extremities
- Select the best protective equipment to fit the hazard involved

When a hazard is identified in the workplace, every effort should be made to eliminate it so that employees are not harmed. Elimination of hazards normally occurs through the application of some type of engineering controls. As an example, consider a milling operation. Assume that the processing results in large quantities of steel chips that fly all over an aisle used by employees. The most effective approach is to engineer this hazard out of the operation. Examples of engineering controls would be to either totally enclose the operation or provide a deflector to keep the chips from reaching the aisle. Preventing the chips from reaching the workers would likely prove to be a far more effective solution in preventing eye injuries than simply requiring the workers to wear protective eyewear.

Some engineering controls seek to reduce or control the hazard by isolating the process. Can that dip tank be relocated? Can the spray booth be enclosed? Can that noisy machine operation be isolated or separated from the rest of the facility? When engineering controls are not feasible, a second and often less effective approach is to control the hazard by use of administrative controls. For example, a milling operation might be done only at certain times of the day when few workers are present. This reduces the flying chip hazard to employees by reducing the incidental traffic in the aisle. But if the milling operation must be done at a time when any employees will be present and engineering controls are not in place, it will be necessary to use personal protective equipment (PPE).

Safety professionals consider PPE the least effective means of controlling exposure to a hazard. PPE does nothing to eliminate the hazard in the workplace and depends entirely on the worker correctly and consistently using the PPE in order to reduce the chance of injury. Even OSHA recognizes that PPE should be used as the last option in controlling exposure to hazards—not the first. One of the first paragraphs in the OSHA standard on respirators (29 CFR 1910.134 (a) (1) reads:

> In the control of those occupational diseases caused by breathing air contaminated with harmful dusts, fogs, fumes, mists, gases, smokes, sprays, or vapors, the primary objective shall be to prevent atmospheric contamination. This shall be accomplished as far as feasible by accepted engineering control measures (for example, enclosure or confinement of the operation, general and local ventilation, and substitution of less toxic materials). When effective engineering controls are not feasible, or while they are being instituted, appropriate respirators shall be used pursuant to this section.

OSHA requires that every employer assess its workplace to determine if hazards are present that may be controlled through the use of PPE (29 CFR 1910.132 (d)(1)). Even worksites that are not regulated by OSHA can benefit from such an assessment. Effectively identifying, eliminating, and controlling hazards can result in fewer workplace injuries and higher productivity that result from having a healthy workforce.

A personal protective equipment survey is most often completed by the safety coordinator or safety committee. Often times, line supervisors are very helpful in identifying hazards specific to their department or their process.

The personal protective equipment assessment usually consists of a walk-through survey in which basic hazard categories such as impact, penetration, compression, chemical, heat, harmful dust, or harmful radiation from sources such as welding or lasers are identified. Any hazards that cannot be eliminated through engineering methods, or controlled through administrative controls, may require the use of PPE.

Workplaces regulated by OSHA are required to certify that this personal protective equipment hazard assessment has been completed (29 CFR 1901.132(d)(2)). OSHA has published an excellent guideline for completing a personal protective equipment assessment. It is titled, *Non-mandatory Compliance Guidelines for Hazard Assessment and Personal Protective Equipment Selection.* The guideline appears in the general industry standards (29 CFR 1910) in Subpart I, Appendix B. This may also be viewed at www.osha.gov – Regulations – Part 1910 – Subpart I. Your local OSHA office can also provide guidance on how to locate the information on its web site. This guideline provides useful information even for worksites that are not regulated by OSHA.

In some situations, the only available protection will be the use of PPE. Often, in emergencies, PPE will be required. Use of PPE should be considered as one aspect of the overall safety plan.

## CONTROLLING HAZARDS

People who must work in areas with uncontrolled hazards should use PPE. To develop an effective PPE program, supervisors should:

- Be familiar with requirements of government standards and company policies.
- Understand the results of the workplace hazard assessment and the need for PPE.

- Be familiar with the safety equipment on the market to protect against specific hazards.
- Know the company procedures for paying for and maintaining the equipment.
- Develop an effective education and training program for employees required to wear the proper protective equipment.
- Review all Material Safety Data Sheets (MSDSs) that require PPE for protection against hazardous chemicals and materials.
- Establish an industrial hygiene evaluation procedure to determine whether PPE meets MSDS recommendations.

As the supervisor, once you know what and when PPE is needed, you need to make sure employees understand the need for PPE and when to use it. If workers fail to use the required protective equipment, you must impress on them its value and help them to recognize the need for it. Getting some workers to use protective equipment may be one of the toughest jobs you face. The more you involve employees in the PPE assessment and selection process, the greater the likelihood they will cooperate and comply with the requirements.

Supervisors should be given an understanding of applicable government regulations and company policies and trained to recognize hazards. You should also know the uses of the equipment and its limitations. PPE vendors will often provide training and information so that you can become thoroughly familiar with all types of protective equipment. Safety and hygiene personnel can also help keep you up to date on changes in PPE requirements.

Local safety conferences are a good source of PPE information. Equipment is usually on display and manufacturers' or distributors' representatives are handy. Hundreds of vendors exhibit at the National Safety Congress & Expo held each fall (www.nsc.org ). A wealth of information on safety equipment is available online or published in manufacturers' catalogs and other trade literature such as the annual PPE issue of *Safety + Health* magazine, published by the National Safety Council.

A good rule to follow when specifying or buying safety equipment is to deal only with reputable vendors. Protective equipment should conform to ANSI Standards. Standards for protective equipment are discussed later in this chapter.

Check with safety or industrial hygiene personnel who can specify the appropriate equipment for the hazards present in the workplace. There should be a method to ensure that PPE fits correctly. There is also a need to develop a good cleaning and maintenance program so that equipment is always sanitary.

By understanding the potential injuries from hazards present in the workplace, empathizing with the users, and understanding the basic principles about protective equipment, you can overcome any objections to PPE usage.

## Cost of Equipment

Companies differ in their personal PPE policies. Personal items that can be used off the job, such as safety shoes and prescription safety glasses, are often provided on a shared-cost basis with employees paying part of the cost. Some firms maintain their own stores of PPE and others have routine scheduled vendor visits, such as safety shoe mobiles, to make equipment available to employees.

Items considered necessary to the job must be supplied by the employer at no cost to the employee (see 29 CFR 1910.132(h)). Examples of such equipment would include respirators, protective aprons, and chemical-resistant gloves. Some companies provide protective uniforms that are laundered by a service provider and then returned to the worksite. Companies should be concerned with controlling costs, but not at the expense of safety. In the long run, it can be less expensive to install engineering controls that eliminate the need for PPE than to maintain necessary equipment and associated compliance requirements such as periodic medical testing and PPE program auditing.

## Employee Involvement in Selection of PPE

The level of employee involvement in the safety management system is often a key indicator of success. Workplaces that have high levels of employee involvement in the safety system tend to have less safety-related disciplinary issues.

It is in everyone's best interest to have employees actively involved in the PPE selection process. This involvement should begin with the workplace hazard assessment that determined the need for PPE. Experienced employees can often identify specific processes or area hazards that even the best trained safety professional may miss. Experienced employees can also be a great help to engineers and safety professionals as they seek to apply engineering or administrative controls that eliminate the need for PPE. All employees should be aware of the attempts the employer has made to eliminate the need for PPE. Employee acceptance of PPE is often higher when employees understand that the company has investigated other methods of control before requiring the use of PPE.

When a decision has been reached that PPE must be used to protect the health and safety of the work force, employees should be involved in the selection of the PPE.

One of the biggest obstacles supervisors face is overcoming the objections some workers have to wearing PPE. Many of the objections encountered are quite similar, whether employees are talking about eye, ear, or head protection; respiratory protective equipment; or protective clothing such as vests, aprons, gloves, and even safety shoes.

Proper fit and comfort are important. Make sure that safety glasses, ear protectors, respirators, aprons, gloves, and shoes are properly fitted. Proper fit is facilitated by having more than one size and more than one brand available for employees' selection of the best fit. No one wants to wear equipment that does not feel comfortable, that is not clean, or that constantly needs adjusting

Another important consideration is availability and convenience of equipment. Make sure it is clean, easily used, and readily available.

Appearance should be considered. If the piece of equipment does not look attractive—for example, big, clumsy safety shoes as opposed to snug steel-toed moccasins or wing-tip dress shoes—workers may not wear it. Significant improvements in the appearance of safety glasses and hard hats have been made. These design improvements provide equipment with the same level of protection and a higher level of employee acceptance.

Involve the affected employees in every step of the PPE process to generate a sense of ownership in the program.

## HEAD PROTECTION

Industrial helmets, commonly known as hard hats, are needed to provide protection from impact and penetration hazards caused by falling objects (Figure 9–1). An example of a situation where a PPE hazard assessment might indicate the need for head protection would include workers who are working below other workers using tools or materials that could fall. Warehouses and elevated mezzanines are common workplace environments where such hazards are present. Other workplace examples that may mandate the use of head protection would include working under overhead conveyors that carry parts or materials, working near piping or storage racks where head bumping is a possibility, or working around exposed electric hazards.

The ANSI/ISEA Z89.1-2009 standard, *Industrial Head Protection*, provides performance and testing requirements for hard hats. It establishes the types and classes of protective helmets, depending on the type of hazard encountered. It includes specifications for helmets designed to offer protection from lateral (side) impact, or top-only impact, giving employers and users the flexibility to specify the helmet that best meets the needs of their specific workplace.

Impact resistance is essential. Industrial head protection meeting the requirements of the ANSI/ISEA standard are classified as Type I for top protection or Type II for lateral impact protection. Both types are tested for impact attenuation and penetration resistance.

There are three classes of helmets, based on electrical insulation rating. Class C (conductive) helmets provide no electrical protection. Class G (general) helmets are rated for 2200 volts or less. Where contact with energized circuits is possible, only helmets that meet the requirements of Class E (electrical), ANSI/ISEA Z89.1-2009, should be worn. These helmets should have no conductive fittings passing through the shell. Class E hard helmets are tested at 20,000 volts.

**Figure 9–1.** Standard protective headgear. *(Courtesy of E.D. Bullard Co.)*

A brim around the helmet provides the most complete protection for the head, face, and back of the neck. In a situation where the brim would get in the way (i.e., welding), it may be desirable for the helmet to be worn in the reverse position. Helmets should not be worn backwards unless approved by the manufacturer because the bill is designed to deflect potential hazards.

Another type of head protection is known as "bump hats" or "bump caps." Bump caps protect the scalp by reducing exposure to abrasions and lacerations caused by minor bumps to the head. This type of PPE is designed for applications in workplaces such as meatpacking, food processing, pest control, meter reading and automotive repair. Bump caps should never be used on construction sites, shipyards, or other locations where more dangerous hazards are present. This type of headgear does not meet the OSHA requirements for head protection and should not be used as a sport safety helmet. Bump caps do not meet requirements of ANSI/ISEA Z89.1-2009 and should not be used when that level of head protection is required or needed.

### Types and Materials

Plastic (molded under high pressure) is most frequently used for safety helmets. It resists impact, water, oil, and electricity. Fiberglass impregnated with resin is preferred because of its high strength-to-weight ratio, high dielectric strength, and resistance to moisture. Helmets with reflective trim are available on special order. They give added protection against incidents at night or in darkened areas.

The hard outer shell of the helmet is supported by a suspension. A cradle is attached to a headband that keeps the shell away from the head and provides some degree of protection against falling objects. This suspension must be adjusted to ensure a good fit. Keep extra cradles and headbands on hand, and replace a cradle or headband whenever it shows signs of deterioration or soiling.

Bands, cradles, and shells can be washed in warm, soapy water and rinsed or steam cleaned. Thirty days is the recommended maximum interval between cleanings. Make sure helmets and suspensions are thoroughly cleaned and inspected before reissuing them. The suspension should be replaced at least once a year.

Never attempt to repair the shell of a helmet once it has been broken or punctured. Do not let people drill holes in their safety helmets to improve ventilation or let them cut notches in the brims. Such practices destroy the ability of the helmet to protect the wearer. A few extra shells should be kept on hand as replacements. Information from the manufacturer should be reviewed for a useful life of the outer shell. Many manufacturers recommend replacing the outer shell at an interval not to exceed five years, regardless of the shell's appearance.

### Auxiliary Features

Liners for safety headgear are available for cold weather use. Do not let workers remove the safety helmet suspension in order to wear the helmet over parka hoods. This practice completely destroys the protection given by the helmet and has led to tragic results.

A chin strap is useful when the wearer may be exposed to strong winds, such as those on oil derricks. An eye shield of transparent plastic can be attached to some helmets. The shield is secured under the brim and lies flat against it when not in use. Full face shields should be used where needed. Brackets to support welding masks or miners' cap lamps are available on some safety helmets. Other mounting accessories for hearing, eye, and face protection are available.

Standard colors of safety helmets are white, yellow, red, green, blue, brown, and black. Other colors are available on special order. Often, distinctive colors or designs are used to designate the wearer's department or trade, especially in companies where certain areas are restricted to selected employees.

Employees should be prohibited from painting the outer shell as paint may contain solvents that can make the shell brittle. Also prohibit applying stickers, which could hide cracks or defects.

### Obtaining User Acceptance

Workers' personal preferences in appearance and dress may clash with safety regulations. For example, hair styles may be incompatible with wearing safety helmets. As the group's leader, you will have to convince people that it is more important (indeed it is required) to wear helmets than to have stylish hairdos.

In certain industries, such as food processing, workers are required not only to wear bump caps but to cover all facial hair—beards and mustaches—with hair nets. This is a requirement of the FDA and is done for sanitary reasons.

In overcoming objections and convincing people to use PPE, you must be prepared to counter all types of complaints. In the case of safety helmets, for instance, workers may complain that in winter the helmets are too cold. A liner or skull cap large enough to come down over the ears can take care of this objection. Some people may complain that helmets give them headaches; however, the suspension must not be removed. This is a possible effect until the employee becomes accustomed to wearing the helmet. In general, however, there is no physiological reason for a properly fitted helmet to cause a headache.

Although safety helmets are slightly heavier than ordinary hats, the difference is unnoticed after a while, particularly if the headband and suspension are properly adjusted. Workers should not be bothered by helmets unless there are unique physiological or psychological reasons for their complaints.

## FACE PROTECTION

Many types of personal protective equipment shield the face (and sometimes the head and neck) against impact, chemical or hot metal splashes, heat radiation, and other hazards.

Face shields of clear plastic protect the face of a person who is sawing or buffing metal, doing sanding or light grinding, or handling chemicals. The shield should be slow burning, and must be replaced if warped or scratched. Eye protection must be worn under the face shield. A regular face shield replacement

**Figure 9–2.** This face shield allows 160-degree peripheral vision and can be worn with prescription glasses. Its curved surfaces divert chemical splashes away from the face. *(Courtesy Millennium Safety Products, Inc.)*

schedule must be set up because plastic tends to become brittle with age.

A face shield that protects against light grinding will not protect against many chemicals. Like all PPE, face shields should be selected based on the specific hazard to which the wearer will be exposed (Figure 9–2). The headgear and shield should be easy to clean and be adjustable to size and contour of the head. Many types permit the wearer to raise the shield without removing the headgear.

Plastic shields with dichromic coating or metal screen face shields deflect heat and still permit good visibility. They are used around blast furnaces, soaking pits, heating furnaces, and other sources of radiant heat.

## Helmets

Welding helmets protect eyes and face against splashes of molten metal and radiation produced by arc welding. Helmets should have proper filter glass to keep ultraviolet and visible rays from harming the eyes. A list of shade numbers recommended by OSHA can be found in 29 CFR 1910.133(a)(5) or at www.osha.gov. Examples of such recommendations are listed in Table 9–A.

Workers' eyes vary due to age, general health, and the care given them. Two people may require different shaded lenses when doing the same job. Cracked or chipped filter lenses must be replaced. Otherwise, they may permit harmful rays to reach the welder's eyes.

The shell of the helmet must resist sparks, molten metal, and flying particles. It should be a poor heat conductor and a nonconductor of electricity. Helmets that develop pinholes or cracks must be discarded. Headgear should permit workers to use both hands and raise the helmet to position their work.

Most types of helmets have a replaceable, heat-treated glass or plastic covering to protect the filter lens against pitting and scratching. Some have a lift-front glass holder that permits the worker to inspect work without lifting or removing the helmet. Many newer versions have lenses that automatically darken when a welding arc is struck, returning to a light state when welding ceases. This allows the wearer to continue work without removing the helmet.

## Table 9–A. Selection of Shade Number for Welding Filters

| Welding Operation | Suggested Shade Number* |
|---|---|
| Shielded Metal-Arc Welding, up to 5/32 in (4 mm) electrodes | 10 |
| Shielded Metal-Arc Welding, 3/16 to 1/4 in (4.8 to 6.4 mm) electrodes | 12 |
| Shielded Metal-Arc Welding, over 1/4 in (6.4 mm) electrodes | 14 |
| Gas Metal-Arc Welding (Nonferrous) | 11 |
| Gas Metal-Arc Welding (Ferrous) | 12 |
| Gas Tungsten-Arc Welding | 12 |
| Atomic Hydrogen Welding | 12 |
| Carbon Arc Welding | 14 |
| Torch Soldering | 2 |
| Torch Brazing | 3 or 4 |
| Light Cutting, up to 1 in (25 mm) | 3 or 4 |
| Medium Cutting, over 1 to 6 in (26 to 150 mm) | 4 or 5 |
| Heavy Cutting, over 6 in (150 mm) | 5 or 6 |
| Gas Welding (Light), up to 1/8 in (3.2mm) | 4 or 5 |
| Gas Welding (Medium), 1/8 to 1/2 in (3.2 to 12.7 mm) | 5 or 6 |
| Gas Welding (Heavy) over 1/2 in (12.7 mm) | 6 or 8 |

*The choice of filter shade may be made on the basis of visual acuity and may therefore vary widely from one individual to another, particularly under different current densities, materials, and welding processes. However, the degree of protection from radiant energy afforded by the filter plate or lens when chosen to allow visual acuity will still remain in excess of the needs of eye filter protection. Filter plate shades as low as shade 8 have proven suitably radiation-absorbent for protection from the arc-welding processes.

In gas welding or oxygen cutting where the torch produces a high yellow light, it is desirable to use a filter lens that absorbs the yellow or sodium line in the visible light of the operation (spectrum).

*Notes:*

1. All filter lenses and plates shall meet the test for transmission of radiant energy prescribed in ANSI Z87.1-1989, *Practice for Occupational and Educational Eye and Face Protection.*

2. All glass for lenses shall be tempered, substantially free from striate, air bubbles, waves, and other flaws. Except when a lens is ground to provide proper optical correction for defective vision, the front and rear surfaces of lenses and windows shall be smooth and parallel.

3. Lenses shall bear some permanent distinctive marking by which the source and shade may be readily identified.

Source: *U.S. Code of Federal Regulations*, Title 29–Labor, Chapter XVII, Part 1910, Occupational Safety and Health Standards, §1910.252(e)(2)(i), and Part 1926, Safety and Health Regulations for Construction, §1926.102(a)(5).

Proper eye protection should be worn under helmets to protect welders from flying particles when the helmets are raised. The spectacle type with side shields is recommended as minimum protection from flying materials, adjacent work, or from the popping scale of a fresh weld. Welders' helpers should also wear the required eye protection while assisting in a welding operation or while chipping flux away after a bead has been run over a joint.

### Welding Shields and Goggles

A hand-held shield can be used where helmet protection is not needed, such as for inspection work, tack welding, and other operations requiring little or no welding. Frame and lens construction are similar to that of the helmet.

Welding goggles are available with filter glass shades up to No. 8. If darker shades are required, then complete face protection is needed because of the danger of skin burns. When shades darker than No. 3 must be used, side shields or cup goggles are recommended.

### Hoods and Face Shields

Acid-proof hoods that cover head, face, and neck are worn to protect against splashes from corrosive chemicals. This type of hood has a window of glass or plastic that is securely joined to the hood to prevent acid from seeping through. Chemical-resistant hoods made of vinyl, fiberglass, rubber, neoprene, plastic film, or impregnated fabrics are available for resistance to different chemicals. Consult manufacturers to find the protective properties of each material.

Hoods with an air supply should be worn when levels of toxic vapors or dusts are unknown or exceed recommended exposure limits. These hoods provide a supply of clean, breathing-quality air, which excludes toxic materials and maintains worker comfort. To support the air hose, the worker should wear a harness or belt.

Transparent face shields, supported from a head harness or headband, are used when there is limited exposure to direct splashes of corrosive chemicals. Splash-resistant cup goggles should be worn under the shield for eye protection.

Ensure that you choose the correct type of face shield based on the hazard assessment completed for the task. A chemical-resistant face shield will offer no protection if the task hazard is potential molten metal splashes or arc flash from an electrical source. Many times PPE, vendors can help assess the hazard and help you choose the correct protection level.

## EYE PROTECTION

Job tasks often expose the eyes to a variety of hazards, such as flying objects, splashes of corrosive liquids or molten metals, dusts, and harmful radiation. Eye injuries are very common, and not only disable a person but they often disfigure the face. More than 2,000 people injure their eyes at work each day. About 1 in

10 injuries require one or more missed workdays to recover. Of the total amount of work-related injuries, 10–20 % will cause temporary or permanent vision loss. Per-injury cost is high to both employee and employer.

Flying objects such as metal or stone chips, nails, or abrasive grits cause most injuries. Prevent Blindness America (www.preventblindness.org) lists the other chief causes of eye injury as:

- abrasive wheels (small flying particles)
- tools
- corrosive substances
- damaging visible or thermal radiation
- splashing metal
- poisonous gases or vapors

Operations in which striking actions occur, such as processes where equipment or material is struck by metal hand tools, or where cutting action may cause particles to fly, will require the operator and other workers nearby to wear eye protection. The flying particle hazard can be minimized by using nonferrous, "soft" striking tools, and by shielding the job with metal, wood, or canvas deflectors. Safety goggles or face shields should be worn when woodworking or cutting tools are used. Occasionally the need for eye protection may be overlooked on jobs where the potential for eye injury is less obvious. Examples of such tasks include cutting wire and cable, striking wrenches, using hand drills, chipping concrete, removing nails from scrap lumber, shoveling material to head level, working on the leeward side of the job, using wrenches and hammers overhead, areas where compressed air is used for cleaning, and other jobs where particles or debris may fall. Evaluate all tasks to determine the need for proper protective eye wear.

## Contact Lenses

Use of contact lenses in the workplace needs to be evaluated for potential safety problems. Where there are appreciable amounts of dust, smoke, irritating vapors, or liquid irritants that could splash into the eyes, contact lenses usually are not recommended. A judgment must be made as to the degree of potential hazard. Workers should wear proper safety glasses or goggles over their contacts where any potential eye hazard exists. One rarely noticed regulation under OSHA prohibits the wearing of contact lenses in contaminated areas with respirators (see 29 CFR 1910.134(e)(5)(ii)).

Prevent Blindness America, through its Advisory Committee on Industrial Eye Health and Safety, has issued the following position statement:

> Except for work environments in which there are significant risks of ocular injury, employees may be allowed to wear contacts in the workplace. However, contact lens wearers must conform to the regulations and directions of management regarding contact lens use. When the work environment involves exposure to chemicals, vapors, splashes, radiant or intense heat, molten metals, or airborne particles, contact lens use considerations should be made on a case-by-case basis in conjunction with the guidelines of the Occupational Safety and Health Administration (OSHA) and the National Institute for Occupational Safety and Health (NIOSH).

For further information, contact Prevent Blindness America at 211 West Wacker Drive, Suite 1700, Chicago, IL 60606, www.preventblindness.org, 800-331-2020.

## Goggles

Goggles and other kinds of eye protection are available in many styles, along with the protective medium of heat-treated or chemically treated glass, plastic, wire screen, or light-filtering glass. Supervisors should be educated about the various forms of eye protection and should be trained to identify the best eye protection for each task.

**Types and Materials.** Cover goggles are frequently worn over ordinary glasses. Lenses that have not been heat- or chemically treated are easily broken. Cover goggles protect against pitting and breakage.

These goggles include the cup type with heat-treated lenses and the wide-vision type with plastic lenses. Both are used for heavy grinding, machining, shipping, riveting, working with molten metals, and similar heavy operations. They offer the advantage of being wide enough to protect the eye socket and to distribute a blow over a wide area.

Glasses without side shields may be worn only when it is unlikely that particles will fly toward the side of the face. However, glasses with side shields are recommended for all industrial uses. Frames must be rigid enough to hold lenses directly in front of the eyes. These glasses should be fitted by a specialist.

Goggles with soft vinyl or rubber frames protect eyes against splashes of corrosive chemicals and exposure to fine dusts or mists. Lenses can be heat-treated glass or acid-resistant plastic. For exposures involving chemical splashes, these goggles are equipped with baffled ventilators on the sides. For vapor or gas exposures, they must be nonventilated. Some types are made to fit over glasses.

Dust goggles should be worn by employees who work around noncorrosive dusts—for example, in cement and flour mills. The goggles have heat-treated or filter lenses; wire-screen ventilators around the eye cup provide air circulation.

Welders' goggles with filter lenses are available for such operations as oxyacetylene welding, cutting, lead burning, and brazing. Table 9–A is a guide for the selection of the proper shade numbers. These recommendations may be varied to suit individual needs.

All goggle frames should be of corrosion-resistant material that will neither irritate nor discolor the skin, that can withstand sterilization, and that are flame-resistant or nonflammable. Metal frames should not be worn around electrical equipment or near intense heat.

To permit the widest range of vision, goggles should be fitted as close to the eyes as possible without the eyelashes touching the lens. The lenses should have no appreciable distortion or prism effect. Some specifications are covered by ANSI Z87.1–2003, *Occupational and Educational Personal Eye and Face Protective Devices.*

## Care of Eye Protection Equipment

Eye protection equipment must be sterilized before being reissued to different employees. Manufacturers' specific directions should be followed. The common procedure is to disassemble the equipment and wash it with soap or detergent in warm water. Rinse thoroughly. Immerse all parts in a solution containing germicide, deodorant, and fungicide. Do not rinse, but hang to dry in air. When dry, place parts in a clean, dust-proof container—a plastic bag is acceptable. Many companies have stations that dispense cleaning liquid and tissues to encourage frequent cleaning after goggles get smudged or dirty. Carrying cases or storage cabinets also help promote care of eye protection equipment.

**Replace Defective Parts.** If a lens has more than just the most superficial scratch, nick, or pit, it should be replaced. Such damage can materially reduce protection afforded to the wearer.

Some companies keep extra goggles in the main supply department, while others keep them in individual departments, along with a stock of parts.

Glasses and goggles should never be placed with the lenses facing down because they could become scratched, pitted, or dirty. They should not be kept in a bench drawer or toolbox, unless they are in a sturdy case.

Cleaning stations or materials should be easily available. A lens "fog" problem can usually be eliminated by use of one of the many commercial anti-fog preparations. In hot weather, workers can wear a sweatband to keep perspiration off their goggles. Although it takes a little effort to keep the goggles clean, this should not be an excuse for workers to discard them and risk losing an eye.

Visitors to areas requiring eye protection should use the same ANSI-approved eye protection used by employees. To avoid confusion and the potential for eyewear with inappropriate protection being worn on the worksite, most employers require all visitors and employees to wear company issued eyewear. Some exceptions may be made if the visitor's eyewear has appropriate ANSI markings.

## EAR PROTECTION

As outlined in Chapter 8, Industrial Hygiene, excessive noise levels should be reduced by engineering changes and administrative controls whenever possible. Hearing protection should be used only as a last resort.

Under 29 CFR 1910.95, where the sound levels exceed an eight-hour time-weighted average of 85 dB, measured on the A scale, a continuing, effective hearing conservation program shall be administered. This level may be increased slightly as the duration of exposure decreases. (Review the discussion in Chapter 8, Industrial Hygiene.)

It takes specialized equipment and trained personnel to analyze a noise exposure and to recommend the types of hearing protection that will be most effective. The best hearing protection is the one that is accepted by the individual and is worn properly. Properly fitted protectors can be worn continuously by most people and will provide adequate protection against most industrial noise exposures. See Figure 9–3 for a sampling of three common types of ear protection devices.

Earmuffs

PVC Earplugs    Foam Earplugs

**Figure 9–3.** Three common types of hearing protection devices.

### Insert Ear Protectors

Insert-type protectors are inserted into ear canals. They vary considerably in design and material. Materials used range from pliable rubber and soft or medium plastic to expandable plastic foam. Rubber and plastic are popular because they are inexpensive, easy to keep clean, and give good performance.

Another type of earplug popular with workers is one that is molded to fit each ear. After being allowed to set, it will hold its form. Because each person's ear canal is shaped differently, these plugs become the property of the individual to whom they were fitted. The plugs, of course, must be molded and fitted by a trained, qualified professional.

Some pressure is required to fit a rubber or plastic earplug into the ear canal so that the noise does not leak around the edges (as much protection as 15 dB could be lost). As a result, points of pressure develop, and they may cause discomfort. Even though plugs can be fitted individually for each ear, a good seal cannot be obtained without some initial discomfort. However, if the earplugs are made of soft material and kept clean, they should not create any lasting problems. Avoid earplugs composed of hard, rigid materials as they could injure the ear canal.

Skin irritations, injured ear drums, or other harmful side effects of PPE are rare when hearing protection is properly designed, well fitted, and kept clean. They should cause no more difficulty than a pair of well-fitted safety goggles. If fitting continues to be a problem, casts can be made of the ear canals, and a plastic plug can be custom-made for each.

## Muff Devices

Cup or muff devices cover the external ear to provide an acoustic barrier. The effectiveness of these devices varies with the size, seal material, shell mass, and suspension of the muff, as well as with the size and shape of workers' heads. Muffs are made in a universal type or in specific head, neck, or chin sizes. Hearing protection kits that can be used with hard helmets are also available.

Many employers prohibit the use of radios or other individual music delivery systems that can be worn by the individual worker. In addition to the possibility of damage to the ears from excessive noise levels, radios or personal music delivery systems worn in the ear may not be safe in a working environment. Employees may not be able to hear oncoming facility trucks, vehicle traffic, other workers, or any number of industrial noises.

As supervisor, you should be educated concerning any noise hazards in your department/area especially if any task or area requires the use of hearing protection. You should also be trained to demonstrate the correct use of hearing protection when other forms of control are not practical. You should also be educated concerning company policy about radios and other personal music delivery systems.

# RESPIRATORY PROTECTION

If hazardous air contaminants are present, engineering controls should be applied to remove the contaminants at the source, or otherwise isolate the contaminants. Because leaks and breakdowns do occur, and some operations expose a person, if only briefly and infrequently, respiratory protection should be available. Respirators may be regarded as emergency equipment or equipment for routine or occasional use. Workers must be medically fit, fit-tested, and instructed and trained in its proper use and limitations.

## Types of Equipment

Respiratory equipment includes air-purifying devices (mechanical filter respirators, chemical cartridge respirators, combination mechanical filter and chemical cartridge respirators, and masks with canisters) and air-supplied devices.

**Air-Purifying Devices.** Air-purifying devices remove contaminants from air prior to its entering the respiratory system. They can be used only in environments containing sufficient oxygen to sustain life. Air-purifying devices are only effective in the limited concentration ranges for which they are designed and must never be used where contaminant levels exceed the respirator manufacturer's accepted protection factor. These respirators generally consist of a soft, resilient facepiece and some kind of replaceable filtering element. Several types of air-purifying respirators, however, are available as completely disposable units. Various chemical filters can be employed to remove specific gases and vapors, while mechanical filters remove particulate contaminants. Air-purifying devices are never used in environments that are immediately dangerous to life or health (IDLH).

**Particulate-Filtering Facepiece Respirators.** Particulate respirators (also known as dust masks) are considered air-purifying devices because they clean particles out of the air you breathe. They are the simplest, least expensive, and least protective of the respirator types available. These respirators only protect against particles. They do not protect against chemicals, gases, or vapors, and are intended only for low-hazard levels. Even if you can't see the particles, there may be too many in the air for this respirator to provide adequate protection. There are nine classes of nonpowered, air-purifying particulate filters: three levels of filter efficiency, each with categories of resistance to filter efficiency degradation due to the presence of oil aerosols (see table below).

| N = for *N*ot resistant to oil | R = for *R*esistant to oil | P = for oil *P*roof |
| --- | --- | --- |
| 100 | 100 | 100 |
| 99 | 99 | 99 |
| 95 | 95 | 95 |

If no oil particles are present, use any series (N, R, or P). If oil particles are present, use only R or P series. Follow the respirator filter manufacturer's service-time-limit recommendations.

**Chemical Cartridge Respirators.** Chemical cartridge respirators have either a half-mask facepiece or a full-mask facepiece connected to one or more small containers (cartridges) or sorbent, typically activated charcoal or soda lime (a mixture of calcium hydroxide with sodium or potassium hydroxide) for absorption of low concentrations of certain vapors and gases.

The life of the cartridges can be relatively short. Check with the manufacturer on what is considered the end service life. Cartridges should be equipped with an end-of-service-life indicator certified by NIOSH for the contaminant, or the employer should develop and implement a change schedule for canisters and cartridges based on objective information or data that will ensure that they are changed before the end of their service life. After use, the cartridges must be discarded. These respirators must not be used in atmospheres immediately dangerous to life or health, such as those deficient in oxygen.

**Gas Masks.** Gas masks consist of a facepiece or mouthpiece connected by a flexible tube to a canister (a very large cartridge). Inhaled air, drawn through the canister, is cleaned chemically. Unfortunately, no one chemical has been found that removes all contaminants.

Gas masks have definite limitations on their effectiveness. Both gas concentration and length of time influence this. Gas masks, like chemical cartridge respirators, do not protect against oxygen deficiency. When the canister is used up, it should be removed and replaced by a fresh one. Even if they have not been used, canisters should be replaced periodically. Manufacturers will indicate the maximum effective life. Masks should be stored away from moisture, heat, and direct sunlight. They should be inspected regularly.

**Supplied-Air Devices.** Supplied-air devices deliver breathing air through a hose connected to the wearer's facepiece. The air source used is monitored frequently to make sure it does not become contaminated—for example, with carbon monoxide.

The air-line respirator can be used in atmospheres not IDLH, especially where working conditions demand continuous use of a respirator. Each person should be assigned his or her own respirator.

Air-line respirators are connected to a compressed air line. A trap and filter should be installed in the compressed air line ahead of the mask to separate oil, water, grit, scale, or other matter from the air stream. When line pressures are over 25 psi (170 kPa), a pressure regulator is required. A pressure release valve set to operate if the regulator fails should be installed.

Keep the compressor intake away from any source of contamination such as a motor vehicle or internal combustion engine exhaust. The compressor should have a carbon monoxide alarm in order to guard against the carbon monoxide hazard either from overheated lubricating oil or from engine exhaust. The most desirable air supply is provided by a nonlubricated or externally lubricated medium-pressure blower—for example, a rotary compressor.

If workers need to move from place to place, they may find that the air hose is a nuisance. Careful planning and job task evaluation may help minimize air hose problems. Care must also be exercised to prevent damage to the hose. The hose should not be permitted to be in liquids such as oil or solvents.

**Abrasive Blasting Helmets.** Abrasive blasting helmets are one example in a variety of air-line respirators designed to protect the head, neck, and eyes against the impact of the abrasive particles as well as give a supply of breathing air. The air quality requirements are the same as those described for air-line respirators.

The helmet should be covered both inside and out with a tough, resilient material. This increases comfort and still offers impact resistance. Some helmets have an outer hood of impregnated material and a zippered inner cape for quick removal. The helmet should contain a glass window, protected by a 30- to 60-mesh fine wire screen or plastic cover plate. Safety glass, used to prevent shattering under a heavy blow, should be free of color and glass defects.

A lighter weight helmet is approved for less hazardous work, such as spraying paint, buffing, bagging, and working in clean rooms.

Self-contained breathing apparatus (SCBA) offers protection for various periods of time by providing a portable, clean, tested air supply, which is worn by the user. Pressure-demand is the most common SCBA type.

The wearer of a SCBA is independent of the surrounding atmosphere; therefore, this kind of respiratory protective equipment can be used in environments where air contaminants are IDLH, or of unknown concentrations. This equipment is frequently used in mine rescue work and in firefighting (Figure 9–4).

**Figure 9–4.** Self-contained breathing apparatus. *(Courtesy Mine Safety Appliances Co.)*

## Selecting the Respirator

Air contaminants range from relatively harmless substances to dusts, fumes, vapors, mists, and gases that may be extremely toxic. Respiratory protection is required in certain areas or during certain operations when engineering controls are not available to reduce airborne concentrations of contaminants to a safe level. Engineering controls might not be present because they are technically unfeasible, or the hazardous operation might be done infrequently, making controls impractical. Respiratory protection is also needed while engineering and administrative controls are being implemented.

Your first step in selecting a respirator is to determine the chemical or other offending substance in the environment and to evaluate the quantity of the hazard. Conducting a hazard assessment should help you with this information and will help you choose the respirator that best protects against the particular hazard. Before respiratory equipment is ordered, discuss the type of exposure you have identified with the company safety or industrial hygiene department, and with manufacturers and dealers.

Protection factors are a measure of the overall effectiveness of a respirator. These factors, based on tests and on professional judgment, range from 5 to 10,000. The maximum use concentration for a respirator is determined by multiplying the Threshold Limit Values (TLVs) of the substance to be protected against by the assigned protection factor (APF) for the respirator. For example, a respirator with an APF of 10 for acetic acid would protect a worker against concentrations up to 10 times its TLV. Because the TLV of acetic acid is 10 ppm, the worker would be protected in atmospheres containing acetic acid concentrations as high as 100 ppm. (TLVs were explained in Chapter 8, Industrial Hygiene.)

The second step is to make sure workers are medically fit to wear a respirator. Routine medical surveillance is required to ensure that workers have adequate lung and heart functioning to wear a respirator that can place physical demands on the worker. See 29 CFR 1910.134, Appendix C, *OSHA Respirator Medical Evaluation Questionnaire* (mandatory) for the minimum medical requirements.

A respirator must be sized so that it properly fits the wearer. Fitting respirators to employees involves trying on several types and sizes and conducting fit checks and fit tests to determine the right respirator for the worker. This process could take one to two hours. In addition, training and education are needed. You should explain why it is essential to wear the respiratory protection, how it works, and, in general, why workers should wear the proper respirator for the operation in question. Get the affected employees involved in the entire process to facilitate their compliance.

Only respirators certified by the National Institute for Occupational Safety and Health (NIOSH) are acceptable. A NIOSH-approved respirator is assigned an approval number prefixed by the letters TC, indicating that it has been tested and certified by NIOSH for the air contaminant at the concentration range stated.

This approval ensures that the design, durability, and workmanship of the equipment meet minimum standards. Technicians have tested the respirator for worker safety, freedom of movement, vision, fit, and comfort of facepiece and headpiece; the ease with which the filter and other parts can be replaced; tightness of the seal against dust; freedom from leakage; and resistance to airflow when the wearer is inhaling and exhaling.

### Cleaning the Respirator

Respirator facepieces and harnesses should be cleaned and inspected regularly. If several persons must use the same respirator, it should be disinfected after each use in order to comply with OSHA regulations (see 29 CFR 1910.134 Appendix B-2, *Respirator cleaning procedures* (mandatory)). Appropriate methods of disinfection and cleaning should be obtained from the equipment manufacturer.

The supervisor or a designated employee should inspect respirators at intervals to check for damage or improperly functioning parts, such as headbands or valve seats. OSHA requires that the respirator program be periodically evaluated to determine if the employees are experiencing any difficulties that may interfere with the proper usage. These periodic surveys can be used as an opportunity to reinforce the company's commitment to protecting the health of the worker.

## BODY PROTECTION

The most common protection for the abdomen and trunk is the full apron. Aprons are made of various materials. Leather or fabric aprons, with padding or stays, offer protection against light impact and against sharp objects. Heat-resistant coats and aprons are often used by those who work around molten metal or other sources of intense conductive heat.

An apron worn near moving machinery should fit snugly around the waist. Neck and waist straps should be either light strings or instant-release fasteners in case the garment is caught. There should be a fastener at each end of the strap to prevent severe friction burns should the strap be caught or drawn across the back of the neck. Split aprons should be worn on jobs that require mobility on the part of the worker. Fasteners draw each section snugly around the legs.

Welders are often required to wear leather vests or capes and sleeves, especially when doing overhead welding, for protection against hot sparks and bits of molten metal. On jobs where employees must carry heavy, angular loads, pads of cushioned leather or padded duck are used to protect the shoulders and back from injury.

### Fall Protection

The use of full-body harnesses with lifelines attached should be considered for workers who labor at high levels or in closed spaces or in tasks that have the potential to bury a worker by loose material. This discussion does not pertain to vehicular seat belts.

Normal full-body harness use involves comparatively light stresses applied during regular work—stresses that rarely exceed the static weight of the user. Emergency use means stopping an individual when he or she falls. Every part of the harness may be subjected to an impact loading many times the weight of the wearer.

Full-body harnesses distribute evenly the shock of an arrested fall. This shock is distributed over shoulders, back, and waist, instead of being concentrated at the waist. The harness permits a person to be lifted with a straight back rather than bent over a waist strap. This makes rescue easier if the victim is unconscious, buried, or must be lifted out through a manhole. Wherever a job requires use of a self-contained breathing apparatus or supplied-air device, a harness and lifeline should also be considered for use in emergency rescue.

In the event that a worker falls from an elevated working surface, the full-body harness is designed to distribute the impact force over the legs and chest as well as the waist. The harness is designed to be used in conjunction with a lanyard or decelerating device, which brings the falling person to a more gradual stop, lessening the impact load on both the equipment and the person. To minimize potential fall distance, the line should be tied off overhead and should be as short as movements of the worker will permit.

All good fall protection programs have a detailed fall rescue plan so as to take as little time as possible to bring a fallen worker to safety. When a fall occurs, any number of factors can create challenges to the effective rescue of the victim. Weather conditions, physical obstacles, and the condition of the victim can consume time and create hindrances for the rescue personnel.

Any rescue plan should be regularly reviewed to ensure that the procedures are manageable and realistic in their time estimates. Employers should act in everyone's best interests by implementing the safest and quickest rescue plan and by practicing procedures to maximize preparation for a real emergency.

**Materials.** Harnesses may be made of natural or synthetic materials. Nylon lanyards and harnesses, although widely used, may not be satisfactory under some chemical conditions. Special types of harnesses are available for certain environments (these include wax-treated to resist paint solvents or mildew, and neoprene-impregnated to resist acids and oil products, to name a few). Be sure to consult with your safety department or safety supply vendor to obtain the right equipment.

Harnesses are designed to give a little when stress loaded. If a person does fall and stresses a harness, the harness should be taken out of service and replaced. Several kinds of devices that attach the harness to a lifeline serve to reduce the stress of a fall; these minimize both the injury to the user and the damage to the harness.

**Inspection.** Harnesses and lines should be examined before each use. Inspectors should be trained to recognize potential flaws in the harness, such as worn and torn fibers. Hardware should be examined, and if worn parts are found the entire harness should be replaced. If the harness is riveted, each rivet should be carefully inspected for wear around the rivet hole. Dirt and dust should be brushed carefully from harnesses to prevent fiber damage. Broken or damaged stitching may require replacement. Most equipment suppliers will provide this harness inspection training at no charge.

Harnesses may be washed in warm, soapy water, rinsed with clear water, and allowed to air dry without excess heat. Follow the manufacturer's instruction for cleaning and maintenance.

The outer surface of rope lines should be examined for cuts and for worn or broken fibers. Rope should be examined for breaks, discoloration, and deterioration. If the rope shows any of these signs, it should be taken out of service and rendered unusable before discarding.

A steel wire rope should be examined for broken strands, rust, and kinks that may weaken it. Ropes must be kept clean, dry, and rust free. They should be lubricated frequently, especially before use in acid atmospheres or before exposure to salt water. After such use, wire rope should be carefully cleaned and again coated with oil.

Wire rope should not be used for lifelines where falls are possible, unless a shock-absorbing device is also used. The rigidity of steel greatly increases the impact loading. Steel is, furthermore, hazardous around electricity.

Knots reduce the strength of all ropes. The degree of loss depends on the type of knot and the amount of moisture in the rope. More information is given in Chapter 15, Materials Handling and Storage.

The recently approved fall protection standard, ANSI/ASSE Z359, was adopted to incorporate basic fall safety principles, including hazard survey, hazard elimination and control, and education and training. For more information on the family of ANSI/ASSE fall protection standards, visit www.asse.org.

## Protection against Chemicals

A workplace assessment is required by OSHA. This evaluation includes an assessment of chemicals that may be inhaled, ingested, injected or absorbed through the skin. If you need help for this task, contact your safety department, insurance company safety engineer, industrial hygiene department, protective equipment supplier, or a chemist.

The best way to protect people is by implementing engineering controls that either contain the toxic chemical or isolate the worker. If the chemical contamination cannot be eliminated, replaced, or isolated, then your only recourse is protective clothing. Bear in mind that the protection is only effective if the protective clothing is worn diligently and properly. See Chapter 11, Hazard Communication, for additional information on chemicals.

### Protection against Ionizing Radiation

Ionizing radiation, although dangerous because of its serious biological effects, need not be feared if it is respected and if suitable precautionary measures are taken. The National Safety Council's *Fundamentals of Industrial Hygiene* text has additional information on the topic and many references. (Also see Chapter 8, Industrial Hygiene, for a more detailed discussion of avoiding radiation hazards.)

Standards for protection against radiation, specified in OSHA regulation 1910.97, reference Nuclear Regulatory Commission (NRC) regulations. These are spelled out in the *Code of Federal Regulations*, Title 10, "Energy," Part 20.

If you need help in protecting employees who may come in contact with this hazard, contact your safety department, insurance company, industrial hygiene department, protective equipment supplier, or your company's radiation safety officer.

### Protective Clothing

Ordinary work clothing, if clean, in good repair, and suited to the job, may be considered safe. "Protective clothing" refers to garments designed for specific, hazardous jobs where ordinary work clothes do not give enough protection against such injuries as abrasions, burns, and scratches.

**Guidelines.** Good fit is important. Trousers that are too long must be shortened to proper length, preferably without cuffs. If cuffs are made, they should be securely stitched down so the wearers cannot catch their heels in them. Cuffs should never be worn near operations that produce flying embers, sparks, or other harmful matter.

Long or loose sleeves, gloves, and loose-fitting garments (especially about the waist) create a hazard because they are easily caught in moving (or revolving) machine parts.

Clothing soaked in oil or a flammable solvent is easily ignited and is a definite hazard. Remember that skin irritations are often caused by continued contact with clothing that has been soaked with solvents or oils (see Dermatitis in Chapter 8, Industrial Hygiene).

**Materials.** A number of protective materials are used in making clothes to protect workers against various hazards. Supervisors should be familiar with these materials:

*Aluminized and reflective clothing* has a coating that reflects radiant heat. Aluminized or glass fiber is used for heavy-duty suits, and aluminized fabric for fire-approach suits.

*Flame-resistant cotton fabric* is often worn by people who work near sparks and open flames. Although the fabric is durable, the flame-proofing treatment may have to be repeated after one to four launderings.

*Flame-resistant duck* is used for garments worn around sparks and open flames, is lightweight, strong, and long-lasting. However, it is not considered adequate protection against extreme radiant heat.

*Glass fiber* is used in multilayered construction to insulate clothing. The facing is made of glass cloth or of aluminized fabric.

*Impervious materials* (such as rubber, neoprene, vinyl, and fabrics coated with these materials) protect against dust, vapors, mists, moisture, and corrosives. Rubber is often used because it resists solvents, acids, alkalis, and other corrosives. Neoprene resists petroleum oils, solvents, acids, alkalis, and other corrosives.

*Leather* protects against light impact and against sparks, molten metal splashes, and infrared and ultraviolet radiation.

*Synthetic fibers* (such as acrylics and low-density polyethylene) resist acids, many solvents, mildew, abrasion and tearing, and repeated launderings. Because some fabrics generate static electricity, garments should not be worn in explosive or high-oxygen atmospheres, unless they are treated with an antistatic agent.

*Water-resistant duck* is useful for exposures to water and noncorrosive liquids. When it is aluminum-coated, the material also protects against radiant heat.

*Wool* may be used for clothing that protects against splashes of molten metal and small quantities of acid and small flames.

*Permanent-press* fabrics are used in a wide variety of protective clothing. However, care must be taken to determine whether or not the fabrics are flammable.

## PROTECTING EXTREMITIES

Workers' extremities are highly vulnerable to injury in most work environments. Protective clothing and gloves can reduce the number and severity of injuries workers suffer each year. As supervisor, you should educate and train all employees in the importance of using such protection. Injuries to or losses of fingers, toes, even parts of an arm or leg, can happen with frightening suddenness. A good hazard assessment will define the type of protection needed and help workers understand the choices and that their best protection is prevention. This section discusses protective equipment for arms, hands, fingers, feet, and legs.

Fingers and hands may be subject to cuts, scratches, bruises, chemical exposure, and burns. Although fingers are hard to protect (because they are needed for practically all work), they can be shielded from any common injuries with such proper protective equipment as the following:

1. Heat-resistant gloves protect against burns and discomfort when the hands are exposed to sustained conductive heat.
2. Metal mesh gloves, used by those who work constantly with knives, protect against cuts and blows from sharp or rough objects.
3. Rubber gloves are worn by electricians. They must be tested regularly for dielectric strength.
4. Rubber, neoprene, and vinyl gloves are used when handling chemicals, etc. Neoprene and vinyl are particularly useful when petroleum products are handled.
5. Leather gloves are able to resist sparks, moderate heat, chips, and rough objects. They provide some cushioning against blows. They are generally used for heavy-duty work. Chrome-tanned leather or horsehide gloves are used by welders (Figure 9–5).
6. Chrome-tanned cowhide leather gloves with steel-stapled leather patches or steel staples on palms and fingers are often used in foundries and steel mills.
7. Cotton fabric gloves are suitable for protection against dirt, slivers, chafing, or abrasion. They are not heavy enough to use in handling rough, sharp, or heavy materials.
8. Heated gloves are designed for use in cold environments, such as deep freezers, and can be part of a heated-clothing system. Other types designed for such work are insulated with foam, and most such gloves are waterproof.

Specially made electrically tested rubber gloves, worn under leather gloves to prevent punctures, are used by linemen and electricians who work with high-voltage equipment. A daily visual inspection and air test must be conducted. Make sure that rubber gloves extend well above the wrist so that there is no gap between the coat or shirt sleeve and glove.

In some instances, however, gloves may be a hazard, particularly when worn around certain machining operations, such as a drill press. The material may become caught in machine parts. Use of gloves in these areas should be prohibited.

Hand leathers or hand pads are often more satisfactory than gloves for protecting against heat, abrasion, and splinters. Wristlets or arm protectors are available in the same materials as gloves.

## Foot Protection

About a quarter of a million disabling occupational foot injuries take place each year. A PPE assessment should be conducted to evaluate the need for protective footwear.

**Figure 9–5.** Leather or leather-reinforced work gloves protect against sharp or abrasive surfaces. *(Courtesy Ansell Edmont Industrial, Inc.)*

In 29 CFR 1910 Subpart I Appendix B, *Compliance Guidelines for Hazard Assessment and Personal Protective Equipment Selection* (non-mandatory), OSHA recommends that employers consider:

Safety shoes and boots which meet the ANSI Z41-1991 Standard provide both impact and compression protection. Where necessary, safety shoes can be obtained which provide puncture protection. In some work situations, metatarsal protection should be provided, and in other special situations electrical conductive or insulating safety shoes would be appropriate.

Safety shoes or boots with impact protection would be required for carrying or handling materials such as packages, objects, parts or heavy tools, which could be dropped; and, for other activities where objects might fall onto the feet. Safety shoes or boots with compression protection would be required for work activities involving skid trucks (manual material handling carts) around bulk rolls (such as paper rolls) and around heavy pipes, all of which could potentially roll over an employee's feet. Safety shoes or boots with puncture protection would be required where sharp objects such as nails, wire, tacks, screws, large staples, scrap metal etc., could be stepped on by employees causing a foot injury.

The responsibility for proper care of protective footwear ordinarily rests with the employees.

*Metal-free shoes*, boots, and other footwear are available for use where there are specific electrical hazards or fire and explosion hazards.

*"Congress"* or *gaiter-type shoes* are used to protect people from splashes of molten metal or from welding sparks. This type can be removed quickly to avoid serious burns. These shoes have no laces or eyelets to catch molten metal.

*Reinforced* or inner soles of flexible metal are built into shoes worn in areas where there are hazards from protruding nails and when the likelihood of contact with energized electrical equipment is remote, as in the construction industry.

*For wet work* conditions, in dairies and breweries, rubber boots and shoes, leather shoes with wood soles, or wood-soled sandals are effective. Wood soles have been so commonly used by workers handling hot asphalt that they are sometimes called "pavers' sandals."

*Foot protection with metatarsal guards* should always be worn during operations where heavy materials, such as drums, pig iron, heavy castings, and timbers, are handled. They are recommended whenever there is a possibility of objects falling and striking the foot above the toe cap. Metal foot guards are long enough to protect the foot back to the ankle, and may be made of heavy gauge or corrugated sheet metal. They are usually built onto the shoe.

Supervisors must often overcome workers' objections to wearing foot protection. Protective footwear used to be hot and heavy, and people often complained that it was uncomfortable. Current designs now make protective footwear comfortable, practical, and attractive. The steel cap weighs about as much as a pair of rimless eyeglasses or a wristwatch. The toe box is insulated with felt to keep the feet from getting too hot or cold. Some shoes have inner soles of foam latex with tiny "breathing cells."

## Leg Protection

Leggings that encircle the leg from ankle to knee, and have a flap at the bottom to protect the instep, protect the entire leg. The front may be reinforced to give impact protection. Such guards are worn by persons who work with molten metal. Leggings should allow fast removal in case of emergency. Hard fiber or metal guards are available to protect shins against impact.

Padding can protect parts of the legs from injury. Knee pads protect employees whose work, like cement finishing or tile setting, requires much kneeling.

Ballistic nylon pads are often used to shield thighs and upper legs against injury from chain saws.

Foot and leg protectors are available in many different materials. The type selected depends on the work being done. Where molten metals, sparks, and heat are the major hazards, heat- or flame-resistant materials or leather is best. Where acids, alkalis, and hot water are encountered, natural or synthetic rubber or plastic, resistant to the specific exposure, can be used.

## SUMMARY OF KEY POINTS

Key points covered in this chapter include:
- Personal protective equipment should be used only after engineering controls or administrative controls to manage hazards are evaluated.
- A personal protective equipment hazard assessment is required by OSHA. The OSHA web site, (www.osha.gov/pls/oshaweb/owadisp.show_document? p_table = STANDARDS&p_id = 10120) has an appendix that offers step by step guidance on how to perform this assessment.
- Supervisors must be familiar with government regulations regarding PPE, be able to recognize and assess hazards requiring PPE use, know the best types of equipment for particular hazards workers face, and educate and train employees in proper use and maintenance of PPE.
- Supervisors can overcome employee objections to PPE by making sure equipment fits properly, looks appealing, and is easy to clean and repair or replace. Employees should be involved in the selection of PPE.
- PPE required by the company must be provided by the employer at no cost to the employee. See 29 CFR 1910.132(h) for the allowable exceptions.
- Head protection is worn to prevent head injuries from falling or flying objects or from bumps against stationary objects. Hard hats can be insulated against electric shock and contain a hard outer shell with inner suspension or cradle and headband. Hard hats should be kept clean and replaced if damaged. Auxiliary features include liners, chin straps, eye shields, brackets, and standard or special-order colors. Hard hats should be adapted to fit comfortably.
- Common causes of eye injuries are flying objects, particles from abrasive wheels, corrosive substances, visible or thermal radiation, splashing molten metal, and poisonous gases or fumes. Workers can be protected either by shielding from the hazard and/or by providing safety goggles and face shields. Use of contact

lenses should be carefully evaluated. Supervisors should be trained to recognize hazards in the work tasks that may require the use of protective eyewear. All eyewear should be sterilized after use and defective parts replaced. Workers should be trained in proper use, handling, and care of protective eyewear.

- Face protection should shield the face (and sometimes head and neck) from impact, chemical or hot metal splashes, heat radiation, and other hazards. Equipment includes hard hats, face shields and goggles, and hoods (covering head, face, and neck). Some hoods are provided with independent air supplies for work around toxic fumes or dusts. Face protection should be selected for the particular hazard, be easy to clean, and be adjustable to fit in size and contour. Supervisors should set up regular replacement schedules, because plastic gear tends to become brittle with age.

- Hearing protection equipment must be provided to workers exposed to noise levels above 85 dBA on an eight-hour time-weighted average to reduce noise to an acceptable level. Equipment includes insert ear protectors and muff devices. These protectors must be designed and fitted properly and kept clean or they can cause ear infections and hearing loss. With cup or muff devices, which cover the external ear, their effectiveness depends on size, shape, seal material, shell mass, suspension of the muff, and how well they fit. Workers should be thoroughly trained in use and care of hearing protection.

- Respiratory equipment can be regarded as emergency equipment or PPE for occasional use on the job. This equipment includes air-purifying devices and supplied-air devices. To select proper respiratory equipment, supervisors must first determine the hazard in the environment, evaluate the extent of hazard, and choose the respirator that best protects against that hazard. Workers should be trained in proper use and maintenance and respirators should be inspected prior to use.

- The most common protection for the abdomen and trunk is full aprons, harnesses, lifelines, suits, equipment designed for radiation and chemical hazards, and protective clothing. Full-body harnesses can contain the worker or help to distribute the shock of an arrested fall. Harnesses and lifelines must be cleaned and inspected frequently. Equipment must meet all ANSI/ASSE standards and be worn when required.

- Protective equipment items for arms, hands, and fingers are usually gloves, wristlets, and arm protectors made of various materials designed to protect against cuts, scratches, bruises, and burns. Supervisors should know which type is best for a particular hazard: heat-resistant, metal mesh, rubber, leather, chrome-tanned cowhide, cotton, thermal, or chemical-resistant gloves. Gloves should be inspected regularly and kept clean.

- Suits and equipment for radiation and chemical hazards include gloves, disposable clothing, suits with supplied air, and respiratory devices. Employees exposed to radiation hazards must be monitored continuously to ensure safety levels. This clothing is designed for specific, hazardous jobs where ordinary work clothes do not give enough protection.

- Protective shoes, leggings, knee, thigh, and leg pads provide protection for the feet and legs. All protective shoes are reinforced with a toe cap. Supervisors should see that protective shoes are chosen for the particular hazard and that they fit properly. They include metal-free footwear; "congress" or gaiter-type shoes; reinforced shoes, boots, and other footwear for wet work; and shoes with metatarsal guards. Leggings must protect workers' legs and permit rapid removal in case of emergency. Hard fiber or metal guards protect shins, and nylon or cotton padding protects knees, thighs, and upper legs. Supervisors must educate employees in proper use and maintenance of protective foot and leg wear.

# 10
# Ergonomics

**After reading this chapter, you will be able to:**

- Describe various issues related to ergonomics and be alert for symptoms among your workers
- Explain the principles and goals of ergonomics and how physiology, anthropometrics, and biomechanics are used to solve work-stress problems
- Analyze how the workspace can be engineered to match workers' body characteristics
- Recognize risk factors related to musculoskeletal disorders
- Give examples of how to reduce ergonomic problems related to hand work, hand tools, whole-body vibration, video display terminals, and other work-related stresses

As a group leader or supervisor, you may be asked to understand and use some of the principles of ergonomics. Ergonomics is the study of how people interact with their work—in other words, how to create a good match between employees and the workstations and tools with which they work. The goal of an ergonomics program is to minimize incidents and illnesses due to chronic physical and psychological stresses, while maximizing productivity, quality, and efficiency. Many corporations have found that effective ergonomics programs have reduced operating costs, improved quality, and increased production and worker morale.

You should be familiar with ways to recognize problems that may be resolved using ergonomic principles, evaluate them, and suggest methods to control the risks. Many types of controls can be applied, from eliminating the problem, to changing work practices, to designing new tools, to training workers in how to use tools and equipment more effectively.

This chapter covers causes and symptoms of ergonomic problems, understanding the field of ergonomics, and applying basic principles to eliminate or reduce work-stress problems.

## WHAT ARE POSSIBLE CAUSES OF ERGONOMIC PROBLEMS?

Physical stress may arise when workstations, equipment, or tools do not fit the worker well. These stresses can cause long-term damage to muscles, nerves, and joints. Most illnesses due to ergonomic causes occur because of forceful or repetitive work activities, mechanical stress, temperature, lighting, or because workers are required to assume awkward postures over a period of time. Although it may take a physician to diagnose a worker's illness, you should be alert to any employee complaints of pain, tingling, numbness, swelling, or other discomforts. These symptoms may indicate a work-related problem.

### Cumulative Trauma Disorders

Back, shoulder, and neck strains and sprains can be results of exposure to stresses. Hand, wrist, and other arm symptoms are also often referred to as cumulative trauma disorders (CTDs). The following are some of the more common types of CTDs. (You may also use the term "repetitive strain injuries.")

**Tenosynovitis.** In this disorder, tendons in the wrist and fingers become sore and inflamed because of repetitive motion or awkward postures. Symptoms include pain, swelling, cracking sounds, tenderness, and loss of function. This disorder is caused by poor workstation design, problems with tool design, or changes in work habits.

**Tendinitis.** Tendinitis is an inflammation of the tendon that occurs when a muscle/tendon unit is repeatedly tensed. Tendinitis causes symptoms similar to those of tenosynovitis. This disorder typically occurs in the shoulder, wrist, hands, or elbow.

**Thoracic Outlet Syndrome.** A disorder of the shoulder that affects nerves in the upper arm, thoracic outlet syndrome is usually caused by doing tasks overhead for long periods of time. Loss of feeling on the little finger side of the hand and arm, with pain, weakness, and deep, dull aching, may occur.

**Ganglion Cysts.** These cysts are associated with cumulative trauma or repetitive motion. They appear on the wrist as bumps containing synovial fluid and can be surgically removed.

**Carpal Tunnel Syndrome.** This condition is caused by excessive overextension or twisting of the wrists, especially where force is used. It affects the median nerve, which runs through a bony channel in the wrist called the carpal tunnel. Symptoms include burning, itching, prickling or tingling feelings in the wrist or thumb and first three fingers. When the condition is severe, some of the thumb muscles may become weaker, and there may be overall weakness in the hand.

**De Quervain's Disease.** This is an inflammation of the sheath of the tendons to the thumb. Scarring in the sheaths restricts the thumb's movement.

**Trigger Finger Syndrome.** This is another form of tendinitis caused by repetitive flexing of the fingers against vibrating resistance. Eventually the tendons in the fingers become inflamed, causing pain, swelling, and a loss of dexterity. Trigger finger may result from using a power tool with too large a handle.

**Epicondylitis.** Also known as tennis elbow, epicondylitis is the inflammation of tissues on the inner (thumb) side of the elbow. It may be caused by violent or highly repetitive action that requires downward rotation of the forearm and wrist.

# UNDERSTANDING ERGONOMICS

Many different fields of study have contributed to industry's understanding of ergonomics. A few of these are:
- anatomy and physiology of the human body
- anthropometrics: study of differences in the size of bodies and body parts in different groups of humans
- biomechanics: study of the way work activities produce forces on muscles, nerves, and bones
- psychology: how people recognize and respond to signals in the environment
- industrial design and engineering: design of workplaces, tools, and processes

In approaching an ergonomic problem, an organization may bring together specialists in these areas to analyze the ways people interact with machines, tools, work methods, and work spaces. Safety professionals, industrial hygienists, and occupational physicians and nurses may add their expertise to help create solutions. Management and employees also have important roles in providing information, helping to evaluate work situations, and giving feedback on changes.

One way that you, as the supervisor, can help in the effort is by keeping track of possible indicators of workplace problems. These include:
- apparent trends in injuries, illnesses, and incident record keeping
- incidence of cumulative trauma disorders
- excessive use of sick days, high turnover rate
- employees changing their workstations (e.g., adding padding to hand tools or equipment, using makeshift platforms to stand on, making changes in the work flow)
- poor product quality

Some of the causative factors you should look for include:
- reliance on incentive-pay systems that motivate people to work faster than is safely possible
- high overtime and increased work rate
- manual materials-handling and other repetitive tasks
- work requiring awkward postures
- high amount of hand force required for tasks
- mechanical stress (e.g., hands or forearms resting on sharp table edges)

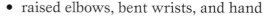

- raised elbows, bent wrists, and hand
- grasping or pinching objects
- exposure to temperature extremes
- use of vibrating tools

As supervisor, you should assist in the review of any new manufacturing processes, designs, and major equipment purchases. In this way, you will be able to point out potential difficulties in how workers may actually use equipment in the workplace and work with the person doing the job.

## Physiology

Work tasks need to be designed to match employees' physical capacities. A person's ability to do work at a given level is usually determined by the ability of his or her respiratory and cardiovascular systems to deliver oxygen to working muscles and to make use of energy derived from food.

One measure of capacity for manual work is maximum oxygen uptake. In industry, maximum effort is usually required only for brief periods, such as when an employee must lift heavy loads onto a hand truck. Over an eight-hour shift, however, the amount of energy used is usually well below the maximum capacity of the worker.

For instance, light work is generally associated with a heart rate of 90 or fewer beats/minute; and heavy work, about 120 beats/minute. Strenuous work produces a heart rate of 140–160 beats/minute or more and cannot be sustained without rest or cessation of effort. At strenuous work loads, waste products of energy consumption build up in the blood and muscles, producing fatigue or a sensation of soreness.

A person's capacity for manual work is also limited by muscle strength; by flexibility, or lack of it, in the joints; and by the strength of the spinal column. The science of biomechanics provides measures of the amount of force put on these parts when employees work in different positions. It helps to determine which positions make best use of muscular strength.

The muscle's ability to perform is also affected by the way it is used. Activity can be either static or dynamic. A dynamic activity, like walking, is created by rhythmic contraction and relaxation of the muscles. Dynamic work allows muscles to rest during the relaxation phase, so that the blood can supply oxygen and remove waste products. In contrast, static work is caused by the worker holding one object or body part in one position for an extended period. (Standing still is a static posture, for example.) The muscle is in a fixed and locked position, blood vessels are compressed, and blood flow is reduced; the muscle tires rapidly. A static muscle consumes more energy. It can also place more stress on tendons and joints, which may result in back, shoulder, and neck pain or inflammation of the tendons.

It is important to remember that heart rate (and the worker's ability to do the job) may also be affected by temperature, humidity, and the worker's age and fitness level. While engineers and work physiologists usually are responsible for assessing the physiological demands of a job, you need to understand the basics of how work demands affect body functioning.

## Anthropometrics

Anthropometric studies provide information that can be used to design workspaces to match body dimensions. The idea that there is an average-sized person or worker is obsolete. Even two people who are of "average" height, say 5 ft 8 in (1.7 m), may have very different body dimensions. Also, a person who has an average arm length might have "long" legs, or one with average leg length might have "long" arms.

Studies of military and civilian personnel have been used as the basis for tables of information on body sizes. These have defined the extremes of body and body part sizes in a particular population. This information is described by percentiles. For example, regarding height, a 5th-percentile female is taller than 4% of the female population and shorter than 95% of the females in the group. A 95th-percentile male is taller than 94% of the male population and shorter than the tallest 5%.

These percentiles are useful to engineers who try to design a workspace to fit as many in the population as possible. Usually they try to make the workstation adjustable so that it can fit anyone from the 5th-percentile female to the 95th-percentile male.

## Biomechanics

The science of biomechanics explains human body characteristics in mechanical terms. This allows us to describe the amount of force produced when a muscle works to move an object of a specific weight. The biochemical models may describe the arm as a level, for example. They are often illustrated as simple line drawings which show the different forces acting on the muscle and on the object. While these seem oversimplified, they are useful in helping to identify stressful postures. Calculations are derived from the laws of physics and are complex, because often more than one muscle is involved.

## Looking at the Workplace

Many workplace operations have potential for causing injury or illness in different ways. The amount of physical stress a job produces is often determined by the following:

- posture the worker must assume to do his or her job
- force needed to perform the task
- number of repetitions of a stressful task
- length of time doing the task

The next sections will discuss these principles of ergonomics and ways to apply them in the workplace.

# MATERIALS MOVEMENT

When materials are moved, the appropriate equipment should be used. This can include mechanical systems such as conveyors, hand trucks, and carts to reduce manual lifting and carrying. Wherever possible, people should be engineered out of the flow of materials handling. Chapter 15, Materials Handling and Storage, discusses this topic in more detail.

## Manual Materials Handling

Manual materials handling refers to the human activity necessary to move an object. This often means lifting and carrying but can also describe pushing, pulling, and shoving. Look for the following situations that may cause injury:

- lifting from the floor or lifting while twisting
- lifting heavy weights or bulky objects
- repetitive lifts
- lifting above shoulder height or while seated
- pushing or pulling loads
- bending while moving objects

## Guidelines for Safe Materials Handling

Making manual materials handling easier may mean changing one or more of the following factors:

- Where possible, modify the object to make movement easier. Handholds should be provided to allow use of a power grip that will make carrying more efficient. In a power grip, the object is clasped between the flexed fingers. This technique allows less muscle force to be used than a pinch grip, in which the hands cannot be placed entirely around the object. Cardboard cartons, for example, should have openings that can accommodate large gloved hands if needed. Surfaces of openings or handles should be smooth to avoid mechanical irritation of the skin (Figure 10–1). The center of gravity should be below the handholds and close to the body.

- Avoid lifting or carrying large, unwieldy, or heavy objects. Mechanical lifting and transporting means should be used wherever possible. These can include equipment such as conveyors, automatic or gravity-feed devices, load-leveling pallets, hoists, and lift trucks.

- Where mechanical means cannot be used, seek assistance from another person in lifting heavy or unwieldy objects.

- Where feasible, split loads in containers into smaller loads. Containers can be made of lighter materials. Shapes can be changed to allow objects to be handled closer to the body.

The weight workers can lift is influenced by many factors. Various calculations have been developed to help determine acceptable and maximum loads for each worker. These calculations are usually based on *Applications Manual for the Revised NIOSH Lifting Equation 94-110*, published by the National Institute

**Figure 10–1.** Handles should be smooth and shaped to permit a power grip.

for Occupational Safety and Health (NIOSH). The guidelines are modeled on lifting an object with two hands, in front of the body, with good handholds and footing, and no twisting required. The NIOSH formula takes into account the distance of the load from the body, the location of the object before and at the end of the lift, and the frequency of lifting.

Many industrial lifts do not meet these criteria. In general, the following guidelines apply to all lifts:

- Lifts should be kept between knuckle and shoulder height.
- The lift should begin as close to the body as possible. Sometimes objects are bulky or must be placed over or in containers or on shelving far back on counters. Job redesign should be considered in these cases, especially if a heavy object is involved. The longer the reach required to grasp the object, the less weight can be lifted. Lifting with arms outstretched puts far more stress on the back than lifting with the object close to the body.
- The height of the object above the floor at the start of the lift should be between shoulder and knuckle (mid-thigh). Placing objects on the floor or other location that requires stooping should be minimized. Rearrange the workplace so starting and ending points of the lift are close together. Avoid stacking items above shoulder height. Store loads so they will be easy to retrieve, and avoid deep shelves. Use spring-loaded bottoms for bins or gravity-feed bins to bring the load into easy reach.
- Avoid or minimize repetitive lifting tasks. Design tasks so that twisting while lifting or carrying is not required.
- The location in which the task is performed should be set up for optimal materials handling. Footing should be comfortable and should not be slippery or sticky. The space should have adequate room to work, but be small enough to limit the need to reach, stretch, or carry. Make sure workers have good lighting and clear sight lines.
- Eliminate the need for pushing or pulling by using conveyors, slides, or chutes. When objects must be pushed or pulled, push whenever possible.
- Carts should have large coasters to make the job easier. Limit ramp slope to 10 degrees or less.
- The worker must be considered as part of the system. Training in correct methods using ergonomics can sometimes help to reduce the risk of injury or illness. Training will probably not be the entire answer, however, especially where jobs with moderate or severe risk of injury due to lack of ergonomics are involved. Also consider a comprehensive back injury-prevention program

that includes ergonomic principles of engineering, work methods, and administrative controls.

It is therefore generally better to design a job for safe lifting and to train people to work safely.

Some experts believe that lifting training is appropriate, although there is still controversy over what constitutes proper lifting techniques. The older rule was to lift with a straight back from a bent-knee position. However, some workers' leg muscles and knee joints may not be strong enough to lift this way repetitively.

Some experts advocate other styles of lifting, such as the "kinetic lift," the "one-handed lift," or the "two-hand squat lift," but it is not clear that these will work for all people. Another approach includes improving workers' physical fitness through stretches and warm-ups.

Therefore, while there are no comprehensive rules for safe lifting, rules described in the previous section may apply. Additional guidelines are summarized below:

- Eliminate manual lifting and lowering from the task, where feasible.
- Be in good physical shape. Workers who are not used to lifting and vigorous exercise should not attempt difficult lifting tasks.
- Think before acting. Place material in a convenient place and within easy reach. Use the handling aids available.
- Get a good grip on the load. Test the weight before trying to lift it. If it is too heavy or bulky, get an assistant, mechanical lifting aid, or both.
- Get the load close to the body with feet close to the load. Stand in a stable position with feet pointing in the direction of movement. Lift mostly by straightening the legs.
- Don't lift with sudden or jerky movements.
- Don't twist the back or bend sideways.
- Don't lift or lower with arms extended.
- Don't heave a heavy object.
- Don't lift a load over an obstacle.

## WORK SPACE AND BODY CHARACTERISTICS

Because each person has unique body characteristics (anthropometrics), the work space must be flexible or adjustable to allow for these differences. Ideally, all work platforms, tables, and carts should be adjustable. If that is not possible, the work space should allow for the tallest person, with adjustable chairs, foot rests, or adjustable platforms used to accommodate shorter workers.

Equipment controls should be placed between shoulder and waist level. The normal work surface should be just below the elbow; the surface should be higher for precise work, lower for heavy work.

As discussed previously, workstation surfaces and leg room need to be designed for the largest potential user, with adjustments to accommodate smaller individuals. However, reaching distances in the workstation need to be designed for the smaller worker. For example, at a seated assembly operation that involves reaching into bins for parts, the chairs and work-surface adjustments should be provided so that short individuals with short arms will not have to overreach, putting stress on shoulders and arm muscles. Infrequent reaching behind or to the sides is acceptable but should not be required on a routine basis.

The work space for the hands should be between hip and chest height in front of the body. Most work should be done just below elbow height. In heavy work, the hands should be lower, while in fine assembly work that involves precise visual inspection, the worker will need a higher surface, perhaps with rests for forearms. Work-surface height should allow the arms to be kept low and close to the body.

Work objects should be located close to the front edge of the work surface to prevent the employee from having to bend over and lean across the surface to grasp items. Surfaces with no sharp edges but rounded or padded edges are preferred. Sufficient leg room must be allowed for seated operators.

Any visual symbols or displays should be placed in front of the body and below the eye level, between 10 degrees and 40 degrees below the horizontal line of sight. The most important information for the task should be placed in the worker's viewing area.

## Posture

The workstation should be designed to reduce static effort. Workers should be reminded to change positions frequently, and to keep the body in what is called the neutral posture:

- body is relaxed, with arms hanging loosely at sides
- wrists are neutral
- shoulders are relaxed
- elbows are close to the body

Working surfaces and seats should be designed to eliminate the need to work with a bent spine. In using ergonomics to identify problems, the supervisor should investigate workstations where workers must:

- bend necks forward more than 15 degrees
- lean forward or twist sideways
- crouch over their work
- work with arms above head or out, away from body
- use awkward wrist positions, especially if work is repetitive or requires forceful movements

## Standing Work

Many jobs can be performed with less effort when the worker is standing. However, prolonged standing can create stresses on the legs and lower back. Workers who stand for a long time on hard concrete surfaces should be supplied with padded antifatigue mats. The height of the work surface should usually be 2 to 6 in (5 to 15 cm) below the level of the worker's elbow when the arm is hanging in a relaxed posture (Figure 10–2).

Resting one foot on a foot rest while standing can help reduce low-back stress. Provide adjustable footrests where possible. When workers alternate between sitting and standing, a stool that leans with the body is sometimes useful.

**Figure 10–2.** Stand-seats help relieve leg and lower back stress. *(Courtesy American Industrial Hygiene Association Journal.)*

## Seating

Industrial seating is often poorly designed or uncomfortable. On the assembly line, providing a chair or stool is frequently an afterthought. As a result, workers tend to rig their own stools, or salvage discarded chairs, cushions, or car seats to use at workstations.

The same care should be given to providing comfortable seating on the shop floor as to providing correct chairs and equipment to office workers. Obviously, not all jobs will allow a worker to sit, but whenever possible, the following guidelines should be considered:

1. The weight-bearing seat area should be large enough to accommodate the

person without putting pressure on the buttocks or thighs (at least 18.2 in (46.2 cm) wide). Seat surface should be comfortably but firmly upholstered to distribute pressure evenly. Upholstery should be a woven fabric to reduce sweating. The seat should allow the worker to change his or her seated posture. The front of the seat should be rounded. It should not produce pressure on the back of the thighs or calves, which can occur when the seat is too deep. To relieve subsequent discomfort, workers may sit too far forward, causing them to lean forward in a static posture. They may slouch, which will put stress on the back. Generally, seats should be between 15 and 17 in (38 and 43 cm) deep.

2. The seat height should be adjustable within a range of approximately 15 to 20 in (38 to 51 cm) if the work surface is at a standard desk height. Adjustments of seat height will have to be made upward where the work surface is higher. For example, on an assembly line where the work surface is 32 in (81 cm) high, the seat should be adjustable from 20 to 26 in (51 to 66 cm) off the floor. For higher seats, foot rests must be provided to prevent legs from dangling, which can cause circulatory problems. Foot rests must also be used for shorter workers at any seat height where their legs may dangle.

3. A minimum 8-in (20-cm) vertical clearance should be provided under the work surface, with 26 in (66 cm) forward clearance for feet and legs. Foot rests should be between 1 and 9 in (2.5 and 23 cm) high, and be slanted, with a depth of about 12 in (30 cm).

4. The back should be supported in the lumbar region. Preferable, the backrest should be adjustable both horizontally and vertically. The support should maintain the back in it natural S-shaped position. The backrest should be at least 6 to 9 in (15 to 23 cm) high and 12 to 14 in (30 to 36 cm) wide. Too small a backrest will provide inadequate support and too large a backrest may prevent workers from moving as required by their work cycle.

## Displays and Controls

A worker's interface with the job may be through a display console. On these jobs, workers are required to receive information through continual sensory signals, to process the information mentally, and to make a decision or take action. The display must be arranged so communication is clear, concise, and easily understood. The controls must also be arranged to minimize error.

Basic ergonomic and human-factors principles of control and display design help ensure that this critical process is performed in the best manner. Controls should be within easy reach of the operator while he or she views displays or the field of operation. They should be easily located by touch, with adequate spacing between controls.

Good sensory input (primarily involving sight and hearing) is a key principle, and requires proper lighting, visual stimuli, sound levels, and sound patterns. Urgent information is best conveyed by auditory signals, such as bells or buzzers. Complicated information can be displayed visually. A combination of visual and auditory signals often improves the effectiveness of alerting the operator. For example, a buzzer could alert an operator to an emergency situation while a visual display would give details of the emergency.

Because a person's capacity for perceiving information is limited, it is important to group and simplify displays and controls. The following design concepts can help an operator to interpret and make decisions:

- Displays (gauges, meters, etc.) should be grouped so that normal ranges are all in one area and pointers are in the same direction. The most frequently used displays and controls should be in the center of the visual field. A deviation from normal is then easy to detect because it stands out. Controls should be clearly labeled and color coded, if appropriate, to distinguish machines, operation, and department.
- Symbols should be bold and simple. Use broad, rotating pointers on dials. The pointer should touch the dial gradations but should not obscure the numbers. The simpler and fewer the dial gradations, the better.
- Controls should be grouped with the related display and should follow the worker's usual expectations. In the United States, for instance, switches are usually moved *up* to activate electrical equipment and *down* to turn them off. Use distinctive shapes to aid in identification by touch.
- Controls should have adequate space between them for efficient handling.
- Controls must be arranged so workers will be able to shift their bodily positions during the workday.
- Switches must be guarded with cowls to prevent accidental operation.
- Hand controls should be used where precision and speed are required; foot controls may be used where greater strength is needed.
- Workers should have time to make decisions about their work. For more complicated jobs, special training in decision making may be needed. Give operators immediate feedback on their decisions and actions.

# HAND WORK AND USE OF TOOLS

Most jobs require some manipulation of objects with the hand. Although the human hand is a highly evolved grasping tool, it cannot always withstand the stresses of rapid, difficult, and/or repetitive occupational tasks. Cumulative trauma problems are especially likely to occur where there are less than 30 seconds per cycle, or where more than 1,000 parts per shift are handled or processed.

## Principles of Reducing Job Stress

The following principles should be considered when determining how much stress the job may produce:

- Activities should be performed in what ergonomists call the "natural range of movement." For the upper limbs, a work posture that allows the arms and wrists to remain in a natural posture as in ordinary activities is ideal. This means avoiding awkward wrists and, when necessary, keeping elbows at right angles when using the arms to apply force to the proper application. It also means avoiding extreme inward or outward rotation of the arms and wrists; instead, keep the wrist and hands in the neutral or "handshake" position (Figure 10–3).
- Repetitive motion, duration, and the amount of force needed to perform work should also be minimized. Jobs that require extremes of these conditions may put the worker at greatly increased risk of cumulative trauma disorders.

Ways to reduce the effect of task repetitiveness and forcefulness include:

- Reducing the weight of the object, picking up fewer objects at a time, lifting the object with two hands instead of one, or changing the size and shape of the objects.
- Planning rotation to jobs where demands on muscles and nervous system are different.
- Alternating use of limbs.
- Allowing workers to pace themselves, taking short breaks as needed.
- Enlarging the job to include a variety of tasks
- Using mechanical aids that perform

**Figure 10–3.** Workers should keep wrists and hands in a neutral or "handshake" position when working with hand tools.

in one motion the many tasks a worker might perform. A pneumatically oper-ated lug wrench at a tire installation operation on an auto assembly line, for example, can reduce repetitive forceful wrist motion for the operator.

- Using jigs and fixtures to reduce the amount of holding required by assembly line workers.
- Modifying the work surface to angle the object so that it is easier to work on.
- Using foot pedals that do not require the worker to continually depress them during operations.
- Avoiding impulse or impact forces on parts of the body. For example, the worker should use the right tool rather than hammer small parts together with his or her palms.

## Good Tool Design and Use

Most workers use hand tools for part of their jobs. As the time spent using tools increases, good tool design becomes more important. Improper design can result in inefficiency and cumulative trauma disorders. Tools that require effort to hold or to operate, have awkward handles that cause the wrist to be bent, or produce excessive pressure or vibration may injure the user.

Static loading of the muscles will occur where the tools or arms have to be held without support. This can lead to fatigue and sometimes cause a disorder in susceptible individuals. At times, bending a tool, such as a straight soldering iron, will reduce the need to keep the arm lifted. To reduce forces on the fingers, the tool should be activated by a bar or grip switch instead of a single-finger trigger.

Minimize awkward postures, especially bent wrists. Because grip strength is greatest when the wrist is straight, tools designed with bent handles may allow the worker to keep the wrist in a more natural posture. However, each job must be analyzed to ensure that the right tool is being used.

Tool handles should be long enough to extend past the palm of the user's hand. The surfaces should be broad enough to distribute pressure evenly, and should be padded or slip-resistant.

Tools such as hammers and screwdrivers should have handles about 1.5 in (3.8 cm) in diameter. Those for fine work should be about half an inch in diam-eter. Handle spans, on two-handled tools, should be in the 2.5 to 3.5 in (6.3 to 8.9 cm) range. They should be long enough to avoid pressure on the flesh of the palm or at the base of the thumb. Textured, but not highly grooved, surfaces are preferred for hand tools. This is all dependent on the individual using the tool.

Purchase tools that allow the worker to use a power rather than a pinch grip. Avoid using tools that expose workers to mechanical pinch points. Also, avoid tools with form-fitting handles that may pinch a larger worker's flesh and may reduce a smaller worker's efficiency. Hand-tool handles should be insulated against electricity, heat, and cold and fit the worker's hand. Avoid repetitive use of hand tools that put pressure on the back of the hand. For example, conventional scissors have been redesigned to remove the oval handle that creates pressure on the backs of the fingers.

Whenever possible, use power tools that are balanced with a neutral buoyancy tool balancer. These items may be better than a balancer that automatically reels the tool upward because they reduce the amount of reaching for and pulling the tool.

All surfaces with which arms, elbows, and other body parts are in contact should be rounded or padded to avoid mechanical contact stresses on the nerves.

Use of power tools that vibrate excessively can produce a disease known as Raynaud's syndrome or vibration white finger. This disease affects the ability of the blood to circulate in the hands, leading to numbness, stiffness, pain, and loss of strength. The affected fingers appear white. If the worker is not removed from exposure, severe tissue damage can occur.

Hand-arm vibration at frequencies below 1,000 hertz (cycles per second) should be avoided. Where workers use hand tools that vibrate, hazards may be reduced by decreasing exposure time or by dampening or isolating equipment vibration.

Driving a fastener into material with a power tool transfers torque to the worker's hand when the fastener bottoms out. The worker experiences this as a snapping action, which is a potential cause of stress. This stress can be reduced by using slip clutches, torque limiters, or torque-absorbing tools; by keeping the torque setting low, by mounting the tool on an articulating arm, or by providing an extra handle so the worker can use two hands to help counter the torque effort.

The grip of a tool can be dampened by making it out of flexible materials. Vibration-damping gloves may be used, but if they are too large or bulky, the worker will need to use extra force to hold onto the tool. Other tool-handling guidelines include:

- reducing tool weight and handle size so the worker does not need to use excessive grip strength to hold onto the tool (or increasing handle size—depending on the worker)

- counterbalancing tools so their weight will not twist the tools out of the workers' hands
- using tools that can be held in one hand and guided by the other hand
- balancing and replacing worn-out parts
- using vibrating tools less frequently or rotating personnel to other jobs

## WHOLE-BODY VIBRATION

Whole-body vibration can also create physical stress. Driving a truck, bus, or car, or operating large vibrating production machinery can produce discomfort and health stresses. When in contact with an object that vibrates at very low frequencies, the body, or body parts, will resonate (pick up the vibration). Workers exposed to very low frequency (1–20 Hz) vibrations may experience difficulty breathing, pains in the chest and abdomen, backache, headache, muscular tension, and other problems. Workers exposed to chronic excessive vibration may suffer spinal problems and intestinal complaints. Strong vibration may impair vision, mental processes, and ability to complete skilled movements (Grandjean, p. 299).

Heavy vibrating tools such as jackhammers can cause damage to bones, joints, and tendons, as well as Raynaud's syndrome. The vibration from large equipment can be reduced by:

- Mounting the tool on springs or compression pads.
- Using structural materials that produce less vibration. These will also possibly reduce noise generated by the equipment.

The person can also be isolated from the vibrating source. For instance, for truck driving or other seated tasks, you can provide cushioning or springs to insulate the seat from the vibrating surface. For standing tasks at large industrial equipment, make sure workers have a vibration-absorbing (rubber or vinyl) floor mat.

## VIDEO DISPLAY TERMINALS (COMPUTER MONITORS)

Video display terminals (VDTs) are widely used in offices and are becoming more common in industrial situations. They provide an example of the need for good displays and controls.

The VDT screen conveys information for the worker. It is important that the screen be free from glare. Shielding windows with drapes or blinds can help to reduce glare. Reflected glare from shiny objects can be decreased by using dark and/or matte finishes. Glare from lighting can be reduced by avoiding overhead lighting, lowering the light level where possible, and adding task lighting on documents as needed.

- Screens should tilt in order for the worker to comfortably look down on the screen. The top of the screen should not be higher than eye level. The screen should be adjustable in placement on the desk. A comfortable viewing distance is usually between 6 and 20 in (15–51 cm), but this varies with each individual.
- Keyboard height and placement should be adjustable, as should seat and desk height. The employee should not remain in the same position for long periods—several breaks in keyboard work should be encouraged.

## LIGHTING, NOISE, AND HEAT

Lighting must be appropriate to the task. For example, low light may be acceptable in storage areas, while high lighting may be needed for fine visual tasks. The amount of light needed will depend on the job, the operator's age and eyesight, and other factors. A full discussion of the technical aspects of lighting levels is beyond the scope of this chapter. However, the same types of guidelines apply to the general workplace as to the VDT operations:

- Provide adequate but not excessive light. Excessive light can create eye fatigue.
- Provide adjustable blinds at windows.
- Avoid direct or reflected light sources in the worker's field of vision.
- Shade or diffuse all light sources. Light should be reflected down from the ceiling, not up from the floor.
- Ensure sufficient contrast for written material (i.e., use black ink rather than a pencil).
- Increase the size of small critical details.
- Put the task material perpendicular to the operator's line of sight (Kodak, vol. 1, p. 225).

Noise and heat stress are discussed in Chapter 8, Industrial Hygiene. When considering workplace ergonomics, remember that too little heat can be as detrimental as too much. In cold, workers' muscles stiffen, making them less able to make precise movements, and making muscles and nerves more vulnerable to injury.

## SUMMARY OF KEY POINTS

Key points covered in this chapter include:
- Ergonomics is the study of how people interact with their work, or how to create a good match between the employee and the workstation and tools with which the person works. The goal of ergonomics is to minimize accidents and illnesses due to chronic physical and psychological stresses on the job and maximize productivity and efficiency. Supervisors must be familiar with the principles of ergonomics to help recognize problems involving ergonomics, evaluate them, and suggest methods to control the risks.
- Problems may arise when workstations, equipment, or tools do not fit the workers well. These stresses may cause immediate or long-term damage to muscles, nerves, and joints. Most injuries and illnesses are caused by forceful or repetitive motion activities or because workers are required to assume awkward postures over a period of time. These disorders are referred to as cumulative trauma disorders and include tenosynovitis and tendinitis (an inflammation of the tendons in shoulder, arm, elbow, wrist, or hands).
- Ergonomics draws on anatomy and physiology, anthropometrics, biomechanics, and psychology to solve work-related problems. Anthropometrics provides information used to design workspaces to match body dimensions. Workspaces are generally designed for the tallest or shortest worker and adjusted for individual differences. Biomechanics explains characteristics of the human body in mechanical terms. These models help to identify stressful postures and to develop countermeasures to relieve stress. In looking at the workspace, supervisors should be aware that the amount of job-produced physical stress is often determined by the posture a worker must assume, the force needed to perform the task, and the number of repetitions required by a stressful task.
- To help companies reduce or eliminate problems, supervisors can provide key information by keeping track of trends in injuries and accidents, incidence of cumulative trauma disorders, sick leave and turnover rates, employee adjustments to workstations and equipment, and product quality levels. Ergonomics

can alert supervisors to possible causes of problems, including high overtime and increased work rate, manual materials handling, repetitive tasks, mechanical stress on the body, temperature extremes, and vibrating tools.

- According to ergonomic principles, work tasks should be designed to match the employee's physical capacity for the job. This capacity can be measured by maximum oxygen uptake or by heat produced by the body. Job demands that exceed an employee's maximum capacity; that put strain on tendons, muscles, or joints; or that disregard environmental factors (heat, humidity, etc.) can cause injuries.

- Ergonomic principles can be applied to manual materials handling to eliminate or reduce stress-related illnesses. Workers should be taught the correct ways to lift, push, or pull objects; to use mechanical means wherever feasible; to ask for assistance; and to modify workloads to make moving objects safer.

- Because each person has unique body characteristics, the workspace must be flexible to allow for these differences. Ergonomic principles applied to workplace conditions include designing furnishings and equipment to keep the body in a neutral posture as much as possible; providing relief when employees must stand to do their work; designing proper seating to prevent lower back and leg problems; and providing display consoles that allow easy viewing, maximize information obtained, and permit rapid, nonstressful operation.

- Ergonomic principles can also be applied to reduce job stress in hand work and when using hand tools. Activities should be performed as much as possible in the "natural range of motion," avoiding awkward postures and bent wrists, and keeping elbows at right angles. Repeated motions should be minimized, and the amount of force required to do the work reduced as much as possible. Using the proper tool for the job may minimize cumulative trauma disorders, and tools should always be kept in good condition.

- Whole-body vibration, such as riding a truck or operating large vibrating tools, can create physical stress. Ergonomic principles can be used to reduce this stress by dampening the vibration and/or isolating the person from the vibrating source.In addition, video display terminals must be free of glare and set at the proper height and angle to prevent neck and shoulder problems. Keyboard height and placement should be adjustable, as should seat and desk height.

- Lighting, noise, and heat must all be controlled and appropriate to the work environment or workers may suffer from work-related problems. Proper lighting is particularly important to prevent fatigue and chronic eye disorders.

# 11

# Hazard Communication

**After reading this chapter, you will be able to:**

- Describe the intent and definitions of the hazard communication standard (HCS), OSHA 29 CFR 1910.1200
- Recognize various types of warning labels that must be used on all containers of hazardous materials
- Describe Material Safety Data Sheets (MSDSs) and know how to generate and maintain a chemical inventory
- List the employer's and employee's training and safety obligations under the HCS
- Discuss the elements of a hazard communication program
- Better understand the more routine actions that a first-line supervisor should take to protect his or her employees from hazardous chemical exposure and to comply with the HCS

The purpose of this chapter is to introduce the requirements of the hazard communication standard (HCS) and methods of compliance. In many locations, the HCS is also known as the "right to know" law. Best practices that exceed the standard as well as the more routine duties of a first-line supervisor in meeting this standard are also discussed.

## HAZARD COMMUNICATION STANDARD REQUIREMENTS

The federal HCS (29 CFR 1910.1200) was issued by OSHA in 1988.

The HCS imposes two duties on U.S. employers. The first duty is to see that all chemicals produced or imported into the United States be evaluated to determine their physical and health hazards. The second duty is to see that information about these hazards be transmitted to employees who may be exposed to such chemicals under normal operations or in a foreseeable emergency. The first-line supervisor is often the critical link in fulfilling the employer's second duty under the hazard communication regulations.

The overall goal of this regulation is to reduce the number and frequency of occupational illnesses and injuries caused by unprotected exposure to hazardous chemicals.

### Definitions

Figure 11–1 gives the definitions that OSHA uses in interpreting the HCS.

The HCS is a "performance-oriented" standard. This means that the standard tells an employer what must be accomplished but leaves up to each workplace to decide exactly how to accomplish the goal. This approach allows the employer great flexibility when setting up workplace programs designed to meet the following mandates (intents) of the standard.

The main sections of the standard cover:

- hazard determination/inventory
- Material Safety Data Sheet (MSDS) availability
- hazard labeling and warning signs
- training and communication of information to employees
- a written hazard communication program for the specific workplace

| Term | Definition |
|---|---|
| ACUTE EXPOSURE | Exposure is usually of short duration and high concentration. Symptoms will usually appear within 24 hours. |
| ASSISTANT SECRETARY | The Assistant Secretary of Labor for Occupational Safety and Health, U.S. Department of Labor, or designee. |
| ARTICLE | A manufactured item formed to a specific shape or design that has an end-use function that depends wholly or partly on its shape or use during end use; and that does not release, or otherwise result in exposure to, a hazardous chemical, under normal conditions of use. |
| CHEMICAL | Any element, chemical compound, or mixture of elements and/or compounds. |
| CHEMICAL MANUFACTURER | An employer with a workplace where chemical(s) are produced for use or distribution. |
| CHEMICAL NAME | The scientific designation of a chemical according to the naming system developed by the International Union of Pure and Applied Chemistry (IUPAC) or the Chemical Abstracts Service (CAS), or a name that will clearly identify the chemical for the purpose of conducting a hazard evaluation. |
| CHRONIC EXPOSURE | Symptoms are usually delayed, cumulative, and result from repeated exposure to low-level concentrations of hazardous chemicals. |
| COMBUSTIBLE LIQUID | Any liquid having a flashpoint at or above 100 F (37.8 C), but below 200 F (93.3 C). The exception is a mixture having components with flashpoints of 200 F, or higher, the total volume of which makes up 99 % or more of the total volume of the mixture. |
| COMMON NAME | Any designation or identification such as a code, trade, brand, or generic name, or a code number used to identify a chemical other than by its chemical name. |
| COMPRESSED GAS | A gas or mixture of gases having, in a container, an absolute pressure exceeding 40 psi at 70 F (21.1 C); or a gas or mixture of gases having, in a container, an absolute pressure exceeding 104 psi at 130 F (54.4 C), regardless of the pressure exceeding 40 psi at 100 F (37.8 C) as determined by ASTM D–323–72. |
| CONTAINER | Any bag, barrel, drum, bottle, box, can, cylinder, reaction vessel, storage tank, or the like, that contains a hazardous chemical. For the purposes of the HCS, pipes, piping systems, engines, fuel tanks, and other operating systems in a vehicle are not considered to be containers. |
| DESIGNATED REPRESENTATIVE | Any individual or organization to whom an employee gives written authorization to exercise said employee's rights under this standard. A recognized or certified collective bargaining agent (such as a union representative) shall be treated automatically as a designated representative without regard to written employee authorization. |
| DIRECTOR | The Director, National Institute for Occupational Safety and Health, U.S. Department of Health and Human Services, or designee. |
| DISTRIBUTOR | A business, other than a chemical manufacturer or importer, that supplies hazardous chemicals to other distributors or employees. |
| EMPLOYEE | A worker who may be exposed to hazardous chemicals under normal operating conditions, or in foreseeable emergencies. Workers such as office workers or bank tellers who encounter hazardous chemicals only in non-routine, isolated instances, are not included in this definition of "employee." |
| EMPLOYER | A person engaged in a business where chemicals are either used, distributed, or are produced for use or distribution, including a contractor or subcontractor. |
| EXPLOSIVE | A chemical that causes a sudden, almost instantaneous release of gas, pressure, or heat, when subjected to sudden shock, pressure, or high temperature. |
| EXPOSURE OR EXPOSED | The instance of being exposed, or an employee who is exposed to, or has the potential to be exposed to, hazardous chemicals in the course of employment, through any route of entry (absorption, inhalation, ingestion, skin contact, injection) to the body. |

**Figure 11–1.** Key terms of the hazard communication standard (HCS).

| Term | Definition |
|------|------------|
| FLAMMABLE | A chemical that falls into one of the following categories: aerosol flammable, gas flammable (includes forming a flammable mixture with air at a 13 % concentration of the chemical, or less, AND forming a flammable range, when mixed with air, that is wider than 12 % by volume, regardless of the lower limit); liquid flammable; and solid flammable. |
| FLAMMABLE LIQUID | Any liquid having a flashpoint below 100 F (38 C). |
| FLASHPOINT | The minimum temperature at which a liquid gives off vapor in sufficient concentration to ignite when tested by the Tagliabue Closed Tester, Pensky-Martens Closed Tester, or Setaflash Closed Tester methods. Organic peroxides, which undergo autoacceleration thermal decomposition, are excluded from any of the flashpoint determination methods mentioned here. |
| FORESEEABLE EMERGENCY | Any potential occurrence such as, but not limited to, equipment failure, rupture of containers, or failure of control equipment, which could result in an uncontrolled release of a hazardous chemical into the workplace. |
| HAZARDOUS CHEMICAL | Any chemical that is a physical or health hazard. |
| HAZARD WARNING | Any words, pictures, symbols, or combination thereof, appearing on a label or other appropriate form of warning, which conveys the hazard(s) of the chemical(s) in the container(s). |
| HEALTH HAZARD | A chemical for which there is statistically significant evidence, based on at least one study conducted in accordance with established scientific principles, that acute or chronic health effects may occur in exposed employees. The term "health hazard" includes chemicals that are carcinogenic, toxic, or highly toxic, reproductive toxins, irritants, corrosives, sensitizers, hepatotoxins, nephrotoxins, neurotoxins, hematopoietic agents, and agents that can damage the lungs, skin, mucous membranes, or eyes. |
| IDENTITY | Any chemical or common name for the chemical that is indicated on the MSDS. The identity used shall permit cross-references to be made among the required list of hazardous chemicals, the label, and the MSDS. |
| IMMEDIATE USE | The hazardous chemical will be used, or under the control of, only the person who transfers it from a labeled container, and only within the work shift in which it was transferred. |
| IMMEDIATELY DANGEROUS TO LIFE AND HEALTH (IDLH) | The maximum level of a substance from which, in the event of respirator failures, a person can escape within 30 minutes without any impairing symptoms or irreversible health effects. |
| IMPORTER | The first business, with employees and within the Customs Territory of the United States that receives hazardous chemicals produced in other countries for the purpose of supplying them to distributors or employees within the United States. |
| LABEL | Any written, printed, or graphic material, displayed on or affixed to, containers of hazardous chemicals. |
| MATERIAL SAFETY DATA SHEETS (MSDSs) | Written or printed material concerning a hazardous chemical that is prepared in accordance with Paragraph (g) of the HCS. |
| MIXTURE | A combination of two or more chemicals, in which the combination is not wholly or partly the result of a chemical reaction. |
| ORGANIC PEROXIDE | An organic, highly reactive compound that contains the bivalent -O-O- structure, and which may be considered to be a structural derivative of hydrogen peroxide, where one or both of the hydrogen atoms has been replaced by an organic radical. |
| OXIDIZER | A chemical, other than a blasting agent or explosive, that initiates or promotes combustion in other materials, thereby causing a fire either of itself, or through the release of oxygen or other gases. |
| PERMISSIBLE EXPOSURE LIMIT (PEL) | Allowable air concentration of a substance in the workplace 8 hours a day, 40 hours a week, as established by NIOSH and enforced by OSHA. |

**Figure 11–1.** Continued.

| Term | Definition |
|------|------------|
| PHYSICAL HAZARD | A chemical for which there is scientifically valid evidence that it is a combustible liquid, a compressed gas, explosive, flammable, an organic peroxide, an oxidizer, pyrophoric, unstable, or water reactive. |
| PRODUCE | To manufacture, process, formulate, or repackage. |
| PYROPHORIC | A chemical that will ignite spontaneously in air at a temperature of 130 F (54.4 C) or below. |
| RESPONSIBLE PARTY | A person who can, if necessary, provide additional information on the hazardous chemical and appropriate emergency procedures. |
| SPECIFIC CHEMICAL IDENTITY | The chemical name, Chemical Abstracts Service (CAS) Registry Number, or any other information that reveals the precise chemical designation of the substance. |
| THRESHOLD LIMIT VALUE (TLV) | Refers to airborne concentrations of substances, and represents conditions under which is believed that nearly all workers may be repeatedly exposed day after day without adverse effect. TLVs are recommended guidelines established by the American Conference of Governmental Industrial Hygienists. |
| THRESHOLD LIMIT VALUE CEILING (TLV-C) | The ceiling level of the exposure that should never be exceeded. This value has been established as the maximum level to be used in computing the TWA and STEL limits. The ceiling value is under the IDLH limit for a given substance. |
| THRESHOLD LIMIT VALUE SHORT-TERM EXPOSURE LIMIT (TLV-STEL) | An exposure level safe to work in for short periods of time (15 minutes) 4 times a day maximum, with at least 60 minutes between exposures. No irritation or other adverse effects should be experienced. This level is higher than the TLV-TWA for a given substance. |
| THRESHOLD LIMIT VALUE TIME-WEIGHTED AVERAGE (TLV-TWA) | The time-weighted average concentration of a substance for a normal 8-hour workday and a 40-hour workweek, to which nearly all workers may be repeatedly exposed, day after day, without adverse effect. |
| TRADE SECRET | Any confidential formula, pattern, process, device, information, or compilation of information that is used in an employer's business, and that gives the employer an opportunity to obtain an advantage over competitors who do not know or use it. |
| UNSTABLE (REACTIVE) | A chemical that, in the pure state or as produced or transported, will vigorously polymerize, decompose, condense, or become self-reactive under conditions of shocks, pressure, or temperature |
| USE | To package, handle, react, or transfer |
| WATER-REACTIVE | A chemical that reacts with water to release a gas that is either flammable or presents a health hazard. |
| WORK AREA | A room or defined space in a workplace where hazardous chemicals are produced or used, and where employees are present. |
| WORKPLACE | An establishment, job site, or project at one geographical location, containing one or more work areas. |

**Figure 11–1.** Concluded.

## Employer and Supervisor Responsibilities

Employers have a responsibility to comply with all OSHA regulations, such as the HCS. A good starting point in attaining compliance is to become familiar with the requirements of the standard.

The first definition that the employer must understand in complying with the HCS is the definition of a hazardous chemical. As identified in Figure 11–1, a hazardous chemical is a chemical that has either a physical or health hazard. If a chemical does not have either a physical hazard or a health hazard, the HCS does not apply.

Examples of physical hazards include a chemical that is flammable, a compressed gas, an oxidizer, as well as other physical hazards listed in the standard. Examples of health hazards include chemicals that are corrosive or irritants, chemicals that cause cancer, or chemicals that have a toxic effect on organs within the body.

OSHA recommends that employers conduct a physical inventory and list all chemicals used in the workplace. Employers are cautioned to think of all chemicals to which an employee may be exposed, such as gases and dusts, and not to just focus on liquid chemicals.

First-line supervisors are often critical in helping develop an accurate chemical inventory. Supervisors are often aware of chemicals that are only used on occasion in the workplace. It is critical that such incidental uses be included within the initial chemical inventory. The first-line supervisor is also critical in alerting employers that employees may want to use a new chemical for cleaning or other purposes.

**Best Practices.** As mentioned earlier, the HCS applies only to chemicals that have either a physical hazard or a health hazard. If no employee is exposed to such a chemical in normal operations or in a foreseeable emergency, the standard does not apply. However, companies that exhibit excellence in safety develop a hazard communication program for all chemicals used in the workplace.

## HAZARD DETERMINATION

All manufacturers are required to determine if the chemicals they produce or import have physical or health hazards. Hazard determination is usually conducted at research laboratories or scientific laboratories using a number of specific scientific methodologies. The process to determine if a chemical has a physical hazard or a health hazard does not normally involve a supervisor or team leader.

Most companies do not conduct hazard determination on chemicals that they purchase. Instead, they rely on the information supplied to them through the manufacturer's MSDS, which is explained in more detail later in the chapter. The first-line supervisor often plays a critical role with regards to the use and maintenance of MSDSs in the workplace.

OSHA's hazard communication standard specifically requires that any workplace in which a worker may be exposed to a hazardous chemical (physical hazard or health hazard) under routine operating conditions or in a foreseeable emergency have a written communication plan.

# ELEMENTS OF A HAZARD COMMUNICATION PROGRAM

A hazard communication program should:
- maintain a list of hazardous chemicals currently on site
- ensure all hazardous chemicals have proper labeling
- maintain MSDSs
- conduct employee training
- include a written hazard communication plan

Each aspect of this hazard communication program must meet or exceed federal HCS requirements. Best-practices suggestions and guidelines on developing and implementing an effective program will be addressed in this section.

## Written Hazard Communication Program

The HCS requires that the employer develop, implement, and maintain a written hazard communication program for the workplace. This written document must be kept at the workplace, and must, at the very least, describe how the employer will meet the responsibilities of establishing the following elements:
- employee training program
- labeling and warning-sign procedures or systems
- MSDS collection, storage, and review
- list of the hazardous chemicals known to be present in the workplace, using an identity referenced on the appropriate MSDS
- methods the employer will use to inform employees of the hazards of nonroutine tasks
- methods the employer will use to inform employees of the hazards associated with chemicals contained in unlabeled pipes in work areas

- methods the employer will use to provide outside employees (who work for another employer) with a copy of MSDSs for the work area
- methods the employer will use to inform outside employees about any precautions that must be taken in the workplace
- methods the employer will use to inform outside employees of the labeling and warning-sign system used in the workplace

The company's written program also provides employers and employees with specific safe operating procedures to follow when handling hazardous chemicals in the workplace. The first-line supervisor or team leader plays a critical role in the success of a hazard communication plan in preventing occupational illnesses and injuries among the work force.

A copy of the written HazCom program must be kept at the workplace and be available on request to:

- an OSHA compliance officer or other OSHA representative
- an investigator from the National Institute for Occupational Safety and Health (NIOSH) or employees and their designated representatives
- all employees

As a supervisor, you should know where the hazard communication plan is located and who is responsible for maintaining the plan. You should also review the plan and make sure that you understand your roles and responsibilities as outlined in the hazard communication plan.

When developing the hazard communication written document, employers must ensure that the policies and procedures cover the exact chemicals and processes that exist at their particular workplace. It is not acceptable to take a "generic" HCS outline, put the company's name on it, and call it "the company's written HazCom program." However, many employers can use a professional safety guideline to ensure that all mandatory sections are addressed in the company-specific HazCom program. (There is a guideline available on OSHA's web site at www.osha.gov/pls/oshaweb/owadisp.show_document?p_table = STANDARDS&p_id = 10104. Another resource is the OSHA booklet *Hazard Communication Guidelines for Compliance,* which is available through the government printing offices.)

The written HazCom program must be reviewed and updated whenever a new hazardous chemical exposure is possible in the workplace. Likewise, anytime the written program is altered significantly, the company must hold information or training sessions to explain to all affected employees the changes in the written program.

As a first-line supervisor, you should alert management if new chemicals have been brought into the workplace, or if you believe that changes in work processes or procedures may be causing exposure to a hazardous chemical. You should also be aware of the process by which you will be notified that a new chemical hazard is being introduced to your work area.

**Best Practices.** When designing the training program to accompany the HazCom program, the employer may have to develop different levels of training for different areas of the workplace.

For example, a warehouse worker may not be exposed to a hazardous chemical during his or her routine job assignments because the chemicals are sealed within shipping drums and containers. However, that warehouse worker would still need to be trained on the procedures to follow in the event that a drum of hazardous chemicals was pierced or a spill, fire, or other emergency involving the hazardous chemicals occurred.

An employee who is directly involved with handling or using hazardous chemicals must receive training covering the physical hazards associated with the chemical; how to safely use the chemical, MSDSs, and personal protective equipment; labeling requirements; and other practical aspects of a complete HazCom program.

Supervisors are often responsible for training exposed employees on the hazardous chemicals in the work area, the company labeling system, how to obtain MSDSs, and the proper work procedures. The first-line supervisor is often responsible for verifying that any new or transferred employee is properly trained prior to being exposed to any hazardous chemical. Typically, supervisors are expected to ensure that all employees are wearing personal protective equipment and following the procedures designed to minimize their exposure to a hazardous chemical.

Implementing the HazCom program usually requires several key steps. The company should begin by taking inventory of all chemicals in the workplace (Figure 11–2). The following chemicals and items are exempt from the HCS because these chemicals are regulated by a federal agency other than OSHA:

- any hazardous waste subject to the EPA Resource Conservation and Recovery Act
- tobacco or tobacco products

## CHEMICAL INVENTORY

| Chemical Name | Toxic/Hazardous Components | Area Where Used/Produced | Dates Use/Production Began | Ceased | Quantity in Area | Quantity in Storage and Location | Substance Category |
|---|---|---|---|---|---|---|---|
| | | | | | | | |
| | | | | | | | |
| | | | | | | | |
| | | | | | | | |
| | | | | | | | |
| | | | | | | | |
| | | | | | | | |
| | | | | | | | |

**INVENTORY**
An inventory of hazardous chemicals should be prepared on a master chart such as the one shown here.

**Figure 11-2.** Sample chemical inventory form.

- wood or wood products (but not wood dust)
- food, drugs, or cosmetics
- food, drugs, or cosmetics intended for personal consumption by employees while in the workplace
- any consumer product or hazardous substance (as defined in the Consumer Product Safety Act and Federal Hazardous Substances Act, respectively) that an employer can demonstrate is used in the workplace in the same manner as in normal consumer use, so that workers are not exposed to the substance any longer or more often than any other consumers
- any drug (as defined in the Federal Food, Drug, and Cosmetic Act) that is in solid, final form for direct administration to a patient
- A manufactured article, formed to a precise shape or design during manufacturing, whose end use function is completely or partially dependent on its precise shape or design. Under normal conditions of use, the item must not be able to release or otherwise expose workers to hazardous chemicals.

In the process of completing the list of chemicals on site, best practice would include documenting the information required for environmental reporting, such as specific chemical name, storage location, annual quantity used, and maximum quantity stored. The inventory process should include procedures for updating the inventory and notification of affected employees and management.

The supervisor is often involved with doing a physical inventory to determine what chemicals are used in his or her work area. It is critical for the supervisor to include any chemical that is used only occasionally, such as during annual maintenance procedures, as well as chemicals used routinely.

All supervisors should know where the company hazard communication plan is located, as it contains the current list of hazardous chemicals on site. All supervisors should be aware of any hazardous chemical that is used in their work area.

After completing the inventory of all chemicals on site, the company should review the available information to determine which chemicals have a physical or health hazard. These are the chemicals that must meet all the requirements of the HCS, such as hazard warning labeling and MSDS availability. The company should decide at this point what its policy will be regarding labeling and retaining MSDSs for chemicals that do not have a physical or health hazard. Best practice would dictate that all chemicals be treated in accordance with the requirements of the HCS.

After determining the extent of the HazCom program, the company should assign roles and responsibilities for each of the three main elements of the standard:

- Who will be responsible for maintaining and distributing MSDSs within the company or facility?
- Who will determine and maintain appropriate hazard labels on the regulated chemical used on site?
- Who will train the employees to protect themselves from hazardous chemical exposures?

The company must also determine how it will address nonroutine exposures, such as those that may occur from bringing in outside contractors, and nonroutine tasks. The role of the first-line supervisor in each of these areas will be examined in more detail in the remainder of the chapter.

## Material Safety Data Sheets

The HCS requires that an employer have a material safety data sheet (MSDS) for any chemical that has a physical or a health hazard. Many chemical manufacturers will routinely supply customers with a material safety data sheet for all their chemicals, even those with no known health or physical hazards.

The HCS requires that each MSDS contain at least the following information:

- the chemical identity/name used on the label—with a few exceptions, including trade secret
- physical and chemical characteristics of the hazardous chemical
- physical hazards of the chemical
- health hazards of the chemical
- the primary routes of entry into the body
- the OSHA permissible exposure limit (PEL), American Conference of Governmental Industrial Hygienists (ACGIH) Threshold Limit Value, and any other exposure limit used or recommended
- whether the chemical is listed in the National Toxicology Program (NTP) Annual Report on Carcinogens (latest edition) or has been found to be a potential carcinogen in the International Agency for Research on Cancer (IARC) Monographs (latest edition), or by OSHA or ACGIH
- any generally applicable precautions for safe handling and use known to the chemical manufacturer, importer, or employer preparing the MSDS
- any generally applicable control measures known to the chemical manufac-

turer, importer, or employer preparing the MSDS

- emergency and first-aid procedures
- the date of preparation of the MSDS or the last change to it
- the name, address, and telephone number of the chemical manufacturer, importer, employer, or other responsible part preparing or distributing the MSDS, who can provide additional information on the hazardous chemical and appropriate emergency procedures

If no relevant information is found for any given category on the MSDS, then the preparer should mark the category to indicate that no information was found. When complex mixtures have similar hazards and contents, one MSDS may apply to all these similar mixtures. The chemical manufacturer, importer, or employer preparing the MSDS must ensure that the recorded information accurately reflects the scientific evidence used in making the hazard determination.

Chemical manufacturers or importers must provide distributors and employers with an appropriate MSDS with the initial shipment and with the first shipment after an MSDS is updated. In turn, distributors must provide the MSDSs and update information to other distributors and employers.

The supervisor is often responsible for explaining the relevant sections of the MSDS to his or her exposed employees.

As a supervisor, you should know what person inside the company is responsible for receiving the MSDSs. This person is required to review each new MSDS to see if the revision contains any new health hazards or physical hazards that may require a change in your own operating procedures.

The hazard communication standard requires that an employer maintain copies of the required MSDSs for each hazardous chemical in the workplace. These MSDSs must be readily accessible to employees during each workshift when workers are in their job areas.

The supervisor is often responsible for making sure that employees know how to access MSDSs as well as for maintaining the material safety data sheets for the chemical used within his or her work area. MSDSs must also be readily available, upon request, to employees' designated agency representatives, to the Assistant Secretary of OSHA, and to the NIOSH director.

**Best-Practices Suggestions.** The HCS requires that MSDSs be written in English. Many companies will provide MSDSs in other languages. Even though the standard requires that the MSDS be in English, the standard is very clear that all employees must understand the health and physical hazards associated with a chemical to which they are exposed. The supervisor is often responsible for making sure that all employees understand the MSDS information regardless of their ability to read English.

MSDSs must be current, updated as necessary, and easily accessible to employees at all times. It is common practice to keep MSDSs in a brightly colored binder, attached to a wall, or stored on a special shelf in the work area. All MSDSs for the chemicals used in a specific work area should be kept in that area. Complete sets of all MSDSs for the workplace should be kept in at least one central location in the facility.

Most MSDS information is broken into nine to 12 sections on the data sheet. MSDS interpretation and familiarization training should address the contents of each section and the various types of MSDSs that exist in the specific workplace or work area.

If an inventory check reveals missing MSDSs, the employer should send a letter to the manufacturer, importer, or distributor, requesting a copy of the missing MSDS. By checking for MSDSs each time a chemical inventory is performed and each time a chemical is received in the workplace, the employer should always have the correct MSDS available for the workplace and work areas.

It is primarily the responsibility of the manufacturer, importer, or distributor to provide timely updates of MSDSs to the employer. However, the employer also should ensure that the MSDS master files, and all area files, contain the most recently updated MSDSs for the hazardous chemicals used in the workplace. This step can be done by sending an annual form letter to the manufacturer, distributor, or importer, requesting updated MSDSs for the chemicals used by the employer. These letters are part of the documentation showing that the employer is complying with the intent of the HCS, and copies should be kept permanently on file.

Several current computer programs and on-line services have a database containing the most common MSDSs. However, these programs and services can be costly and may not cover all the mixtures and compounds an employer might use, store, or transfer. Whether MSDSs are supplied by the manufacturer, distributor, or importer; developed in-house; or accessed through a database or on-line service, the employer should write a policy that describes how MSDSs will be acquired, updated, integrated with inventory and receiving procedures, distributed, and accessed by employees.

The HCS does not specifically require that MSDSs be kept on site for a particular length of time after the chemical is no longer in use. According to OSHA 29 CFR 1910.120, access to the records, however, is required for 30 years. Some companies insist that all materials must be stored in accordance with the medical record retention procedures. At a minimum, it is suggested that a record of the identity of substances used be kept indefinitely to comply with other federal regulations regarding retention of employee exposure and medical records. Employers should consult their legal department to decide on the retention period for MSDSs.

## Labeling

The HCS requires that any container of a hazardous chemical be labeled with an appropriate hazard warning. The warnings must be prominently displayed, in English, and maintained in a legible condition as long as the container holds any amount of the hazardous chemicals.

First-line supervisors are often responsible for ensuring that all containers of hazardous chemicals received or used in their department have hazard warning labels on them. Supervisors are often a critical link in ensuring that any employee exposed to a hazardous chemical understands the labeling system being used by the facility. We will look at these labeling requirements in more detail.

**HCS Requirements.** The HCS establishes specific labeling requirements to ensure the proper labeling of hazardous chemicals (Figure 11–3). The requirement places responsibility for creating an appropriate label at the source of the hazardous chemical, namely, the chemical manufacturer, importer, or distributor (Figures 11–2 and 11–4).

Chemical manufacturers, importers, and distributors of hazardous chemicals are required to ensure that, unless specifically exempt, each container of hazardous chemical that they ship is labeled, tagged, or marked with:

- identity of the hazardous chemical(s)
- appropriate hazard warnings
- name and address of chemical manufacturer, importer, or other responsible party

The following types of materials are exempt, or partially exempt, from the labeling requirements. (It should be noted that even if a hazardous material is exempt from the labeling requirements, it may still need to meet other requirements specified by the HCS.)

- Pesticides (as defined by the Federal Insecticide, Fungicide, and Rodenticide Act—

**Figure 11–3.** Sample hazardous waste label. *(Label courtesy of Lab Supply Inc., Janesville, WI.)*

FIFRA), when the product is subject to FIFRA and EPA labeling requirements. Containers used for mixing at a loading site do not require labels under the HCS labeling requirements, as long as the mixture is to be used within 12 hours, and the FIFRA labeling and warning sign system is in use in the mixing/loading area.

• Any food, food additive, color additive, drug, cosmetic, or medical or veterinary device (including articles that are components of these devices), as defined in the Food and Drug Act, that are subject to the labeling requirements under the Food and Drug Act (labeling requirements apply only when the drug is packaged, and not before).

• Any distilled spirits, wine, or malt (as defined in the Federal Alcohol Administration Act and regulations issued under the act) when subject to labeling requirements issued under the act by the Bureau of Alcohol, Tobacco, and Firearms.

• Any consumer product or hazardous substance (as defined in the Consumer Product Safety Act and the Federal Toxic Substances Act, respectively) when subject to a consumer product safety standard or labeling requirement of those acts by the Consumer Product Safety Commission.

As stated earlier, chemical manufacturers, importers, and distributors of hazardous chemicals are required to ensure that each container of hazardous chemical that they ship be labeled with an appropriate hazard warning. In many companies, the shipping supervisor is responsible for seeing that all containers leaving the workplace have the required labels.

In companies that simply use but do not produce a hazardous chemical, the employer is responsible for maintaining the appropriate warnings on every container of hazardous chemicals used or stored in the workplace (Figure 11–5). The supervisor is normally held responsible for seeing that the labels are maintained in his or her work area. The supervisor should also understand what procedure to follow in the event that an unlabeled container is discovered.

There are three specific circumstances in which the HCS exempts the employer from labeling a hazardous chemical container in the workplace:

## LABEL REQUEST LETTER

(Date)

(Name)
(Title)
(Company Name)
(Address)
(City, State, ZIP Code)

Dear Sir:

The OSHA Hazard Communication Standard (29 CFR 1910.1200) requires companies that sell hazardous materials to label containers with the identity of the chemical, the name and address of the manufacturer, importer, or other responsible party, as well as appropriate hazard warning(s). Please provide a quantity of labels for _____ which we purchase
(Name of Product)
from your company.

Your prompt attention to this request will be appreciated.

Sincerely,

---

Any supplier improperly labeling hazardous chemicals should be notified in writing and all correspondence dated and retained in an appropriate file.

**Figure 11–4.** Sample label request letter.

1. Portable containers do not have to be labeled when they contain hazardous chemicals that were transferred from labeled containers, and the chemicals in the unlabeled containers are intended to be used immediately (during the same workshift) by the person who made the transfer. Best practice would include labeling all containers to avoid confusion and mistakes, especially in the event of an emergency.

2. Employers may affix signs, placards, process sheets, batch tickets, operating procedures, or other such written materials, to individual process containers instead of affixing labels, provided:
   ○ The container to be labeled by this alternate method is clearly identified.
   ○ The alternative labeling method contains the same chemical identity and hazard information found on the appropriate MSDS and on the list of in-house hazardous chemicals.

3. If existing labels already convey the required information, chemical manufacturers, distributors, importers, and employers do not have to relabel the chemical to comply with the HCS.

Employers are required to ensure that, unless specifically exempted as described above, all containers, including those used to transfer hazardous chemicals, as well as original containers, are labeled, tagged, or marked with:

- the identity of the chemical
- the appropriate hazard warning(s) including target organ effects

All labels must be legible and in English (labels in another language may be added to, but not substituted for the English language label).

Employers do not have to relabel an original container unless the container label is obscured, hard to read, defaced, or missing. Employers are not to remove or deface existing labels on incoming containers of hazardous materials.

A description of the methods chosen to meet the labeling requirements (employers can order printed labels or produce their own) should become a part of the written HazCom program. If the employer uses a numerical coding system for the in-house identification of hazardous materials (often used when the labeling system is integrated with the chemical inventory system), then he or she must make sure employees are trained to understand and use the numerical system, and to obtain more information from the MSDSs.

If the hazardous chemical is regulated by OSHA in a chemical-specific health standard, the chemical manufacturer, importer, distributor, or employer must

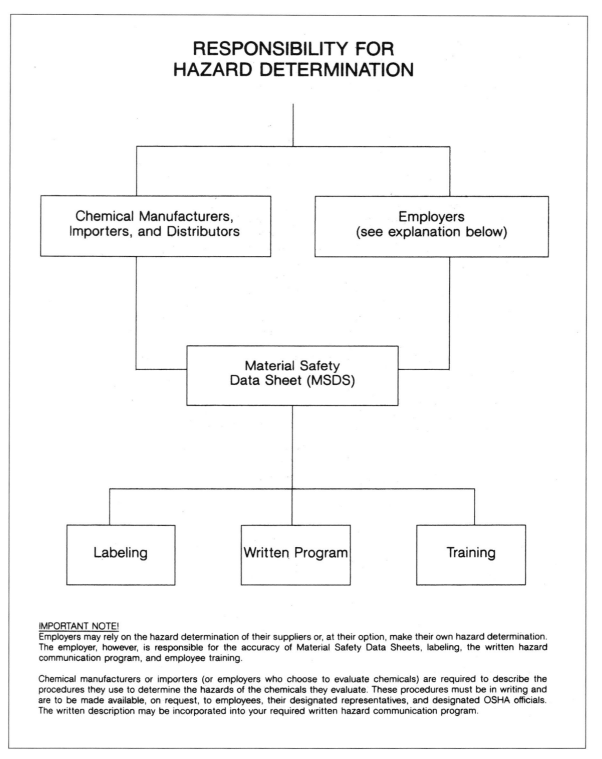

# RESPONSIBILITY FOR HAZARD DETERMINATION

Chemical Manufacturers, Importers, and Distributors

Employers (see explanation below)

Material Safety Data Sheet (MSDS)

Labeling

Written Program

Training

IMPORTANT NOTE!
Employers may rely on the hazard determination of their suppliers or, at their option, make their own hazard determination. The employer, however, is responsible for the accuracy of Material Safety Data Sheets, labeling, the written hazard communication program, and employee training.

Chemical manufacturers or importers (or employers who choose to evaluate chemicals) are required to describe the procedures they use to determine the hazards of the chemicals they evaluate. These procedures must be in writing and are to be made available, on request, to employees, their designated representatives, and designated OSHA officials. The written description may be incorporated into your required written hazard communication program.

**Figure 11–5.** Responsibility for hazard determination.

ensure that the labels or other forms of warning used are in accordance with the requirements of that standard. Competent individuals, familiar with the hazards of the chemical and with the language and interpretation of the law, should prepare the information to appear on the labels. Sources of information for creating in-house labels include MSDSs, ANSI guidelines, and software programs.

**Best-Practices Suggestions.** Employers can use any in-house labeling and hazard warning-sign method(s) they want, as long as they comply with the HCS requirements. Two systems commonly used are the National Fire Protection Association (NFPA) program and the Hazardous Material Identification System (HMIS). These two labeling and warning-sign systems are easy to use. They usually refer to the information on the MSDS and require no special printing or manufacturing procedures.

The HMIS is considered an appropriate method for complying with HCS labeling provisions. The NFPA system is used primarily as a guide by emergency response teams when choosing fire-fighting equipment, techniques, and evacuation procedures at fires involving hazardous chemicals.

In both NFPA and HMIS systems, the category "health hazards" refers to the ability of the material to cause personal injury from absorption, ingestion, inhalation, or injection into the body. The NFPA health hazards were developed mainly for reference to acute exposures (i.e., fire fighting), not everyday chronic exposure. The category "flammability hazards" indicates how easily the material burns. The category "reactivity" (instability) is concerned with the ability of the material to release energy when it reacts with itself or other materials.

*NFPA Labeling System.* This system uses three main categories to identify hazards of a chemical: health, flammability, and reactivity (stability). The degree of severity is indicated numerically by five divisions ranging from 0 to 4. A "4" rating indicates a severe hazard, while a "0" means that no special hazard exists. The NFPA label is diamond-shaped, and includes four smaller different-colored diamonds within the large diamond. Flammability information appears in the red diamond at the top of the large diamond label. Health appears in the blue diamond on the left side (as it faces the reader). Reactivity (stability) appears in the yellow diamond on the right. The white diamond at the bottom is used to indicate whether the chemical is water-reactive, or whether other special conditions exist (Figure 11–6).

*HMIS Labeling System.* This is a comprehensive and uniform labeling system designed to alert employees to chemical hazards to which they are, or may be, exposed in the workplace. The hazard communication portion of the HMIS system communicates information regarding chemical identity, health, flammability, and reactivity hazards; and proper personal protective equipment.

The chemical identity is conveyed by chemical or common name, a code number, or other descriptive term. The identity process in use should serve as a direct link to the appropriate MSDS. According to the HCS, English must be

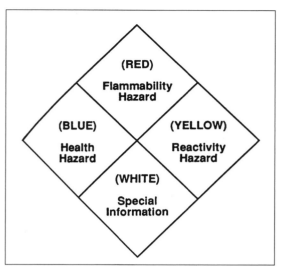

**Figure 11–6.** NFPA 704 M hazard identification system.

the primary language appearing on the label, although additional languages also can be used. As in all labeling systems, the same name or identifier used the chemical on the HMIS label must also appear on the corresponding MSDS label.

Numerical ratings communicate the acute health, flammability, and reactivity hazards of the chemical. An alphabetic or picture code designates the type of personal protective equipment that should be used. An asterisk (*), or other symbol, is used to communicate health hazards on the HMIS label.

The same numerical identification system as described in the NFPA system is used in the HMIS (a rating of "0" indicates fairly low to nonexistent hazard; a rating of "4" indicates severe or extreme hazard). As a rule, any category with a "2" or higher rating is considered very dangerous, or as having the potential to create a very dangerous situation. The basic HMIS label is a colored rectangle that provides:

- identification of the chemical
- acute health, flammability, and reactivity ratings
- chronic health hazard information
- personal protective equipment designation

The rectangle is broken into five rectangles, with each rectangle identified by its specific color and written category name. The top white rectangle is blank and is for writing in the chemical name/identification. Next is the blue rectangle for health. The next is red for flammability. Next is yellow for reactivity/stability. The last rectangle is white for personal protective equipment.

**More Best-Practices Suggestions.** Like any other method, a labeling system is only good if employees use it consistently. Supervisors should make visual checks of all process, storage, transfer, and use areas at regular intervals. Employees must be encouraged to express their concerns and preferences about the labeling system so that any changes can incorporate employees' observations.

A labeling program is useless if employees do not know the meaning of the hazard information on the labels. Employees should be required to demonstrate competency in using the labeling system by participating in training scenarios that require them to explain the hazard being conveyed as well as any company procedures for dealing with the hazard.

Training for use of the labeling system is critical and should consist of at least the following elements:

- Audiovisual aids to explain and illustrate the labeling system being used in the workplace.
- Wallet cards that provide basic information needed to understand the hazard ratings used on the workplace hazard labels. Employees should be encouraged to carry the cards at all times.
- Wall posters to explain the numerical identification system being used. The wall posters can also display personal protective equipment codes and be hung in appropriate work areas when in use for training sessions.
- Written material that explains the workplace labeling system in easily understood terms and format. This information should be supplied in a second language also, if needed, as long as the primary language is English.
- Raw materials information for employees that explains the procedure and criteria for assigning chemical hazard ratings.
- Implementation procedures describing the necessary steps for understanding and implementing the hazard label system.
- Training sessions on how to read and interpret the labels, to give specific information regarding product hazards, and information on personal protective equipment.

It is common for employers to require that the first-line supervisor ascertain through observation and discussion whether all employees can properly interpret the hazard label warning system. As mentioned earlier, it is common to require that the first-line supervisor verify that any container of a hazardous chemical in his or her work be properly labeled.

## Employee Information and Training

Given today's complex job demands and diverse work force, employee information and training regarding hazardous materials has become even more critical. Employers must make sure that workers have the proper information and training, and that the information is provided in a way they can understand and interpret correctly.

The HCS mandates that employers give their employees information and training regarding hazardous chemicals in their work areas. This information and training must be provided with the employee's first work assignment, and whenever a new hazard (including a process change) is introduced.

All employees who are, or may potentially be, exposed to hazardous chemicals under normal work conditions or in foreseeable emergencies, must be provided with required information and training before beginning work in the area that contains hazardous chemicals.

Employees must be informed of:

- HCS requirements..
- Any operations in their area where hazardous materials are present.
- Location and availability of the employer's written HazCom program. The program must include required lists of hazardous chemicals and associated MSDSs.

Employers must be trained and demonstrate competency in:

- All methods and observations that may be used by the employee to detect hazardous chemical presence or release in the work area.
- Physical and health hazards of chemicals used in the work areas.
- Details of the HazCom program developed by the employer. This part of the training must include explanation of the labeling system, MSDS use, and information content. Employees must demonstrate competency in correctly using the hazard information sheets and area controls, such as PPE, to prevent or minimize employee exposure.
- Measures employees can take to protect themselves from chemical hazards in their work areas. This includes instruction in appropriate work practices; emergency procedures; and use, maintenance, and selection of PPE.

The employer is responsible for:

- providing the correct types of PPE for chemicals being used
- developing a written policy that addresses PPE use

- providing employees with training and medical surveillance needed when PPE is used in the workplace.

**Best-Practices Suggestions.** The management group should define the employee training program. The written HazCom training plan should include how employees will be informed of the requirements of the HCS, the title of the person(s) doing the training, the availability and location of the written HazCom program, a training schedule, the method of communicating the training schedule to the employees, and documentation of the content of, and attendees at, the various training sessions.

First-line supervisors are normally held responsible for ensuring that any employee exposed to a hazardous chemical in the workplace has received proper training. Supervisors should be particularly aware of any employee who has transferred into his or her work area from other areas in the facility. It is possible that transferred employees may require additional hazard communication training if a hazardous chemical is only used in certain departments. Supervisors should also document any new job assignments where an employee might be exposed to hazardous chemicals in pipes, or while performing a nonroutine job task. Appropriate HCS training should then be given.

Product and employee liability concerns make record keeping an essential part of an employee training program. The record-keeping documents can include:

- Notices posted for each training session (Figure 11–7).
- Topics discussed in each training session.
- Minutes of each training session.
- An attendance sheet that contains the session date, session topic, printed employee name, and employee signature (Figure 11–8). It is recommended that these attendance records be kept for an employee's employment duration, and for at least five years after.
- A document signed by the employee stating that he or she understands the session content and that continued employment with the company depends on attending all training sessions and following all company policies and procedures (Figure 11–9).
- Documentation of qualifications of instructor(s) teaching the training sessions.

Some companies may want to create a learning lab (usually an area that is set up for supervised, unsupervised audiovisual, or computer-based instruction

## HAZARD COMMUNICATION EMPLOYEE TRAINING MEMO

TO:_____DATE:_____

DEPARTMENT:_____ _____

CLOCK, EMPLOYEE OR SOC. SEC. #_____

All employees exposed or potentially exposed to hazardous chemicals under normal operating conditions are required to attend a training session at the time of initial assignment or when a new hazard is introduced into the workplace. If the date assigned conflicts with another important activity, please have your supervisor contact this office. Otherwise, please sign and return this form to:

_____

You have been scheduled to attend the next hazard communication training program:

DATE:_____

STARTING TIME:_____

COMPLETION TIME:_____

LOCATION:_____

☐ Yes, I will be able to attend.

Employee's
Signature_____

Supervisor's
Approval_____

☐ No, I cannot attend. My supervisor will explain.

The Hazard Communication Employee Training Memo can be used to notify the employee of a training session. It should be returned to the Compliance Manager and will become part of the permanent record of the employee's training.

**Figure 11–7.** Sample hazard communication employee training memo.

# HAZARD COMMUNICATION TRAINING LOG

SUBJECT:_____      COMPANY: _____
A/V _____      ADDRESS: _____
Handouts _____      TRAINER'S NAME _____
                                                TITLE _____
                            DATE OF TRAINING _____

| EMPLOYEE NAME | EMPLOYEE NO | DEPARTMENT | EMPLOYEE SIGNATURE |
|---|---|---|---|
| | | | |
| | | | |
| | | | |
| | | | |
| | | | |
| | | | |
| | | | |
| | | | |
| | | | |
| | | | |
| | | | |
| | | | |
| | | | |
| | | | |
| | | | |
| | | | |
| | | | |
| | | | |
| | | | |
| | | | |
| | | | |
| | | | |
| | | | |
| | | | |
| | | | |
| | | | |
| | | | |
| | | | |
| | | | |

Use a separate training log for each lesson. Employees attending the lesson should signify their presence with their signature. Note absentees, and schedule makeup classes.

**Figure 11–8.** Sample hazard communication training log.

## EMPLOYEE TRAINING EVALUATION SHEET

Date: _____

1. List the hazardous chemicals you use in your work area: _____
_____
_____
_____

2. Were all of these chemicals covered in the training program? ____ Yes   ____ No
   If NO, list those chemicals not covered.
_____
_____
_____

3. Did the program explain the protective clothing/equipment needed to work with the hazardous chemicals you use? ____ Yes   ____ No. If NO, which chemicals were omitted?
_____
_____
_____

4. Do you have and use the required protective clothing/equipment? ____ Yes   ____ No
   If NO, explain.
_____
_____
_____

5. Did the training program cover all the questions you had about the safe and proper use of hazardous chemicals?
   ____ Yes   ____ No. If NO, explain.
_____
_____
_____

6. Use the back of this sheet to offer any comments on how to improve this training program.

NAME: _____

DEPARTMENT: _____

> In addition to training on my rights and obligations under the law, I have...
>
> • been instructed in how to read a MSDS and a label.
>
> • I have knowledge of where the company written hazard communication program and MSDSs are kept and that I have access to these.
>
> • I understand the necessary precautions to be taken when dealing with hazardous chemicals in my work area.
>
> • I know where medical supplies and safety equipment are kept.
>
> _____        DATE: _____
> (employee signature)

> The Employee Training Evaluation Sheet is used for student opinion and is also documentary evidence of training attendance.

**Figure 11–9.** Sample employee training evaluation sheet.

on company policies and procedures) for training employees. This type of training program is viewed as a time and personnel saver by many companies, but it has some major flaws.

For example, the HCS requires that all employees be trained before they begin to work with hazardous chemicals. Unless a supervisor or trainer is assigned to ensure that employees report to the learning lab before beginning to work with hazardous chemicals, the company is in violation of the HCS.

Another flaw common to an unsupervised learning lab, a home study program, or an interactive computer-based training program, is that often no one is available to answer questions, a practice that can lead to dangerous situations in the workplace. Similarly, because the HCS requires that all employees understand the company HazCom program and the dangers of the specific workplace chemicals, even the best interactive computer learning program is limited because it cannot answer unprogrammed questions.

Language and literacy barriers also may pose problems for the company training and documentation program. OSHA requirements for employee training and understanding of information presented in formal training programs do not differentiate between employers with more or less formal education or those who speak English as their second language or do not understand it, and any other employees. All employees must meet the HCS requirements for training, and the employer must prove that all employees understand the policies and procedures of the company's HazCom program.

The training program for employees with little formal education, or with English as a second or foreign language, may require more class time, translation of English materials into another language, or the use of international pictograms.

Whatever is done to facilitate the employees' understanding of hazards associated with the workplace, the training should be shown to be effective and in compliance with regulations, through audits and other documentation processes.

Proving that the training program is effective and operational will be an ongoing job for supervisors, trainers, and management staff. The employer can conduct periodic walk-through inspections to assess workplace conditions, job performance, and adherence to the company's HazCom program.

Another way to assess success of the training program is to conduct OSHA-like audits during walk throughs. This type of audit could entail random questioning of individual employees regarding the company's HazCom program, checking on availability of MSDSs, having employees find certain sections in the HazCom program and explaining their content to the "inspector," or asking employees if they have experienced problems performing their job tasks safely and within the HazCom program guidelines. Information obtained during walk-throughs can be used to correct and update the HazCom program, and can serve as the basis for a reward system (e.g., the department with the most correct answers and least deficiencies within a six-month period could receive a monetary or other award).

The success of the training program also depends on the completeness of the chemical inventory. Once the initial inventory is finished and all chemicals have been identified as hazardous or nonhazardous, the hazardous chemicals can be categorized further by the type(s) of hazard they pose to the employee and the workplace, and then incorporated into the training program. It is recommended that the chemical inventory be continually updated as needed.

It is important that employees understand that there are between 5 million and 6 million currently known chemicals, and that only some of them are considered to be hazardous. Chemicals considered to be hazardous are referenced in OSHA, EPA, NIOSH, DOT, and ACGIH standards, guidelines, and lists (reference sources for these lists and other information sources appear at the end of this chapter). Copies of chemical lists and information sources can be kept on file for reference and training purposes.

In addition, the employer should incorporate information into the HazCom program that addresses types of symptoms usually associated with acute and chronic exposure to hazardous chemicals. The employer may choose to have a medical professional, toxicologist, safety expert, or industrial hygienist conduct these information and training sessions.

There are generally four methods that the employer can adopt to reduce unnecessary employee exposure to hazardous chemicals:

1. **Product Substitution.** The employer should try to choose a different, less harmful chemical for a specific process or operation, or change the process to avoid using any chemicals. This is the preferred method of employee protection. Care must be taken to completely remove the more harmful chemical from the work area to prevent employees from accidentally using it. The employer is responsible for constantly reviewing the types of chemicals used in work processes and in the workplace. Reduce the number and type of chemicals by consolidating the use of oils, cleaning agents, etc.

2. **Use of Engineering Controls.** These controls include proper equipment when handling containers of hazardous chemicals, process enclosures, worker isolation, and local exhaust and general ventilation. The employer

should develop written policies and training programs for appropriate use of these types of engineering controls. Safety eyewash stations and deluge showers are also engineering controls employees can use.

3. **Use of Administrative Controls.** These controls include rotating employees through areas or jobs with high probability of exposure to chemicals. This method is not acceptable with highly harmful chemicals, such as carcinogens. Establishing proper work practices and housekeeping measures are other administrative controls.

4. **Use of Personal Protective Equipment.** When a less hazardous chemical cannot replace a more hazardous chemical, and engineering controls do not adequately reduce exposure, employees usually must wear PPE specifically chosen for use with the hazardous chemical.

The written HazCom program should include procedures and requirements for documenting these methods and steps taken to control exposure. In addition, employees should receive training in basic first aid for chemical exposures.

The employer should post emergency phone numbers at all phone locations and develop a written policy that addresses what the employees are to do in case of accident, fire, or explosion. Employees should be trained in these procedures, and the employer should hold several fire drills, plus other drills each year, to test the emergency response policies.

A spill, fire, or explosion involving hazardous chemicals may require action from employees who have received special training in responding to and abating hazardous chemical emergency situations. (This training refers to the requirements of the OSHA standard [29 CFR, 1910.120] of March 6, 1989, for Hazardous Waste Operations known as Emergency Response and HAZWOPER.) The employer should then develop a policy that addresses these types of emergencies and supply the extra training courses and equipment required to teach employees how to handle the dangers involved in a hazardous chemicals incident.

Hazardous waste site operations fall under many specific regulations and jurisdictions, including the HCS. The basics of the HCS apply to all hazardous waste site personnel, including use of MSDSs, training, PPE, hazard identification, and written policies and procedures. However, other regulations, standards, and agencies also affect the hazardous waste site employee (such as SARA Title III, RCRA, and DOT).

# NONROUTINE ELEMENTS OF A HAZARD COMMUNICATION PROGRAM

A facility-specific procedure should be developed for the transfer of hazardous chemical information to and from outside contractors who enter the worksite. The policy should have two clear goals. First, the procedure should describe how any contract employee entering the facility will be advised of exposure to a hazardous chemical in the work area. Second, the procedure should describe how the employer will ensure that no employee has a new exposure to a hazardous chemical that is being brought on site by the outside contractor.

The procedure should identify the specific person(s) responsible for distributing the hazardous chemical information to outside contractors, and for ensuring that the contractor signs a document stating that the necessary information has been received. The facility supervisor should also request, and approve from the contractor, a list of chemicals (and copies of MSDSs) that the contractor will be using at the facility. Any hazard information should be shared with all affected employees.

The first-line supervisor is often required to audit contractors' operating procedures as they complete the contracted tasks in their work area.

## Temporary and Contract Employees

In a general sense, the employer is responsible for the safety, health, and well-being of all people who enter the workplace or work area. To this end, the employer should establish procedures covering all employees and visitors as a part of the company's written HazCom program. These procedures should specifically address the type of hazardous chemical training and information that maintenance workers, outside contractors, office personnel, and cleaning crews should receive before they enter workplaces where they may be exposed to hazardous chemicals.

**Best-Practices Suggestions.** These procedures should contain at least the following:

- list of hazardous chemicals located in assigned work areas and in any areas through which workers must pass to reach their work area
- floor plan showing areas where workers are not permitted
- location of written HazCom program

- location of MSDS file for the assigned work area
- review of contractor's PPE program, and if necessary, training in use of specific PPE and engineering controls to be used in assigned work area
- emergency procedure training for fires, spills, explosions, and other disasters
- conditions to avoid in work area to reduce possibility of exposure to hazardous chemicals

To close the communication circle, employers should request a list of the chemicals to be used in their workplace, a description of how they will be used, and copies of the MSDSs. Proper disposal information should also be shared with the contractors for the facility. Some companies schedule a prework meeting to discuss this information and approve/reject chemical usage at their facility.

All temporary and nonfacility employees, with no exceptions, should receive formal training before they begin work in the employer's workplace. The reasons for this rule are simple: the potential liability to the employer and the risk of a serious incident to an uninformed worker or "outsider" are too great. Even if outside crews have worked in similar situations, with the same or similar chemicals, they should be trained regarding specifics of each new employer's facility and HazCom program.

### Visitor Orientation

Visitors can represent a serious liability risk to employers, particularly if supervisors do not provide visitors with appropriate orientation and protective equipment. Supervisors should always know where visitors are in their areas, and either assign someone to escort them or accompany the visitors themselves. There are no specific HSC requirements for temporary and contract employees.

**Best-Practices Suggestions.** If the visitor will be in areas where hazardous chemicals are used, stored, processed, or transferred, or must pass through these areas, it is suggested that appropriate training and PPE be supplied before the tour begins.

The employer should create a policy that addresses visitor restrictions, training, identification, access, and PPE and include it in the company's written HazCom program.

### New Employee Orientation

Orientation is a critical part of new employee training. Supervisors must ensure that all new employees have been properly oriented to work requirements, safety procedures, and emergency measures before beginning their job assignment.

**HCS Requirements.** The OSHA HCS requires that all new employees receive initial training on workplace hazards, MSDSs, the company's written HazCom program, definitions, labeling and warning signs, chemical inventory, and the intent of the HCS.

**Best Practices Suggestions.** All new employees, even if they have worked previously with hazardous chemicals, must receive HazCom training before they begin to work with, or are potentially exposed to, hazardous chemicals used in the new employer's workplace. The employer can select the training methods to be used (provided they meet HCS requirements and intent), are in written form, and are part of the company's HazCom program. It is suggested that the training should:

- include hands-on and classroom sessions
- include examples of actual labels, MSDSs, PPE, and engineering controls used in the workplace
- be performed in a location that promotes and stimulates learning (a boiler room, busy lunchroom, or other cramped or noisy area would not usually be a good place to hold the training session)
- include an orientation tour of entire workplace
- be conducted by a knowledgeable person
- include provisions for immediate training if a new chemical or process is introduced to the workplace (mandatory under the HCS)
- include a written test to audit training effectiveness and level of employee understanding
- be scheduled so employees in training sessions are not interrupted by work problems that require them to leave
- include periodic refresher courses

Periodically, management should review the effectiveness of the written HazCom program. Regularly, all changes in chemicals, processes, emergency response procedures, equipment, and MSDSs should be entered into the written program and the new information communicated to all employees. This new information may be communicated through training sessions and written summaries posted on employee bulletin boards.

The company can test program effectiveness through the use of drills, management walk-throughs, and question sessions. The results of all drills, walk-throughs, and question sessions can be critiqued by supervisory staff and employee representatives. Summaries of these critiques can be posted on employee bulletin boards. Management should immediately include any changes to the HazCom plan adopted during the critiques in the written plan. This information should be passed along to all employees.

The HCS helps employers provide a safe, healthy work environment for all employees and gives them the tools they need to perform assigned tasks safely and effectively. The best-practices suggestions are offered to help employers meet the HCS's intent. They are designed as examples and tools the reader can either use, modify, or discard when developing a company HazCom program.

## SUMMARY OF KEY POINTS

Key points covered in this chapter include:
- The federal hazard communication standard, known as OSHA 29 CFR, 1910.1200, regulates an employer's duty to communicate chemical hazard information to employees who may be exposed on the job. Federal or state agencies and OSHA enforce this standard.
- Companies should understand the differences between HCS requirements and best-practices suggestions. The best-practices suggestions are meant for use as guidelines for implementing HCS requirements or when no HCS requirements have been developed to cover specific situations. These suggestions must never replace HCS requirements.
- The HazCom program must include written information on hazardous chemicals, proper labeling, employee training, chemical inventory, and MSDS distribution. Each aspect of the HazCom program must meet or exceed federal HCS requirements.
- A written hazard communication document provides employers and employees with specific policies and procedures to follow when training employees to handle hazardous substances. A copy of the written program must be kept at the workplace and be available to employees and to OSHA compliance officers and NIOSH investigators. The policies and procedures developed must cover the exact chemicals and processes that exist at a company.
- All containers and drums of hazardous materials must be labeled accurately

and the labels prominently displayed. Workers must be trained to understand the labels and interpret chemical hazards, first aid, or emergency instructions. Supervisors must know what substances are exempt from HCS requirements and what circumstances exempt employers from labeling hazardous chemical containers.

- MSDSs, prepared by chemical manufacturers and importers, provide information on hazardous chemicals, including chemical name; characteristics; physical and health hazards; exposure limits; precautions and control measures; medical treatment procedures; and name, address, and telephone number of a person who can provide additional information on the chemical. MSDSs should be kept in the workplace, be accessible to employees at all times, and be regularly updated.

- Under HCS regulations, employers must provide employees with information about HCS requirements, hazardous operations in their work area, and location and availability of the written HazCom program. Training sessions should inform workers of methods they can use to detect the presence or release of a hazardous chemical, the physical and health hazards of chemicals, protective measures for safe use, and details of the HazCom program as developed by the employer.

- Employers are also responsible for the safety and health of visitors and of new, temporary, or outside employees working on a company property. These employees and visitors should receive orientation sessions and be provided with protective gear before they are allowed into work areas. The company's HCS program should include policies and procedures to cover these nonroutine elements of hazard communication.

## RESOURCES

*Annual Report on Carcinogens.* National Toxicology Program (NTP).

*Chemical Data Sheets.* Chemical Cards and Water Information Sheets. Chemical Manufacturing Association, 1825 Connecticut Avenue, N.W., Washington, DC 20009.

*Chemical Safety Slide Rule.* National Safety Council, 1121 Spring Lake Drive, Itasca, IL 60143–3201.

*Code of Federal Regulations.* General Industry Safety and Health Standards: 29 CFR 1910; Toxic and Hazardous Substances. 29 CFR, Part 1910, Subpart

Z, Transportation: 49 CFR Parts 100–199. Washington, DC 20402: Superintendent of Documents, U.S. Government Printing Office.

Computer Programs:

*OHMTADS:* Accessed through Chemical Information System, Inc., (800) 247–8737.

*CHRIS* (Chemical Hazard Response Information System): Accessed through Chemical Information System.

*CAMEO* (Computer-Aided Management of Emergency Operations): Widely used database. U.S. Department of Commerce. NOAA Hazardous Materials Response Branch, 7600 Sandy Point Way NE, Seattle, WA 98115, (206) 526–6317. The database can also be purchased through the National Safety Council.

*TOXLINE* (Toxicology Information On-Line).

MEDLARS, Management Section, Specialized Information Systems, National Library of Medicine, 8600 Rockville Pike, Bethesda, MD 20814; (301) 496–6193; (800)638–8480.

*CHEMTREC* (Chemical Transportation Emergency Center): Access via telephone, (800) 424–9300; (202) 483–7616.

*Guidelines to the Handling of Hazardous Materials.*

Source of Safety, Inc., 8303 East Kenyon Drive, Denver, CO 80237.

International Agency for Research on Cancer (IARC) Monographs.

*National Fire Codes.* National Fire Protection Association, 1 Batterymarch Park, Quincy, MA 02269.

*NIOSH Pocket Guide to Chemical Hazards.* Washington, DC: U.S. Department of Health and Human Services.

NIOSH Registry of Toxic Effects of Chemical Substances. Washington, DC: U.S. Department of Health and Human Services.

Threshold Limit Value. American Conference of Governmental Industrial Hygienists, P.O. Box 1937, Cincinnati, OH 45201.

Bronstein, AC, Currance, PL. *Emergency Care for Hazardous Materials Exposure, No. 2.* St. Louis, MO: C.V. Mosby, 1994.

Budavari, S, Ed. *The Merck Index of Chemicals, Drugs, and Biologicals.* 11th Edition. Rahway, NJ: Merck and Company, Inc., 1989.

*Hazard Communication Guidelines for Compliance.* OSHA Report No. 3111. U.S. Department of Labor, Occupational Safety and Health Administration. Washington, DC: U.S. Government Printing Office.

Lewis, RJ, Sr. *Hawley's Condensed Chemical Dictionary*. 12th Edition. New York, NY: Van Nostrand Reinhold Publishing Company. 1993.

Meyer, E. *Chemistry of Hazardous Materials*. Englewood Cliffs, NJ: Prentice-Hall, Inc., 1997.

# 12

# Regulatory Issues

**After reading this chapter, you will be able to:**

- Understand the scope of OSHA regulations
- Describe the General Duty Clause and how it is applied by OSHA inspectors
- Determine whether written plans or documentation are required for regulatory compliance
- Understand key elements of OSHA regulations that have the most significant regulatory impact on the workplace

This chapter was contributed by Philip E. Hagan, Director of Safety & Environmental Management, Department of Human Sciences, Georgetown University.

# SCOPE OF THE OCCUPATIONAL SAFETY AND HEALTH ACT

In general, the Occupational Safety and Health Act of 1970 (OSHAct) extends to all employers and their employees in the 50 states, the District of Columbia, Puerto Rico, and all other territories under federal government jurisdiction. Coverage is provided either directly by federal OSHA or through an OSHA-approved state occupational safety and health program in states that have approved programs.

All supervisors should know which set of regulations govern their workplace. Twenty-six states and the District of Columbia are covered by federal OSHA regulations. The other 24 states (see Table 12–A), Puerto Rico, and the Virgin Islands, have their own regulatory plans that have been approved by federal OSHA and cover the private and government sectors.

The Connecticut, New Jersey, New York and Virgin Islands plans cover public sector (state and local government) employment only. This means that in these states, government workers would be covered by the state plan and private sector employees would be covered by federal OSHA regulations.

In some cases, there are regulatory requirements from other sources that need to be followed. The EPA, DOT, and the Nuclear Regulatory Commission (NRC) are examples of potential sources of other regulatory requirements, in addition to OSHA, that may govern a workplace. As a supervisor, it is a good idea to ask your employer for support in determining which regulations are applicable if this has not been done. This chapter deals primarily with OSHA regulations governing general industry.

OSHA regulations can be found in Title 29 of the Code of Federal Regulations (CFR), Parts 1902-1990. These regulations cover general industry, maritime, construction and agricultural workplaces. These regulations can be found at www.osha.gov.

Workplaces covered by these regulations should ensure that that the OSHA 3165 poster, *Job Safety and Health: It's the Law*, be displayed in a conspicuous place visible to all employees and potential employees so that workers know they are covered by the protections outlined in the OSHAct. Federal government agencies must post a Federal Agency Poster. The poster is available for free from the OSHA Office of Publications. Reproductions or facsimiles of the poster shall be at least 8 ½ by 14 inches with 10-point type. Employers do not need to replace previous versions of the poster.

## Table 12–A. Sources for Regulatory Information for States with Their Own Plans

| | |
|---|---|
| Alaska | Alaska Division of Labor Standards and Safety – http://labor.state.ak.us/lss/home.htm |
| Arizona | Arizona Secretary of State Website – www.ica.state.az.us/Divisions/osha/index.html |
| Connecticut | CONN-OSHA adopts Federal OSHA regulations and standards identically. |
| California | California DOSH – www.dir.ca.gov/dosh/ |
| Hawaii | Hawaii Standards – http://hawaii.gov/labor/hiosh/ |
| Indiana | Indiana OSHA adopts Federal OSHA regulations and standards identically. |
| Iowa | Iowa OSH Administrative Rules – www.iowaworkforce.org/labor/iosh/index.html |
| Kentucky | OSH Regulations – www.labor.ky.gov/ows/osh/osharegulations |
| Maryland | Maryland Standards – www.dllr.state.md.us/labor/mosh.html |
| Michigan | Standards & Legislation – www.michigan.gov/miosha |
| Minnesota | Minnesota Standards & Regulations – www.dli.mn.gov/MnOsha.asp |
| Nevada | Nevada Standards – http://dirweb.state.nv.us/OSHA/osha.htm |
| New Jersey | See Regulations and Standards section – http://lwd.dol.state.nj.us/labor/lsse/employer/Public_Employees_OSH.html          www.nj.gov/health/eoh/peoshweb |
| New Mexico | New Mexico Regulations – www.nmenv.state.nm.us/Ohsb_Website/index.htm |
| New York | See Regulations and Standards section – www.labor.state.ny.us/workerprotection/safetyhealth/DOSH_PESH.shtm |
| North Carolina | NCDOL Standards – www.nclabor.com/osha/osh.htm |
| Oregon | Oregon OSHA Rules and Laws – www.orosha.org |
| Puerto Rico | Spanish translations of safety and health standards/Normas traducidas al español radicadas – www.dtrh.gobierno.pr/PROSHA.asp?m=6 |
| South Carolina | South Carolina – www.llr.state.sc.us/Labor/Osha/index.asp |
| Tennessee | Tennessee Department of Labor & Workforce Development – www.state.tn.us/labor-wfd/tosha.html |
| Utah | R614 – Labor Commission, Occupational Safety and Health – www.laborcommission.utah.gov/UOSH/index.html |
| Vermont | Vermont Standards – www.labor.vermont.gov/Default.aspx?tabid=74 |
| Virgin Islands | VIDOSH adopts Federal OSHA regulations and standards identically. |
| Virginia | Virginia Standards – www.doli.virginia.gov |
| Washington | Washington Standards – www.lni.wa.gov/Safety/default.asp |
| Wyoming | Wyoming OSHA Standards – http://wydoe.state.wy.us/osha |

## GENERAL DUTY CLAUSE

In many cases, there are specific regulations that govern different workplace activities. For those cases where there are no specific regulations, there is Section 5(a)(1) of the OSHAct, often referred to as the "General Duty Clause." The General Duty Clause requires employers to "furnish a place of employment that is free from recognized hazards which are causing or are likely to cause death or serious physical harm to employees." The General Duty Clause is used by OSHA inspectors to cite employers when hazards are identified that are not covered by other regulations. However, for the General Duty Clause to be used, the identified hazard needs to meet several criteria.

OSHA looks at the following when evaluating whether the General Duty Clause is applicable:

- There is no applicable OSHA standard for an identified hazard.
- The employer failed to keep the workplace free of a hazard to which employees of that employer were exposed.
- The hazard was recognized or should have been recognized:
  ○ The employer knows about the hazard as shown by written or oral statements made during or before an OSHA inspection.
  ○ The hazard is recognized by others in the same industry.
  ○ Common sense indicates that any reasonable person would recognize the hazard.
- The hazard was causing or was likely to cause death or serious physical harm.
- There was a feasible and useful method to correct the hazard.

A general duty citation must involve both the presence of a serious hazard and exposure of the cited employer's own employees.

## REQUIRED WRITTEN PLANS AND OTHER DOCUMENTATION

Regulations can take different forms and their requirements can vary widely. Some require written plans or other forms of documentation that will serve as a guide for complying with the regulation. Common examples of regulations requiring written plans are 29 CFR 1910.1030(c)(1)(i) and 1910.1030(c)(1)(iii) (bloodborne pathogens), 29 CFR 1910.1200(e)(1) and 1910.1200(e)(4) (hazard communication), and 29 CFR 1910.146(c)(4) (permit-required confined spaces). Traditionally, these written programs have been kept in separate binders in appropriate work areas in order to comply with the standards. Generally, these written plans need to be readily available to employees or their representatives.

These written plans can be general in nature (the hazard communication standard) or they can have very specific requirements like those found in the bloodborne pathogens standard. Sometimes the only requirement is that written records be maintained. Because computers are more common in the workplace now that in years past, OSHA will allow a written program to be in either paper or electronic format, as long as the program meets all other requirements of the standard in question. This allows an employer to obtain significant benefits in consistency, ease of use, and accuracy in maintaining and updating the materials in a timely manner.

A partial listing of OSHA regulations with written components required in many workplaces includes the following:

OSHA Log 300—Recording and Reporting Occupational Injuries and Illness § 1904

Emergency Action Plan § 1910.38(a)

Fire Prevention Plan § 1910.38(a) § 1910.38(a)

Hearing Conservation § 1910.95

Process Safety Management § 1910.119

HAZWOPER § 1910.120

Respiratory Protection § 1910.134

Confined Spaces § 1910.146

Control of Hazardous Energy § 1910.147

Fire Extinguishers § 1910.157

Toxic/Hazardous Substances (Subpart Z) § 1910.1000

Bloodborne Pathogens § 1910.1030

Hazard Communication § 1910.1200

Laboratory Standard § 1910.1450

Supervisors must determine if regulations governing their workplace require written documentation. It is very difficult to comply with a regulation if there are written components and there is nothing in writing. When an OSHA inspector visits, quite often the first area evaluated deals with those regulations with written requirements.

# MOST FREQUENTLY CITED OSHA STANDARDS

This chapter addresses some of the standards that have resulted in frequent OSHA citations and large monetary penalties.

Recently, the following standards were among the top 10 most frequently cited by OSHA inspectors:

1. Hazard communication standard, general industry (29 CFR 1910.1200)
2. Control of hazardous energy (lockout/tagout), general industry (29 CFR 1910.147)
3. Respiratory protection, general industry (29 CFR 1910.134)
4. Electrical, wiring methods, components and equipment, general industry (29 CFR 1910.305)
5. Powered industrial trucks, general industry (29 CFR 1910.178)
6. Machines, general requirements, general industry (29 CFR 1910.212)
7. Electrical systems design, general requirements, general industry (29 CFR 1910.303)

The following standards are among those for which OSHA has recently assessed the highest penalties:

1. Electrical, hazardous (classified) locations (29 CFR 1910.307)
2. Control of hazardous energy (lockout/tagout), general industry (29 CFR 1910.147)
3. Machines, general requirements, general industry (29 CFR 1910.212)
4. General duty clause (Section 5(a)(1) of the OSH Act)
5. Powered industrial trucks, general industry (29 CFR 1910.178)
6. Walking-working surfaces, general requirements (29 CFR 1910.22)
7. Process safety management of highly hazardous chemicals (29 CFR 1910.119)

Each supervisor should determine if any of these regulations are applicable in their workplace and if so, ensure that a compliance program is in place.

## Hazard Communication Standard

OSHA's web site indicates that more than 30 million American workers are exposed to hazardous chemicals in the workplaces. The OSHA hazard communication standard (HCS) is intended to ensure that these workers and their employers are informed of the identities of these hazardous chemicals, associated health and safety hazards, and appropriate protective measures. Hazard communication is addressed in specific standards for the general industry, shipyard employment, marine terminals, longshoring, and the construction industry. The HCS covers some 650,000 hazardous chemical products found in more than 3 million facilities. As a supervisor, it is important to determine if employees in your workplace could be exposed to hazardous chemicals. Given the right (or wrong) conditions, most chemicals can be hazardous.

In general, the HCS works like this: chemical manufacturers or importers determine the hazards of each chemical product and communicate the hazard information to customers through labels and through Material Safety Data Sheets (MSDSs). The employer (and supervisor) has the following responsibilities:

- Identify and list hazardous chemicals in the workplaces.
- Obtain MSDSs and labels for each hazardous chemical.
- Develop and implement a written hazard communication program, which should include information on container labels, MSDSs and their availability, and employee training.

**Written Program.** The written program does not have to be lengthy or complicated, and must be available by request to employees and employees' designated representatives. The written hazard communication program should contain a list of the hazardous chemicals in each work area, how the employer will inform employees of nonroutine task hazard, and the hazards associated with chemicals in unlabeled pipes. If the workplace has multiple employers on site, it is necessary to ensure that information regarding hazards and protective measures be made available to the other employers as appropriate.

**Labels.** All containers of hazardous chemicals must be labeled, tagged, or marked with the identity of the material and must show hazard warnings appropriate for protecting employees. As long as the hazards of the chemicals are conveyed, the hazard warning can be any type of message using words, pictures, or symbols. Labels must be legible (they need to be clear to anyone who would be using the chemical).

There are some exceptions regarding container labels:

- Signs or placards can be posted that convey hazard information if there are stationary containers within a work area that have similar contents and hazards.
- Standard operating procedures, batch tickets, blend tickets, process sheets,

and other similar written materials can be used instead of container labels on stationary process equipment if the same information is covered and readily available to employees in the work area.

- Portable containers into which hazardous chemicals are transferred from labeled containers and that are intended only for the immediate (within the immediate work shift) use of the employee who makes the transfer.
- Pipes or piping systems are not required to be labeled.

**Material Safety Data Sheets.** Each MSDS must be in English (they can also be in other languages) and include information regarding the specific chemical identity of the hazardous chemical(s) involved and the common names. In addition, information must be provided on:

- identification and contact information for the organization responsible for preparing the MSDS
- names, synonyms, and other identification information
- physical and chemical characteristics of the hazardous chemical
- fire and exposure data
- health hazard information (including whether it is considered a carcinogen)
- exposure limits
- precautions for safe handling and use
- control methods
- emergency and first-aid procedures

Copies of the MSDSs need to be readily accessible to employees in their work areas. This means they have to be available during each workshift in case someone either wants or needs to review the information (see Chapter 11).

**List of Hazardous Chemicals.** The written plan needs to contain a list of all hazardous chemicals in the workplace. There should also be an MSDS for each chemical on the list. If there are hazardous chemicals without an MSDS, the supplier, manufacturer, or importer needs to be contacted to obtain the missing MSDS. If the MSDS is not received within a reasonable period of time, OSHA should be contacted. It is a good idea to document any actions in trying to obtain an MSDS.

**Employee Training.** All employees exposed to hazardous chemicals in their work area need to receive training. This needs to occur at the time of initial assignment and/or whenever a new hazard is introduced into the work area. The training plan should cover the following elements:

- how the hazard communication program is implemented, how to read and interpret information in MSDSs and on labels
- the hazards of the chemicals in the work area—hazards may be discussed by each chemical or by hazard categories, such as corrositivity or flammability
- measures that can be used to protect oneself from hazards
- procedures used to provide protection, such as engineering controls, work practices, and the use of personal protective equipment (PPE)
- methods and observations workers can use to detect the presence of a hazardous chemical to which they may be exposed—this could be from a smell or something as simple as identifying a spill from a container

**Trade Secrets—Medical Emergency.** Sometimes chemicals will come under the trade secret provision and the identity and ingredients will not be known. However, employers must immediately disclose the specific chemical identity of a hazardous chemical to a treating physician or nurse when the information is needed for proper emergency or first-aid treatment. If you need the information from the MSDS, the chemical manufacturer, importer, or employer may obtain a written statement of need and a confidentiality agreement that will protect them for having to disclose the trade secret.

## Control of Hazardous Energy (Lockout/Tagout) Standard

Approximately 3 million workers service equipment and face the greatest risk of injury if a lockout/tagout program is not properly implemented. Lockout/tagout (LOTO) refers to specific practices and procedures to safeguard employees from the unexpected energization or startup of machinery and equipment, or the release of hazardous energy during service or maintenance activities.

Compliance with the lockout/tagout standard (29 CFR 1910.147) prevents an estimated 120 fatalities and 50,000 injuries each year. Workers injured on the job from exposure to hazardous energy lose an average of 24 workdays for recuperation. In a study conducted by the United Auto Workers (UAW), 20% of the fatalities (83 of 414) that occurred among their members between 1973 and 1995 were attributed to inadequate hazardous energy control procedures, specifically lockout/tagout procedures. Based on these statistics, it is clear that a supervisor should pay close attention to ensuring that an effective LOTO program is implemented throughout his or her designated work areas.

**Lockout/Tagout.** Workers performing service or maintenance on machinery and equipment may be exposed to injuries from the unexpected energization, startup of the machinery or equipment, or release of stored energy in the equipment.

The LOTO standard requires the adoption and implementation of practices and procedures to shut down equipment, isolate it from its energy source(s), and prevent the release of potentially hazardous energy while maintenance and servicing activities are being performed. As long as minimum performance requirements are met, employers have the flexibility to develop LOTO programs that are suitable for their respective facilities.

**Program Elements.** A LOTO program should cover the following elements: energy control procedures, employee training, and periodic inspections to ensure that before service and maintenance is performed, machines and equipment that could unexpectedly start up, become energized, or release stored energy are isolated from their energy source(s) and rendered safe.

This program should cover servicing and/or maintenance on machines or equipment and who may be exposed to the unexpected energization, startup, or release of hazardous energy. "Unexpected" covers situations in which the servicing and/or maintenance is performed during ongoing normal production operations if an employee is required to remove or bypass machine guards or other safety devices. It also covers situations in which an employee is required to place any part of his or her body into a point of operation or into an area on a machine or piece of equipment where work is performed, or into the danger zone associated with the machine's operation.

Activities and operations covered by this standard include any servicing and/or maintenance of machines or equipment when the source of energy to the machines or equipment is electrical, mechanical, hydraulic, pneumatic, chemical, thermal, or other energy.

This also includes activities like constructing, installing, setting up, adjusting, inspecting, modifying, maintaining and/or servicing machines or equipment—including lubrication, cleaning or unjamming of machines or equipment—and making adjustments or tool changes where employees could be exposed to the unexpected energization or startup of the equipment or release of hazardous energy.

**Exceptions.** This regulation does not cover work on cord- and plug-connected electrical equipment if the equipment is unplugged from the energy source and the authorized employee has exclusive control of the plug. Another exception to this standard is when continuity of service is essential, shutdown of the system is impractical, documented procedures are followed, and employees are effectively protected by special equipment.

**Lockout.** Lockout is placement of a lockout device on an energy-isolating device in accordance with an established procedure, ensuring that the energy-isolating device and the equipment being controlled cannot be operated until the lockout device is removed. This would include the use of any device that uses positive means, such as a lock, blank flanges, and bolted slip blinds, to hold an energy-isolating device in a safe position, thereby preventing the energizing of machinery or equipment.

**Tagout.** Tagout is used where lockout is not feasible and placement of a tagout device on an energy-isolating device is used in accordance with an established procedure. Tagout indicates that the energy-isolating device and the equipment being controlled may not be operated until the tagout device is removed. This would include the use of any prominent warning device, such as a tag and a means of attachment, that can be securely fastened to an energy-isolating device indicating that the machine or equipment may not be operated until the tagout device is removed.

**Required Procedures.** Employers must develop, document, and use specific procedures to control potentially hazardous energy when employees are servicing equipment or machinery. The procedures must outline the scope, purpose, authorization, rules, and techniques that the employer will use to control hazardous energy.

At a minimum, the procedures must include:

- a specific statement of the intended use of the procedure
- specific procedural steps for shutting down, isolating, blocking, and securing machines or equipment to control hazardous energy
- specific procedural steps for the placement, removal, and transfer of lockout devices or tagout devices, and a description of who has responsibility for them
- Specific requirements for testing a machine or piece of equipment to determine and verify the effectiveness of lockout devices, tagout devices, and other energy control measures.

Periodic inspections should be performed at least annually to identify any deficiencies or deviations and correct them. Where lockout is used, the inspector must review each authorized employee's responsibilities under the procedure with that employee (group meetings are acceptable). Where tagout is used, the inspector must review both the authorized and affected employee's responsibilities with those employees for the energy control procedure being inspected, and any additional training responsibilities. The inspection report should include the following information:

- identity of the machine on which the procedure was used
- date of inspection
- identity of the employees included in the inspection
- identity of the inspector who performed the inspection

**Removing Lockout or Tagout Devices.** Before lockout or tagout devices are removed and energy is restored, the work area must be inspected to ensure that nonessential items (e.g., tools, spare parts) have been removed and that all of the machine or equipment components are operationally intact.

The work area must be checked to ensure that all employees have been safely positioned or have cleared the area. In addition, all affected employees must be notified that the lockout or tagout devices have been removed before the equipment is energized.

Each lockout or tagout device must be removed from the energy-isolating device by the employee who applied the device. When the authorized employee who applied the lockout or tagout device is not available to remove it, that device may be removed under the direction of the employer, provided that specific procedures and training for such removal have been developed, documented, and incorporated into the employer's energy control program.

Employers must ensure the continuity of employee protection by providing for the orderly transfer of lockout or tagout device protection between off-going and incoming employees. This will help to minimize exposure to hazards from the unexpected energization or startup of the machine or equipment or the release of stored energy.

**Training for Lockout/Tagout.** The standard requires different levels of training for three categories of employees:

- Authorized employees must receive training on the recognition of applicable hazardous energy sources, the type and magnitude of the energy available in the workplace, and the methods and means necessary for energy isolation and control.

- Affected employees must receive training on the purpose and use of the energy control procedure.

- Other employees (those whose work activities are or may be in an area where energy control procedures may be used) must be instructed about the procedure and the prohibition relating to attempts to restart or reenergize machines or equipment that are locked out or tagged out. Employers and supervisors must train employees in the limitations of tags. If changes occur in job procedures or problems are identified, then employees should be re-trained. Training should be documented with a certificate that includes each employee's name and the dates of training and/or retraining. See Figure 12–1 for a sample LOTO checklist.

## Respiratory Protection Standard

An estimated 5 million workers are required to wear respirators in 1.3 million workplaces throughout the United States. Respirators protect workers against insufficient oxygen environments, harmful dusts, fogs, smokes, mists, gases, vapors, and sprays. These hazards may cause cancer, lung impairment, other diseases, or death.

Whenever possible, engineering controls should be used to control employee exposure to airborne hazards. When this is not feasible, a respiratory protection program should be instituted that complies with OSHA's respirator standard, 29 CFR 1910.134. This standard requires the use of respirators to protect employees from breathing contaminated and/or oxygen-deficient air when effective engineering controls are not feasible, or while they are being instituted. Several other OSHA regulations also require the use of respirators.

Respirators shall be selected on the basis of hazards to which the worker is exposed (i.e., particulates, vapors, oxygen-deficiency, or combination). In addition, only respirators certified by the National Institute for Occupational Safety and Health (NIOSH) shall be used. Certified respirators will be marked with "NIOSH," the manufacturer's name and part number, and an abbreviation to indicate the cartridge or filter type (see Figures 12–2a and 12–2b).

When respirators are required in the workplace, a written respiratory protection program with worksite-specific procedures and elements for required respirator use is also required. The provisions of the program shall include

## LOTO Checklist

❏ Is all machinery or equipment capable of movement required to be de-energized or disengaged and blocked or locked out during cleaning, servicing, adjusting, or setting up operations?

❏ If the power disconnect for equipment does not also disconnect the electrical control circuit, are the appropriate electrical enclosures identified and is a means provided to ensure that the control circuit can also be disconnected and locked out?

❏ Is the locking out of control circuits, instead of locking out main power disconnects, prohibited?

❏ Are all equipment control valve handles provided with a means for locking out?

❏ Does the lockout procedure require that stored energy (mechanical, hydraulic, air, etc.) be released or blocked before equipment is locked out for repairs?

❏ Are appropriate employees provided with individually keyed personal safety locks?

❏ Are employees required to keep personal control of their key(s) while they have safety locks in use?

❏ Is it required that only the employee exposed to the hazard can place or remove the safety lock?

❏ Is it required that employees check the safety of the lockout by attempting a startup after making sure no one is exposed?

❏ Are employees instructed to always push the control circuit stop button prior to reenergizing the main power switch?

❏ Is there a means provided to identify any or all employees who are working on locked-out equipment by their locks or accompanying tags?

❏ Are a sufficient number of incident prevention signs or tags and safety padlocks provided for any reasonably forseeable repair emergency?

❏ When machine operations, configuration, or size require an operator to leave the control station and part of the machine could move if accidentally activated, is the part required to be separately locked out or blocked?

❏ If equipment or lines cannot be shut down, locked out, and tagged, is a safe job procedure established and rigidly followed?

**Figure 12–1.** This checklist can be used in conjunction with training employees on lockout/tagout procedures.

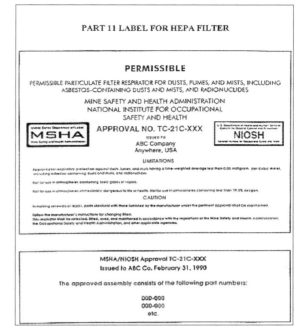

**Figure 12–2a and b.** Respirators should be certified by NIOSH and carry labels marked with NIOSH, the manufacturer's name and part number, as well as some indication of the cartridge or filter type.

procedures for selection, medical evaluation, fit testing, training, use and care of respirators.

**Fit Testing.** Proper respirator size is determined through a fit test. A fit test is a method used to select the right-size respirator for the user. Employees using negative or positive pressure tight-fitting facepiece respirators must pass an appropriate fit test using the procedures detailed in OSHA's respirator standard.

Fit testing of all negative or positive pressure tight-fitting facepiece respirators is required prior to initial use, whenever a different respirator facepiece is used, and at least annually thereafter. An additional fit test is required whenever there are changes in the user's physical condition that could affect respirator fit (e.g., facial scarring, dental changes, cosmetic surgery, or an obvious change in body weight). The worker must be fit tested with the same make, model, style, and size of respirator that will be used.

Workers using tight-fitting facepiece respirators are required to perform a user seal check each time they put on the respirator. A *user seal check* is a method to verify that the user has correctly put on the respirator and adjusted it to fit properly.

Manufacturers make several different sizes of respirators. Respirators may also vary in size from one manufacturer to another. Users may be able to get a better fit by trying a respirator made by another manufacturer. In some cases, the use of powered air-purifying respirators may be appropriate. Employers must help employees find a suitable respirator.

Tight-fitting facepiece respirators must not be worn by employees who have facial hair that comes between the sealing surface of the facepiece and the face or that interferes with valve function. Respirators that do not rely on a tight face seal, such as hoods or helmets, may be used by bearded individuals.

**Maintenance, Storage, and Replacement.** The employer must provide for the cleaning and disinfecting, storage, inspection, and repair of respirators used by employees, according to the procedures.

Some respirators are disposable and cannot be disinfected, and are therefore assigned to only one person. Disposable respirators must be discarded if they are soiled, physically damaged, or reach the end of their service life. Replaceable filter respirators may be shared, but must be thoroughly cleaned and disinfected after each use and before being worn by a different person.

Respirators with replaceable filters are reusable, and a respirator classified as disposable may be reused by the same worker as long as it functions properly. All filters must be replaced whenever they are damaged, soiled, or causing noticeably increased breathing resistance (e.g., causing discomfort to the wearer). Before each use, the outside of the filter material should be inspected. If the filter material is physically damaged or soiled, the filter should be changed (in the case of respirators with replaceable filters) or the respirator discarded (in the case of disposable respirators). Always follow the respirator filter manufacturer's service-time-limit recommendations.

Standard operating procedures for storing, reusing, and disposing of respirators that have been designated as disposable and for disposing of replaceable filter elements have to be developed.

Respirators must be stored to protect them from damage, contamination, dust, sunlight, extreme temperatures, excessive moisture, and damaging chemicals. They must also be packed or stored to prevent the facepiece and exhalation valve from deformation or damage. A good method is to place them in individual storage bins. Keep in mind that respirator facepieces can become distorted and the straps will lose their elasticity if hung on a peg for a long time. Check for these problems before each use.

Storing the respirator in a plastic sealable bag after use is not considered a good practice. The respirator may be damp after use and sealing prevents drying and encourages microbial growth. If plastic bags are used, respirators must be allowed to dry before storage.

Employers are not required to include in a written respiratory program those employees whose only use of respirators involves the voluntary use of filtering facepieces (dust masks).

## Machine Guarding Standard

Moving machine parts have the potential to cause severe workplace injuries, such as crushed fingers or hands, amputations, burns, or blindness. Safeguards should be used for any machine part, function, or process that presents a danger that may result in an injury. When the operation of a machine or accidental contact could injure the operator or others in the vicinity, the hazards must be eliminated or controlled.

Machine guarding hazards are addressed in specific standards for general industry, marine terminals, longshoring, and the construction and agriculture industries. The machine guarding standard, 29 CFR 1910.212, provides general guarding requirements for machines and provides a blueprint for machine safety.

All machines consist of three fundamental areas: the point of operation, the power transmission device, and the operating controls. Despite the fact that all machines have the same basic components, safeguarding needs often differ due to varying physical characteristics and operator involvement.

The point of operation is where work is performed on the material, such as cutting, shaping, boring, or forming of stock. When a machine is in use, the point of operation is a place where an employee is potentially exposed to injury and needs to be guarded. The guarding device needs to conform with any appropriate standard or, in the absence of applicable specific standards, be designed and constructed so as to prevent the operator from having any body parts in the danger zone during the operating cycle.

The following is a list of machines that usually require point of operation guarding:

- guillotine cutters
- shears
- alligator shears
- power presses

- milling machines
- power saws
- jointers
- portable power tools
- forming rolls and calenders
- press brake
- barrels, containers, and drums—revolving barrels, containers, and drums must be guarded by an enclosure interlocked with the drive mechanism so the barrel, gun, or container cannot revolve unless the guard enclosure is in place.
- exposure of blades—when the periphery of the blades of a fan is less than 7 ft (2.1 m) above the floor or working level, the blades must be guarded. the guard must not have openings larger than $\frac{1}{2}$ in (1.8 cm).
- anchoring fixed machinery—a machine designed for a fixed location must be securely anchored to prevent walking or moving.

The power transmission apparatus is all of the components of the mechanical system that transmit energy to the part of the machine performing the work. These components include flywheels, pulleys, belts, connecting rods, couplings, cams, spindles, chains, cranks, and gears. Operating controls refers to all of the parts of the machine that move while the machine is working. These can include reciprocating, rotating, and transverse moving parts, as well as feed mechanisms and auxiliary parts of the machine.

Operating controls are the input devices that the operator uses to activate or control the mechanical system. They need to be arranged so that a worker is protected from danger while operating the machinery. Often additional safety devices are configured as part of the operating control process in order to enhance the safety of an operator.

One or more methods of machine guarding needs to be provided to protect the operator and other employees in the machine area from hazards such as those created by point of operation, ingoing nip points, rotating parts, flying chips, and sparks. Examples of guarding methods include barrier guards, two-hand tripping devices, and electronic safety devices, to name a few.

**Machine Guarding.** The purpose of machine guarding is to protect the machine operator and other employees in the work area from potential hazards. Guards shall be affixed to the machine where possible and secured elsewhere if, for any reason, attachment to the machine is not possible.

Safeguards must meet minimum general requirements: The safeguard must prevent hands, arms, and any other part of a worker's body from making contact with dangerous moving parts. Workers should not be able to easily remove or tamper with the safeguard. The safeguard should ensure that no objects can fall into moving parts. If possible, one should be able to lubricate the machine without removing the safeguards.

A fixed guard is a permanent part of the machine and is usually preferable to all other types because of its relative simplicity. It is not dependent on moving parts to function. It may be constructed of sheet metal, screen, wire cloth, bars, plastic, or any other material that is substantial enough to withstand whatever impact it may receive and to endure prolonged use.

The safeguard must create no new hazards and create no interference for the worker. The guard should not be an incident hazard in itself.

**Servicing or Maintenance.** The employer must establish an energy control program consisting of energy control procedures, employee training, and periodic inspections to ensure that before any employee performs any servicing or maintenance on a machine or equipment, the machine or equipment is isolated from the energy source and rendered inoperative. The maintenance and repair crew must never fail to replace the guards before the job is considered finished and the machine released from lockout. In order to prevent hazards while servicing machines, each machine or piece of equipment should be safeguarded during the conduct of servicing or maintenance by:

- notifying all affected employees (usually machine or equipment operators or users) that the machine or equipment must be shut down to perform some maintenance or servicing
- stopping the machine
- isolating the machine or piece of equipment from its energy source
- locking out or tagging out the energy source
- relieving any stored or residual energy
- verifying that the machine or equipment is isolated from the energy source

When the servicing or maintenance is completed, there are specific steps that must be taken to return the machine or piece of equipment to service. These steps include:

- inspection of the machine or equipment to ensure that all guards and other safety devices are in place and functional

- checking the area to ensure that energization and start-up of the machine or equipment will not endanger employees
- removal of the lockout devices
- reenergization of the machine or equipment
- notification of affected employees that the machine or equipment may be returned to service

**Training.** A worker needs to knows how and why to use a safeguarding system. The training should be specific and detailed. Operator training should involve instruction or hands-on training in the following:
- a description and identification of the hazards associated with particular machines
- the safeguards themselves, how they provide protection, and the hazards for which they are intended
- how to use the safeguards and why
- how and under what circumstances safeguards can be removed, and by whom (in most cases, repair or maintenance personnel only)
- when a lockout/tagout program is required
- what to do (e.g., contact the supervisor) if a safeguard is damaged, missing, or unable to provide adequate protection

This type of safety training is necessary for new operators, and maintenance or setup personnel, when any new or altered safeguards are put in service, or when workers are assigned to a new machine or operation. The machine safeguarding checklist in Figure 12–3 can be used to help determine the safeguarding needs of a wokplace.

## Electrical Safety Standards

Electricity has long been recognized as a serious workplace hazard. The Department of Labor (DOL) considers electrical safety a high priority. The Bureau of Labor Statistics indicates that from 1992 to 2002, deaths due to electrical contact averaged nearly 300 per year and lost-time injuries numbered more than 4,000 per year. OSHA's electrical standards are designed to protect employees exposed to dangers such as electric shock, electrocution, fires, and explosions. Electrical hazards are addressed in specific standards for the general industry, shipyard employment, and marine terminals.

OSHA has a brought additional requirements to the workplace to govern electrical work by revising 29 CFR 1910 Subpart S, Electrical Standard, which went into effect on August 15, 2007.

The final rule of this revision is partially based on Part I of "NFPA 70E Standard for Electrical Safety in the Workplace 2000 Edition" and "NFPA 70 National Electrical Code 2002 Edition." This final rule revises OSHA's existing standard for electrical installations, and focuses on safety in the design and installation of electric equipment in the workplace. It applies, as the previous standard did, to employers in general industry, shipyard employment, longshoring, and marine terminals. It is important for supervisors who work with electricity to become familiar with relevant parts of "NFPA 70E Standard for Electrical Safety in the Workplace 2000 Edition" and "NFPA 70 National Electrical Code 2002 Edition."

It must be considered that "NFPA 70 National Electric Code" and "NFPA 70E Standard for Electrical Safety in the Workplace" are revised and updated periodically; and these updates become the standard of practice to which OSHA will hold an employer for compliance purposes. For a listing of the codes and requirements for electrical installation, see Appendix 12–A at the end of this chapter.

**Frequently Cited Standards Related to Electricity.** In recent years, the following standards have resulted in either frequent or costly citations:

- 29 CFR 1910.303, Electrical systems design, general requirements, general industry
- 29 CFR 1910.305, Electrical, wiring methods, components and equipment, general industry
- 29 CFR 1910.307, Electrical, hazardous (classified) locations

In addition, OSHA has extended the ground-fault protection requirement to temporary receptacles used in construction activities performed in general industry. Ground-fault circuit-interrupter (GFCI) protection is required for all receptacle outlets on temporary wiring installations used during maintenance, remodeling, repair of buildings, structures, or equipment, or during other construction activities. A GFCI device will protect a worker by opening a circuit if current going into and coming out of an electric circuit is not equal. This will prevent another current pathway, such as flow through a worker's body, causing injury to the worker: (For more on GFCI devices, see Chapter 16, Electrical Safety.)

| Requirements for All Safeguards | YES | NO |
|---|:---:|:---:|
| 1. Do the safeguards provided meet the minimum OSHA requirements? | ❏ | ❏ |
| 2. Do the safeguards prevent workers' hands, arms, and other body parts from making contact with dangerous moving parts? | ❏ | ❏ |
| 3. Are the safeguards firmly secured and not easily removable? | ❏ | ❏ |
| 4. Do the safeguards ensure that no object will fall into the moving parts? | ❏ | ❏ |
| 5. Do the safeguards permit safe, comfortable, and relatively easy operation of the machine? | ❏ | ❏ |
| 6. Can the machine be oiled without removing the safeguard? | ❏ | ❏ |
| 7. Is there a system for shutting down the machinery before safeguards are removed? | ❏ | ❏ |
| 8. Can the existing safeguards be improved? | ❏ | ❏ |

## Mechanical Hazards

The point of operation:

| | YES | NO |
|---|:---:|:---:|
| 1. Is there a point-of-operation safeguard provided for the machine? | ❏ | ❏ |
| 2. Does it keep the operator's hands, fingers, body out of the danger area? | ❏ | ❏ |
| 3. Is there evidence that the safeguards have been tampered with or removed? | ❏ | ❏ |
| 4. Could you suggest a more practical, effective safeguard? | ❏ | ❏ |
| 5. Could changes be made on the machine to eliminate the point-of-operation hazard entirely? | ❏ | ❏ |

Power transmission apparatus:

| | YES | NO |
|---|:---:|:---:|
| 1. Are there any unguarded gears, sprockets, pulleys, or flywheels on the apparatus? | ❏ | ❏ |
| 2. Are there any exposed belts or chain drives? | ❏ | ❏ |
| 3. Are there any exposed set screws, key ways, collars, etc.? | ❏ | ❏ |
| 4. Are starting and stopping controls within easy reach of the operator? | ❏ | ❏ |
| 5. If there is more than one operator, are separate controls provided? | ❏ | ❏ |

Other moving parts:

| | YES | NO |
|---|:---:|:---:|
| 1. Are safeguards provided for all hazardous moving parts of the machine, including auxiliary parts? | ❏ | ❏ |

## Nonmechanical Hazards

| | YES | NO |
|---|:---:|:---:|
| 1. Have appropriate measures been taken to safeguard workers against noise hazards? | ❏ | ❏ |
| 2. Have special guards, enclosures, or personal protective equipment been provided, where necessary, to protect workers from exposure to harmful substances used in machine operation? | ❏ | ❏ |

## Electric Hazards

| | YES | NO |
|---|:---:|:---:|
| 1. Is the machine installed in accordance with National Fire Protection Association and National Electrical requirements? | ❏ | ❏ |
| 2. Are there loose conduit fittings? | ❏ | ❏ |
| 3. Is the machine properly grounded? | ❏ | ❏ |

| | | |
|---|---|---|
| 4. Is the power supply correctly fused and protected? | ❏ | ❏ |
| 5. Do workers occasionally receive minor shocks while operating any of the machines? | ❏ | ❏ |
| **Training** | | |
| 1. Do operators and maintenance workers have the necessary training in how to use the safeguards and why? | ❏ | ❏ |
| 2. Have operators and maintenance workers been trained in where the safeguards are located, how they provide protection, and what hazards they protect against? | ❏ | ❏ |
| 3. Have operators and maintenance workers been trained in how and under what circumstances guards can be removed? | ❏ | ❏ |
| 4. Have workers been trained in the procedures to follow if they notice guards that are damaged, missing, or inadequate? | ❏ | ❏ |
| **Protective Equipment and Proper Clothing** | | |
| 1. Is protective equipment required? | ❏ | ❏ |
| 2. If protective equipment is required, is it appropriate for the job, in good condition, kept clean and sanitary, and stored carefully when not in use? | ❏ | ❏ |
| 3. Is the operator dressed safely for the job (i.e., no loose-fitting clothing or jewelry)? | ❏ | ❏ |
| **Machinery Maintenance and Repair** | | |
| 1. Have maintenance workers received up-to-date instruction on the machines they service? | ❏ | ❏ |
| 2. Do maintenance workers lock out the machine from its power sources before beginning repairs? | ❏ | ❏ |
| 3. Where several maintenance persons work on the same machine, are multiple lockout devices used? | ❏ | ❏ |
| 4. Do maintenance persons use appropriate and safe equipment in their repair work? | ❏ | ❏ |
| 5. Is the maintenance equipment itself properly guarded? | ❏ | ❏ |
| 6. Are maintenance and servicing workers trained in the requirements of 29 CFR 1910.147, lockout/tagout hazard, and do the procedures for lockout/tagout exist *before* they attempt their tasks? | ❏ | ❏ |

**Figure 12–3.** Answers to the questions in this checklist should help determine the machine safeguarding needs of a workplace.

To summarize, a supervisor should understand that electrical safety is a complex issue:

- There are new requirements based on a revised standard.
- It is important to ensure that workers receive specialized education and training when working with electrical systems.
- Consensus standards can be used to help in the selection of the best method to achieve compliance with OSHA regulations.
- Some OSHA state plans are more restrictive than federal OSHA require-

ments and as such may have adopted or incorporated consensus standards. But this is on a state-by-state basis and should be evaluated for each employer location.

- In most cases, electrical shock will be a recognized hazard that could cause or likely cause death or serious physical harm to employees. This means that electrical shock hazards not covered by existing regulations can be cited by use of the General Duty Clause described earlier.

### Powered Industrial Trucks Standard

OSHA requires that all forklifts be examined at least daily before being placed in service. Forklifts used on a round-the-clock basis must be examined after each shift.

The operator should conduct a pre-start visual check with the key off and then perform an operational check with the engine running. The forklift should not be placed in service if the examinations show that the vehicle may not be safe to operate. If a vehicle is in need of repair, defective, or in any way unsafe, it should not be driven and should be taken out of service immediately. Any problems should be recorded on the appropriate documents and reported to a supervisor.

**Pre-operation inspection.** Before a vehicle is started, a pre-operation (or pre-start) inspection that checks a variety of items should occur, including:
- fluid levels—oil, water, and hydraulic fluid
- visual leaks, cracks, or any other visible defect including hydraulic hoses and mast chains *(NOTE: Operators should not place their hands inside the mast. Use a stick or other device to check chain tension.)*
- tire condition and pressure including cuts and gouges
- condition of the forks and the top clip retaining pin and heel
- load backrest extension
- finger guards
- safety decals and nameplates—all warning decals and plates are in place and legible, and that information on the nameplate matches the model, serial numbers, and attachments
- operator manual on truck and legible
- operator compartment—check for grease and debris
- all safety devices are working properly, including the seat belt

In addition to this general inspection, additional items should be checked depending on the forklift type (electric, internal combustion, or liquid propane).

### Electric forklifts

- cables and connectors for frayed or exposed wires
- battery restraints
- electrolyte levels
- hood latch

*NOTE: Always use personal protective equipment, such as a face shield and goggles, rubber apron, and rubber gloves, when checking electrolyte levels.*

### Internal combustion forklifts

- engine oil
- brake reservoir
- engine coolant
- air filter
- belts and hoses
- radiator
- hood latch

### Liquid propane forklifts

- properly mounted tank
- pressure relief valve pointing up
- hose and connectors
- tank restraint brackets
- tank for dents and cracks
- tank fits within profile of truck
- leaks (use a soapy solution)

*NOTE: Always use personal protective equipment such as a face shield, long sleeves, and gauntlet gloves when checking liquid propane tanks and fittings.*

**Operational Inspection.** After completing the pre-operation inspection, operators should conduct an operational inspection with the engine running. This inspection includes:

- accelerator linkage

- inch accelerator (if equipped)
- brakes
- steering
- drive control (forward and reverse)
- tilt control (forward and back)
- hoist and lowering control
- attachment control
- horn
- lights
- back-up alarm (if equipped)
- hour meter

*NOTE: Unusual noises or vibrations should be reported immediately.*

The OSHA powered industrial truck standard lists a number of conditions under which a forklift must be removed from service. If the operator notes these conditions while driving, the driver must stop, park the vehicle and get assistance.

- Any powered industrial truck not in safe operating condition shall be removed from service. All repairs shall be made by authorized personnel.
- Defects when found must be immediately reported and corrected.
- Any vehicle that emits hazardous sparks or flames from the exhaust system shall immediately be removed from service and not returned to service until the cause for the emission of such sparks and flames has been eliminated.
- When the temperature of any part of any truck is found to be in excess of its normal operating temperature, thus creating a hazardous condition, the vehicle shall be removed from service and not returned to service until the cause for such overheating has been eliminated.
- No truck shall be operated with a leak in the fuel system until the leak has been corrected.

Scheduled maintenance is critically important to the safe operation of a vehicle. Never operate a forklift requiring maintenance, and always ensure repair problems are reported to a responsible party.

Never operate a vehicle that requires maintenance or is in any way thought to be unsafe.

- Keep industrial trucks in clean condition, free of lint, excess oil, and grease.
- Use noncombustible agents for cleaning trucks.

- ○ High flashpoint (at or above 100 F) solvents may be used. Do not use low flashpoint (below 100 F) solvents.
- ○ Take recommended precautions regarding toxicity, ventilation, and fire hazards.

**Training.** The standard requires employers to develop and implement a training program based on the general principles of safe truck operation, the types of vehicle(s) being used in the workplace, the hazards of the workplace created by the use of the vehicle(s), and the general safety requirements of the OSHA standard. Trained operators must know how to do the job properly and do it safely as demonstrated by workplace evaluation. Formal (lecture, video, etc.) and practical (demonstration and practical exercises) training must be provided.

Employers must also certify that each operator has received the training and evaluate each operator at least once every three years. Prior to operating the truck in the workplace, the supervisor must evaluate the operator's performance and determine whether the operator is competent to operate a powered industrial truck safely.

Refresher training is needed whenever an operator demonstrates a deficiency in the safe operation of the truck. Training shall consist of a combination of formal instruction (e.g., lecture, discussion, interactive computer learning, videotape, written material), practical training (demonstrations performed by the trainer and practical exercises performed by the trainee), and evaluation of the operator's performance in the workplace.

**Training Program Content.** Powered industrial truck operators shall receive initial training in applicable truck-related topics (i.e., operating instructions, warnings, controls, instrumentations, etc.) and workplace-related topics (i.e., load manipulation, pedestrian traffic, surface conditions where the truck will be operated, etc.).

## Walking-Working Surfaces Standard

Slips, trips, and falls constitute the majority of general industry accidents. They cause 15 % of all accidental deaths, and are second only to motor vehicles as a cause of fatalities. The OSHA standards for walking-working surfaces apply to all permanent places of employment, except where only domestic, mining, or agricultural work is performed.

Walking-working surfaces are addressed in specific standards for the general industry, shipyard employment, marine terminals, longshoring, and the construction industry. For general industry, the following sections apply:

1910 Subpart D—Walking-Working Surfaces

1910.21—Definitions

1910.22—General requirements

1910.23—Guarding floor and wall openings and holes

1910.24—Fixed industrial stairs

1910.25—Portable wood ladders

1910.26—Portable metal ladders

1910.27—Fixed ladders

1910.28—Safety requirements for scaffolding

1910.29—Manually propelled mobile ladder stands and scaffolds (towers)

**Housekeeping.** Housekeeping is a very important aspect of OSHA's walking-working surfaces standard. OSHA requires that all places of employment, passageways, storerooms, and service rooms shall be kept clean and orderly and in a sanitary and dry condition. Where wet processes are used, drainage shall be maintained, and false floors, platforms, mats, or other dry standing places should be provided where practicable. To facilitate cleaning, every floor, working place, and passageway shall be kept free from protruding nails, splinters, holes, or loose boards.

**Aisles and Passageways.** Where mechanical handling equipment is used, sufficient safe clearances shall be allowed for aisles, at loading docks, through doorways, and wherever turns or passage must be made. Aisles and passageways shall be kept clear and in good repairs, with no obstruction across or in aisles that could create a hazard. Permanent aisles and passageways shall be appropriately marked.

**Covers and Guardrails.** Covers and/or guardrails shall be provided to protect workers from the hazards of open pits, tanks, vats, ditches, holes, and such.

**Floor Loading Protection.** In every building or other structure, or part thereof, used for mercantile, business, industrial, or storage purposes, the loads approved by the building official shall be marked on plates of approved design, which shall be supplied and securely affixed by the owner of the building or his authorized agent, in a conspicuous place and in each space to which they relate. Such plates shall not be removed or defaced, but if lost, removed, or defaced, shall be replaced by the owner or his agent.

It is unlawful to place, cause, or permit to be placed—on any floor or roof of a building or other structure—a load greater than that for which such floor or roof is approved by the building official.

**Protecting Openings.** OSHA's 29 CFR 1910.23 provides extensive guidelines for many situations in which there is an opening that could result in a fall if the opening were to be left unprotected. Included in their extensive list are stairway floor openings, ladderway floor openings or platforms, hatchway and chute floor openings, skylight floor openings and holes, pits and trapdoor floor openings, infrequently used, wall openings where there is a drop of more than four feet, manhole floor openings, every temporary floor opening, every floor hole into which persons can accidentally walk, places where doors or gates open directly on a stairway and those situation where there is a hazard of materials falling through a wall hole. It would be a good idea for each supervisor to be familiar with both the unsafe conditions described by OSHA and the remedies for addressing those same conditions.

## Process Safety Management of Highly Hazardous Chemicals Standard

OSHA Fact Sheet 93-45 outlines and clearly explains the general requirements for complying with the process safety management (PSM) of the highly hazardous chemicals (HHCs) standard, 29 CFR 1910.119. This standard is intended to prevent or minimize the consequences of a catastrophic release of toxic, reactive, flammable or explosive HHCs from a process. A process is any activity or combination of activities, including any use, storage, manufacturing, handling, or the on-site movement of HHCs. A process includes any group of vessels that are interconnected as well as separate vessels that are located such that a HHC could be involved in a potential release.

**Application.** The standard applies to a process that contains a threshold quantity or greater amount of a toxic or reactive HHC as specified in Appendix 12–B at the end of this chapter. Also, it applies to 10,000 pounds or greater amounts of flammable liquids and gases and to the process activity of manufacturing explosives and pyrotechnics.

**Exceptions.** The standard does not apply to retail facilities, normally unoccupied remote facilities, and oil or gas well drilling or servicing activities. Hydrocarbon fuels used solely for workplace consumption as a fuel are not covered, if such fuels are not part of a process containing another HHC covered by the standard. Atmospheric tank storage and associated transfer of flammable liquids, which are kept below their normal boiling point without benefit of chilling or refrigeration, are not covered by the PSM standard unless the atmospheric tank is connected to a process or is in close proximity to a covered process such that an incident in a covered process could involve the atmospheric tank.

**Process Safety Information.** The standard constitutes a compilation of written process safety information (PSI), including hazard information on HHCs, technology information, and equipment information on covered processes.

**Employee Involvement.** The standard requires developing a written plan of action regarding employee participation; consulting with employees and their representatives on the conduct and development of process hazard analyses (PHAs) and on the development of other elements of process safety management required under the rule; and providing to employees and their representatives access to process hazard analyses and to all other information required to be developed under the rule. Employees include worksite and contractor employees.

**Process Hazard Analysis.** The standard specifies that PHAs must be conducted as soon as possible for each covered process using compiled PSI in an order based on a set of required considerations.

**Operating Procedures.** Operating procedures must be in writing and provide clear instructions for safely conducting activities involving covered process consistent with PSI; must include steps for each operating phase, operating limits, safety and health considerations and safety systems and their functions; be readily accessible to employees who work on or maintain a covered process, and be reviewed as often as necessary to ensure they reflect current operating practice; and must implement safe work practices to provide for special circumstances such as lockout/tagout and confined space entry.

**Training.** Employees operating a covered process must be trained in the overview of the process and in the operating procedures addressed previously. This training must emphasize specific safety and health hazards, emergency operations, and safe work practices. Documented refresher training is required at least every three years.

**Contractors.** The standard identifies the responsibilities of worksite employer and contract employers with respect to contract employees involved in maintenance, repair, turnaround, major renovation or specialty work, on or near covered processes. Contract employers are required to train their employees to safely perform their jobs. They must document that employees received and understood training and ensure that contract employees know about potential process hazards and the worksite employer's emergency action plan. Assurance must be made that employees follow safety rules of the facility and advise the worksite employer of hazards the contract work poses.

**Pre-start up Safety Review.** The standard mandates a safety review for new facilities and significantly modified worksites to confirm that the construction and equipment of a process are in accordance with design specifications; to ensure that adequate safety, operating, maintenance, and emergency procedures are in place; and to ensure that process operator training has been completed. Also, for new facilities, the PHA must be performed and recommendations resolved and implemented before start up. Modified facilities must meet management of change requirement.

**Mechanical Integrity.** The standard requires that the on-site employer establish and implement written procedures for the ongoing integrity of process equipment, particularly those components that contain and control a covered process.

**Hot Work.** Hot work permits must be issued for hot work operations conducted on or near a covered process.

**Management of Change.** The worksite employer must establish and implement written procedures to manage changes, except "replacements in kind," to facilities that affect a covered process. The standard requires that the worksite employer and contract employers inform and train their employees on the changes prior to start-up. Process safety information and operating procedures must be updated as necessary.

**Incident Investigation.** The standard requires employers to investigate as soon as possible (but no later than 48 hours after) incidents that did result or could reasonably have resulted in catastrophic releases of covered chemicals. The standard calls for an investigation team, including at least one person knowledgeable (a contract employee when the incident involved contract work) in the process and others with knowledge and experience to investigate and analyze the incident, and to develop a written report on the incident. Reports must be retained for five years.

**Emergency Planning and Response.** The standard requires employers to develop and implement an emergency action plan. The emergency action plan must include procedures for handling small releases.

**Compliance Audits.** This requires employers to certify that they have evaluated compliance with process safety requirements at least every three years. Prompt response to audit findings and documentation that deficiencies are corrected is required. Employers must retain the two most recent audit reports.

**Trade Secrets.** This sets requirements similar to trade secret provisions of the 1910.1200 hazard communication standard requiring information required by the PSM standard to be available to employees (and employees' representatives). Employers may enter into a confidentiality agreement with employees to prevent disclosure of trade secrets.

Although information from OSHA Fact Sheet No. 93-45 provides a good overview of the regulatory requirements for complying with the process safety management of highly hazardous chemicals standard, compliance can only be ascertained by carefully addressing each of the provisions in the standard.

## Confined Space Entry Standard

Because the confined space standard—29 CFR 1910.146—applies to all of general industry, a performance-oriented standard was developed rather than a specification standard. Currently, construction, marine terminal, shipyard employment, and agriculture are not subject to the OSHA general industry permit-required confined spaces regulation. However, employers in those industries should be aware that their workers are covered when they perform work that falls under the general industry category. For example, maintenance, repair, and refurbishing work are covered under general industry rules even though done by "construction" contractors.

It is an employer's obligation to evaluate a workplace to determine if any spaces are permit-required confined spaces. It must first be determined whether a space is a confined space. If it is a confined space, then it must be determined if it is a permit-required confined space. If it is a permit-required confined space, then the question is whether full-permit entry rules apply or less restrictive alternative entry rules apply.

A confined space is characterized by restricted means of entry or exit, size sufficient to contain a worker, and not specifically designed for worker occupancy.

A permit-required space is a confined space that has a potential hazard to health or life associated with it. Hazards may be the result of atmosphere or materials in the space or the result of the shape of the space.

In general, the permit-required confined spaces standard requires that the employer evaluate the workplace to determine if any spaces are permit-required confined spaces. If permit spaces are present and workers are ever authorized to enter such spaces, a comprehensive permit spaces program must be developed, which is an overall plan/policy for protecting employees from permit space hazards and for regulating employee entry into permit spaces.

Permit spaces must be identified by signs, and entry must be controlled and limited to authorized persons. An important element of the requirements is that entry be regulated by a written entry permit system, and that entry permits be recorded and issued for each entry into a permit space. The standard specifies strict procedures for evaluation and atmospheric testing of a space before and during entry by workers.

The standard requires that entry be monitored by an attendant outside the space and that provisions be made for rescue in the event of an emergency. The standard specifies training requirements and specific duties for authorized entrants, attendants, and supervisors. Rescue service provisions are required, and where feasible, rescue must be facilitated by a non-entry retrieval system, such as a harness and cable attached to a mechanical hoist. Many would-be rescuers have become trapped by the hazard requiring the rescue in the first place.

The OSHA permit-required confined spaces standard provides for alternative (less stringent than full-permit procedures) entry procedures in cases where the only hazard in a space is atmospheric and the hazard can be controlled by forced air. The alternative procedure is allowed only in cases where specified requirements for substantiation and notification are met.

Many hazards related to uncontrolled releases of energy may be located in confined spaces, and any program should ensure that measures are available to assess potential hazards from the following:

- high-pressure fluids and gases
- mechanical energy
- oxygen deficiency
- toxins
- flammable materials
- engulfment
- wet or slick surfaces
- electrical energy

OSHA requires the following elements be covered in a permit-required confined space entry program:
- identification and evaluation of confined spaces and permit-required confined spaces
- identification of hazards in those spaces
- procedures for controlling hazards
- written program
- permit system
- employee training
- record keeping
- rescue provisions

To be defined as a confined space means a space:
- is large enough and so configured that an employee can bodily enter and perform assigned work
- has limited or restricted means for entry or exit (e.g., tanks, vessels, silos, storage bins, hoppers, vaults, and pits are spaces that may have limited means of entry)
- is not designed for continuous employee occupancy

A permit-required confined space has to meet any of the following four requirements:
- contains or has a potential to contain a hazardous atmosphere
- contains a material that has the potential for engulfing an entrant
- has an internal configuration such that an entrant could be trapped or asphyxiated by inwardly converging walls or by a floor which slopes downward and tapers to a smaller cross-section
- contains any other recognized serious safety or health hazard

The next step is to evaluate potential hazards, which includes (but is not limited to) the following:

- oxygen deficiency, toxics, flammable/explosive
- engulfment, mechanical and electrical energy sources, release of materials, noise, wet or slick surfaces, falling objects, hot and cold temperatures

Oxygen deficiency, which refers to oxygen levels below 19.5%, is a major potential hazard in confined spaces and is described in Table 12–B. Confined space entry requires that specific procedures are developed covering both pre-entry and entry activities. Pre-entry activities consist of evaluating potential hazards, organizing entry, and completing the written permit.

### Table 12–B. Effects of Oxygen Deficiency

| % Oxygen | Effects |
|---|---|
| 16 to 12 % | Deep breathing, accelerated heartbeat, impaired attention, impaired thinking, impaired coordination |
| 14 to 10 % | Very faulty judgment, very poor coordination, rapid fatigue from exertion that may cause permanent heart damage, intermittent breathing |
| 10 % or Below | Nausea, vomiting, inability to perform vigorous movement or loss of all movement, unconsciousness followed by death |
| Less than 6 % | Spasmodic breathing, convulsive movements, death in minutes |

An entry process/system needs to include the following information and actions:

- work to be performed
- identity of individuals performing work
- time period
- identification and evaluation of potential hazards in the space
- complete written permit
- obtain equipment
- review permit with team members
- rescue procedures
- communication processes
- contact rescue team if not at entry
- LOTO procedures completed

- clean/purge atmosphere of any hazardous constituents
- ventilation system in place based on identified requirements
- documentation of atmospheric conditions
- entry to perform work
- exit wit all entrants
- cancel permit by appropriate party
- entry equipment maintenance

Procedures for multiple operations in the same space also need to be developed that would cover incompatible jobs, oversight for the operation, notification procedures, scheduling, and an entry-permit system. As needed, air monitoring devices, portable ventilation, appropriate personal protective equipment, communication devices, and retrieval equipment should be obtained prior to entry.

Additionally, other equipment might be specified when indicated, such as:
- explosion proof lighting
- GFCI electrical protection
- nonsparking tools
- barriers to protect the opening
- entry harnesses
- non-entry retrieval

Specific duties must be assigned to the entry team, including entry supervisor, authorized entrant, and attendant. The authorized entrant must be trained to understand hazards, possible health symptoms and consequences, proper use of equipment, the importance to maintain ongoing communication with attendant, and to immediately leave the space when directed by a team member.

The entry supervisor must follow the procedures to end an entry, including notifying all team members that the entry has ended; ensuring that equipment and materials have been removed from the space; and securing the space to prevent unauthorized entry.

Training for all confined entry personnel must be completed and documented. Employees should be retrained when there are changes in the program, workplace or equipment, deviations during an entry, or inadequate knowledge is observed. No program, permit system, entrant, or rescue procedures are required if:
- the only hazard is a hazardous atmosphere
- ventilation will be maintained to provide a safe atmosphere

&#9675; documented through monitoring and inspections

&#9675; is permitted only to verify atmosphere

Reclassification to a "non-permit" space is allowed under the following:
- No actual or potential atmospheric hazards
- Entry space to eliminate hazards must follow 1910.146(d)–(k) of standard
- Reclassification process needs to be documented

**Contractors.** Contractors who are hired to enter confined spaces must be provided with the following information regarding the confined space by the host employer:
- the nature of the hazards involved
- the necessary precautions to be taken
- the use of protective and emergency equipment required

Dual entry missions should be coordinated between contractors and the employer. Contractors should be debriefed after entry.

Before allowing any employees to enter a confined space, the contractor should make sure that they have the following information from the host company: "dual-entry" coordination, contractor program elements, hazards confronted or created during entry. This information should be shared with the contractor through a briefing and prior to entry.

Further guidance may be obtained from American National Standard ANSI Z117.1-1989, Safety Requirements for Confined Spaces. This standard provides minimum safety requirements to be followed while entering, exiting, and working in confined spaces at normal atmospheric pressure. This standard does not pertain to underground mining, tunneling, caisson work or other similar tasks that have established national consensus standards.

## RECORDING AND REPORTING OCCUPATIONAL INJURIES AND ILLNESSES STANDARD

Reporting a work-related injury, illness, or death does not mean someone was at fault or that an OSHA regulation has been violated. When an employee dies, or three or more employees are admitted to a hospital from a work-related incident, an oral report to OSHA must occur within eight hours. This can be done in person or by phone (800-321-OSHA/800-321-6742).

Illnesses or injuries related to the workplace are recorded on the OSHA Log 300. A current OSHA Log 300 can be downloaded from www.osha.gov. For an injury or illness to be recorded, it should be new, work-related, and serious. New injuries and illnesses are those that are not the same type and same part of the body that was recorded previously unless the injury or illness had completely recovered.

**OSHA Record Keeping.** The purpose of this section is to provide, the supervisor with information about the uses and benefits of record-keeping for occupational injuries and illnesses, as well as how to comply with OSHA injury and illness record-keeping requirements. Types of records and reports, record-keeping requirements, and determination of recordability will be reviewed. Although the supervisor may not be the individual that maintains the OSHA record-keeping log, it is important he or she fully understands the requirements to assist the individual who makes the final determination for record keeping.

The Occupational Safety and Health Act of 1970 (OSHAct) requires covered employers to prepare and maintain occupational injury and illness records. The OSHAct and record-keeping regulations in 29 CFR 1904 provide specific recording and reporting requirements.

There are many specific OSHA standards and regulations that require record maintenance and retention of medical surveillance, inspections, exposure monitoring, and other activities not covered here. Companies are responsible for keeping informed of current OSHA regulations at all times. The purpose of the rule Part 1904 is to require employers to record and report work-related fatalities, injuries, and illnesses.

All employers covered by the OSHAct are covered by these Part 1904 regulations. States that operate their own job safety and health programs have adopted comparable record-keeping regulations that have been effective since Jan. 1, 2002. States must have the same requirements for which injuries and illnesses are recordable and how they are recorded. Other provisions may be different as long as they are as stringent as the federal requirements. Employers in some state-plan states may be subject to more stringent reporting requirements (e.g., California requires every case of "serious injury or illness" to be reported).

This rule promotes improved employee awareness and involvement in the record-keeping process, providing workers and their representatives access to information on record-keeping forms and increasing awareness of potential workplace hazards. Employee privacy concerns have been addressed—the former rule had no privacy protections covering the log used to record work-related injuries and illnesses.

The current rule uses a question-and-answer format written in plain language with checklists and flowcharts to provide easier interpretations of recordkeeping requirements. Employers are afforded more flexibility in using computers and telecommunications technology to meet record-keeping requirements. The term "lost workdays" is eliminated and the record keeping focuses on days away or days restricted or transferred. Calendar days are used for counting instead of workdays. A company executive must sign and certify that the 300 A form is accurate.

**Exemptions for Recording Injuries and Illnesses.** Some employers do not have to keep OSHA injury and illness records unless OSHA or the BLS informs them in writing that they must keep records. For example, employers with 10 or fewer employees, and business establishments in certain industry classifications, are partially exempt from keeping OSHA injury and illness records.

OSHA has identified two categories of exemptions that may impact a company's obligation to record injuries and illnesses sustained by their employees: exemption for size and exemption for low hazard industries. However, remember that all employers must report to OSHA any workplace incident that results in a fatality or the hospitalization of three or more employees.

- **Exemption for size.** A company with 10 or fewer employees at all times during the last calendar year does not need to keep OSHA injury and illness records unless OSHA or BLS informs the company in writing.
  - The partial exemption for size is based on the number of employees in the entire company.
  - To determine if a company is exempt because of size, determine the company's peak employment during the last calendar year. If it had no more than 10 employees at any time in the last calendar year, the company qualifies for the partial exemption for size.
- **Exemption for Low-Hazard Industries.** If a business establishment is classified in a specific low-hazard retail, service, finance, insurance, or real estate industry, it does not need to keep OSHA injury and illness records unless the government asks to keep the records.
  - All employers, however, must report any workplace incident to OSHA that results in a fatality or the hospitalization of three or more employees.

○ The partial industry classification exemption applies to individual business establishments. If a company has several business establishments engaged in different classes of business activities, some of the company's establishments may be required to keep records, while others may be exempt. Table 12–C lists some partially exempt industries.

## OSHA Record-keeping Forms

The injury and illness records required by OSHA's record-keeping rule are an important source of information for OSHA, employers, and employees. There are penalties for failure to comply with OSHA record-keeping obligations. OSHA requires completion of the following forms:

- OSHA 300 Log: Log of Work-Related Injuries and Illnesses
- OSHA 300-A Summary Form: Summary of Work-Related Injuries and Illnesses
- OSHA 301 Incident Report: Injury and Illness Incident Report

The OSHA 300 Log (Form 300) is used to classify work-related injuries and illnesses and to note the extent and severity of each case. When an incident occurs, a supervisor should use the log to record specific details about what happened and how it happened.

The OSHA 300-A Summary Form is a separate form used to show total work-related injuries and illnesses recorded in a single year in each category. Each covered employer must complete the summary at the end of the year and post it for three months (February 1 to April 30). The employer must review the records at year-end for accuracy before summarizing them. Additional certification of accuracy by a company executive and additional data on the average employment and hours worked at the establishment are also required.

The OSHA 301 Incident Report (Form 301) or equivalent form must be completed for each injury or illness recorded on Form 300 as follows:

- This form or an equivalent must be filled out within seven calendar days after a company receives information that a recordable work-related injury or illness has occurred. Some state workers' compensation, insurance, or other reports may be acceptable substitutes.
- To be considered an equivalent form, any substitute must contain all the information asked for on this form.
- According to Public Law 91-596 and 29 CFR 1904, OSHA's record-keeping rule, companies must keep this form on file for five years following the year to which it pertains.

## Table 12–C. Partially Exempt Industries

| SIC Code | Industry Description | SIC Code | Industry Description |
|---|---|---|---|
| 525 | Hardware Stores | 725 | Shoe Repair and Shoeshine Parlors |
| 542 | Meat and Fish Markets | 726 | Funeral Service and Crematories |
| 544 | Candy, Nut, and Confectionery Stores | 729 | Miscellaneous Personal Services |
| 545 | Dairy Products Stores | 731 | Advertising Services |
| 546 | Retail Bakeries | 732 | Credit Reporting and Collection Services |
| 549 | Miscellaneous Food Stores | 733 | Mailing, Reproduction, and Stenographic Services |
| 551 | New and Used Car Dealers | 737 | Computer and Data Processing Services |
| 552 | Used Car Dealers | 738 | Miscellaneous Business Services |
| 554 | Gasoline Service Stations | 764 | Reupholstery and Furniture Repair |
| 557 | Motorcycle Dealers | 78 | Motion Picture |
| 56 | Apparel and Accessory Stores | 791 | Dance Studios, Schools, and Halls |
| 573 | Radio, Television, and Computer Stores | 792 | Producers, Orchestras, and Entertainers |
| 58 | Eating and Drinking Places | 793 | Bowling Centers |
| 591 | Drug Stores and Proprietary Stores | 801 | Offices & Clinics Of Medical Doctors |
| 592 | Liquor Stores | 802 | Offices and Clinics Of Dentists |
| 594 | Miscellaneous Shopping Goods Stores | 803 | Offices Of Osteopathic Physicians |
| 599 | Retail Stores, Not Elsewhere Classified | 804 | Offices Of Other Health Practitioners |
| 60 | Depository Institutions (banks and savings institutions) | 807 | Medical and Dental Laboratories |
| 61 | Nondepository Institutions (credit institutions) | 809 | Health and Allied Services, Not Elsewhere Classified |
| 62 | Security and Commodity Brokers | 81 | Legal Services |
| 63 | Insurance Carriers | 82 | Educational Services (schools, colleges, universities, and libraries) |
| 64 | Insurance Agents, Brokers, and Services | 832 | Individual and Family Services |
| 653 | Real Estate Agents and Managers | 835 | Child Day Care Services |
| 654 | Title Abstract Offices | 839 | Social Services, Not Elsewhere Classified |
| 67 | Holding and Other Investment Offices | 841 | Museums and Art Galleries |
| 722 | Photographic Studios, Portrait | 86 | Membership Organizations |
| 723 | Beauty Shops | 87 | Engineering, Accounting, Research, Management, and Related Services |
| 724 | Barber Shops | 899 | Services, not elsewhere classified |

- Companies not sure whether a case is recordable can call the local OSHA office for help. When in doubt, companies should record it. They can always line it out later.

*NOTE: Federal OSHA requires other types of record keeping for many other reasons. For example, some OSHA regulations require training records documenting that training was conducted and understood. It is beyond the scope of this chapter to discuss those various recordkeeping requirements.*

Each employer that is required to keep records of fatalities, injuries, and illnesses must record each fatality, injury, and illness that:

- is work-related
- is a new case
- meets one or more of the general recording criteria

## Determining Work-Relatedness and OSHA Recordability

There are a number of points to consider in the process of accurately determining whether an injury or illness is work-related. An injury or illness is considered work-related if an event or exposure in the work environment caused or contributed to the condition or significantly aggravated a preexisting condition. Work-relatedness is presumed for injuries and illnesses resulting from events or exposures occurring in the workplace, unless an exception specifically applies. OSHA defines the work environment as:

> …the establishment and other locations where one or more employees are working or are present as a condition of their employment. The work environment includes not only physical locations, but also the equipment or materials used by the employee during the course of his or her work.

An injury or illness occurring in the work environment that falls under one of the following exceptions is not work-related, and therefore is not recordable:

- The injury or illness occurred when the employee was present in the work environment as a member of the general public rather than as an employee.
- The injury or illness involves signs or symptoms that surface at work, but result solely from a non-work-related event or exposure that occurs outside the work environment.

An injury or illness that occurs in the work environment is not considered work-related when the injury or illness:

- results solely from voluntary participation in a wellness program or in a medical, fitness, or recreational activity such as blood donation, physical examination, flu shot, exercise class, racquetball, baseball, and such
- is solely the result of an employee eating, drinking, or preparing food or drink for personal consumption. (e.g., if the employee is injured by choking on a sandwich while in the employer's establishment, the case would not be considered work-related.)

*NOTE: If the employee is made ill by ingesting food contaminated by workplace contaminants (such as lead), or gets food poisoning from food supplied by the employer, the case would be considered work-related.*

- is solely the result of an employee doing personal tasks (unrelated to employment) at the establishment outside of the employee's assigned working hours
- solely results from personal grooming, self medication for a non-work-related condition, or is intentionally self-inflicted
- is caused by a motor vehicle accident and occurs on a company parking lot or company access road while the employee is commuting to or from work
- is the common cold or flu

*NOTE: Contagious diseases such as tuberculosis, brucellosis, hepatitis A, or plague are considered work-related if the employee is infected at work.*

- is a mental illness, which will not be considered work-related unless the employee voluntarily provides the employer with an opinion from a physician or other licensed health care professional with appropriate training and experience (psychiatrist, psychologist, psychiatric nurse practitioner, etc.) stating that the employee has a mental illness that is work-related

**New Cases.** An employer must consider an injury or illness a new case to be evaluated for recordability if the employee:

- has not previously experienced a recorded injury or illness of the same type that affects the same part of the body
- previously experienced a recorded injury or illness of the same type that affected the same part of the body but had recovered completely (all signs/symptoms of the previous injury or illness had disappeared) and an event or exposure in the work environment caused the injury or illness, or its signs or symptoms, to reappear

### Determining Recordability

Record work-related injuries and illnesses that result in:

- death
- loss of consciousness
- days away from work
- restricted work activity or job transfer
- medical treatment beyond first aid

You must record any significant work-related:
- injury or illness that is diagnosed by a physician or other licensed health care professional
- case involving cancer, chronic irreversible disease, a fractured or cracked bone, or a punctured eardrum

**Additional Criteria.** You must also record work-related injuries and illnesses that meet any of the additional criteria listed below:
- any needlestick injury or cut from a sharp object that is contaminated with another person's blood or other potentially infectious material
- any case requiring an employee to be medically removed under the requirements of an OSHA health standard
- tuberculosis infection evidenced by a positive skin test or diagnosed by a physician/other licensed health care professional after exposure to a known case of active tuberculosis
- An employer's hearing test (audiogram) reveals:
  - that the employee has experienced a Standard Threshold Shift (STS) in hearing in one or both ears (averaged at 2,000, 3,000, and 4,000 Hz)
  - the employee's total hearing level is 25 decibels or more above audiometric zero (also averaged at 2,000, 3,000, and 4,000 Hz) in the same ear(s) as the STS.

Definitions for medical treatment, first aid, and what constitutes restricted work activity, as defined by OSHA, can be found at www.osha.gov. OSHA also provides a variety of examples that will assist supervisors in classifying injuries and illnesses.

## SUMMARY OF KEY POINTS

Key points covered in this chapter include:

- The Occupational Safety and Health Act of 1970 (OSHAct) extends coverage through OSHA, or an OSHA-approved safety and health program, to all employers and their employees in the 50 states, the District of Columbia, Puerto Rico, and all other territories under federal government jurisdiction. All supervisors should know and understand which set of regulations govern their workplace. OSHA regulations that cover general industry, maritime, construction and agricultural workplaces can be found in Title 29 of the Code of Federal Regulations (CFR), Parts 1902–1990.

- For the most part, there are specific regulations governing different workplace activities. However, there are some cases where no specific OSHA standard applies. For those cases, there is a section of the OSHAct (Section 5(a)(1)) commonly referred to as the General Duty Clause. This clause states that employers must provide a place of employment that is free from recognized hazards that cause or are likely to cause serious physical harm to employees. This clause is used by OSHA inspectors to cite employers when hazards are identified that are not covered by specific regulations. Supervisors should know that a general duty citation must involve both the presence of a serious hazard and exposure of the cited employer's own employees.

- Supervisors should be aware that some regulations, depending on the industry, require written plans and/or documentation for compliance with the standard. For example, regulations that cover bloodborne pathogens, hazard communications, and permit-required confined spaces, to name a few, require documentation that must be kept in the work area, accessible to employees or their representatives, in order for the employer to be in compliance with the regulation for that industry. Written plans may be general or very specific depending on the industry. Often, an OSHA inspector will begin his or her evaluation of a facility by requesting the appropriate written documentation first.

- Some regulations are more frequently cited by OSHA inspectors than others and, as a result, have a greater regulatory impact on workplaces in terms of large monetary penalties. The OSHA hazard communication standard (HCS) is one of the most frequently cited standards. According to OSHA, more than 30 million workers are exposed to hazardous chemicals in the workplace. The HCS covers approximately 650,000 hazardous chemicals and is intended to ensure that exposed workers and their employers are informed of the identities of these chemicals, associated health and safety hazards, and appropriate protective measures. A supervisor (or employer) is responsible for (1) identifying and listing hazardous chemicals in the workplace, (2) obtaining Material

Safety Data Sheets (MSDSs) and labels for each hazardous chemical, and (3) develop and implement a written hazardous communication plan that includes information on container labels, MSDSs and their availability, and employee training.

- Supervisors should understand and be familiar with the record-keeping requirements and standards of their industry. Record-keeping regulations in 29 CFR 1904 provide specific recording and reporting requirements of occupational injuries and illnesses. The regulation covers types of records and reports to be kept, record-keeping requirements, and determination of recordability. It is important for the supervisor to know that all employers covered by the OSHAct must report work-related injuries, illnesses, and fatalities unless exempted by OSHA. The OSHA web site provides specific information, including forms, definitions, and classifications of injuries and illnesses to ensure compliance with recordkeeping standards.

## APPENDIX 12–A. CURRENT REQUIREMENTS FOR ELECTRICAL INSTALLATION AND UTILIZATION EQUIPMENT

### Coverage

OSHA has brought additional requirements to the workplace to govern electrical work by revising 29 CFR 1910 Subpart S, Electrical Standard, which went into effect in August 2007.

The final rule, as does the current standard, exempts older electrical installations from meeting some of the provisions of the design safety standards for electrical systems. Older installations must meet fewer requirements than newer ones. The few requirements in this short list are so essential to employee safety that even the oldest electrical installations must be modified, if necessary, to meet them.

The following requirements apply to all electrical installations and utilization equipment, regardless of when they were designed or installed:

§ 1910.303(b)—Examination, installation, and use of equipment

§ 1910.303(c)(3)—Electrical connections—Splices

§ 1910.303(d)—Arcing parts

§ 1910.303(e)—Marking

§ 1910.303(f), except (f)(4) and (f)(5)—Disconnecting means and circuits

§ 1910.303(g)(2)—600 volts or less—Guarding of live parts

§ 1910.304(a)(3)—Use of grounding terminals and devices

§ 1910.304(f)(1)(i), (f)(1)(iv), and (f)(1)(v)—Overcurrent protection—600 volts, nominal, or less

§ 1910.304(g)(1)(ii), (g)(1)(iii), (g)(1)(iv), and (g)(1)(v)—Grounding—Systems to be grounded

§ 1910.304(g)(4)—Grounding—Grounding connections

§ 1910.304(g)(5)—Grounding—Grounding path

§ 1910.304(g)(6)(iv)(A) through (g)(6)(iv)(D), and (g)(6)(vi)—Grounding—Supports, enclosures, and equipment to be grounded

§ 1910.304(g)(7)—Grounding—Nonelectrical equipment

§ 1910.304(g)(8)(i)—Grounding—Methods of grounding fixed equipment

§ 1910.305(g)(1)—Flexible cords and cables—Use of flexible cords and cables

§ 1910.305(g)(2)(ii) and (g)(2)(iii)—Flexible cords and cables—Identification, splices, and terminations

§ 1910.307, except as specified in § 1910.307(b)—Hazardous (classified) locations

The list is unchanged from the current standard, except for the following:
- It is prohibited to use grounding terminals and devices for purposes other than grounding.
- Documentation is required for hazardous locations made under the zone classification system.
- Zone classification system requirements

The following requirements apply only to electrical installations and utilization equipment installed after April 16, 1981:

§ 1910.303(h)(4)—Over 600 volts, nominal—Entrance and access to work space

§ 1910.304(f)(1)(vii) and (f)(1)(viii)—Overcurrent protection—600 volts, nominal, or less

§ 1910.304(g)(9)(i)—Grounding—Grounding of systems and circuits of 1000 volts and over (high voltage)

§ 1910.305(j)(6)(ii)(D)—Equipment for general use—Capacitors

§ 1910.306(c)(9)—Elevators, dumbwaiters, escalators, moving walks, wheelchair lifts, and stairway chair lifts—Interconnection between multicar controllers

§ 1910.306(i)—Electrically driven or controlled irrigation machines

§ 1910.306(j)(5)—Swimming pools, fountains, and similar installations—Fountains

§ 1910.308(a)(1)(ii)—Systems over 600 volts, nominal—Aboveground wiring methods

§ 1910.308(c)(2)—Class 1, Class 2, and Class 3 remote control, signaling, and power-limited circuits—Marking

§ 1910.308(d)—Fire alarm systems

The following requirements apply only to electrical installations and utilization equipment installed after August 13, 2007:

§ 1910.303(f)(4)—Disconnecting means and circuits—Capable of accepting a lock

§ 1910.303(f)(5)—Disconnecting means and circuits—Marking for series combination ratings

§ 1910.303(g)(1)(iv) and (g)(1)(vii)—600 Volts, nominal, or less—Space about electric equipment

§ 1910.303(h)(5)(vi)—Over 600 volts, nominal—Working space and guarding

§ 1910.304(b)(1)—Branch circuits—Identification of multiwire branch circuits

§ 1910.304(b)(3)(i)—Branch circuits—Ground-fault circuit interrupter protection for personnel

§ 1910.304(f)(2)(i)(A), (f)(2)(i)(B) (but not the introductory text to § 1910.304(f)(2)(i)), and (f)(2)(iv)(A)—Overcurrent protection—Feeders and branch circuits over 600 volts, nominal

§ 1910.305(c)(3)(ii)—Switches—Connection of switches

§ 1910.305(c)(5)—Switches—Grounding

§ 1910.306(a)(1)(ii)—Electric signs and outline lighting—Disconnecting means

§ 1910.306(c)(4)—Elevators, dumbwaiters, escalators, moving walks, wheelchair lifts, and stairway chair lifts—Operation

§ 1910.306(c)(5)—Elevators, dumbwaiters, escalators, moving walks, wheelchair lifts, and stairway chair lifts—Location

§ 1910.306(c)(6)—Elevators, dumbwaiters, escalators, moving walks, wheelchair lifts, and stairway chair lifts—Identification and signs

§ 1910.306(c)(7)—Elevators, dumbwaiters, escalators, moving walks, wheelchair lifts, and stairway chair lifts—Single-car and multicar installations

§ 1910.306(j)(1)(iii)—Swimming pools, fountains, and similar installations—Receptacles

§ 1910.306(k)—Carnivals, circuses, fairs, and similar events

§ 1910.308(a)(5)(v) and (a)(5)(vi)(B)—Systems over 600 volts, nominal— Interrupting and isolating devices

§ 1910.308(a)(7)(vi)—Systems over 600 volts, nominal—Tunnel installations

§ 1910.308(b)(3)—Emergency power systems—Signs

§ 1910.308(c)(3)—Class 1, Class 2, and Class 3 remote control, signaling, and power-limited circuits—Separation from conductors of other circuits

§ 1910.308(f)—Solar photovoltaic systems

## APPENDIX 12–B. HIGHLY HAZARDOUS CHEMICALS (HHCS)

This listing of toxic and reactive highly hazardous chemicals presents a potential for a catastrophic event at or above the threshold quantity.

| CHEMICAL NAME | CAS* | TQ** |
|---|---|---|
| Acetaldehyde | 75-07-0 | 2500 |
| Acrolein (2-Popenal) | 107-02-8 | 150 |
| Acrylyl Chlorde | 814-68-6 | 250 |
| Allyl Chlorid | 107-05-1 | 1000 |
| Allylamine | 107-11-9 | 1000 |
| Alkylaluminum | Varies | 5000 |
| Ammonia, Anhydrous | 7664-41-7 | 10000 |
| Ammonia solutions (greater than 44% ammonia by weight) | 7664-41-7 | 15000 |
| Ammonium Perchlorate | 7790-98-9 | 7500 |
| Ammonium Permanganate | 7787-36-2 | 7500 |
| Arsine (also called Arsenic Hydride) | 7784-42-1 | 100 |
| Bis(Chloromethyl) Ether | 542-88-1 | 100 |
| Boron Trichloride | 10294-34-5 | 2500 |
| Boron Trifluoride | 7637-07-2 | 250 |
| Bromine | 7726-95-6 | 1500 |
| Bromine Chloride | 13863-41-7 | 1500 |

| | | |
|---|---|---|
| Bromine Pentafluoride | 7789-30-2 | 2500 |
| Bromine Trifluoride | 7787-71-5 | 15000 |
| 3-Bromopropyne (also called Propargyl Bromide) | 106-96-7 | 100 |
| Butyl Hydroperoxide (Tertiary) | 75-91-2 | 5000 |
| Butyl Perbenzoate (Tertiary) | 614-45-9 | 7500 |
| Carbonyl Chloride (see Phosgene) | 75-44-5 | 100 |
| Carbonyl Fluoride | 353-50-4 | 2500 |
| Cellulose Nitrate (concentration greater than 12.6% nitrogen | 9004-70-0 | 2500 |
| Chlorine | 7782-50-5 | 1500 |
| Chlorine Dioxide | 10049-04-4 | 1000 |
| Chlorine Pentrafluoride | 13637-63-3 | 1000 |
| Chlorine Trifluoride | 7790-91-2 | 1000 |
| Chlorodiethylaluminum (also called Diethylaluminum Chloride) | 96-10-6 | 5000 |
| 1-Chloro-2,4-Dinitrobenzene | 97-00-7 | 5000 |
| Chloromethyl Methyl Ether | 107-30-2 | 500 |
| Chloropicrin | 76-06-2 | 500 |
| Chloropicrin and Methyl Bromide mixture | None | 1500 |
| Chloropicrin and Methyl Chloride mixture | None | 1500 |
| Cumene Hydroperoxide | 80-15-9 | 5000 |
| Cyanogen | 460-19-5 | 2500 |
| Cyanogen Chloride | 506-77-4 | 500 |
| Cyanuric Fluoride | 675-14-9 | 100 |
| Diacetyl Peroxide (concentration greater than 70%) | 110-22-5 | 5000 |
| Diazomethane | 334-88-3 | 500 |
| Dibenzoyl Peroxide | 94-36-0 | 7500 |
| Diborane | 19287-45-7 | 100 |
| Dibutyl Peroxide (Tertiary) | 110-05-4 | 5000 |
| Dichloro Acetylene | 7572-29-4 | 250 |
| Dichlorosilane | 4109-96-0 | 2500 |
| Diethylzinc | 557-20-0 | 10000 |
| Diisopropyl Peroxydicarbonate | 105-64-6 | 7500 |
| Dilauroyl Peroxide | 105-74-8 | 7500 |
| Dimethyldichlorosilane | 75-78-5 | 1000 |
| Dimethylhydrazine, 1,1- | 57-14-7 | 1000 |
| Dimethylamine, Anhydrous | 124-40-3 | 2500 |
| 2,4-Dinitroaniline | 97-02-9 | 5000 |
| Ethyl Methyl Ketone Peroxide (also Methyl Ethyl Ketone Peroxide; concentration greater than 60%) | 1338-23-4 | 5000 |

| | | |
|---|---|---|
| Ethyl Nitrite | 109-95-5 | 5000 |
| Ethylamine | 75-04-7 | 7500 |
| Ethylene Fluorohydrin | 371-62-0 | 100 |
| Ethylene Oxide | 75-21-8 | 5000 |
| Ethyleneimine | 151-56-4 | 1000 |
| Fluorine | 7782-41-4 | 1000 |
| Formaldehyde (Formalin) | 50-00-0 | 1000 |
| Furan | 110-00-9 | 500 |
| Hexafluoroacetone | 684-16-2 | 5000 |
| Hydrochloric Acid, Anhydrous | 7647-01-0 | 5000 |
| Hydrofluoric Acid, Anhydrous | 7664-39-3 | 1000 |
| Hydrogen Bromide | 10035-10-6 | 5000 |
| Hydrogen Chloride | 7647-01-0 | 5000 |
| Hydrogen Cyanide, Anhydrous | 74-90-8 | 1000 |
| Hydrogen Fluoride | 7664-39-3 | 1000 |
| Hydrogen Peroxide (52% by weight or greater) | 7722-84-1 | 7500 |
| Hydrogen Selenide | 7783-07-5 | 150 |
| Hydrogen Sulfide | 7783-06-4 | 1500 |
| Hydroxylamine | 7803-49-8 | 2500 |
| Iron, Pentacarbonyl | 13463-40-6 | 250 |
| Isopropylamine | 75-31-0 | 5000 |
| Ketene | 463-51-4 | 100 |
| Methacrylaldehyde | 78-85-3 | 1000 |
| Methacryloyl Chloride | 920-46-7 | 150 |
| Methacryloyloxyethyl Isocyanate | 30674-80-7 | 100 |
| Methyl Acrylonitrile | 126-98-7 | 250 |
| Methylamine, Anhydrous | 74-89-5 | 1000 |
| Methyl Bromide | 74-83-9 | 2500 |
| Methyl Chloride | 74-87-3 | 15000 |
| Methyl Chloroformate | 79-22-1 | 500 |
| Methyl Ethyl Ketone Peroxide (concentration greater than 60%) | 1338-23-4 | 5000 |
| Methyl Fluoroacetate | 453-18-9 | 100 |
| Methyl Fluorosulfate | 421-20-5 | 100 |
| Methyl Hydrazine | 60-34-4 | 100 |
| Methyl Iodide | 74-88-4 | 7500 |
| Methyl Isocyanate | 624-83-9 | 250 |
| Methyl Mercaptan | 74-93-1 | 5000 |
| Methyl Vinyl Ketone | 79-84-4 | 100 |
| Methyltrichlorosilane | 75-79-6 | 500 |
| Nickel Carbonly (Nickel Tetracarbonyl) | 13463-39-3 | 150 |
| Nitric Acid (94.5% by weight or greater) | 7697-37-2 | 500 |
| Nitric Oxide | 10102-43-9 | 250 |
| Nitroaniline (para Nitroaniline | 100-01-6 | 5000 |
| Nitromethane | 75-52-5 | 2500 |
| Nitrogen Dioxide | 10102-44-0 | 250 |
| Nitrogen Oxides (NO; NO(2); | | |

| | | |
|---|---|---|
| N2O4; N2O3) | 10102-44-0 | 250 |
| Nitrogen Tetroxide (also called Nitrogen Peroxide) | 10544-72-6 | 250 |
| Nitrogen Trifluoride | 7783-54-2 | 5000 |
| Nitrogen Trioxide | 10544-73-7 | 250 |
| Oleum (65% to 80% by weight; also called Fuming Sulfuric Acid) | 8014-94-7 | 1000 |
| Osmium Tetroxide | 20816-12-0 | 100 |
| Oxygen Difluoride (Fluorine Monoxide) | 7783-41-7 | 100 |
| Ozone | 10028-15-6 | 100 |
| Pentaborane | 19624-22-7 | 100 |
| Peracetic Acid (concentration greater 60% Acetic Acid; also called Peroxyacetic Acid) | 79-21-0 | 1000 |
| Perchloric Acid (concentration greater than 60% by weight) | 7601-90-3 | 5000 |
| Perchloromethyl Mercaptan | 594-42-3 | 150 |
| Perchloryl Fluoride | 7616-94-6 | 5000 |
| Peroxyacetic Acid (concentration greater than 60% Acetic Acid; also called Peracetic Acid) | 79-21-0 | 1000 |
| Phosgene (also called Carbonyl Chloride) | 75-44-5 | 100 |
| Phosphine (Hydrogen Phosphide) | 7803-51-2 | 100 |
| Phosphorus Oxychloride (also called Phosphoryl Chloride) | 10025-87-3 | 1000 |
| Phosphorus Trichloride | 7719-12-2 | 1000 |
| Phosphoryl Chloride (also called Phosphorus Oxychloride) | 10025-87-3 | 1000 |
| Propargyl Bromide | 106-96-7 | 100 |
| Propyl Nitrate | 627-3-4 | 2500 |
| Sarin | 107-44-8 | 100 |
| Selenium Hexafluoride | 7783-79-1 | 1000 |
| Stibine (Antimony Hydride) | 7803-52-3 | 500 |
| Sulfur Dioxide (liquid) | 7446-09-5 | 1000 |
| Sulfur Pentafluoride | 5714-22-7 | 250 |
| Sulfur Tetrafluoride | 7783-60-0 | 250 |
| Sulfur Trioxide (also called Sulfuric Anhydride) | 7446-11-9 | 1000 |
| Sulfuric Anhydride (also called Sulfur Trioxide) | 7446-11-9 | 1000 |
| Tellurium Hexafluoride | 7783-80-4 | 250 |
| Tetrafluoroethylene | 116-14-3 | 5000 |
| Tetrafluorohydrazine | 10036-47-2 | 5000 |
| Tetramethyl Lead | 75-74-1 | 1000 |
| Thionyl Chloride | 7719-09-7 | 250 |
| Trichloro (chloromethyl) Silane | 1558-25-4 | 100 |
| Trichloro (dichlorophenyl) | | |

| | | |
|---|---|---|
| Silane | 27137-85-5 | 2500 |
| Trichlorosilane | 10025-78-2 | 5000 |
| Trifluorochloroethylene | 79-38-9 | 10000 |
| Trimethyoxysilane | 2487-90-3 | 1500 |

\* Chemical Abstract Service Number

\*\* Threshold Quantity in Pounds (Amount necessary to be covered by this standard.)

# 13

# Machine Safeguarding

**After reading this chapter, you will be able to:**

- Explain the principles and benefits of safeguarding machines and equipment
- Describe the basic principles of effective safeguard design
- List the primary hazards of machinery and the safeguard options needed for each hazard
- Explain the safeguards needed for automated machines and equipment
- Plan, install, and maintain a guarding system
- Establish a maintenance and repair program for all safeguards in your department

The first step in any operation is to engineer the hazards out of a job as much as possible, to ensure continued production, employee safety and health, a good profit margin, and reduced equipment damage. At times, however, finding a better way is not feasible—or the technical knowledge is not sufficiently advanced. In these cases, you must then use safeguarding to protect employees from injury. If such precautions are not taken, the consequences can be serious. When an employee is injured while operating machines or equipment, production capacity and cost of operations are adversely affected. Poorly designed, improperly safeguarded, or unguarded machinery or equipment is an ongoing threat to the well-being of employees and to production capacity.

An injury can take an operator away from his or her job for a long time, sometimes permanently. The worker must be replaced to keep the department going. Even if a qualified person is transferred from another department, eventually a new person must be hired. Finding, hiring, and training the right worker takes time and money.

Safeguards are a critical factor in controlling hazards and in preventing accidents. No matter how much training or experience a worker may have, no one can keep focused on work every minute. Also, some machines look so deceptively easy to operate that the supervisor might allow semiskilled or unskilled people to operate them. The results can be disastrous in terms of personal injury and machine or material damage.

The supervisor must have a complete understanding of machine safety principles and communicate these concerns to the workers.

## PRINCIPLES OF SAFEGUARDING: GUARDS, DEVICES, AND SAFE DISTANCE

Safeguarding is frequently thought of as being concerned only with points of operation (Figure 13–1) or the means of power transmission. Although safeguarding against these hazards is required, this step also can prevent injuries from other causes, both on and around machines and from equipment and damaged material.

The point of operation is that point where the work is actually being performed on the material (i.e., cutting, boring, shaping, forming, shearing, ripping, or grinding). The point of operation can be safeguarded by guards (physical barriers that prevent access by the operator and others to the point of operation) or by devices (a mechanism or control designed for safeguarding at the point of operation). Guards can include die enclosures, fixed barriers, adjustable barriers, and interlocked barriers. Devices can include two-hand trips; two-hand controls; pullbacks; and presence-sensing, movable-barrier, holdout, or restraint devices.

Barrier guards can prevent injuries from the following sources:

1. Direct contact with exposed moving parts of a machine—either points of operation on machines (power presses, shears, milling machines, machine tools, or woodworking equipment) or power-transmitting parts of mechanisms (gears, pulleys and sheaves, slides, or couplings). Figure 13–2 shows one type of barrier guard.

2. Work in process: for example, pieces of wood that kick back from a power ripsaw, or metal chips that fly from machining operations.

3. Machine failure, which usually results from lack of preventive maintenance, overloading, metal fatigue, or misuse.

4. Electrical failure, which may cause malfunctioning of the machine or electrical shocks or burns.

5. Operator error or failure caused by lack of knowledge, instructions, training, or skill; emotional distractions; misunderstandings; unsafe operation; illness; fatigue; and so on.

The experience of more than six decades of organized accident prevention proves that it is unwise to rely entirely on education and training of operators, or their cooperation, to avoid mishaps. Many factors affect a person's attitude, judgment, and ability to concentrate. For example, skilled workers with emotional or physical problems may not pay strict attention to their production responsibilities or give their best effort.

**Figure 13–1.** A ¾-in (9-mm) opening permits a small part of the fingers to slip past the guard. A smaller, ¼-in (6-mm) opening would prohibit the fingers crossing into the danger zone.

**Figure 13–2.** These expanded metal covers safeguard protruding moving parts and power transmissions.

### Safeguarding Hazards Before Incidents

Guarding the hazard is a fundamental principle of incident prevention, and is not limited to machinery. When you survey your department from the point of view of safeguarding against all hazards, you will list many potential sources of injury that should be protected by barricades, rails, toeboards, enclosures, or other means. As you list electric switches and equipment, motors, engines, fixed ladders, stairs, platforms, and pits, you might ask, "Can an incident occur here, or here … or here?" See Chapter 1, *Safety Management*, and Chapter 5, *Employee Involvement*, for details on finding and analyzing hazards.

### Benefits of Safeguarding

When operators are preoccupied with being injured, or afraid of their machines or of getting close to moving parts, they cannot pay strict attention to production responsibilities. Once their fears have been removed, workers can concentrate on the operation and often are more productive. In addition, when moving parts are left exposed, other objects may fall into them or get caught and damage the machine or material in process.

Well-designed and carefully maintained safeguards assure workers that management is committed to preventing incidents and ensuring safe production. When employees realize this fact, they are more inclined to contribute to safety efforts.

An added benefit of a proper safeguarding program is raising employee morale, especially when workers who must use the safeguards are consulted before the safeguards are made or bought. Employees often have good ideas that contribute to both safety and economy. Even if they have no suggestions, workers will usually feel satisfied at being consulted about their jobs. As a result, they may take a greater interest in the project, and be less likely to remove guards or barriers, or fail to replace them. Finally, a proper safeguarding program will help to reduce injury, damage, downtime, and workers' compensation expenses resulting from injuries. Proper safeguarding also helps to ensure compliance with state and federal regulations.

### Types of Safeguards

Many standardized guards and devices, and specially designed safeguards, are available from the manufacturer to protect machine operators, particularly their hands, at the point of operation. Table 13–A presents concise descriptions of the nature, action, advantages, and limitations of a variety of such guards and devices.

## Table 13–A. Point-of-Operation Protection

| Type of Guarding Methods | Action of Guard | Advantages | Limitations | Typical Machines on Which Used |
|---|---|---|---|---|
| **Enclosures or Barriers** | | | | |
| Complete, simple fixed enclosure | Barrier or enclosure admits stock but not hands into danger zone because of feed-opening size, remote location, or unusual shape | Provides complete enclosure if kept in place<br>Both hands free<br>Generally permits increased production<br>Easy to install<br>Ideal for blanking on power presses<br>Can be combined with automatic or semiautomatic feeds | Limited to specific operations<br>May require special tools to remove jammed stock<br>May interfere with visibility | Bread slicers<br>Embossing presses<br>Meat grinders<br>Metal square shears<br>Nip points of inrunning rubber, paper, and textile rolls<br>Paper corner cutters<br>Power presses |
| Warning enclosures (usually adjustable to stock being fed) | Barrier or enclosure admits the operator's hand but gives warning before danger zone is reached | Makes "hard to guard" machines safer<br>Generally does not interfere with production<br>Easy to install<br>Admits varying sizes of stock | Hands may enter danger zone—enclosure not complete at all times<br>Danger of operator not using guard<br>Often requires frequent adjustment and careful maintenance | Band saws<br>Circular saws<br>Cloth cutters<br>Dough brakes<br>Ice crushers<br>Jointers<br>Leather strippers<br>Rock crushers<br>Wood shapers |
| Barrier with electrical contact or mechanical stop activating mechanical or electrical brake | Barrier quickly stops machine or prevents application of injurious pressure when any part of the operator's body contacts it or approaches danger zone | Makes "hard to guard" machines safer<br>Does not interfere with production | Requires careful adjustment and maintenance<br>Possibility of minor injury before guard operates<br>Operator can make guard inoperative | Dough brakes<br>Flat roll ironers<br>Paper box corner stayers<br>Paper box enders<br>Power presses<br>Paper calenders<br>Rubber mills |
| Enclosure with electrical or mechanical interlock | Enclosure or barrier shuts off or disengages power and prevents starting of machine when guard is open; prevents opening of guard while machine is under power or coasting (Interlocks should not prevent manual operation or "inching" by remote control) | Does not interfere with production<br>Hands are free; operation of guard is automatic<br>Provides complete and positive enclosure | Requires careful adjustment and maintenance<br>Operator may be able to make guard inoperative<br>Does not protect in event of mechanical repeat | Dough brakes and mixers<br>Foundry tumblers<br>Laundry extractors, driers, and tumblers<br>Power presses<br>Tanning drums<br>Textile pickers, cards |

| Type of Guarding Methods | Action of Guard | Advantages | Limitations | Typical Machines on Which Used |
|---|---|---|---|---|
| **Automatic or Semiautomatic Feed** | | | | |
| Non-manual or partly manual loading of feed mechanism with point of operation enclosed | Stock fed by chutes, hoppers, conveyors, movable dies, dial feed, rolls, etc. Enclosure will not admit any part of body | Generally increases production<br>Operator cannot place hands in danger zone | Excessive installation cost for short run<br>Requires skilled maintenance<br>Not adaptable to variations in stock | Baking and candy machines<br>Circular saws<br>Power presses<br>Textile pickers<br>Wood planers<br>Wood shapers |
| **Hand-Removal Devices** | | | | |
| Hand restraints | A fixed bar and cord or strap with head attachments which, when worn and adjusted, do not permit an operator to reach into the point of operation | Operator cannot place hands in danger zone<br>Permits maximum hand feeding; can be used on higher-speed machines<br>No obstruction to feeding a variety of stock<br>Easy to install | Requires frequent inspection, maintenance, and adjustment to each operator<br>Limits movement of operator<br>May obstruct work space around operator<br>Does not permit blanking from hand-fed strip stock | Embossing presses<br>Power presses |
| Hand pull-away device | A cable-operated attachment on slide, connected to the operator's hands or arms to pull the | Acts even in event of repeat<br>Permits maximum hand feeding; can be used on higher-speed machines<br>No obstruction to feeding a variety of stock<br>Easy to install | Requires unusually good maintenance and adjustment to each operator<br>Frequent inspection necessary<br>Limits movement of operator<br>May obstruct work space around operator<br>Does not permit blanking from hand-fed strip stock | Embossing presses<br>Power presses |
| **Two-Hand Trip** | | | | |
| Electrical | Simultaneous pressure of two hands on switch buttons in series actuates machine | Can be adapted to multiple operation<br>Operator's hands away from danger zone<br>No obstructions to hand feeding<br>Does not require adjustment<br>Can be equipped with continuous pressure remote controls to permit "inching"<br>Generally easy to install | Operator may try to reach into danger zone after tripping machine<br>Some trips can be rendered unsafe by holding with the arm, blocking or tying down one control, thereby permitting one-hand operation<br>Not use for some blanking operations | Dough mixers<br>Embossing presses<br>Paper cutters<br>Pressing machines<br>Power presses<br>Washing tumblers |

| Type of Guarding Methods | Action of Guard | Advantages | Limitations | Typical Machines on Which Used |
|---|---|---|---|---|
| Mechanical | Simultaneous pressure of two hands on air control valves, mechanical levers, controls interlocked with foot control or removal of solid blocks or stops permits normal operation of machine | | | |
| **Miscellaneous** | | | | |
| Limited slide travel | Slide travel limited to ¼ in or less; fingers cannot enter between pressure points | Provides positive protection<br>Requires no maintenance or adjustment | Small opening limits size of stock | Foot power (kick) presses<br>Power presses |
| Presence-sensing device | Sensing field and brake quickly stop machine or prevent its starting if hands are in danger zone | Does not interfere with normal feeding or production<br>No obstruction on machine or around operator | Expensive to install<br>Does not protect against mechanical repeat<br>Generally limited to use on slow-speed machines with friction clutches or other means to stop the machine during the operating cycle<br>Can be circumvented | Embossing presses<br>Power presses<br>Rubber mills<br>Squaring shears<br>Press brakes |
| Light curtains | Sense presence of person through electric beam | Provides positive protection<br>Does not interfere with work | Must be closely maintained | |
| Special tools or handles on dies | Long-handled tongs, vacuum lifters, or hand die holders which avoid need for operator's putting hand in the danger zone | Inexpensive and adaptable to different types of stock<br>Sometimes increases protection of other guards | Operator must keep hands out of danger zone<br>Requires good employee training, close supervision<br>Only one part of prevention needed<br>Will not bring employer into OSHA compliance | Dough brakes<br>Leather die cutters<br>Power presses<br>Forging hammers |
| Special jigs or feeding devices | Hand-operated feeding devices of metal or wood which keep the operator's hands at a safe distance from the danger zone | May speed production as well as safeguard machine<br>Generally economical for long jobs | Machine itself not guarded; safe operation depends upon correct use of device<br>Requires good employee training, close supervision<br>Suitable for limited types of work | Circular saws<br>Dough brakes<br>Jointers<br>Meat grinders<br>Paper cutters<br>Power presses<br>Drill presses |

Sheet metal, perforated metal, expanded metal, heavy wire mesh or steel stock may be used as material for the construction of guards. The best practice is to follow the requirements of OSHA and ANSI standards when selecting material for the design and construction of guards.

If moving parts must be visible, transparent impact plastic or safety glass can be used where strength of metal is not required. Guards or barriers may be aluminum or other soft metal where resistance to rust is essential or where iron or steel guards can cause damage to machinery. Wood, plastic, and glass fiber barriers are usually inexpensive, but are low in strength compared with steel. However, these materials effectively resist effects of splashes, vapors, and fumes from corrosive substances that would react with iron or steel and reduce its strength and effectiveness.

## SAFEGUARD DESIGN

It is easier to establish effective methods for safeguarding power transmissions than it is for safeguarding points of operation because power transmissions are more standardized. OSHA and ANSI standards for mechanical power-transmission apparatus (ANSI/ASME B15.1 and 29 CFR 1910.219) apply to all moving parts of mechanical equipment. Point-of-operation guarding can be found in ANSI standards for specific machinery, such as in the following safety requirements:

- Bakery Equipment, Z50.1
- Construction, Care, and Use of Drilling, Milling, and Boring Machines, B11.8
- Construction, Care, and Use of Grinding Machines, B11.9
- Construction, Care, and Use of Lathes, B11.6
- Construction, Care, and Use of Mechanical Power Presses, B11.1
- Construction, Care, and Use of Metal Sawing Machines, B11.10
- Construction, Care, and Use of Power Press Brakes, B11.3
- Safety Standard for Stationary and Fixed Electric Tools, ANSI/UL 987
- Safety Standard for Mechanical Power Transmission Apparatus, ANSI/ASME B15.1

OSHA standards for the safeguarding of mechanical equipment, 29 CFR, Part 1910.212–219, Subpart O, cover safeguarding of machinery and machine guarding. It is a violation of law to ignore these regulations.

For safeguards to be effective, whether built by the user or purchased from manufacturers, they should meet certain performance or design standards. If a company does not have a safety professional, it should get the advice of the insurance company safety engineer, manufacturer, consulting engineer, or the National Safety Council consulting department as to what designs are available in machinery and machine guarding. The National Safety Council offers guidance to members—the council's *Accident Prevention Manual for Business & Industry, Engineering & Technology*; *Safeguarding Concepts Illustrated*; and *Power Press Safety Manual* give many ideas in addition to those presented here.

To be generally acceptable, a safeguard should:

1. Conform to or exceed applicable ANSI standards and requirements of OSHA and/or OSHA-approved state plans.
2. Be considered a permanent part of the machine or equipment.
3. Afford maximum protection for operators and those performing maintenance, repairs, and servicing, and also for passersby.
4. Prevent access to danger zones and points of operation (hazards created by moving parts of the machine and equipment).
5. Not weaken machine structure.
6. Be convenient. Guards must not interfere with efficient machine operation, cause operator discomfort, or complicate light maintenance and/or cleaning the machine area.
7. Be designed for the specific job, piece part, and specific machine. (Provisions must be made for oiling, inspecting, adjusting, and repairing machine parts.)
8. Be resistant to fire and corrosion. Be easy to repair.
9. Be strong enough to resist normal wear and shock, and durable enough to serve over a long period with minimal maintenance (Figure 13–3).
10. Not be a source of additional hazards, such as splinters, pinch points, sharp corners, rough edges, or other injury sources. If possible, a safeguard covering rotating parts should be interlocked with the machine itself so the machine cannot be operated unless the safeguard is in place.

**Figure 13–3.** Safeguards should be made of sturdy materials. *(Courtesy of International Stamping Co./Midas International Corp.)*

## A Safeguarding Program

When installations of new machinery or the relocation of existing machinery or machine guarding and retrofitting are contemplated, people in the organization who have a specific interest in the machinery should be consulted and should function as a team. Management should seek the opinions of facility engineers, the safety department, and the purchasing department. They should also include machine operators, supervisors, setup and maintenance personnel, millwrights, oilers, electricians, and manufacturers. In regards to safeguarding, the manufacturers have safeguards available that are superior to any others that might be installed (see discussion in this chapter under Built-In Safeguards). Remember to consult manufacturers if the proposed safeguarding changes or interferes with the original equipment design or safety function.

OSHA and ANSI standards should be studied before design, construction, and installation of safeguarding takes place. Every safeguard, whether installed when machinery is purchased or constructed or purchased later, must meet or exceed OSHA requirements. The team, supervisor, or person responsible for purchasing, constructing, or maintaining safeguards should be familiar with applicable OSHA 29 CFR 1910, Subpart O (or other applicable portion) and ANSI standards.

A safeguarding program should include all equipment in the establishment. When several pieces of equipment are connected to form a manufacturing "cell"—they must be guarded as a cell.

A good safeguarding program can also help reinforce observance of other preventive measures. If hands and arms are protected by safeguarding, it makes sense to protect other parts of the body. Like providing eye and foot protection where applicable, a safeguard is a warning signal—a constant reminder to workers that there is a potential for hazard and danger.

## Maximum-Sized Openings

An important factor to consider in safeguard design is the maximum size of openings. If a barrier is to provide complete protection, openings must be small enough to prevent the operator or others from reaching into the danger zone. If, for operational purposes, an opening must be larger than that specified by the standard (Figure 13–4), further protection must be provided.

## Built-In Safeguards

Although sometimes a safeguard made in the facility can do a good job, often the best guard or barrier is the one provided by the manufacturer of a machine.

**Figure 13–4.** These point-of-operation guard locations show where barriers should be placed to prevent hands or forearms from contacting a hazard. The danger line is the point of operation. The clearance line marks the distance required to prevent contact between the guard and the moving parts. Minimum guarding line is ½ in. from the danger line. *(From OSHA requirements 29 CFR, 1910.217.)*

Many standard machinery and safeguarding manufacturers design first-class, serviceable safeguards for equipment, available when contacted or specified on a purchase order. The manufacturer's safeguards are usually designed to be an integral part of the machine and are sometimes superior in appearance, placement, and function to those made in house.

Companies may sometimes believe partial or makeshift barriers can do the job well enough.

Disadvantages of makeshift safeguards are obvious. Because they give a false sense of security, they may be more harmful than no safeguard at all. They require operators to be constantly alert in order to make up for the guard's inadequacy and thus give little protection against human failure. Makeshift guards are often flimsy and are certain to become damaged and ineffective, sometimes shortly after installation. At times, makeshift safeguards may not be used by workers who feel they serve no useful purpose.

A completely effective safeguard is one that eliminates the hazard completely and permanently and can withstand handling and normal wear and tear.

Too often, machinery is purchased and installed without necessary safeguards to protect operators and other workers. Frequently, an excuse given is that purchase of a machine as a capital expense is closely budgeted, while in-house constructed safeguarding is categorized as operating cost and is more acceptable. As a result, some companies buy machinery and make safeguards after the machinery has been installed.

This tactic is poor economy. Machine safeguards can be provided by the manufacturer more cheaply because costs are spread over a large number of machines. In addition, the employer increases its liability.

Advantages of securing as much built-in protection as possible from the manufacturer are as follows:

1. The cost of safeguards designed and installed by the manufacturer is usually lower than the cost of installing safeguards after a machine has been purchased.
2. Built-in safeguards conform more closely to the machine contours.
3. A built-in safeguard can strengthen a machine, act as an exhaust duct or oil retainer, or serve some other functional purpose, thereby simplifying the design and reducing the cost of the machine.
4. Built-in safeguards generally improve production and prevent damage to equipment or material in process.

### Substitution as a Safeguard

Substituting one type of machine for another sometimes can mean elimination or reduction of hazards. For example, substitution of direct-drive machines or individual motors for overhead line-shaft transmission decreases hazards inherent in transmission equipment (assuming that some older line-shaft installations still remain). Speed reducers can replace multi-cone pulleys. Remote-controlled automatic lubrication can eliminate need for employees to work dangerously close to moving parts.

### Matching Machine or Equipment to Operator

This chapter discussion has emphasized safeguarding transmission parts or points of operation. Safe operation of machinery, however, involves more than eliminating or covering hazardous moving parts. Overall hazard potential of the machine operation must be considered. These basic questions should be asked:

- Is there a materials-handling hazard?
- Are the limitations of a person's manual effort—lifting, reaching, bending, pushing, and pulling—recognized?
- Is the design of existing or proposed safeguards based on physiological factors and human body dimensions?

All physical or design features of a production machine and the workplace station should be evaluated as though the machine were an extension of a person's body and can do only what that person wants it to do. To match the machine or equipment to the operator, consider the following factors:

- **The workplace.** Machines and equipment should be arranged so the operator does minimal lifting and traveling. Conveyors and skids to feed raw stock, and chutes or gravity feeds to remove finished stock, should be considered.
- **The work height.** The workstation should be of optimal height in relation to stand-up or sit-down methods of operation. Proper height and type of chair or stool must be determined. Elbow height is the determining factor in minimizing worker fatigue. Effective work level is generally 41 in (1 m) from floor to work surface, with chair height from 25 to 31 in (0.6 to 0.8 m).
- **Controls.** Machine speed and ON-OFF controls should be readily accessible. Position and design of machine controls—such as dials, push buttons, and levels—are important. Controls should be standardized on similar machines. Operators then can be switched back and forth, as necessary, without having to use different controls. E-STOP locations must be available for both left and right hands.
- **Materials-handling aids.** Aids should be provided to minimize manual handling of raw materials and in-process or finished parts, both to and from machines. Overhead chain hoists, belts or roller conveyors, and work positioners are typical examples.
- **Operator fatigue.** Workers become fatigued at a machine workstation, often as the result of combined physical and mental activities, not simply from expending energy. Excessive speed-up, boredom from monotonous operations, and awkward work motion or operator position also contribute to fatigue.
- **Adequate lighting and other environmental considerations.**
- **Excessive noise.** More than an annoyance, excessive noise can be a real hazard because it can cause permanent hearing damage. Methods of protecting workers are given in Chapters 8 and 9 and are covered by OSHA regulations.

## SAFEGUARDING MECHANISMS

Machines have certain basic mechanisms that, if exposed, always need safeguarding. These mechanisms, which incorporate the primary hazards involved in machinery, can be grouped under the following headings:

- rotating mechanisms
- cutting or shearing mechanisms
- in-running nip points
- screw or worm mechanisms
- forming or bending mechanisms
- impact mechanisms

A piece of equipment may involve more than one type of hazardous exposure. For instance, a belt-and-sheave drive is a hazardous rotating mechanism. It also has hazardous in-running nip points.

### Rotating Mechanisms

A rotating part (Figure 13–5) is dangerous unless it is safeguarded. Mechanical power transmission apparatus represent most of this type of hazardous mechanism. Although relatively few injuries are caused by such apparatus, the injuries often are permanently disabling.

Small burrs or projections on a shaft can easily snag hair or clothing, or

**Figure 13–5.** Rotating mechanisms can seize and wind up loose clothing, belts, hair, etc. They should, therefore, be guarded. Left to right, they are (A) projecting key and set screw, (B) spokes and burrs, (C) coupling bolts, (D) bit and chuck, (E) turning bar stock, and (F) rotating shaft.

catch a cleaning rag or apron and drag a person against and around the shafting. Vertical or horizontal transmission shafts, rod or bar stock projecting from lathes, set screws, flywheels and their cross members, drills, rotating couplings, and clutches are common hazardous rotating machine parts.

Shafting, flywheels, pulleys, gears, belts, clutches, prime movers, and other types of power-transmission apparatus may seem safe by virtue of their location. However, many accidents happen where such apparatus is located—"where no one ever goes." The supervisor, oiler, and maintenance people go into these seldom-entered places. For their safety, hazards in such areas must also be safeguarded.

When a flywheel, shaft, or coupling is located so that any part of it is less than 8 ft (2.4 m) (ANSI/ASME B15.1; or 7 ft in OSHA 29 CFR 1910.212 and 219) above a floor or work platform, it must be safeguarded in one of several methods set forth in the federal standards. For example, exposed pulleys, belts, and shafting with rotating parts within 8 ft of the floor should have substantial safeguards designed to protect employees from all possible contact. The underside of belts running over passageways or work areas should be protected by heavy screening so they can cause no harm if they break.

Enclosures should be removable, or provided with hinged panels and interlocked, to facilitate inspection, oiling, or repairing of the mechanism parts. However, enclosures should not be easily removed and should be taken off only when the machine is stopped and all energy sources isolated and lockout procedures implemented, and then only by the supervisor's special order. Machines should never be operated until they are placed back in service by the person(s) performing the work and the enclosures have been properly replaced and secured.

Ends of rotating shafts projecting into passageways or work areas should be cut off or protected by nonrotating caps, safety sleeves, or guarded. Hangers should be securely fastened and well lubricated. Setscrews with slotted or hollow heads should be used instead of the projecting type. Couplings should be protected to prevent contact not only with coupling, but with the exposed rotating shafts also.

Exposed gears should be entirely enclosed by substantial fixed barriers or protection in some other equally effective way. Sheet metal is preferable. The in-running side of the gearing should be treated with special care. Removable barriers should be equipped with interlocks and the power disconnect locked in the OFF position for inspection and maintenance. Other materials can be used where excessive exposure to water or chemicals might cause rapid deterioration of metal barriers.

It is essential to enclose the gearing flywheel in power presses because the large forces involved may cause overloading, which, in turn, can cause fatigue cracks in the shaft or gear. The fact that gears have split in half, and have fallen, indicates the need for strong enclosures.

Clutches, cutoff couplings, and clutch pulleys with projecting parts 7 ft (2 m) or less from the floor should also be enclosed by fixed barrier safeguards (29 CFR 1910.212 and 219). Some clutches within the machine may be considered "guarded by location" because they are out of normal reach. But if any possibility of contact exists, a complete enclosure should be provided.

## Cutting or Shearing Mechanisms

The hazards of cutting or shearing mechanisms (Figure 13–6) lie at points where the work is being done, and where the movable parts of the machine approach or cross the fixed parts of the piece or machine. Guillotine cutters, shear presses, band and circular saws, milling machines, lathes, shapers, and abrasive wheels are typical machines that present cutting or shearing hazards.

**Saws.** A circular saw must be shielded by a guard that covers the blade at all times to at least the depth of the teeth. The hood must adjust itself automatically to the

**Figure 13–6.** Protection should be provided for all variations of common cutting or shearing mechanisms.

thickness of the material being cut to remain in contact with it. The hood should be made so that it protects the operator from flying splinters or broken saw teeth. A table saw should be equipped with a spreader or splitter. It should also have an anti-kickback device to prevent materials from being thrown back at the operator.

All band saw blade parts must be enclosed or otherwise safeguarded except the working portion of the blade between the bottom of the guide rolls and the table. The enclosure for the blade part between the sliding guide and upper saw wheel guard should be self-adjusting, if possible. Band saw and band knife wheels should be completely enclosed. The barrier should be constructed of heavy material, preferably metal.

The upper half of a swing cutoff saw should be equipped with a complete enclosure. The lower half must be designed so that it will ride over the fence and drop down on the table or the work being cut. A counterweight or other device should automatically return the saw to the back of the table when the saw is released at any point in its travel. Limit chains or other equally effective devices should be used to prevent the saw from swinging past the table toward the operator. Saw blades should be kept sharp at all times to prevent force feeding and development of cracks.

**Cutters.** On a milling machine, cutters should be shielded by an enclosure that provides positive protection for the operator's hands. A rotary cutter or slitter should have a barrier that completely encloses the cutting disks or knives. This makes it impossible for the operator to come into contact with the cutting edges while the machine is in motion.

**Shears.** The knife head on both hand- and power-operated shears should be equipped with a barrier to keep the operator's fingers away from the cutting edge. The barrier should extend across the full width of the table and in front of the hold-down. This barrier may be fixed or automatically adjusted to the thickness of the material to be cut.

If material being cut is narrower than the knife blade width, adjustable finger barriers should be installed. These will protect the open area at the sides of the material being sheared so the operators cannot get their hands caught at the sides under the blade. The barrier may be slotted or perforated to allow the operator to watch the knife, but the openings should not exceed ¼ in (6 mm), according to ANSI B11.1, and should preferably be slotted vertically to provide maximum visibility. A cover should be provided over the entire treadle length on the shears, leaving only enough room between the cover and treadle for the operator's foot. This device prevents accidental tripping of the machine if something should fall on the treadle.

**Grinding Wheels.** Because the abrasive grinding wheel is a common power tool and often used by untrained workers, it is the source of many injuries. Stands for grinding wheels should be heavy and rigid enough to prevent vibration. They should be securely mounted on a substantial foundation. Wheels should neither be forced on the spindle nor be too loose.

The supervisor must ensure that a worker trained in correct and safe procedures should be in charge of wheel installations. Before a wheel is mounted, it should be carefully examined for cracks or other imperfections that might cause it to disintegrate. A "ring" test must be performed prior to mounting (29 CFR 1910.215). The work rest should be rigid and set no farther away than ¼ in (6 mm) from the wheel face so that the material cannot be caught between the wheel and the rest. Wheels should be kept true and in balance.

Each wheel should be enclosed with a substantial hood made of steel plate to protect the operator in case the wheel breaks. The threaded ends of the spindle should be covered so that clothing cannot get caught in them. Hoods should be connected to effective exhaust systems to remove particles and prevent harmful dust from entering the grinding operation.

The manufacturer's recommended wheel speed should be adhered to at all times. Eye and face protection should be provided and properly fitted. Supervisors should instruct workers that respirators should be used if considerable dust is generated, or if a hazardous or dangerous material is present and the air contaminants cannot be properly exhausted. Respirators should be of the type required to reduce the exposure to the acceptable level and cleaned in accordance with the manufacturer's instructions. Respirators should not be issued for use by more than one employee.

It is important that grinding wheels be properly stored and handled to prevent cracking or other damage. They should be stored in a dry place where they will not absorb moisture. They should be kept in racks, preferably in a vertical position. The wheel manufacturers can supply additional information on proper storage and handling methods.

**Buffing and Polishing Wheels.** This equipment should be equipped with exhaust hoods to catch particles the wheels throw off. Not only do hoods protect the operator, but they also prevent accumulation of particles on the floor and in the work area. Supervisors should provide eye or face protection for operators. Workers should not wear gloves or loose clothing. Protruding nuts or the ends of spindles, which might catch on operators' hands or clothing, should be covered by caps or sleeves.

**Wire Brush Wheels.** The same machine setup and conditions that apply to polishing and buffing wheels apply to brushes. The speed recommended by the manufacturer should be followed strictly.

The hood on scratch wheels should enclose the wheel as completely as the nature of the work allows. It should be adjustable so protection is maintained as the diameter of the wheel decreases. The hood should cover the spindle end, nut, and flange protection. Personal protective equipment is especially important in operating scratch wheels because wires can break off and fly into an operator's face.

The materials should be held at the horizontal center of the brush. The wire tips of the brush should do the work. Forcing the work into the brush only results in (1) merely wiping or dragging the wires across the material, with no increase in cutting action, (2) increasing wire breakage, and (3) snagging the work. Small pieces should be held in a jig or fixture. More details about abrasive wheels, buffers, and scratch brushes are given in Chapter 14, Hand Tools and Portable Power Tools.

### In-Running Nip Points

Whenever two or more parallel shafts that are close together rotate in opposite directions (Figure 13–7), an in-running nip point is formed. Objects or parts of the body may be drawn into this nip point and be crushed or mangled. Typical examples of nip points are found on intermeshing gears, rolling mills, calenders, chains and sprockets, conveyors, belts and sheaves, racks and pinions, and at points of contact between any moving body and a corresponding stationary body.

Nip points should be made inaccessible by fixed barrier guards or should be protected by instantaneous body-contact cutoff switches with automatic braking devices. The in-running side of rolls like those used for corrugating, crimping, embossing, printing, or metal graining should be protected. Supervisors should ensure that operators can feed material to the machine without catching their fingers between the rolls or between the barriers and the rolls.

Protect calenders and similar rolls with a device that the operator can use to

**Figure 13–7.** These typical in-running nip points require guarding.

shut off the rolls immediately at the feed point by means of a lever, rod, or treadle. The emergency stop must be fully accessible at front and back of machines (29 CFR 1910.218). Otherwise, the nip should be guarded by an automatic electronic device that will stop the rolls if anything but stock approaches the intake points. Enclosures are usually the best way to protect chains and sprockets, racks and pinions, belts and pulleys (or sheaves), and drive mechanisms for conveyors.

## Screw or Worm Mechanisms

The hazard involved in the operation of screw or worm mechanisms is the shearing action set up between the moving screw and the fixed housing parts. Screw or worm mechanisms are generally used for conveying, mixing, or grinding materials. Examples are food mixers, meat grinders, screw conveyors, dry material mixers, and grinders of various types.

Screw conveyor covers should not be used as a walkway. If they must be walked on, additional protection should be provided. Corrosion and/or abrasive material in this conveyor type can erode the metal from the conveyor cover underside so that a person might, without thinking, step through it. When screw conveyors are constantly fed while in motion and use of an interlocked cover is impractical, install a heavy screen or mesh guard or grid. In this way, when the cover is removed for product inspection, the mesh will prevent anyone from falling through the opening.

Covers should also be provided for all mixers. The cover should be hinged to prevent removal and should have an interlock that will cut off the power source to stop action immediately when the cover is raised. The supervisor should see to it that power switches for such machines be locked out during maintenance or cleaning operations.

Rigid grids should cover openings on grinder hoppers. Such coverings should be large enough to permit materials to be fed to the grinder, but small enough to prevent any part of the operator's body from touching the cutting knives or the worm. Removable hoppers should be interlocked so that the grinders cannot be operated when the hoppers are removed.

Other safeguards for screw or worm mechanisms may include guards and devices that require the operator to use both hands on the controls. Operating controls should be located so that they cannot be activated while any body part is in a position to be caught in the machine. Points of operation often can be guarded by regulating the size, shape, and location of the feed opening.

## Metal Forming, Shearing, Punching, Bending Mechanisms

Use of metal-forming machines for stamping and forming pieces of metal and other materials has grown rapidly. Hand and finger injuries on these presses have become frequent. Misuse or abuse of these machines constitutes one of the most serious sources of mechanical problems in incident prevention.

Factors that make the problem difficult are variations in operations and operating conditions. Size, speed, and type of press; size, thickness, and kind of pieces to be worked; construction of dies; degree of accuracy required in the finished work; and length of run vary greatly. It is unwise to depend solely on operator skill for protection.

To make certain that operating methods are safe, insist that the die setter not only set the dies for a new run and test the machine for proper operation, but also set and adjust all other safeguards to the piece part and the operator. The supervisor should then ask the operator to run the machine a few strokes to make sure the safeguards are in place, and any adjustments made are correct. The supervisor must see to it that die setters or setup persons and operators know that they must complete each of these steps before starting a production run. This procedure applies equally to shearing mechanisms: for example, guillotine cutters, veneer chippers, paper-box corner cutters, and leather dinking machines. As a group, these and similar machines present the worst kind of point-of-operation hazards, yet supervisors often believe that safeguarding them would slow down production. The right safeguards, properly designed and utilized, will not only diminish the hazard, but typically will increase production.

A procedure that omits any of these steps provides no guarantee that the dies are in alignment, that the kickout is working properly, that the clutch is in proper condition, or that the guard (if it is removable) is in place and operating. Supervisors with long experience in hazard control customarily require the machine setup crew to fill out a tag as written evidence that the procedure was thorough and complete for each setting. According to 29 CFR 1910.217 (e)(1):

> Inspection, maintenance, and modification of presses—inspection and maintenance records. It shall be the responsibility of the employer to establish and follow a program of periodic and regular inspections of his power presses to insure that all their parts, auxiliary equipment, and safeguards are in a safe operating condition and adjustment. The employer shall maintain records of these inspections and the maintenance work performed.

Each job poses questions. Should fixed guards, pull-back devices, two-hand control devices, or presence-sensing devices be used? The answer depends on such factors as the type of press, full- or part-revolution; kind of stock (piece, strip, roll); type of feed (hand, slide, dial, automatic); and type of knockout or ejection (mechanical or pneumatic) used to complete the operation.

## Anti-Repeat Protection

Power presses sometimes repeat (an uninitiated or unexpected successive powered stroke of the press, resulting from a malfunction). Safeguarding, therefore, must be designed to prevent injuries that could result from such occurrences. Safeguarding measures include anti-repeat mechanisms, control component failure (control reliability), and brake monitors. Workers have sustained injuries when they reached impulsively for misplaced stock after the press has been tripped. Details of how to prevent these incidents are given later in this chapter in the section on Automatic Protection Devices. Even the best training, experience, and supervision cannot substitute for well-designed safeguarding and the constant observance of safety practices.

Power-press operations consist of primary and secondary operations. Primary operations (blanking) are the easiest to safeguard at the point of operation. Because flat material is used, guards can be constructed or adjusted to allow only sufficient opening for the material to pass through the guard into the die.

The trailing edge of strips can often be processed (1) by holding the scrap skeleton and pulling the stock through, (2) by pushing it through with the leading edge of the next skeleton or (3) by pushing it through with the leading edge of the next strip. When none of these is possible, inward shaping (a funnel shape) of a guard to meet the die at the point of stock entry to the die can help to fully use a manually advanced strip. If this customizing of the guard is not possible, the remaining tail end of the material should be discarded as scrap.

In secondary operations, parts or finished products are produced using blanks from primary operations. Safeguarding generally is more difficult than in primary operations, because manipulation of parts is required at or near the point of operation. The operator should be able to work efficiently without reaching into the point of operation. Ways to achieve safe, efficient operation include (1) providing a choice of operator-controlled feeding mechanisms, (2) using special tools, and (3) designing in positive part ejection and scrap removal.

Wherever possible, high-production or long-run dies should be guarded individually. This procedure saves setup time, and a permanent guard will not be detached or lost.

If die enclosures or fixed barriers cannot be used on secondary operations, use a presence-sensing device, a gate or movable barrier device, a two-handed trip device or two-handed control device, or a pull-back device that removes the hands from the point of operation. Unless the press has automatic or semiautomatic feeding and ejection with a die-enclosure or a fixed-barrier guard, hand tools should be used. However, a hand-feeding tool is not a point-of-operation guard or protection device and should never be used instead of it.

Safety devices must be maintained. Their use must be supervised strictly in accordance with the manufacturer's instructions; otherwise, they may fail, resulting in injuries and loss of workers' confidence. Typically, your safety department, machine manufacturer, or an insurance company safety engineering department is willing to assist in the design and installation of the best guard for a specific operation.

## Feeding Methods

Automatic or semiautomatic feeds, if they are used in conjunction with a fixed enclosure guard, may increase production, or reduce costs, or both. With these feeds and safeguarding, operators do not need to place their hands into the point of operation during ordinary feeding. An air-ejection jet can help in removing parts from the die. The enclosure should be interlocked with the clutch brake mechanism.

The choice of feed will depend on the design of the die, the shape and quantity of parts being processed, and the type of equipment available. Feed types include gravity or chute, push, follow, magazine, automatic magazine, dial, roll, reciprocating, hitch, and transfer. Feeds can be automatic, semiautomatic, or manual.

Among the advantages gained with automatic feeding are (1) the operator does not have to reach into the point of operation to feed the press; (2) the feeding method usually makes it possible to enclose the die completely; and (3) the operator can load the feed mechanism, start the press, and then leave the vicinity of the press for a considerable number of strokes. Sometimes, the operator may be able to run several presses at once.

With semiautomatic feeds, the operator does not have to reach into the point of operation, but usually must manually load the feed mechanism repeatedly or at frequent intervals. The feeding method usually makes it possible to enclose the die completely. Semiautomatic feeding is not adaptable for certain blanking operations or for nesting off-shaped pieces.

When manual feeding is required, eliminate the need for operators to place their hands or fingers within the point of operation. If this cannot be done, some method must be employed to protect the employee should the ram descend. Special tools have been developed and used successfully on operations where automatic feeds or enclosure guards are impractical. Such tools include pushers, pickers, pliers, tweezers, forks, magnets, and suction disks. These are usually made of soft metal to protect the die. Strict discipline is necessary to force operators to use them consistently. Such hand tools are not substitutes for guards, but should be used in conjunction with guard devices.

Bear in mind that few kinds of press guards provide complete protection. An automatic or semiautomatic feed may make it unnecessary for operators to place their hands in the danger zone, but may not prevent their doing so. If the operator attempts to straighten a part just before it passes under the slide, this method of safeguarding provides no real protection. Therefore, good practice combines such guards with a two-hand control device that requires constant pressure or control during the downward stroke of the press. (A two-handed control device can only be used when the press has a part revolution clutch or direct drive capable of being stopped during the stroke.)

Automatic or semiautomatic methods of feeding usually can be installed on jobs that are fed manually. Automatic feeds should be supplemented by a substantial enclosure at the point of operation, especially on slow-moving equipment, for complete protection. If possible, machine parts should be adjusted to reduce the hazard. For instance, the stroke on a press may be limited so that fingers cannot enter between the dies. More details are given later in this chapter.

## Ejecting Material

Safe removal of material is as necessary as safe feeding. Because the way the finished piece is removed from the machine may influence the choice of feeding method, the removal method must be considered when an automatic or semiautomatic feed is selected.

Various methods of ejection may be used: compressed air, punch, knockouts, strippers, and gravity. Operators should not be required to remove finished parts or scrap material from the die manually because of the hazards involved. Air blowoff systems, crankshaft-operated scrap cutters, and other devices may be used.

## Controls

Power presses should have actuating devices that prevent hands from getting under the slide or ram when the press is operated. Such devices include two-hand control devices or levers, treadle bars, pedals, and switches, located away from the point of operation. If two-hand controls or levers are used, relays or interlocks should be installed so that one switch or lever cannot easily be made inoperable and permit the press to be controlled with one hand, thus defeating the safeguard.

Press brakes are the source of many incidents. The gaging stops may be too low, and the piece being processed may slip beyond them. Because the ram motion is slow, some operators reach through the area between the ram and die to adjust the work. In doing so, their hands can be caught. The proper type of starting device will make it impossible to reach under the ram after the press has begun to operate.

Foot controls should be covered by stirrup-type covers that extend over the entire length of the treadle arm (an inverted U-shaped metal shield above the control) to prevent accidental tripping. When two or more operators run a press, foot controls or hand controls should be connected in series. Each person then is in the clear before the press can be operated.

On die-casting machines, two-hand tripping devices have been widely used. However, many consider it safer to install a sliding door that covers the die area. As the door closes, protecting the hazardous zone, it activates switches that set the machine in motion. Such a door virtually eliminates burns from splashing material.

It is prerequisite for safe die casting that no operator be allowed to place a hand or arm between the dies at any time. Long-handled pliers or tongs or similar tools should be used to place and remove stock; mechanical feeds and ejectors are even safer. Effective safeguards for these machines include two-hand tripping devices, sliding doors, treadle bars, and electrical or mechanical interlocking devices. To eliminate dangers involved in machine operation, enclosures may be built and installed over hazardous areas or equipment may be redesigned to eliminate such exposed parts.

The modern lathe is a good example of machinery made safe through improved design. Its motor drive and gear box are enclosed so that line shafts, pulleys, and belts are dispensed with. The modern power press, in which all the working parts, with the exception of the slide (ram) are enclosed, is another good example.

Safeguards used to make machinery safe include the fixed guard or enclosure, the interlocking guard or barrier, and the automatic protection device. Automatic or semiautomatic feeding and ejection methods are also ways of safeguarding machine operations.

## Fixed Guards or Enclosures

The fixed guard or enclosure is considered preferable to all other types of protection and should be used in every case, unless it has been definitely determined that this type is not at all practical. The principal advantage of the fixed guard is that it prevents access to the dangerous parts of the machine at all times. (If a fixed barrier is to provide complete protection, the openings in it must be small enough to prevent a person from getting into the danger zone.) This type of guard has another advantage. When the production job is finished, the guard remains with the die until it is needed for the next run.

Fixed safeguards may be adjustable to accommodate different sets of tools or various kinds of work. Once they have been adjusted, however, they should remain fixed. Under no circumstances should they be detached or moved.

Typical examples of the application of fixed safeguards are found on power presses, sheet-leveling or flattening machines, milling machines, gear trains, drilling machines, and guillotine cutters. Some remote feeding arrangements make it unnecessary for the operator to approach the danger point, and fixed barriers can be installed at a distance.

Where a fixed safeguard cannot be used, an interlocking guard or barrier should be fitted onto the machine as the first alternative. Interlocking may be mechanical, electrical, pneumatic, or a combination of types. The purpose of the interlock is to prevent operation of the control that sets the machine in motion until the guard or barrier is moved into position. Operators then cannot reach the point of operation, the point of danger.

When the safeguard is open, permitting access to dangerous parts, the starting mechanism is locked to prevent accidental starting. A locking pin or other safety device is used to prevent the basic mechanism from operating; for example, to prevent the main shaft from turning. When the machine is in motion, the enclosure cannot be opened. It can be opened only when the machine has stopped or has reached a fixed position in its travel.

To be effective, an interlocking safeguard must satisfy three requirements:

1. Guard the dangerous part before the machine can be operated.
2. Stay closed until the dangerous part is at rest.
3. Prevent operation of the machine if the interlocking device fails.

Two-hand control devices are used in many types of interlocking controls on mechanical power presses but they can be used only with part-revolution clutch equipment. These devices require simultaneous and sustained pressure of both hands on switch buttons, air-control valves, mechanical levers, or controls interlocked with a foot control, to name just a few. Two-hand operating attachments should be connected, so it is impossible to block, tie down, or hold down one button, handle, or lever, and still operate the machine.

When gate devices or hinged barriers are used with interlocks, they should be arranged so that they completely enclose the pinch point or point of operation before the operating clutch can be engaged.

Interlocking controls are often installed on bakery machinery, guillotine cutters, power presses, dough mixers, some kinds of pressure vessels, centrifugal extractors, tumblers, and other machines on which covers or barricades must be in place before the starting control can be operated.

## Automatic Protection Devices

An automatic protection device may be used, subject to certain restrictions, when neither a fixed barrier nor an interlocking safeguard is practicable. The device must prevent the operator from coming in contact with the dangerous part of the machine while it is in motion, or must be able to stop the machine in case of danger.

An automatic device functions independently of the operator. Its action is repeated as long as the machine is in motion. The advantage of this type of device is that tripping can occur only after the operator's hands, arms, and body have been removed from the danger zone.

An automatic protection device is usually operated by the machine itself through a system of linkage, through levers, or by electronic means. There are many variations. It can also be a hand-restraint or similar device, or a photoelectric relay.

Pull-back devices are attached to the operator's hands or arms and connected to the slide or ram plunger, or outer side of the press so that the operator's hands or fingers will be withdrawn from the danger zone as the slide (or ram plunger) or outer slide descends. These devices should be installed and adjusted for each operator and readjusted for each job change, piece part, and press changes so that they will properly pull the operator's hands clear of the danger zone.

All electronic safety devices (two-handed control devices) for power presses are made to perform the same function when energized. They interrupt the electrical current to the power press and shut off the machine (just as if the STOP button had been pushed). Electronic safety devices are effective only on power presses that have air, hydraulic, or friction clutches. Such devices are not effective on power presses with full-revolution clutch equipment. This is because once the operating cycle of this type of power press starts, nothing can prevent completion of the cycle.

To be effective, the electronic device should be operated from a closed electric circuit. Interruption of the current will automatically prevent the press from tripping. It is the supervisor's responsibility to make sure these devices are properly adjusted and maintained in peak operating condition. Many injuries have occurred because of improper adjustments and because parts have been allowed to become worn or need repair.

One advantage claimed for electronic devices (presence-sensing devices) is the absence of a mechanism in front of the operator. This is particularly advantageous on large presses. The electric-eye device should be installed far enough from the danger zone so that it will stop the slide or ram before the operator's hand can get underneath. Sufficient light beams should cover the bad-break area with a curtain of light.

Indexing is another press-shop term used to describe a mechanical method of feeding stock into press dies. One method is by a dial feeder. The dial feed is constructed, as its name implies, in the form of a dial having multiple stations that progress into the die by the indexing motion of the dial. The indexing of the dial should take place in conjunction with the up-stroke of the ram. When hand feeding, the index circuit should be controlled by dual-run buttons. Release of either button, during indexing, should stop the index cycle by releasing a safety clutch in the table. Safety guards should be connected to switches that stop the crank motion of the press whenever they are bumped. Both of these safety precautions will prevent accidents if the operator is tempted to reach into the die area to correct an improperly positioned part on the dial.

## AUTOMATION

Automation, a somewhat misused term, is defined in this text as the mechanization of processes by the use of automatically controlled conveying equipment. Automation has minimized the hazards associated with manually moving stock in and out of machines and transferring it from one machine to another. It has also minimized exposure to the causes of hernias, back injuries, and foot injuries.

However, experience shows that those who work with automated equipment must have a thorough knowledge of its hazards and must be well-trained in safe work practices to avoid incidents. Most finger and hand injuries result from operator exposure to the closing or working parts of a machine in the process of loading and unloading. The use of indexing fingers, sliding dies, tongs, or similar hand tools reduces the hazard from such exposure. Nevertheless, the supervisor must make sure that these devices are used consistently and correctly.

Like most innovations, automation has brought not only benefits, but hazards. Because each automatic operation is dependent on others, machine breakdown or failure must be corrected quickly. Also, because speed is highly important, maintenance employees may expose themselves, inadvertently perhaps, to working equipment parts. It is therefore imperative to have a mandatory policy that machines, equipment, and processes must be completely de-energized, isolated, and locked out at the power source before servicing the machine. A well-defined lockout procedure should be written down and followed carefully.

Automated handling also has increased the use of stiles or crossovers. These should be constructed and installed in accordance with ANSI A12.1, *Safety*

*Requirements for Floor and Well Openings, Railings, and Toeboards.* Automation eliminates or greatly reduces exposure to mechanical and handling hazards. In the single-operation process, however, the basic principles of safeguarding equipment must still be applied. These principles are:

- Engineer the hazard out of the job as much as possible.
- Guard against the remaining hazards.
- Educate and test the performance of workers.
- Insist on use of safeguards provided.

## Robotics

Robots—machines specifically designed and programmed to perform certain operations—are rapidly becoming a part of the work environment. These machines can cause incidents to unsuspecting people working in the vicinity. Robot installation and safeguarding must conform to *Industrial Robots and Industrial Robot Systems—Safety Requirements*, ANSI/RIA 15.06–1999.

As early as possible, consult the engineer who designed or worked on the robot to learn about its capabilities, features, hazards, and operation. Allow for proper clearances when the robot is working with peripherals, such as machine tools, presses, transfer lines, palletizers, gaging stations, and so on. Even the robot with the best conceptual design may be efficient only if it has these clearances. Providing clearance for personnel, however, is an even more important factor. Although robots perform repetitive tasks for long periods of time, they are still machines and require periodic preventive maintenance. Eventually, the robots will need repairs to keep their devices, modules, and/or tooling working properly. Clear spaces and safe areas must be incorporated in the design stage.

A major hazard of robots involves their computer programming. When an operation is being performed, it may be virtually impossible to shut it off or reverse the cycle because of the automatic programming feature. Therefore, workers should be warned not to get near these machines. In some cases, it may be possible to totally enclose the robot, but in many instances, this may be impractical.

Control systems for robots usually work with a "priority interrupt" scheme. For example, higher priorities are interrupted by special hardware devices to signal that personnel are entering the work area, or that tooling needs to be changed or repaired. The supervisor should follow these basic rules:

1. Get as much information as possible on the operation of the robot.
2. Provide proper clearances and barrier guards around the robot.

3. Discuss all facets of the operation, including safety, with your people. All personnel should be trained.

4. Do not allow workers near robots in operation. Wait for the cycle to be completed and the machine de-energized before approaching.

Some jobs may require work to be done with power "on" (such as alignment and repair of servo systems). This must be realized in the design stage by the user's engineers. It cannot be assumed that work always will be performed in a power-off state. The simplest guarantee that equipment is safe to work on is to insist that repair personnel and the maintenance crew follow their company's lockout procedures. These are discussed later in this chapter.

## Safe Practices

Safeguards are of primary importance in eliminating machine incidents, but they are not enough. Employees who work around mechanical equipment or operate a piece of machinery must have respect for safeguards.

Before being permitted to run equipment, operators should be instructed in all the practices required for safe operation. Even experienced operators should be given refresher training, unless they know the hazards and the necessary precautions to be taken. In addition, employees who do not operate machinery, but who work in machine areas, should also receive instruction in basic safety practices.

Positive safety procedures should be established to prevent employees from misunderstanding instructions. You should enforce the following rules:

1. No guard, barrier, or enclosure should be adjusted or removed for any reason by anyone unless that person has permission from the supervisor, has been specifically trained to do this work, and machine adjustment is considered a normal part of the job.

2. Before safeguards or other guarding devices are removed so that repair or adjustments can be made, or equipment can be lubricated or otherwise serviced, the power for the equipment must be turned off and the main switch locked out and tagged.

3. No machine should be started unless the safeguards are in place and in good condition.

4. Defective or missing safeguards should be reported to the supervisor immediately.

5. Employees should not work on or around mechanical equipment while wearing neckties, loose clothing, watches, rings, or other jewelry.

# MAINTENANCE OF SAFEGUARDS

The supervisor is responsible for scheduling inspection of machine safeguards as a regular part of machine inspection and maintenance. Such inspections are necessary because employees tend to operate their machines without safeguards if they are not functioning properly, if they have been removed for repairs, or if they interfere in any manner with their operations. A guard or enclosure that is difficult to remove or to replace may never be replaced once it has been taken off. An inspection checklist can be developed for each machine to simplify the job and provide a convenient record for follow-up.

## Maintenance and Repairs

Machines are subject to wear and deterioration, and can become unsafe to operate. Wear cannot be prevented, but it can be reduced by controlling loads through proper manufacturing methods, alert supervision, employee attention, and good maintenance.

Lubrication is a basic maintenance function. Centralized lubrication will reduce hazards to oilers when climbing ladders to reach less accessible points. Some changes may be made so that most lubrication can be done at floor level. When oilers or repair crews must lubricate flywheel bearings, motors, and other parts at the tops of presses, good work practice suggests that permanent ladders with sturdy cages or wells should be installed.

Regardless of the type of lubrication necessary or the method used, it is the supervisor's responsibility to ensure that the machinery—including driving mechanisms, gears, motors, shafting hangers, and other parts—is being lubricated properly. Where automatic lubrication is not possible or feasible, extension grease or oil pipes should be attached to machines so that the oiler can avoid coming in contact with moving parts. Automatic lubrication should not be used, however, in cases where the oil or grease can congeal in the pipes.

## Lockout/Tagout Procedures: The Control of Hazardous Energy Sources

Lockout or tagout procedures are designed to isolate or shut off machines, equipment, and processes from their power sources before employees perform any servicing or maintenance work. Employees must be trained in these procedures and instructed to replace safeguards after the work is completed. The supervisor, group leader, or authorized employee must enforce the company's lockout program.

### Replacements

It is generally up to the supervisor to alert management whenever replacement of machinery and/or its safeguards is advisable. This step may seem like a production problem, but it is related to incident prevention. When machinery, safeguards, and safety devices become so worn that repairs cannot restore them to original operating efficiency, they constitute hazards. Often, unfortunately, these are concealed hazards. For example, if a non-repeat device on a press becomes so worn that it does not operate or fails to operate properly, workers can be seriously injured. This hazard can be eliminated only by replacing the machine part.

Several methods are used to determine when a machine needs maintenance or an overhaul to prevent a breakdown, or when replacements are in order. Some supervisors rely on reports of machine operators. Some have occasional inspections made by oilers, machine setters, or similar mechanics. Others set up inspections on a regular schedule and do it themselves, or have it done, routinely. The best plan is an established system of frequent, regularly spaced inspections.

## SUMMARY OF KEY POINTS

Key points covered in this chapter include:
- The basic principles of safeguarding are: (1) engineer the hazard out of the job as much as possible, (2) guard against remaining hazards, (3) educate and test safety performance of workers, and (4) use safeguards provided. If such precautions are not taken, serious injury and property loss can result.
- Safeguarding is often thought of as being concerned only with the point of operation or with the means of power transmission. However, safeguards also prevent incidents from direct contact with exposed moving parts, work in process, machine failure, electrical failure, and operator error or human failure. The primary benefits of safeguarding are reducing the possibility of injury and improving production.
- Many safety devices, as well as barriers, enclosures, and tools, have been developed to protect machine operators. OSHA and ANSI standards require and present detailed guidelines for safeguarding such machines and equipment. Establishing effective methods of safeguarding power transmissions is easier than guarding points of operation because power transmissions are more standardized.

- An acceptable safeguard should conform to or exceed standards, be a permanent machine part, afford maximum protection, prevent access to the danger zone, be convenient, be designed for the specific machine and job, resist fire and corrosion, resist wear and shock, be long lasting, and not be a source of additional hazards. Built-in safeguards are generally more effective and less costly in the long run than "in-house" designed and constructed barriers.

- Safeguarding programs should adhere to pertinent regulatory codes and be instituted before machinery or equipment is installed. At times, substituting a less hazardous machine can solve a safeguarding problem. Regardless, supervisors should make every effort to match equipment to operators in terms of workplace, work height, controls, material handling aids, operator fatigue, adequate lighting, noise levels, and other environmental factors.

- Certain basic machine mechanisms, if exposed, always need safeguarding. These include rotating, cutting or shearing, screw or worm, forming or bending, impact mechanisms, and in-running nip points. These parts must have special safeguarding covers, barriers, or other equipment to shield operators and nearby workers from injury and to prevent equipment or product damage.

- Power presses consist of primary and secondary operations in which workers are exposed to potential injury. Wherever possible, high-production or long-run dies should be guarded individually. These devices should shut off the presses or prevent normal operation if workers' hands, clothing, or tools become caught in a press.

- Automatic or semiautomatic feeds, if they can be used, are generally successful safeguards. They can increase production, reduce costs, or both. The main advantages of these safeguards are that operators do not have to reach into the point of operation to feed the press, the die can be enclosed completely, and operators can leave the press untended for brief periods.

- Safe removal, or ejection, of material is as necessary as safe feeding. Various methods of ejection include compressed air, punch, knockouts, strippers, and gravity. Operators should not be required to manually remove finished parts or scrap material.

- Controls should be designed so that machinery can be operated without endangering workers. Foot controls, for example, should be covered by stirrup-type covers to prevent accidental tripping. Dangers involved in machine operation can be eliminated either by enclosing hazardous machine parts, or by redesigning to eliminate exposed parts.

- Safeguards include fixed guards or enclosures, interlocking guards or barriers, and automatic protection devices. Fixed guards or enclosures are preferred to all other types because workers cannot remove them. When this safeguard is not practicable, interlocking guards or barriers can be fitted onto the machine to prevent operation of the controls until the guard or barrier is positioned.

- When neither the fixed nor interlocking safeguard can be used, an automatic protection device can be applied. This safeguard prevents the operator from coming into contact with the dangerous part of the machine while it is in motion. It can also stop the machine in an emergency.

- Automation has minimized many hazards associated with manually moving stock in and out of machines but has brought its own hazards. Automated machinery must be completely de-energized (all energy sources isolated and locked out) before anyone is allowed to service or repair the machines. To apply effective safeguards on robots, supervisors must learn as much about the operation of the robot as possible, provide proper clearance and barriers around the robot, discuss all facets of operation and safety issues with workers, and not allow workers near robots in operation until the cycle is completed and the machine is de-energized.

- Supervisors must instruct workers in the use of effective safeguards and safe practices to prevent incidents and injuries. All workers, before operating machinery, should be thoroughly trained in how to use safeguards on equipment. Positive safety procedures should be established and vigorously enforced.

- Supervisors must set a schedule for inspecting and maintaining machine safeguards as a regular part of operations. Wear and tear on safeguards can be reduced by proper machine-handling methods, alert supervision, attention of employees, and regular maintenance. Before allowing workers to service machines and equipment, supervisors should be sure that lockout/tagout procedures have been implemented. Worn out, corroded, or damaged machine parts or safeguards should be replaced or repaired immediately.

# 14

# Hand Tools and Portable Power Tools

**After reading this chapter, you will be able to:**

- Establish an incident control program based on four safe work practices: selecting the right tools for the job, using tools correctly and safely, keeping tools in good condition, and storing tools properly
- Describe the major hazards and safeguards for a wide variety of hand and portable power tools
- Explain the need for a centralized tool supply room or other way of controlling access to company tools
- Establish procedures for regular tool inspection, repair, and replacement

A major responsibility of the supervisor is to train people in safe handling of hand and power tools. This chapter discusses ways that hand and portable power tool injuries can be avoided. To pinpoint where major problems lie, first review department or company incident records. Identify all tool-related incidents, try to identify specific hazards, and check to see if any employees have had an unusual number of injuries. Such workers should have additional training and closer supervision. Develop written procedures for any tools used only rarely, to remind workers how to use the equipment properly and safely.

Training and close supervision are also needed when new tools, processes, and equipment are introduced in the operation or changes are made to an existing job. A job safety analysis of each task will serve to clarify tool needs, personal protective equipment (PPE) required, and safe work practices.

## SAFE WORK PRACTICES

To be effective, a program to reduce tool-related injuries must include training in five basic safe practices. Your incident control program should include the following activities:

1. Evaluate and purchase the appropriate tools.
2. Train employees to select the right tools for each job. Good job safety analysis and job instruction training helps here. (See Chapter 4, Safety and Health Training.)
3. Train and supervise employees in the correct use of tools.
4. Set up regular tool inspection procedures, including inspection of employee-owned tools and provide good repair facilities to ensure tools are being maintained in safe condition.
5. Set up a procedure for control of company tools. A checkout system at tool cribs is ideal. Provide proper storage facilities in the tool room and on the job.

### Select the Right Tool

While many tools can be used to do a job, proper selection means choosing the right tool that can be most safely used by the employee.

Using improper tools can result in errors, injuries, and damage to products. For example, an adjustable wrench can tighten a nut, but a box end or a socket wrench of the proper size is better. Ergonomically designed tools that meet job requirements provide the best fit for the employee and the job.

Tool selection is related to other factors, such as job setup, workspace available, work height, and so forth, each related to selection of the safest, most efficient tool for the job. Cost and quality of tools are additional items to consider. Employers generally are expected to provide the correct tool(s) to do the job safely.

## Use Tools Correctly and Safely

The supervisor shows employees how to use tools safely and correctly. Too often workers may select tools that are neither safe nor suited to the task, such as substituting a screwdriver for a crowbar, or using pliers instead of the proper wrench. In many situations, the employee should study a tool's manual, which may suggest the use of proper PPE. Supervisors should ensure that workers understand the correct use of their tools and are aware of effects on other workers in the vicinity. Guards on tools must be checked to be sure they operate effectively.

Every tool should be examined before each use. Employees in the tool crib and workers using tools on the job should be trained to determine wear and damage on all tools used so that worn or damaged items can be repaired or discarded. Cutting edges must be kept sharp. Tools must be kept in good repair through a planned tool maintenance program. Good housekeeping is closely related to good, clean tools.

Tools that have deteriorated should not be used until repaired to meet manufacturers' specifications. Examples are wrenches with cracked or worn jaws, screwdrivers with broken bits (points) or broken handles, hammers with loose heads, dull saws, deteriorated extension or power cords on electrical tools, broken plugs, and improper or removed grounding systems. Procedures should be established to check and remove from service any deteriorated tools.

Employees are usually not expected to do tool maintenance work. If they are, they should be trained and supervised.

## Store Tools in a Safe Place

Tools need to be stored safely either at the work area or in a common tool crib. If employees own their own tools, safe storage is of equal importance.

Make a complete check of operations to determine a need for special tools to do work more safely. Special tools may require unusual handling and storage. For example, tools like powder-actuated hand tools should be kept under lock and key and made available only to personnel trained and certified in their use.

# USE OF HAND TOOLS

The misuse of common hand tools is a major source of injury to industrial workers. Injury often results because it is assumed that "anybody knows how to use" common hand tools. Observation and records of injuries show that this is not the case. The supervisor should study each job, train new workers, and retrain veteran employees on correct procedures for using tools. Where employees have the privilege of selecting or providing their own tools, supervisors should advise them on using the safest equipment for each task. Regardless of who provides or owns the tools, they must be of the type specified by the employer and maintained as required by the employer. This requirement is so important that considerable attention must be given to establishing safe work practices with hand and portable power tools.

The supervisor must enforce all established practices. Check the condition of tools frequently to see that they are maintained and sharpened correctly, and that guards are not altered or removed. Make routine checks regarding the use and condition of PPE used with these tools. Adopt specific practices for using hand tools and associated PPE for each operation. The required assessment of hazards should identify your protective equipment needs.

## Cutting Tools

All edged hand-operated cutting tools should be used in such a way that, if a slip occurs, the direction of force will be away from any part of the user's body. For safety and efficiency, edged tools should be kept sharp, and ground to the proper angle. A dull tool does a poor job and may stick or bind. Any unexpected difference in tool performance may cause the user to lose control of the tool. Thus, all cuts should be made along the grain when possible.

**Metal Chisels.** Factors determining selection of a cold chisel are (1) materials to be cut, (2) size and shape requirements of the tool, and (3) depth of the cut to be made. The chisel should be heavy enough so that it does not buckle or spring when struck. For best results and maximum safety, flat and cape chisels should be ground so that the faces form an angle of 70 degrees for working on cast iron, 60 degrees for steel, 50 degrees for brass, and about 40 degrees for babbitt and other soft metals.

A chisel edge large enough for the job should be selected, so the blade rather than the point or corner, is used. Also, a hammer heavy enough to do the job should be used. Some workers prefer to hold the chisel lightly in the hollow of their hands, supporting the chisel with the thumb and first and second fingers. If the hammer glances from the chisel, it will strike the soft palm rather than the knuckles. Other workers think that a grip with the fist holds the chisel steadier and minimizes the chances of glancing blows. In some positions, this may be the only grip that is natural or even possible.

Hand protection can also be supplied by a rubber pad, forced down over the chisel to provide a hand cushion. Workers should chip in a direction away from the body. When shearing with a cold chisel, workers should hold the tool at the vertical angle that permits one cutting edge bevel to be flat against the shearing plane. Employees should wear eye protection when using chisels and set up a shield or screen to prevent injury to other workers in the vicinity. If a shield does not afford enough protection to all exposed employees, these workers should also wear eye protection.

Bull chisels held by one person and struck by another require tongs or a chisel holder to guide the tool so the worker is not exposed to injury. Both workers should wear eye, head, and foot protection.

The one who swings the sledgehammer should not wear gloves.

Dress all chisel heads at the first sign of mushrooming because mushroomed heads often produce flying chips. It is suggested that when using a ½-in. (1.27 cm) steel chisel, a hammer with a 1½-in. (3.8 cm) diameter face be used.

**Tap and Die Work.** This type of work requires certain safeguards. The work piece should be firmly mounted in a vise. Only a T-handle wrench or adjustable tap wrench should be used. Steady downward pressure should be applied on the taper tap. Excessive pressure causes the tap to enter the hole at an angle or bind the tap, causing it to break. A properly sized hole must be made for the tap, and the tap should be lubricated as necessary.

Keep hands away from broken tap ends. Broken taps should be removed with a tap extractor. If a broken tap is removed by using a prick punch or a chisel and hammer, the worker should wear eye protection. When threads are being cut with a hand die, hands and arms should be kept clear of the sharp threads coming through the die, and metal cuttings should be cleared away with a brush.

**Hacksaws.** These tools should be adjusted in the frame to prevent buckling and breaking, but should not be so tight that the pins that support the blade break. Install blades with teeth pointing forward.

The preferred blade to be used is shown in Table 14–A. A general rule is that at least two teeth should be in the cutting piece. It is advisable to release tension on the blade before storing hacksaws and then re-tension the blades before starting work.

| | **Table 14–A. Selector for Hacksaw Blades** | |
|---|---|---|
| **Pitch of Blade (Teeth per Inch)** | **Stock to Be Cut** | **Explanation** |
| 14 | Machine steel<br>Cold rolled steel<br>Structural steel | The course pitch makes saw free and fast cutting |
| 18 | Aluminum<br>Babbitt<br>Tool steel<br>High-speed steel<br>Cast iron | Recommended for general use |
| 24 | Tubing<br>Tin<br>Brass<br>Copper<br>Channel iron<br>Sheet metal of 18 gage or more | Thin stock will tear and strip teeth on a blade of coarser pitch |
| 32 | Sheet metal of less than 18 gage<br>(18 Gage = 1.27 Inch) | |

**Files.** Selecting the right kind of file can prevent injuries, lengthen the life of the file, and increase production. A file-cleaning card or brush should be used to keep the file in peak condition. Files should not be hammered or used as prying tools. Such abuse frequently results in the file's chipping or breaking, causing injury to the user. A file should not be made into a center punch, chisel, or any other type of tool because the hardened steel can fracture. Use a vise, whenever possible, to hold the object being filed.

The correct way to hold a file is to grasp the handle firmly in one hand and use the thumb and forefinger of the other to guide the point, using smooth file strokes. This technique provides good control and produces better work more safely.

A file should never be used without a smooth, crack-free handle; otherwise, if the file binds, the tang may puncture the palm of the hand, the wrist, or other part of the body. Under some conditions, a clamp-on, raised offset handle can provide extra clearance for the hands. Files should not be used on lathe stock turning at high speeds (faster than three turns per file stroke). The end of the file can strike the chuck, dog, or face plate and throw the file (or a metal chip) back at the operator, inflicting serious injury.

**Tin Snips.** This tool should be heavy enough to cut the material easily so the worker needs to use only one hand on the snips and the other to hold the material. The material should be well supported before the last cut is made so the cut edges do not press against the hands. Push down the sharp edges next to the hand holding the snips when cutting long sheet-metal pieces. The jaws of snips should be kept tight and well lubricated. When not in use, tin snips should be hung up or laid on a shelf. Burrs on the cut usually indicate a need for adjusting or sharpening the snips.

Workers should always wear gloves and use eye protection when trimming corners or slivers of sheet metal. Small particles can fly with considerable force into a worker's face.

Other cutters, such as those used on wire, reinforcing rods, or bolts, should have ample capacity for the stock. Otherwise, the jaws may spring or spread. Also, a chip can fly from the cutting edge and injure the user.

Cutters are designed to cut only at right angles to the plane of the material being cut. They should not be "rocked," hammered, or pushed against the floor to facilitate the cut because they are not designed to withstand excessive strain. This practice can nick the cutting edges, lessening the cutter's ability to perform.

A good rule is, "If it doesn't cut with ease, use a larger cutter or a cutting torch." Cutters require frequent lubrication. To keep cutting edges from becoming nicked or chipped, cutters should not be used as nail pullers or pry bars.

Cutter jaws should have the hardness specified by the manufacturer for the particular kind of material to be cut. By adjusting the bumper stop behind the jaws, workers can set the cutting edges to have a clearance of 0.003 in (0.076 mm) when closed. Always store cutters in a safe place (consult the manufacturer's recommendations).

**Punches.** Punches are used like chisels. They should be held firmly and securely, and be struck squarely. The tip should be kept shaped as the manufacturer has specified. Punches should be held at right angles to the work.

**Wood Chisels.** All employees should be instructed in the proper method of holding and using wood chisels. While molded plastic and metal handles are often available, wood handles are also found on wood chisels. If the handle is wood, it should be free of splinters and cracks. The wood chisel handle that is struck by a mallet should be protected with a metal or leather cap to prevent splitting and mushrooming. *NOTE*: Do not use a metal hammer on wood chisels.

Work to be cut should be free of nails to avoid damage to the blade or to prevent a chip from flying into the user's face or eye. (Eye protection is required.) The steel in a chisel is hard so that the cutting edge will hold, so it is also brittle enough to break if the chisel is used as a prying bar. When not in use, the chisel should be kept in a rack, on a workbench, or in a slotted section of the tool box so that the sharp edges will be protected. Sharp edges can and should be safeguarded.

**Wood Saws.** These tools should be carefully selected for the work they perform do. For fast crosscut work on green wood, a coarse saw (4 to 5 points per inch) is best. A fine saw (more than 10 points per inch) is better for smooth, accurate cutting in dry wood. Saws should be kept sharp and well-set to prevent binding. When saws are not in use, make sure they are placed in racks.

Saw set (the amount of angle or lean of a point from the blade) is needed to cut wood or other materials properly and cleanly. Inspect the material to be cut to avoid sawing into nails or other metal.

Manual sawing should be a one-hand operation. When starting a cut, workers should guide the saw with the thumb of their free hand held high on the side of the saw blade. They should not place a thumb on material being cut. Begin with a short, light stroke toward the body. After the cut begins, increase pressure and increase stroke length. Use light, shorter strokes as the cut is completed.

**Axes.** To use an axe safely, workers must be taught to check the axe head and handle, clear the area for an unobstructed swing, and swing correctly and accurately. Accuracy is attained by practice and proper handhold—and good training. All other workers must be kept a safe distance away from the direct line of the axe swing.

A narrow axe with a thin blade should be used for hardwood, and a wide axe with a thick blade for soft wood. A sharp, well-honed axe gives better chopping speed and is safer to use because it bites into the wood. A dull axe may glance off the wood being cut and strike the user in the foot or leg.

The person using the axe should make sure there is a clear circle in which to swing before starting to chop. Vines, brush, and shrubbery within the range, especially overhead vines that may catch or deflect the axe, should be removed.

Axe blades must be protected with a sheath or metal guard wherever possible. When the blade cannot be covered, it is best to carry the axe at one's side. The blade on a single-edged axe should be pointed down.

**Hatchets.** Hatchets, which are used for many purposes, frequently cause injury. For example, when workers attempt to split a small piece of wood while holding it in their hands, they may strike their fingers, hand, or wrist. Hatchets are dangerous tools in the hands of an inexperienced worker. To start the cut, it is good practice to strike the wood lightly with the hatchet, then force the blade through, by striking the wood against a heavier or larger block of wood.

Hatchets should not be used for striking metal because flying chips may injure the user or others nearby.

**Knives.** These tools cause more disabling injuries than any other hand tool. In the meatpacking industry, for example, hand knives account for more than 15 % of all disabling injuries. The principal hazard in using knives is that the hands may slip from the handle onto the blade, or the knife may strike the body or the free hand. A handle guard or a finger ring (and swivel) on the handle reduces these hazards.

The cutting stroke should be away from the body. If that is not possible, the hands and body should be in the clear; a heavy leather apron, cut-resistant gloves, or other protective clothing should be worn; and, where possible, a rack or holder should be used for the material to be cut. Jerky motions and the use of excessive force should be avoided to help maintain balance and control. Be sure employees are trained and supervised. Training in handling knives is very important in the food-service and similar industries.

Workers who must carry knives with them on the job should keep them in sheaths or holders. Never carry a sheathed knife on the front part of a belt. Always carry it over the right or left hip, toward the back, for added protection by locating it away from vital areas of the body.

Knives should never be left lying on benches or in other places where they could cause hand injuries. When not in use, they should be kept in racks with the edges guarded. Safe placement and storage are important to knife safety.

Ring knives—small, hooked knives attached to a finger ring—are used where string or twine must be cut frequently. Supervisors should make sure that the cutting edge is kept outside the hand, not pointed inside. A wall-mounted cutter or blunt-nose scissors would be safer.

Carton cutters are safer than hooked or pocket knives for opening cartons. They not only protect the user, but limit the depth of cuts that could damage carton contents. Frequently, damage to contents of soft, plastic bottles may not be detected immediately; subsequent leakage may cause chemical burns, damage other products, or start fires.

Make certain that employees who handle knives have ample room, so they are not in danger of being bumped by moving machinery or other employees. For instance, a left-handed worker should not stand to the right of a right-handed person. The left-handed person might be placed at the left end of the bench or otherwise given more room. Workers should be trained to cut away from or out of line with their bodies.

Knives must be kept separate from other tools to protect the cutting edges and workers. Be particularly careful about employees leaving knives hidden under product, scrap paper, or wiping rags, or among other tools in work boxes or drawers.

Work tables should be smooth and free of slivers. Floors and working platforms should have slip-resistant surfaces and be kept unobstructed and clean. If sanitary requirements permit mats or wooden duck boards, they should be in good repair so workers do not trip or stumble. Conditions that cause slippery floors should be controlled as much as possible by good housekeeping.

Careful job analysis and incident investigation may suggest some changes in the work procedures that will make knives safer to use. For instance, on some jobs, special jigs, racks, or holders may be provided so it is not necessary for the operator to stand close to the piece being cut.

Wiping dirty or oily knives on aprons or clothing should be eliminated. The blade should be wiped with a towel or cloth with the sharp edge turned away from the wiping hand. Sharp knives should be washed separately and in a way that they will not be hidden under wash water.

Horseplay should be prohibited around knife operations. Throwing, "fencing," trying to cut objects into smaller and smaller pieces, and similar practices are dangerous.

The supervisor should make sure that nothing is cut that requires excessive pressure on the knife, such as frozen meat. Food should be thawed before it is cut or else frozen food should be sawed. In addition, knives should not be used as substitutes for openers, screwdrivers, or ice picks.

## Miscellaneous Cutting Tools

Planes, scrapers, bits, and draw knives should be used only by experienced personnel. Keep these tools sharp, clean, dry, and in good condition. When not in use, they should be placed in a rack on the bench or in a toolbox to protect the user and prevent damage to cutting edges.

## Torsion Tools

Many tools used to grip materials are available. Proper tool selection and usage are important. Damaged tools should be removed from service.

**Wrenches.** All wrenches should be pulled, not pushed (Figure 14–1). Workers' footing should be secure. Ample work space is needed near the wrench to provide clearance for fingers. Use a short, steady pull. If a nut does not loosen or tighten fully, a larger wrench may be needed. Open-end and box wrenches should be inspected to be sure that they fit properly. Do not mix metric with nonmetric sizes. Wrenches should not be hammered or struck unless they are designed for striking. See Figure 14–2 for a striking-face wrench, which is designed to be hammered.

Socket wrenches (Figure 14–3) offer great flexibility. The use of special types should be encouraged on jobs where workers can be injured. In some situations, socket wrenches can be safer to use than adjustable or open-end wrenches, and they also protect the bolt head or nut.

Wrench jaws that fit prevent damage to nuts and bolt heads because they are less likely to slip.

Adjustable wrenches are not intended to take the place of open-end, box, or socket wrenches. They are used mainly for nuts and bolts that do not fit standard wrench sizes. The correct method

**Figure 14–1.** Adjustable (crescent) wrenches are probably the most often misused wrenches. They are made for use on odd-sized nuts and bolts that other wrenches might not fit. To use, the wrench should be placed on the nut so that when the handle is pulled, the moveable jaw is closer to the user's body. The pulling force will then be applied to the fixed jaw, which is the stronger of the two.

**Figure 14–2.** Using a striking-face wrench designed to be hammered is a safe way to handle large nuts that must be set tight or frozen nuts that must be loosened.

**Figure 14–3.** Socket wrenches have a square hole in the drive end into which various handles can be fitted. The ratchet handle increases turning speed because the socket does not have to be removed from the nut between turns. The universal joint fitting provides flexibility where space is limited.

is to apply the open side of the jaws toward you so that pressure is applied to the fixed jaw as the wrench is pulled toward the body.

**Pipe Wrenches.** Pipe wrenches, both straight and chain tong, should have sharp jaws and be kept clean to prevent slipping.

The adjusting nut of the wrench should be inspected frequently. If it is cracked, the wrench should not be used. A cracked nut may break under strain, causing failure of the wrench and possible injury to the user.

Workers on overhead jobs have been seriously injured when pipe wrenches slipped on pipes or fittings, causing them to lose their balance and fall. Using a wrench of the wrong length is another cause of accidents. A wrench handle too short for the job does not give proper grip or leverage. An oversized wrench handle may strip the threads or break the fitting or the pipe suddenly, causing a slip or fall.

A pipe wrench should never be used on nuts or bolts. The corners will break the teeth of the wrench, making it unsafe to use on pipe and fittings. Also, a pipe wrench used on nuts and bolts can damage their heads. The wrench should not be used on valves, struck with a hammer, or used as a hammer unless, as with specialized types, it is designed for such use.

**Torque Wrenches.** These have a scale to show the amount of force applied to the nut or bolt (usually foot pounds). It is important that a torque wrench be well cared for to ensure accurate measurements. Cleanliness of the bolt or nut threads is also important.

**Tongs.** Although tongs are usually purchased, some companies make their own to do specific jobs. To prevent pinching hands, the end of one handle should be upended toward the other handle, to act as a stop. It is also possible to braze, weld, or bolt bumpers on the handles a short distance behind the pivot point so the handles cannot close against the fingers.

**Pliers.** Side-cutting pliers sometimes cause injuries when short ends of wire are cut. A guard over the cutting edge and the use of eye protection will help to prevent eye injuries.

The handles of electrician's pliers should be insulated. Because pliers do not hold the work securely, they should not be used as a substitute for a wrench.

Vise-grip pliers can be locked on the object, but they should be used with care because they can pinch the fingers or hand.

**Special Cutters.** These tools include those for cutting banding wire and strap. Claw hammers and pry bars should not be used to snap metal banding material. Only cutters designed for the work provide safe and effective results.

**Nail Band Crimpers.** These crimpers make it possible to keep the top band on kegs and wood barrels after nails or staples have been removed. Use of these tools eliminates injury caused by reaching into kegs or barrels that have projecting nails and staples.

**Pipe Tongs.** Such tongs should be placed on the pipe only after the pipe has been lined up and is ready to be fashioned. A 3-in or 4-in. (7.5- or 10-cm) block of wood should be placed near the end of the travel of the tong handle. The block should be parallel to the pipe to prevent injury to the hands or feet if the tongs slip.

Workers should not stand or jump on the tongs, or place extensions on the handles to obtain more leverage. They should use larger tongs, if necessary, to do the job.

**Screwdrivers.** The screwdriver is probably the most commonly abused tool. Supervisors should discourage workers from using screwdrivers as punches, wedges, pinch bars, or pries. If used in such ways, they become unfit for the work they are intended to do. A broken handle; bent shaft; or dull, broken, or twisted blade may cause a screwdriver to slip out of the slot and injure a worker's hand (Figure 14–4).

GROUND RIGHT

A—SHOWS HOW A SCREWDRIVER SHOULD FIT THE SCREW SLOT

B—A POOR FIT DAMAGES SCREWDRIVER AND SCREW SLOT

**Figure 14–4.** Screwdriver blades must be ground flat in that portion of the tip that enters the screw slot, and then gradually taper out to the diameter of the shank (left). When selecting a screwdriver, make sure the blade makes a good fit in the screw slot (right).

The tip must be kept clean and sharp to permit a good grip on the head of the screw. A screwdriver blade should fit the screw snugly. A sharp, square-edged blade will not slip as easily as a dull, rounded one, and requires less pressure. The part to be worked on should never be held in the hands. It should be laid on a bench or flat surface, or held in a vise. This practice will lessen the chance of hand injury if the screwdriver should slip. In addition, workers should observe the following precautions:

- Never use any screwdriver for electrical work unless it is insulated.
- In wood and sheet metal, make a pilot hole for the screw.
- Do not carry screwdrivers in your pockets.
- Keep screwdriver handles clean.
- Keep screwdriver blades properly dressed.

Screwdrivers used in a shop are best stored in a rack. This layout allows easy selection of the proper screwdriver.

**Allen Wrenches.** These wrenches are used like screwdrivers. The key factor is choosing the correct size wrench for the job. It is easy to apply too much force on small Allen wrenches.

**Vises.** Vises in a variety of sizes, jaws, and attachments are used to grip and hold objects being worked on. Vises should be secured solidly to a bench or similar base. When sawing material held in a vise, make the cut as close to the vice jaws as possible. Lightly oil all moving parts of a vise.

**Clamps.** Many kinds of clamps are used in a variety of operations. Nearly all clamps can be used with pads to prevent marring the work. Overtightening a clamp can break it or damage the product. If there is a swivel tip on the moving jaw of the clamp, it must operate or move freely. Moving parts, such as the threads, should be lightly oiled. Clamps should be stored in a rack and not in a drawer.

## Impact Tools

The term impact tools refers to a variety of hammers and related tools used to drive items such as nails into material by the use of manual or powered force. These tools are a common source of injuries, particularly to hands and arms. (See Chapter 10, Ergonomics, for more information on potential injuries.) It may be difficult to get employees to follow safe work practices because people often believe they know how to use hammers. In routine safety inspections, supervisors must make sure employees are not being careless with these tools.

**Hammers.** All hammers should have a securely fitting handle suited to the type of head used. The handle, whether reinforced plastic, wood, or metal, should be smooth, free of oil, shaped to fit the hand, and of the specified size and length. Warn employees against using a steel hammer on hardened steel surfaces. Instead, they should use a soft-head hammer; ball-peen hammer; or one with a plastic, wood, or rawhide head. Protective eyewear should be worn to protect against flying chips, nails, or scales.

Hammers may chip or spall, depending on four factors:

- The more square the corners of the hammer, the easier it chips.
- The harder the hammer is swung, the more likely it is to chip.
- The harder the object is struck, the more chipping increases.
- The greater the angles between the surface of the object and the hammer face, the greater the chances of chipping.

Selecting the proper hammer is important. One that is too light is as unsafe and inefficient as one that is too heavy.

To drive a nail, hold the hammer close to the end of the handle. Use a light blow to start, and increase power after setting the nail. Hold the hammer so the angle of the hammer face and the surface of the object being struck is parallel. The nail will drive straighter, and there will be less chance for damage. Placing the hammer on the nail before drawing it up to swing may increase accuracy of the aim.

**Sledgehammers.** These tools have two common defects that present a hazard to workers: split handles and loose or chipped heads. Because these tools are used infrequently in some industries, a loose or chipped head may not be noticed. Some companies place a steel band around the head and bolt it to the handle to prevent the head from flying off. The heads should be dressed whenever they start to check or mushroom. A sledgehammer so light that it bounces off the work is hazardous. Similarly, one that is too heavy is hard to control and may cause body strain.

**Riveting Hammers.** Often used by sheet metal workers, riveting hammers should have the same kind of use and care as ball-peen hammers. Inspect them often for checked or chipped faces.

**Carpenter's or Claw Hammers.** These hammers are designed primarily for driving and drawing nails. When a nail is to be drawn from a piece of wood, a block of wood may be used under the hammer head to increase leverage and protect the wood surface.

The striking faces should be kept well dressed at all times to reduce the hazard of flying chips or nails. A checker-faced head is sometimes used to reduce this hazard.

Eye protection is recommended for all nailers and all employees working in the same area.

### Spark-Resistant Tools

Spark-resistant tools of nonferrous materials generally should be used where flammable gases, vapors, or explosive materials might be present. These tools will not produce sparks when striking hard surfaces, thus reducing risk of fire or explosion.

# PORTABLE POWER TOOLS

Portable power tools are divided into four primary groups according to power source: (1) electric, (2) pneumatic, (3) internal combustion, and (4) explosive (powder actuated). Several types of tools, such as saws, drills, and grinders, are common to the first three groups, whereas explosive tools are used exclusively for penetration work and cutting.

A portable power tool presents hazards similar to those of a stationary machine of the same kind. Typical injuries caused by portable power tools are burns, cuts, contusions, abrasions, and sprains. Sources of injury include electrical shock, particles in the eyes, fires, falls, explosion of gases, and falling tools. Because of the extreme mobility of power-driven tools, they can easily come in contact with the operator's body, and it is difficult to guard such equipment completely. The operator is also more closely exposed to the source of power (electricity, compressed air, liquid fuel, or explosive cartridge), which creates additional hazards.

When using powder-actuated tools (explosive cartridge equipment) for driving anchors into concrete, or when using air-driven hammers or jacks, it is imperative that hearing protection and eye and face protection be worn by operators, assistants, and nearby personnel when the tool is in use.

All companies and manufacturers of portable power tools should attach to each tool a set of instructions for safely operating the equipment. These are meant to supplement thorough training that each power tool operator should receive. (More details on using these tools are given below.)

Power-driven tools should be kept in safe places and not left in areas where they can be struck or activated accidentally by a passerby. The power cord should always be disconnected before accessories on a portable power tool are changed. Guards should be correctly adjusted before the tool is used.

## Selecting the Proper, Best-Designed Tool

Power tools can be more hazardous than hand tools. Check with your safety professional to make sure the tools you select meet not only current safety standards but also best practices. Tool manufacturers can recommend the best tool to do a job. Describe the job you want accomplished, the material to be worked on, and the space available in your shop or facility work area. Tell the manufacturer if the operation is intermittent or continuous. Power tools are available in a variety of light- to heavy-duty models, and the manufacturers should be able to provide a reliable recommendation.

## Electric Tools

Electrical shock is the chief hazard from electrically powered tools. Injury categories include minor and lethal shock. Serious electrical shock is not entirely dependent on the voltage of the power input. Nearly all portable power tools are powered at 110 volts. As explained in Chapter 16, Electrical Safety, the ratio of voltage to resistance determines the current. The current is regulated by the body's resistance to the ground and by environmental conditions. As little as 100 milliamps can be lethal. It is possible for a tool to operate with a defect or short in the wiring. Use of a ground wire, however, protects the operator under most conditions.

Electrical tools used in damp areas or in metal tanks expose the operator to conditions favorable to current flow through the body, particularly if the worker perspires. Many electrical shocks from tools are caused by failure of insulation between current-carrying parts and metal frames of tools. Only tools in good repair that are listed by an authorized testing laboratory should be used.

Insulating platforms, rubber mats, and rubber gloves provide an additional safety factor when tools are used in damp locations, such as in tanks or boilers or on wet floors. Safe low voltage of 6, 12, or 24 volts, obtained through portable step-down transformers, will reduce shock hazards in damp locations.

If battery-powered tools are selected for the task, the electrical shock hazard is reduced significantly. The one thing that should always be considered is that the charger or charging stations for the battery packs are usually 110-volt systems. The same precautions, such as grounding and ground-fault circuit-interrupter (GFCI) use should always be applied in the charging area or environment, as explained in Chapter 16, Electrical Safety.

**Double-Insulated Tools.** Protection from electrical shock while using portable power tools depends on third-wire protective grounding. However, "double-insulated" tools are available and provide reliable shock protection without third-wire grounding or a GFCI (Figure 14–5). The National Electrical Code permits double insulation for portable tools and appliances. Tools in this category are permanently marked by the words "double insulation" or "double insulated." Units designed in this category that have been tested and listed by Underwriters Laboratories are labeled as such. Many U.S. and European manufacturers use the symbol shown in Figure 14–6 to denote double insulation. Double-insulated tools do not require separate ground connections. The third wire or ground wire is not needed and should not be used.

Failure of insulation is harder to detect than worn or broken external wiring, and points out the need for frequent inspection and thorough maintenance. Care in handling the tool and frequent cleaning will help prevent wear and tear that causes defects. See Chapter 16, Electrical Safety, for more details.

**Grounding.** The easiest way to safeguard the operator against electrical shock is to ground portable electrical tools. If any defect or short circuit inside the tool occurs, the current will flow harmlessly from the metal frame tool housing through a low-resistance ground wire rather than passing through the operator's body. All electrical power tools should be effectively grounded unless they are double insulated or cordless.

**Figure 14–5.** A double-insulated electrical tool has an internal layer of protective insulation that isolates the electrical components from the outer housing.

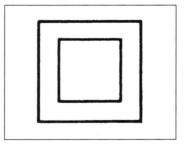

**Figure 14–6.** Symbol for double insulation for electrical appliances and tools.

The noncurrent-carrying metal parts of equipment may be grounded in two ways: first, by a grounding conductor, or second, by enclosing in metal the conductors feeding such equipment—provided an approved grounding-type attachment plug is used. The grounding conductor should be run with the power supply conductors in an insulated cable assembly or flexible cord. It must be properly terminated in an approved grounding-type attachment plug having a fixed grounding contacting member. If the grounding wire is separate, but insulated, it must be colored either in solid green or green with one or more yellow stripes.

A major safety concern is proper maintenance of the electrical power supply system, especially the ground. Periodic tests of the electrical system should be made by an electrician to ensure that it remains grounded.

A ground-fault circuit-interrupter (GFCI) is a device that in this case protects the tool operator if a ground fault occurs. The GFCI should be installed at the origin of the circuit supplying the power tool. GFCIs come in a variety of overcurrent protection values and must be tested periodically. GFCIs are described in detail in Chapter 16, Electrical Safety.

Power tool electrical cords should be inspected frequently and kept in good condition. If the cord has been changed, have a qualified person check that the new cord is the right wire size. Use heavy-duty plugs that clamp to the cord insulation to prevent strain on current-carrying parts under manual use. Employees should be trained not to pull on cords and to protect them from sharp objects, heat, and oil or solvents that might damage or soften the insulation.

**Extension Cords.** Use only three-wire extension cords with three-prong grounding-type plugs and three-pole receptacles that accept the tool's plug. Replace or repair damaged or worn cords immediately. Be sure to select the proper size cord because the flow of current (amperage) demanded by the electrical appliance will cause an undersized conductor to overheat. (See discussion in Chapter 16.) Table 14–B shows the proper gauge wire to use for various lengths of extension cord.

### Table 14–B. Minimum Gage Wire Size for Extension Cords

| Tool Nameplate Amperes | Cord Length in Feet | | | |
|:---:|:---:|:---:|:---:|:---:|
| | 25 | 50 | 100 | 150 |
| 0-6 | 18 | 16 | 16 | 14 |
| 6-10 | 18 | 16 | 14 | 12 |
| 10-12 | 16 | 16 | 14 | 12 |
| 12-16 | 14 | 12 | | |

Source: Power Tool Institute

**Electric Drills.** These power tools can cause injuries in several ways: a part of the drill may be pushed into the hand, the leg, or other parts of the body. The drill may be dropped when the operator is not actually drilling. The eyes can be hit either by material being drilled or by parts of a broken drill bit. Avoid clearing chips using a finger—it can get pulled into the drill and result in serious injury.

Oversized bit spindles should not be ground down to fit electric drills with smaller drill bit chucks. Instead, a reduction adapter can be used that will fit the large bit. However, this adapter is an indication of improper drill size and should be used sparingly to avoid placing excess strain on the smaller drill motor. When drills are used, the pieces of work should be clamped on or anchored to a sturdy base to prevent whipping.

Electric drills should always be of the proper size for the job. A ¼-in. drill means a maximum ¼-in. (diameter) drill bit that can be used for wood and light metal. A ½-in. drill bit would be needed to pierce steel and masonry. To operate, the chuck key should be attached to the cord but removed from the chuck before starting the drill. Install the chuck key at the plug end so the employee must pull the plug before inserting the chuck key. If the drill has a side handle, it should be used. Make a punch mark to facilitate starting the drill and bit. Hold the drill firmly and at the proper angle for the job. Then start slowly. Gradually increase speed as needed.

**Routers.** The widespread use of routers is based on their ability to perform an extensive range of smooth finishing and decorative cuts.

Safety in using a router starts with an understanding that it operates at a very high speed, in the 20,000 RPM range—25 times faster than a drill. It cuts quickly, is noisy, and the cutting blades are unguarded and out of sight when in use.

For work with routers, always wear eye protection against flying particles and airborne wood dust. Use a dust mask in dusty work conditions. Wear hearing protection during extended periods of operation.

Do not wear gloves, loose clothing, jewelry, or any dangling objects, including long hair, which may catch in rotating parts or accessories.

Install router bits securely and according to the operator's manual. Always use the wrenches provided with the tool.

When in use, keep a firm grip with both hands on a router at all times. Failure to do so could result in loss of control, possibly leading to serious injury. Read the operator's manual carefully regarding laminate trimmers and other small routers that are designed for use with one hand. Always face the cutter blade opening away from the body.

When the router is equipped with carbide-tipped bits, start by pointing the bit in a direction that will protect the operator and others against flying cutter pieces if the bit disintegrates.

Hold only those gripping surfaces of the router designated by the manufacturer. Check the operator's manual. If a router is equipped with a chip shield, keep it properly installed. Keep your hands away from the bits or cutter area when the router is plugged in. Do not reach underneath the work while bits are rotating. Never attempt to remove debris while the router is operating.

Your desired cutting depth adjustments should be made only according to the tool manufacturer's recommended procedures for these adjustments. Tighten adjustment locks. Make certain the cutter shaft is engaged in the collet at least ½-in. Check the operator's manual carefully.

Be certain to secure clamping devices on the workpiece before applying the router. The router switch should be in the OFF position before plugging into the power outlet. For greater control, always allow the motor to reach full speed before feeding the router into the work. Never force a router.

When removing a router from the workpiece, always be careful not to turn the base and bit toward the body. Unplug and store the router immediately after use, and remove the cutter or cap before storage.

**Electric Circular Saws.** These saws are usually well guarded by the manufacturer, but employees must be trained to use the guard as intended. The guard should be checked often to be sure it operates freely and encloses the teeth completely when not cutting, and encloses the unused portion of the blade closest to the operator when it is cutting.

Circular saws should not be jammed or crowded into the work. The saw should be started and stopped outside the work. At the beginning and end of the stroke or when the teeth are exposed, the operator must take extra care to keep the body extremities and the power cord away from the cutting line. All portable circular saws have a trigger switch to shut off power when pressure is released.

Injuries that occur when using portable circular saws are caused by contact with the blade; electrical shock or burns; tripping over the electrical cord, saws, or debris; losing balance; and kickbacks resulting from the blade being pinched in the cut.

Important requirements for safe operation of the portable circular saw are proper use, frequent inspection, and a rigid maintenance schedule. The manufacturer's recommendations for operation and maintenance must be followed faithfully and treated as standard procedure.

The following are specific safety "musts" when using any portable circular saw. Not doing so must be considered dangerous.

- Do not use a circular saw that is too heavy for workers to easily control.
- Be sure the trigger switch works properly. It should turn the tool on and return to the OFF position after release.
- Use sharp blades. Dull blades cause binding, stalling, and possible kickback;

waste power; and reduce motor life.

- Use the correct blade for the application. Check these points carefully: Does it have the proper size and shape arbor hole? Is the speed marked on the blade at least as high as the no-load RPM on the saw's nameplate?

- The workpiece must be securely clamped. For maximum control, use both hands to properly and safely guide the saw (Figure 14–7).

**Figure 14–7.** The workpiece must be securely clamped, and for maximum control, the operator must use both hands to properly and safely guide the saw. *(Courtesy: Power Tool Institute.)*

Check blades carefully before each use for proper alignment and possible defects. Be sure blade washers (flanges) are correctly assembled on the shaft and that the blade is properly supported and tightened.

Is the blade guard working? Check for proper operation before each cut. Check the adjustment of anti-kickback dogs. Check often to be sure that guards return to their normal position quickly. If a guard seems slow to return or "hangs up," repair or adjust it immediately. Never defeat the guard to expose the blade—for example, tying back or removing the guard.

Before starting a circular saw, be sure the power cord and any extension cord are out of the blade path and are sufficiently long to complete the cut. Stay constantly aware of the cord location. A sudden jerk or pulling on the cord can cause the operator to lose control of the saw.

For maximum control, hold the saw firmly with both hands after securing or clamping the workpiece. Check frequently to be sure clamps remain secure. Never hold a workpiece in your hand or across your leg when sawing. Avoid cutting small pieces of material that cannot be properly secured, or material on which the saw shoe cannot properly rest.

When making a "blind" cut (you can't see behind what is being cut), be sure that hidden electrical wiring, water pipes, or any mechanical hazards are not in the blade path. If there is any possibility of cutting into electrical conductors, disconnect power to these conductors. Contact with live wires could cause lethal shock or fire.

Set blade depth to no more than ⅛-inch to ¼-inch greater than the thickness of the material being cut. Always hold the tools by the insulated grasping surfaces. When you start the saw, allow the blade to reach full speed before the workpiece is contacted. Be alert to the possibility of the blade binding and kickback occurring. If a fence or guide board is used, be certain the blade is kept parallel with it. Never overreach.

When making a partial cut, or if power is interrupted, release the trigger immediately and don't remove the saw until the blade has come to a complete stop. Never reach under the saw or workpiece.

Portable circular saws are not designed for cutting logs or roots, or trimming trees or shrubs. Attempting this would be extremely hazardous.

Switch the tool off after a cut is completed, and keep the saw away from the body until the blade stops. Unplug, clean, and store the tool in a protected, dry place after use.

Kickback is a sudden reaction to a pinched blade, causing a combination of the saw being lifted up and out of the workpiece toward the operator and the workpiece being propelled in the direction of the blade rotations. Kickback is the result of tool misuse and/or incorrect operating procedures or conditions. Take specific precautions to help prevent kickback when using any type of circular saw:

- Keep saw blades sharp. A sharp blade will tend to cut its way out of a pinching condition.
- Make sure the blade has adequate set in the teeth. Tooth set provides clearance between the sides of the blade and the workpiece, thus minimizing the probability of binding. Some saw blades have hollow-ground sides instead of tooth set to provide clearance.
- Keep saw blades clean. A buildup of pitch or sap on the saw blade surface increases the thickness of the blade and also increases the friction on the blade surface. These conditions cause an increase in the likelihood of a kickback.
- Don't cut wet wood. It produces high friction against the blade. The blade will tend to load up with wet sawdust and create a much greater probability of kickback.
- Be cautious of stock that is pitchy, knotty, or warped. These are most likely to create pinching conditions and possible kickback.
- Always hold the saw firmly with both hands.
- Release the switch immediately if the blade binds or the saw stalls.
- Support large panels so they will not pinch the blade. Use a straightedge as a guide for ripping.

- Never remove the saw from a cut while the blade is rotating.
- Never use a bent, broken, or warped saw blade. The probability of binding and resultant kickback is greatly increased by these conditions.

Overheating a saw blade can cause it to warp and result in a kickback. Buildup of sap on the blades, insufficient set, dullness, and unguided cuts can all cause an overheated blade. Never use more blade protrusion than is required to cut the workpiece—$\frac{1}{8}$-inch. to $\frac{1}{4}$-inch greater than the thickness of the stock is sufficient. This minimizes the amount of saw blade surface exposed, and reduces the probability of kickback and severity if any kickback does occur. Minimize blade pinching by placing the saw shoe on the clamped, supported portion of the workpiece and allowing the cutoff piece to fall away freely.

When the saw is being used in a damp environment or outdoors, the operator should wear electrical insulating footwear and gloves. GFCI protection or a double-insulated saw is necessary for outside use.

**Cutoff Saws.** Do not use a cutoff saw beyond its rated speed. Check catalog RPM against safe saw speed. Never try to cut through thick material in one try; make a series of shallow cuts that gradually deepen to the cut desired. Operators who are exposed to harmful or nuisance dusts should wear approved respirators and eye protection.

**Electric Reciprocating Saws.** The common saber saw and reciprocating saw are two basic types of electric saws. Because the blade is almost fully exposed when in use and storage, the tool must be handled with extreme care. The blade should be removed when the saw is stored. The type of blade needed will be determined by the material being cut.

Hold the shoe as securely as possible against the work when cutting. Turn off the tool when the job is completed and do not remove the saw from the material until the motor has fully stopped.

The versatility of the reciprocating saw in cutting metal, plastic, wood, and other materials has made it a widely used tool. By design, it is a simple tool to handle. Its few demands for safe use, however, are very important:

- Without exception, use the blade specifically recommended for the job. Check your operator's manual carefully concerning this.
- Position yourself to maintain full control of the tool. Avoid cutting above shoulder height.

- To minimize blade flexing and provide a smooth cut, use the shortest blade that will do the job.
- The workpiece must be clamped securely, and the shoe of the saw held firmly against the work, to prevent operator injury and blade breakage.

**Abrasive Wheels, Buffers, and Scratch Brushes.** These tools should be guarded as completely as possible. Portable abrasive wheels must have proper guarding. For portable grinding, the maximum angular exposure of the periphery and sides should not exceed 180 degrees. The top portion of the wheel should always be enclosed. Guards should be adjustable so that operators will be inclined to make the correct adjustment rather than remove the guard. However, the guard should be easily removable to facilitate replacement of the wheel. In addition to this mechanical guarding, the operator must use personal protective equipment (including eye protection) in accordance with the hazard assessment.

A portable grinding wheel is exposed to more abuse than a stationary grinder. The wheel should be kept away from water and oil, protected against dropping blows from other tools, and used carefully to avoid striking the sides of a wheel against objects. Cabinets or racks can help to protect the wheel against damage.

The speed and weight of a grinding wheel, particularly a large one, make it more difficult to handle than some other power tools. Because part of the wheel must necessarily be exposed, it is important that employees be trained in the correct way to hold and use the wheel so that it does not touch their clothes or body.

The wheels should be mounted by trained personnel only, with the wheels and safety guards conforming to ANSI standard B7.1, *Safety Requirements for the Use, Care, and Protection of Abrasive Wheels.* Grinders should be marked to show the maximum abrasive wheel size and speed. Operators must make sure that the correct-speed wheel is used for each job. Abrasive wheels should be sound-tested (ring-tested) before being mounted. A ring test involves tapping the wheel. For details, see the standard, and always follow the manufacturer's recommendations.

**Sanders.** Whether belt or disk type, sanders can cause serious skin abrasions when the rapidly moving abrasive touches the body. Because it is impossible to guard sanders completely, employees must be thoroughly trained to use them safely. The motion of the sander should be away from the body. All clothing should be kept clear of the moving parts.

A vacuum system attached to the sander or an exhaust system can be used to remove the dust from the work area. Eye protection also should be worn. If harmful dusts are created, a respirator certified for the exposure by the National Institute for Occupational Safety and Health (NIOSH) or the Mine Safety and Health Administration (MSHA) should be worn. (See Chapter 9, Personal Protective Equipment.)

Safety precautions for using sanders include the following:

- Do not wear loose clothing, long hair, jewelry, or any dangling objects that may catch in rotating parts or accessories.
- Stay constantly aware of cord location.
- Never lock a portable sander in the ON position when the nature of a job may require stopping the sander quickly, such as using a disc sander on an automobile's fender well, where the rotating disc could get jammed in the well and result in an incident.
- With portable sanders, be careful not to expose the tool to liquids, or to use in damp, wet locations.
- When adjusting the tracking of the belt on a portable belt sander, be certain that you have the sander supported and positioned to avoid accidental contact with yourself or adjacent objects.
- Use jigs or fixtures to hold your workpiece whenever possible.
- Use appropriate PPE.

Sanders require especially careful cleaning because of the dusty nature of the work. If a sander is steadily used, it should be dismantled periodically, as well as thoroughly cleaned every day by being blown out with pressurized air. If compressed air is used, the operator should wear protective eyewear and use an air nozzle with sufficient side openings so that if "dead ended" against the skin, air pressure against the skin will not exceed 30 psig.

Because wood dust presents a fire and explosion hazard, keep dust to a minimum. Sanders can be equipped with a dust collection or vacuum bag. Electrical equipment should be designed to minimize the explosion hazard. Fire extinguishers approved for Class C (fires where there is an associated electrical exposure) should be readily available. Employees should be trained in what to do in case of fire.

**Disc Grinders.** Portable grinders basically remove material by contact with a rotating abrasive wheel or disc and buffing disc. There are safety precautions that apply to all grinders and other specific recommendations that apply to specialized grinding operations.

Before operating a grinder, compare the data on the nameplate with the voltage source. Be sure the voltage and frequency are compatible. Remove material or debris from the area that might be ignited by sparks. Be sure others are not in the path of the sparks or debris. Keep a properly charged fire extinguisher available.

Maintain proper footing and balance. Never attempt to grind from an awkward position. A portable grinder can kick, and glance off the work, if not tightly controlled.

Always disconnect the tool from the power source before installing or changing discs. Use only those discs marked with a rated speed that is at least as high as, or above, the speed rating on the nameplate of the tool. Do not use an unmarked wheel. Handle or store discs carefully to prevent damage or cracking.

Be careful not to over-tighten the spindle nut. Too much pressure will deform the flanges and stress the disc.

After mounting a wheel or disc and replacing the guard, stand to the side and allow a one-minute run-up at no load, to test the integrity of the wheel. Your grinder should come up to full speed each time before you contact the workpiece. Do not apply excessive pressure to the disc because that will stress the wheel, overheat the workpiece, and reduce your control.

Portable straight grinders should be used only with high-strength, bonded wheels. Position this type of grinder away from the body and allow it to run for one minute before contacting work.

Tuck point grinders are a variation of straight grinders and are equipped with reinforced abrasive discs and the appropriate guard. Because using a tuck point grinder is very dusty work, a dust mask, face shield, and eye protection with side shields are recommended. Maintain firm control of the tool. Never overreach. Carefully maintain balance. Do not allow the grinding wheel to bend, pinch, or twist in the cut because kickback may result.

Angle grinders are used primarily with reinforced abrasive discs or wire cup brushes for the removal of metal or masonry. Use of the proper wheel and guard combination is critical. The user must follow the manufacturer's recommendations given in the operator's manual.

Many angle grinders are equipped with guards that can be mounted with the opening in a variety of positions. Take care to position the guard to provide maximum protection. Many angle grinders can be converted for use as sanders.

When guards are removed for a sanding operation, it is essential that they be replaced before the tool is again used for grinding.

**Soldering Irons.** Such tools are often the source of burns and illness resulting from inhalation of fumes. Insulated, noncombustible holders will virtually eliminate the fire hazard and the danger of burns from accidental contact. Ordinary metal covering on wood tables is not sufficient because the metal conducts heat and can ignite the wood.

Holders should be designed so employees cannot accidentally touch the hot irons if they reach for them without looking. The best holder completely encloses the heated surface. It is inclined so the weight of the iron prevents it from falling out. Such holders must be well ventilated to allow the heat to dissipate, otherwise the life of the tip will be reduced, and the wiring or printed circuit may be damaged.

Local exhaust ventilation is required to prevent inhalation exposure to solder and base metal fumes in excess of allowable levels established for each material. Sample the air to verify that the amount of the contaminant in the air is not harmful. Food and beverages should not be consumed around soldering operations.

Particles should not be allowed to accumulate on the floor or work tables. Employees should wear protective eyewear, face shields, and respirators to reduce exposures that are not reduced through engineering.

**Electrically Heated Glue Guns.** These tools are being used increasingly in industry. Although all guns have insulated handles, there is danger in the high temperature of the glue and tip of the gun. Proper holders and storage can minimize the exposure. Electrical shock is also a risk, and all tools should be properly grounded and insulated.

**Percussion Tools.** This family of tools is primarily associated with masonry applications as varied as chipping, drilling, anchor setting, and breaking of pavement. They range from pistol grip types to large demolition hammers. Normal operating modes include hammering, hammering with rotary motion, and rotation or drilling only. Many models incorporate a varied combination of the above types.

Always wear protective eyewear with side shields, and a full face shield. Use a respirator in dusty work conditions. Wear hearing protection during extended periods of operation so that the time-weighted noise exposure does not exceed the established levels. Do not wear loose clothing, long hair, jewelry, or dangling objects that may catch in rotating parts or accessories.

For maximum control, use the auxiliary handles provided with the tool. Do not tamper with clutches on those models that provide them. Have the clutch settings checked at the manufacturer's designated service facility at the intervals recommended in the operator's manual.

Check for subsurface hazards such as electrical conductors or water lines before drilling or breaking blindly into a surface. If wires are present, they must be disconnected at the power source by a qualified person, or be certain they are avoided, to prevent the possibility of lethal shock or fire. Water pipes must be drained and capped.

Always hold the tool by the insulated grasping surfaces. Do not force the tool. Percussion tools are designed to hit with predetermined force. Added pressure by the operator only causes operator fatigue, excessive bit wear, and reduced control.

When cutting, drilling, or driving into walls, floors, or wherever "live" electrical wires may be encountered, workers should never touch any metal parts of the tool. Instead, they should hold the tool only by the insulated gripping surfaces to prevent electrical shock if contact is made with an energized conductor.

## Air-Powered Tools

Operators of air tools should be instructed to:

- Keep hands and clothing away from the working end of the tool.
- Follow safety requirements applicable to the tool being used and to the nature of the work being performed.
- Inspect and test the tool, air hose, and coupling before each use.

Most air-powered tools are difficult to guard. Therefore, the operator must be careful to prevent injury to hands, feet, or the body if the machine slips or the tool breaks. Pistol or doughnut hand grips, or flanges in front of the hand grip, provide protection for the hands and should be used on all twist or percussion tools.

**Air Hose.** An air hose presents the same tripping hazard as power cords on electrical tools. Anything accidentally hitting the hose may throw the operator off balance or cause the tool to fall. To prevent a tripping hazard, an air hose on the floor should be protected against trucks and pedestrians by placing two planks on either side or by building a runway over it.

Workers should be warned against disconnecting the air hose from the tool and using it for cleaning machines. Air hoses should not be used to remove dust from clothing.

Incidents may sometimes occur when the air hose becomes disconnected and whips about. A short chain attached to the hose coupling and to the tool housing will keep the hose from whipping if the coupling breaks. Other couplings in the air line should also be safely pinned or chained. Air should be cut off before attempting to disconnect the air hose from the air line. Air pressure inside the line should be released before disconnecting.

A safety check valve installed in the air line at the manifold will shut off the air supply automatically if a fracture occurs anywhere in the line. If kinking or excessive hose wear is a problem, the hose can be protected by a wrapping of strip metal or wire.

**Air-Powered Grinders.** These tools require the same kind of guarding as electric grinders. Be sure the speed regulator or governor on these machines is carefully maintained to avoid overspeeding the wheel (runaway). Hold grinder properly. Regular inspection by qualified personnel at each wheel change is recommended.

**Pneumatic Impact (Percussion) Tools.** Air percussion tools, such as riveting guns and jackhammers, are essentially the same, in that the tool is fitted into the gun and receives its impact from a rapidly moving reciprocating piston driven by compressed air at about 90 psig (625 kPa).

Two safety devices are needed to protect workers who operate these tools. First, the ON/OFF trigger should be located inside the handle where it is reasonably safe from inadvertent operation. The machine should operate only when the trigger is held in depressed position. Second, make sure the tool has a device holding it in place so that it cannot be shot accidentally from the barrel (Figure 14–8). On small air hammers not designed to use this device, use a spring clip to prevent the tool and piston from falling from the hammer.

It is essential that employees be thoroughly trained in proper use of pneumatic impact tools. Impress on all operators of small air hammers the following safety rule: Do not squeeze the trigger until the tool is on the work.

Air percussion tools either produce heavy blows or, due to rapid pulsating, vibrate strongly. Because tools such as rotary drills, saws, or grinders—especially those rotating at high speeds—produce excessive vibration, continued use of this equipment may cause damage to the body. Supervisors should instruct work-

**Figure 14–8.** A spring clip, like the one shown here, should always be used to prevent a rivet set from falling from the air hammer.

ers to reduce transmission of this vibration by using rubber hand grips, air-cushion devices, and other vibration dampeners wherever possible. Prolonged use should be avoided. (See Chapter 10, Ergonomics, for more information on potential injuries.)

**Jackhammers.** Handling of heavy jackhammers causes fatigue and strain. Jackhammer handles should be provided with heavy rubber grips to reduce vibration and fatigue. Operators should wear protective footwear covering toes and metatarsal areas to reduce the possibility of injury if the hammer falls.

Two chippers should work turned away from each other, back to back, to prevent face cuts from flying chips. Workers should not point a pneumatic hammer at anyone. They should never stand in front of operators handling pneumatic hammers.

Many incidents are caused by the steel drill breaking, causing the operator to lose balance and fall. If the steel is too hard, a particle of metal may break off and strike the operator. The manufacturer's instructions for sharpening and tempering the steel should be followed.

**Air-Operated Nailers and Staplers.** The principal hazard from these tools is accidental discharge of the fastener. In such instances, the fastener can become a dangerous projectile that can inflict serious injury at a considerable distance. Operators should be trained in use of these tools and must follow the manufacturer's operating instructions.

**Personal Protective Equipment.** As with all pneumatic impact tools, there is a hazard from flying chips. Operators should wear protective eyewear. If other employees must be in the vicinity, they should be similarly protected. Where possible, screens should be set up to shield workers near chippers, riveting guns, or air drills in use. Operators and others exposed should also wear hearing protection so that the exposure to noise does not exceed established federal regulations. The sound of pneumatic tools can be extremely loud—similar to firecrackers or gunfire.

Eye and hearing protection should also be provided when using powered saws, grinding wheels, screwdrivers, buffers, scratch brushes, and sanders. The noise levels from some of these tools can reach or exceed 90 dBA (the federal standard time-weighted average for an eight-hour workday).

**Flexible Shaft Tools.** These tools require the same type of PPE as do direct power tools of the same type. Abrasive wheels should be installed and operated in compliance with ANSI B7.1, discussed earlier. The flexible shaft must be protected against denting and kinking, which can damage the inner core and shaft.

It is important that the power be shut off whenever the tool is not in use. When the motor is being started, the tool end should be held with a firm grip to prevent injury from sudden whipping.

The abrasive wheel or buffer of the tool is difficult to guard. Because it is more exposed than the wheel or buffer on a stationary grinder, workers should use extra care to avoid damaging the equipment. When not in use, wheels should be placed on the machine or put on a rack, not on the floor.

**Gasoline Power Tools.** These tools are commonly used in logging, construction, and other heavy industry. The most well-known is the chain saw. (See 29 CFR 1910.266(c)(5), Chain Saw Operations and the National Safety Council Occupational Safety & Health Data Sheet 320, Portable Power Chain Saws.)

Operators of gasoline power tools must be trained in their proper operation and follow the manufacturer's instructions. They must also be familiar with the fuel exhaust, heated surface, and noise hazards (see the discussion in Chapter 15, Materials Handling and Storage).

**Powder-Actuated Fastening Tools.** These tools are used for fastening fixtures and materials to metal; precast or prestressed concrete, masonry block, and to brick, stone, and wood surfaces, tightening rivets, and punching holes. Blank cartridges provide the energy and are ignited by means of a conventional percussion primer.

The hazards encountered in the use of these tools are similar to those encountered with firearms. The handling, storing, and control of explosive cartridges present additional hazards. Therefore, rules for the use, handling, and storage of both tools and cartridges must be just as rigid as those governing blasting caps and firearms (see OSHA regulations 1910.243(d)).

Additional specific hazards associated with these tools are accidental discharge; ricochets; ignition of explosive or combustible atmospheres; projectiles penetrating through the work; and flying dirt, scale, and other particles. In case of misfire, the operator should hold the tool in operating position for at least 30

QUALIFIED OPERATOR OF POWDER-ACTUATED TOOLS

Make(s)_____ Model(s)_____

This certifies that _____
(NAME OF OPERATOR)

Card No._____ Soc. Sec. No._____

Has received the prescribed training in the operation of powder actuated
tools manufactured by
_____
(NAME OF MANUFACTURER)

Trained and issued by _____
(SIGNATURE OF AUTHORIZED INSTRUCTOR)

I have received instruction in the safe operation and maintenance of
powder actuated fastening tools of the makes and models specified
and agree to conform to all rules and regulations governing their use.
Failure to comply shall be cause for immediate revocation of this card.

_____    _____
(SIGNATURE)              (DATE)

**Figure 14–9.** A wallet card for qualified powder-actuated tool operators is available from tool manufacturers. A list of instructors should be maintained by each manufacturer.

seconds and follow the manufacturer's instructions for removing the load.

Powder-actuated tools can be used safely if workers are given special training and proper supervision. Manufacturers of the tools will aid in this training. Only properly qualified personnel should ever be permitted to operate or handle the tools. Workers can become qualified after a few hours of instruction (Figure 14–9).

Powder-assisted, hammer-driven tools are used for the same purposes as powder-actuated tools, and generally, the same precautions should be followed. Workers should be trained to use powder charges of the correct size to drive studs into specific surfaces. They should be made responsible for safe handling and storage of cartridges and tools.

Powder-actuated tools should not be used on concrete with a thickness less than three times the fastener shank penetration depth, or on very hard or brittle materials, including cast iron, glazed tile, hardened steel, glass block, natural rock, hollow tile, or smooth brick. Fasteners should not be driven closer than 3 in. (7.5 cm) from an unsupported edge or corner.

Operators must wear adequate eye and face protection when firing the tool (Figure 14–10). Where the standard shield cannot be used for a particular operation, order special shields from the tool manufacturer. Hearing protection should also be worn to prevent damage from short, powerful bursts of sound. See Chapter 9, Personal Protective Equipment, for more details.

**Propane Torches.** Commonly used in many industries, propane torches are dangerous because of their open

**Figure 14–10.** This powder-actuated tool drives studs into concrete slabs. The operator should wear eye, ear, and head protection. Note the holster (left) for carrying the tool. *(Courtesy of Hilti Inc.)*

flame. Make sure that the work area is clear of flammable and combustible materials. Proper storage of propane cylinders is important to prevent leakage of this highly flammable gas. Operating instructions are available from the manufacturer.

# SUPERVISORY CONSIDERATIONS

As supervisor, you are responsible for controlling access to tools, seeing that employee-owned tools meet company standards, and ensuring that workers follow safe practices in carrying tools from place to place. You must be aware of these aspects of safety within your department, in addition to the hazards associated with particular tools.

## Centralized Tool Control

Centralized tool control helps to ensure uniform inspection and maintenance of tools by a trained person. A toolroom attendant can promote tool safety by recommending or issuing the right type of tool, by encouraging employees to turn in defective or worn tools, and by stressing the safe use of tools. The attendant can be trained to recommend the correct protective equipment, such as welder's safety goggles or ear protection to be used when the tool is issued. Centralized control and careful records of tool failure and other accidents will pinpoint hazardous conditions and unsafe practices.

Some companies issue each employee a set of numbered checks that are exchanged for tools from the toolroom. With this system, the attendant knows where each tool is and can recall it for inspection at regular intervals.

If you have a toolroom attendant you may want to set up a procedure so that the toolroom attendant can send tools in need of repair to a department or to the manufacturer for a thorough reconditioning.

Companies performing work at scattered locations may find it impractical to maintain a central toolroom. In such cases, the job supervisor should inspect all tools frequently and remove defective items from service. Many companies have supervisors check all tools weekly, using a preprinted checklist.

## Personal Tools

In trades or operations where employees are required or prefer to have their own personal tools, supervisors may encounter a serious problem regarding the safety of such items. Personal tools are usually well maintained, but some people may purchase inferior or nonindustrial-rated tools, make inadequate repairs, or attempt to use unsafe tools (such as hammers with broken and taped handles, or electrician's pliers with cracked insulation).

To ensure that personal tools meet company standards, the company should spell out the general requirements that tools must meet before they can be used. First, where applicable, they should meet the appropriate ANSI standards and/or be listed by a nationally recognized testing firm. Second, you should arrange a thorough inventory and initial inspection of personal tools. Inspect and list additional tools employees buy as though the items had been purchased by the company. Do not permit workers to use inferior or unsafe tools.

## Carrying Tools

When climbing a ladder or any structure, workers should never carry tools in any way that might interfere with using both hands freely. A strong bag, bucket, or similar container should be used to hoist tools from the ground to the job. Tools should be returned in the same manner, not brought down by hand, carried in pockets, or dropped to the ground.

Mislaid and loose tools cause a substantial portion of hand-tool injuries. Tools are put down on scaffolds, on overhead piping, on top of stepladders, or in other locations from which they can fall on people below. Leaving tools overhead is especially hazardous in areas of heavy vibration or where people are moving about or walking.

Chisels, screwdrivers, and pointed tools should never be carried in a worker's pocket. They should be carried in a toolbox or cart, in a carrying belt (sharp or pointed end down) like that used by electricians and steel workers, in a pocket tool pouch, or in the hand, with points and cutting edges held away from the body. Tools should be handed from one worker to another, never thrown.

Edged or pointed tools should be passed, preferably in their carrying cases, with the handles toward the receiver. Workers carrying tools on their shoulders should pay close attention to clearances when turning around. They should handle the tools so that they will not strike other people.

# MAINTENANCE AND REPAIR

A toolroom attendant or tool inspector should be qualified by training and experience to pass judgment on the condition of tools for further use. Dull or damaged tools should not be returned to stock. Keep enough tools of each kind on hand so that when a defective or worn tool is removed from service, you can replace it immediately.

Efficient tool control requires periodic inspections of all tool operations. These inspections should cover housekeeping in the tool supply room, tool maintenance, service, inventory, handling routine, and condition of the tools. Responsibility for such periodic inspections is usually given to the department supervisor. This job should not be delegated. However, all employees need to recognize signs indicating when a tool is damaged, dull, or unsafe so that they can report it at once.

Hand tools receiving the heaviest wear—chisels, wrenches, sledges, drills and bits, cold cutters, and screwdrivers—will require regular maintenance. Proper maintenance and repair of tools require adequate facilities: workbenches, vises, a forge or furnace for hardening and tempering, a variety of protective eyewear, repair tools, grinders, and good lighting.

## Tempering Tools

Such tools as chisels, stamps, punches, cutters, hammers, sledges, and rock drills should be made of steel and be heat-treated. This procedure makes them hard enough to withstand blows without excessive mushrooming, and yet not so hard that they chip or crack. Hardening and tempering of tools require special skills in manufacture.

## Metal Fatigue

Metal fatigue is a common problem in cold weather. Metals exposed to subzero temperatures undergo cold-soaking. A molecular change takes place that makes steel brittle and easily broken. This phenomenon shortens the life of such equipment as pneumatic impact tools, drill bits, and dies for threading conduit.

Ferrous metals, including carbon steel, alloy steel, and cast iron, are characterized by decreased toughness and corresponding brittleness at low temperatures. Nonferrous metals, on the other hand, such as aluminum and aluminum alloys, nickel, copper and copper alloys, chromium, zinc and zinc alloys, magnesium alloys, and lead, are more resistant to low temperatures and are often used in areas of extreme cold.

### Safe-Ending Tools

Such tools as chisels, rock drills, flatters, wedges, punches, cold cutters, and stamping dies should have their heads properly hardened by a qualified worker. Short sections of tight-fitting rubber hose can be set flush with the striking ends of chisels, hand drills, mauls, and blacksmith's tools to keep chips from flying, since they usually embed themselves in the rubber sleeve. Chisels, drift punches, cutters, and marking tools that manufacturers claim do not spall or mushroom are available. This feature is achieved by using a combination of alloys and scientific heat treatment to temper the metal.

### Dressing Tools

Tools should have regular maintenance of their cutting edges or striking surfaces. In most cases, once the cutting or striking surfaces have been properly hardened and tempered, only an emery wheel, grindstone, or oilstone is needed to keep the tool in good condition. Be sure to grind in easy stages. Keep the tool as cool as possible with water some or other cooling medium.

Tools that require a soft or medium-soft head should be dressed as soon as they begin to mushroom. A slight radius ground on the edge of the head, when it is dressed, will enable the tool to stand up better under pounding and will reduce the danger of flying chips. A file or oilstone, rather than an abrasive wheel, is recommended for sharpening pike poles and axes.

A wood-cutting tool, because of its fine cutting edge, can be initially dressed on a grinder with a wheel recommended by the manufacturer. An oilstone set securely in a wood block placed on a bench should be used to obtain a fine, sharp cutting edge. The oilstone should never be held in the hand; a slip off the face of the stone could cause a severe hand injury. Often a few finishing strokes on a leather strop will produce a keener edge.

Metal-cutting tools, because they generally have greater body, can be dressed or sharpened (or both) on an abrasive wheel. Do not use too much pressure against the wheel or the tool will overheat.

Make sure workers follow the manufacturer's recommendations for choosing the proper kind of abrasive wheel to sharpen wood or metal-cutting tools.

Each cutting edge should have the correct angle according to its use.

## Handles

The wooden handles of hand tools should be ¾-in. to 1½-in. (19 mm to 38 mm) of the best straight-grained material, preferably hickory, ash, or maple. They should be free from slivers. Make sure that they are properly attached. Poorly fitted or loose handles are a hazard. They can damage the work material and make it difficult for the worker to control the tool.

No matter how tightly a handle may be wedged at the factory, both shrinkage and use will loosen it. Inspect tools regularly for damaged and loose handles. These should be removed from service and repaired, if feasible.

## SUMMARY OF KEY POINTS

Key points covered in this chapter include:
- One of a supervisor's major responsibilities is to train and supervise people in the safe handling of hand and portable power tools. Supervisors must pinpoint where major problems in tool handling lie and devise a program to reduce tool injuries. Four basic safe work practices include (1) select the right tool designed for the job and user, (2) use tools properly and safely, (3) keep tools in good condition, and (4) store tools in a safe place.
- Supervisors must establish procedures for regular inspection, repair, and replacement of tools and set up a central tool supply room or other method of controlling access to company tools. Special tools may need to be kept under lock and key.
- The misuse of common hand tools is a major source of injury to industrial workers. These tools include cutting, torsion, shock, and spark-resistant equipment. Supervisors must make sure that employees are trained in the proper selection and use of these tools. They must see to it that employees follow all safety guidelines and procedures. Personal protective equipment (PPE) such as gloves, protective eyewear and helmets, hearing protection, aprons, and protective footwear should be worn.
- Cutting tools include metal and wood equipment, such as chisels, tap and die, hack saws, files, tin snips, punches, saws, axes, hatchets, knives, and miscellaneous tools. All edged tools should be used so that the direction of force is away from the body. Cuts should be made with the grain when possible. Edged

cutting tools should be kept sharp and ground to the proper angle. Saws must be sharpened and oiled regularly or they may stick in the cut. Workers should wear PPE to shield them from flying chips or metal slivers and from cuts. Cutting tools should never be used frivolously in mock fights or contests of skill. They can inflict serious injuries in a moment of carelessness.

- Torsion tools are used to fasten or grip materials and include various types of wrenches, tongs, pliers, special cutters, nail band crimpers, screwdrivers, Allen wrenches, vises, and clamps. These tools must fit the job precisely to prevent slippage, the jaws or blades should be kept sharp and clean, and pressure should be applied away from the body. Workers must protect their hands, arms, and faces with special care when using these tools.

- Shock tools include regular hammers, sledge hammers, riveting hammers, and carpenter's or claw hammers. All hammers should have secure handles and properly dressed heads. Workers must be sure to select the proper weight and size hammer to perform work safely.

- Spark-resistant tools of nonferrous materials are used where flammable or explosive materials are stored. These tools will not create sparks when striking metal or other hard surfaces.

- Portable power tools are divided into four groups according to the power source: electric, pneumatic, internal combustion, and explosive. Typical injuries caused by portable power tools include burns, cuts, and sprains. These tools must be carefully handled and stored so that they cannot be activated accidently. Workers must be thoroughly trained in proper use. In addition to the hazards associated with the tool itself, the power source represents a risk to workers. All employees must wear hearing protection and eye and/or face protection when operating portable power tools or working near them.

- Electrical shock is a significant hazard from electrically powered tools. Workers can suffer minor or lethal shock. Insulating platforms, rubber mats, rubber gloves, ground fault circuit interrupters, double-insulated tools, grounding wires, and grounded extension cords are all ways of protecting workers who use electrical tools. Workers should wear appropriate personal protective gear to minimize risk of injury and hearing loss. All damaged or worn cords should be replaced immediately. Cords should be inspected frequently to be certain they can carry the required electrical current safely.

- Important requirements for safe operation of electrically powered drills, routers, saws, wheels, sanders, soldering irons, percussion, and other portable power tools are proper use, frequent inspection, and a rigid maintenance schedule.

The manufacturer's recommendations for operation and maintenance must be followed strictly.

- Air-powered tools such as grinders and pneumatic tools (jackhammers, nailers, and staplers) present serious hazards to workers. Operators must be instructed to keep hands and clothing away from the working end of the tool, protect the air hose and check the safety value, follow safety requirements for the tool being used and work performed, and inspect and test the tool, air hose, and coupling before each use. Hands, feet, and torso should be protected in case the machine slips or the tool breaks, and hearing and eye protection should be used. Pistol or doughnut hand grips, or flanges in front of the hand grip, provide protection and should be used on all twist or percussion tools.

- Special power tools include flexible shaft tools, gasoline power tools, powder-actuated fastening tools, and propane torches. This equipment requires the same type of personal protection gear as do electric or air-powered tools. In addition, workers must be trained in fuel hazards and the dangers associated with explosive cartridges used in powder-actuated tools. Workers generally are given special instruction to operate these tools safely. They must know the manufacturer's recommendations for using the tools under various conditions.

- Supervisors must establish routine procedures for controlling access to company tools. For example, centralized tool control helps to ensure uniform inspection and maintenance by a trained person. The attendant can recommend proper safety equipment and issue the right tool for the right job.

- When employees purchase their own tools, supervisors must ensure that such tools meet established company standards, are regularly inspected and repaired or replaced, and fit the work to be done. Supervisors should also instruct employees on the proper way to carry tools. Workers should never carry tools while climbing ladders, should always carry pointed tools with the sharp end covered or held away from the body, and should always hand, never throw, tools to another worker.

- Supervisors must assume responsibility for inspection and maintenance and repair of tools in their department. Tools should be checked for quality construction and manufacture, for signs of metal fatigue in extremely cold environments, for properly hardened heads, for appropriately dressed cutting edges or striking surfaces, and for properly fitting handles.

# 15

# Materials Handling and Storage

**After reading this chapter, you will be able to:**

- Describe common materials-handling problems and precautions
- Explain safe manual handling methods for lifting and carrying loads and handling specific container shapes
- Describe hazards, safeguards, and operating guidelines for common materials-handling equipment
- Establish a training program for workers to ensure that they follow safe practices while moving materials
- Establish a program of regular inspection, repair, and replacement of materials-handling equipment and tools

Materials handling, whether done manually or with mechanical equipment, can be a major source of occupational injuries. Safe materials handling off the job is also important.

This chapter presents basic materials-handling hazards and safeguards to help prevent injuries and destruction of property. The topics covered include materials-handling problems; manual handling methods; equipment used for materials handling; inspection of ropes, chains, and slings; and safe storage of materials. The supervisor must know how to instruct workers to safely move materials.

## MATERIALS-HANDLING PROBLEMS

Manual handling of materials accounts for an estimated 25 % of all occupational injuries. These injuries are not limited to the shipping department or warehouse, but come from all operations because it is impossible to run a business without moving or handling materials. Moreover, such injuries are not limited to back strains, but can affect legs, feet, arms, fingers, hands, and the neck. Given the cost of injuries, the importance of proper materials-handling techniques becomes increasingly apparent.

Common injuries that workers suffer include strains and sprains, cuts, fractures, and bruises. These are caused primarily by unsafe conditions and practices—improper lifting techniques, carrying a load that is too heavy, incorrect gripping, back and leg positioning, etc.

Another major cause of materials-handling accidents can be traced to poor job design. Ask the following questions about your present operating practices:

- Can the job be re-engineered to eliminate manual handling of materials or minimize reaching, bending, and twisting?
- Can employees be given mechanical materials-handling aids—or properly sized boxes, containers, powered industrial trucks, or hoists—that will make their jobs safer (Figure 15–1)?
- Will personal protective clothing, or other equipment, help to prevent injuries?

**Figure 15–1.** Power lift table helps prevent strains in a parts-storage area.

These are not the only questions that might be asked, but they serve as a start for an overall appraisal. It is beneficial for everyone to adhere to safe lifting and handling techniques. These methods are discussed in the next section.

## MANUAL HANDLING METHODS

Because most injuries occur to the back, fingers, and hands, people need to be taught how to pick up and put down heavy, bulky, or long objects. Some general precautions are in order:

1. Inspect materials for slivers, jagged edges, burrs, and rough or slippery surfaces.
2. Get a firm grip on the object.
3. Keep fingers away from pinch points, especially when putting materials down.
4. When handling lumber, pipe, or other long objects, keep hands away from the ends to prevent them from being pinched.
5. Wipe off greasy, wet, slippery, or dirty objects before trying to handle them.
6. Keep hands free of oil and grease.

In most cases, gloves, hand leathers, or other hand protectors can be worn to prevent hand injuries. In other cases, handles, holders, or other handling-assisted aids can be attached to the objects themselves (for example, handles or straps for moving auto batteries, wheeled carts, trays, or baskets for carrying laboratory samples).

Workers receive a large share of injuries to feet, legs, and toes while handling materials. One of the best ways workers can avoid such injuries is to wear footwear with integral protective toes and metatarsal guards.

The eyes, head, and trunk of the body are also vulnerable to injury. Especially when removing steel or plastic strapping from boxes, containers, piping and bales, make sure to follow instructions on safe removal and ensure that workers wear eye protection as well as stout gloves. Care should be taken to prevent strapping ends from flying loose and striking the face or body. The same precaution applies to coiled steel, wire, piping, lumber, and cable. If a material is in bags and the likelihood of airborne dust is created or present, the person handling it should consider the use of a respirator or other suitable personal protective equipment. Workers also need training in handling heavy objects. See the directions below for lifting and carrying.

## LIFTING AND CARRYING

Obviously, the best means to reduce back injuries is to try to eliminate manual lifting. If this cannot be done, another way is to reduce exposure. This can be achieved by cutting weight loads, using mechanical handling aids, rearranging or redesigning the workplace, or asking co-workers for help. Basic rules for manual lifting include:

1. Never let workers overexert themselves when lifting. If the load is thought to be more than one person can handle, assign another person to the job.
2. Lift gradually, without jerking, to minimize the effects of acceleration.
3. Keep the load close to the body.
4. Lift without twisting the body.
5. Follow the safe lifting procedures described below.

In reference to the safe lifting procedure, remember that some researchers working in the area of safe lifting now feel that it is better to let workers choose the lifting position most comfortable for them.

## TRAINING FOR SAFE LIFTING PRACTICES

Numerous attempts have been made to train materials handlers to do their work, particularly lifting, in a safe manner. Unfortunately, hopes for significant and lasting reductions of overexertion injuries through the use of training have been generally disappointing. There are several reasons for the disappointing results:

- If the job requirements are stressful, behavioral modification alone will not eliminate the inherent hazard. Designing a safe job is still necessary.
- People tend to revert to previous habits and customs if practices to replace previous ones are not reinforced and refreshed periodically.
- Safe work practices must be enforced and any unapproved deviation must be corrected immediately.
- Emergency situations, the unusual case, a quick movement, increased body weight, or impaired physical well-being/nutrition may overly strain the body, and training does not allow for these conditions.

Training for safe materials handling (which is not limited to lifting) should not be expected to solve the problem of job overload, lack of proper equipment, etc. If properly applied and periodically reinforced, training and behavior modification can and will help to alleviate some aspects of the basic problem.

Originally, the straight back/bent knees lifting posture was recommended. However, frequency and intensity of back injuries was not reduced during the last 40 years while this lifting method was taught. Biomechanical and physiological research has shown that leg muscles used in this lifting technique do not always have the needed strength. Also, awkward and stressful postures may be assumed if this technique is applied when the object is bulky, for example. Hence, the straight back/bent knees action evolved into the "kinetic" lift. The back is kept mostly straight and the knees are bent, but the position of the feet, chin, arms, hands, and torso are prescribed. Another variant is the "freestyle" lift, which may be better for male (but not female) workers than the straight back/bent knee technique. It appears that no single lifting method is best for all situations.

Training of proper lifting techniques is an unsettled issue. It is unclear what exactly should be taught, who should be taught, and how often a technique should be taught. This uncertainty concerns both objectives and methods, as well as expected results. Claims about effectiveness of one technique or another are frequent but are not usually supported by convincing evidence.

A thorough review of existing literature indicates that the issue of training for prevention of back injuries in manual materials handling is confusing at best. In fact, training may not be effective in injury prevention, or its effect may be so uncertain and inconsistent that the money and effort paid for training programs might be better spent on research and implementation of techniques for worker selection and ergonomic job design.

**Rules for Lifting.** There are no comprehensive and sure-fire rules for "safe" lifting. Manual materials handling is a very complex combination of moving body segments, changing joint angles, tightening muscles, and loading the spinal column. The following DOs and DO NOTs apply, however:

- DO design out manual lifting and lowering in tasks and workplaces. If it must be done by a worker, perform it between knuckle and shoulder height.
- DO be in good physical shape. If you are not accustomed to lifting and vigorous exercise, do not attempt to do difficult lifting or lowering tasks.
- DO think before acting. Place material conveniently within reach. Have handling aids available. Make sure sufficient space is cleared.
- DO get a good grip on the load. Test the weight and balance before trying to

move it. If it is too bulky or heavy, get a mechanical lifting aid or somebody else to help, or both.

- DO get the load close to the body. Place feet close to the load. Stand in a stable position with feet pointing in the direction of movement. Lift mostly by straightening the legs.
- DO NOT twist the back or bend sideways.
- DO NOT lift or lower awkwardly.
- DO NOT hesitate to get mechanical help or help from another person.
- DO NOT lift or lower with the arms extended.
- DO NOT continue lifting when the load is too heavy.
- DO not lift above your shoulders and below your knees.

## HANDLING SPECIFIC SHAPES

**Boxes, Cartons, and Crates.** These are best handled by grasping at alternate top and bottom corners and drawing one corner between the legs (see Figure 15–2). Any box, carton, or crate that appears either too large or too heavy for one person to handle should be handled by two people or with mechanical handling equipment, if possible.

When two people handle a crate, box, or carton, it is preferable that they be nearly the same size so both carry an equal portion of the load. They must be able to see in the direction they are walking and be aware of pedestrians, industrial trucks, machines, walls, and equipment.

Bags or sacks are grasped in the same manner as boxes. If a sack is to be raised to shoulder height, it should be raised to waist height first, and rested against the belly or hip before it is lifted to the shoulder. It should rest on its side. If two people are to lift a large bag or sack, they should be nearly the same size. This makes the job easier and keeps the load balanced. The two should lift at the same time, on an agreed signal.

**Figure 15–2.** When lifting a box or carton, the load should be drawn close to the body, and the arms and elbows tucked into the side of the body. This helps keep the weight centered.

**Barrels and Drums.** Workers need special training to handle barrels and drums safely.

If two people are assigned to upend a full drum, they should use the following procedure:

- Stand on opposite sides of the drum and face one another.
- Grasp both edges near their high points. Lift one end; press down on the other.
- As the drum is upended and brought to balance on the bottom edge, release the grip on the bottom chime and straighten it up with the drum.

When two people are to overturn a full drum, they should use this procedure:

- Make sure there is enough room. Cramped quarters can result in badly injured hands.
- Stand near one another, facing the drum. Grip the closest point of the top edge with both hands. Rest palms against the side of the drum, and push until the drum balances on the lower edge.
- Step forward a short distance, and release one hand from the top chime in order to grip the bottom chime. Ease the drum down to a horizontal position until it rests solidly on its side.

If one person is to overturn a drum, he or she should:

- Make sure there is enough room.
- Stand in front of the drum, reach over it, and grasp the far side of the top chime with both hands. (A short person can grasp the near side of the chime, if this is easier.) If the drum is tight against a wall or against other drums, pull on the edge with one hand and push against the wall (or other drum) with the other hand for additional control.
- Pull the top of the drum toward you, until it is balanced on the edge of the lower chime.
- Transfer both hands to the near side of the top chime. Keep the hands far enough apart to avoid being pinched when the drum touches the floor.
- Lower the drum. Keep the back straight, inclined as necessary. Bend the legs so that the leg muscles take the strain.

If one person is to upend a drum, reverse this procedure. Actually, if one person must handle a drum, he or she should have a lifter bar that hooks over the chime and gives powerful leverage and excellent control. Barrel and drum lifters are commercially available. To roll a barrel or drum, push against the sides with the hands. To change direction of the roll, grip the chime. No one should kick a drum with the feet.

**Figure 15–3.** This spill-containment caddy minimizes strain during drum transport. *(Courtesy of Justrite Manufacturing Co., LLC.)*

To lower a drum or barrel down a skid, turn it and slide it on end. Do not roll it. To raise a drum or barrel up a skid, two workers should stand on opposite sides of the skid (outside the rails, not inside, and not below the object being raised). Then, roll the object up the incline.

Handling drums and barrels can be hazardous, even when using utmost care. Special handling equipment and tools (approved by the safety professional or safety department) is recommended and should be available to make this job safer and easier (Figure 15–3).

Long objects, like ladders, lumber, or pipe, should be carried over the shoulder. The front end should be held as high as possible to prevent it from striking other employees, especially when turning corners. When two or more people carry an object, they should place it on the same shoulder, respectively, and walk in step.

## MATERIALS HANDLING EQUIPMENT

### HAND TOOLS

Many hand tools are available for specific jobs and for specific materials. They should be used only for their designated purpose. Further, hand tools should always be kept in good condition for best results and for safety reasons. For instance, if a chisel is used, the blade should be sharp and free of burrs. A mushroom-headed chisel should be repaired or discarded. (Refer to Chapter 14, Hand Tools and Portable Power Tools, for more details.)

Crowbars are probably the most common hand tools used in materials handling. Select the proper kind and size for the job. Because the bar can slip, people should never work astride it. They should position themselves to avoid being pinched or crushed if the bar slips or the object moves suddenly. The bar should have a good "bite" and not be dull or broken. When not in use, crowbars should be hung on a rack or placed so they cannot fall on or trip workers.

Special rollers are designed for moving heavy or bulky objects. Workers should be careful not to crush fingers or toes between the rollers and the floor.

Workers should use a sledge or bar, not their hands and feet, to change the direction of the object. They should avoid unnecessary turns by making sure the rollers are properly placed under the equipment before moving.

Handles of tongs and pliers should be offset so that hands and fingers will not be pinched.

Hand hooks used for materials-handling should be kept sharp so they will not slip when applied to a box or other object. The handle should be strong, securely attached, and shaped to fit the hand. The handle and the point of long hooks should be bent on the same plane so that the bar will lie flat and not be a tripping hazard. The hook point should be shielded when not in use.

Dipping a shovel into a pail of water occasionally will help to keep it free from sticky material. Greasing or waxing the shovel blade will also prevent some kinds of material from sticking. When not in use, shovels should be secured upright against a wall or kept in racks or boxes.

## NONPOWERED HAND TRUCKS

**Two-Wheeled Manual Hand Trucks.** These hand trucks look as though they would be easy to handle, but workers must follow several safe procedures:

1. Keep the center of gravity of the load as low as possible. Place heavy objects below lighter objects.
2. Balance the load so the weight will be carried by the axle, not the handles.
3. Position the load so it will not slip, shift, or fall. Load only to a height that will allow a clear view ahead.
4. Let the truck carry the load. The operator should only balance and push.
5. Never walk backwards with a hand truck.
6. When going up or down an incline, keep the truck in front of you.
7. Move the truck at a safe walking pace. Do not run.
8. Keep the truck under control at all times.

A truck that is designed for a specific purpose should be used only for that purpose. A curved bed truck should be used for handling drums or other circular materials, where a horizontal platform truck should be used. Except for handling acetylene or compressed gas cylinders, foot brakes can be installed on wheels of two-wheeled trucks so that operators need not place their feet on the wheel or axle to hold the truck. Handles should have knuckle guards, unless they are narrower than the distance between the outside of the wheel axle or the chisel, whichever is greater.

**Manually Operated Four-Wheeled Hand Trucks.** Operating rules for these vehicles are similar to those for two-wheeled trucks. Extra emphasis should be placed on proper loading, however. Four-wheeled trucks should be evenly loaded to prevent tipping. These trucks should be pushed rather than pulled, except for those with a fifth wheel and a handle for pulling.

Trucks should not be loaded so high that operators cannot see where they are going. If there are high racks on the truck, two people should move the vehicle. One should guide the front end, the other the back end. Handles should be placed at protected places on the racks or truck body so that passing traffic, walls, or other objects will not crush or scrape the operator's hands. Truck contents should be arranged so that they will not fall or be damaged if the truck or the load is bumped.

**General Precautions.** Truckers should be warned of four major hazards: (1) running wheels off bridge/dock plates or dock platforms, (2) colliding with other trucks or obstructions, (3) jamming hands between the truck and other objects, and (4) running wheels over feet.

When not in use, trucks should be parked in a designated area, not in aisles or other places where they constitute tripping hazards or traffic obstruction. Trucks with drawbar handles should be parked with handles up and out of the way. Two-wheeled trucks should be stored on the chisel with handles leaning against a wall or the next truck.

## POWERED HAND TRUCKS

Only trained and authorized operators shall be permitted to operate powered hand trucks. Use of powered hand trucks, controlled by a walking operator, is increasing. Principal hazards include the operator being caught between the truck and another object, and possible collisions with objects or people.

The truck should be equipped with a "dead-man" control, wheel guards, and an ignition key that can be taken out when the operator leaves the truck. Workers must be trained not to use trucks unless authorized. Training should follow instructions included in the truck manufacturer's manual. General instructions include:

1. Do not operate the truck with wet or greasy hands.
2. Lead the truck from right or left of the handle. Face the direction of travel

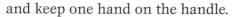

and keep one hand on the handle.

3. Powered hand trucks must enter an elevator or other confined areas with load end forward.

4. Always give pedestrians the right of way.

5. Stop and proceed with caution at blind corners, doorways, and aisle intersections to prevent collisions.

6. Never operate the truck faster than normal walking pace.

7. Handle flammable or corrosive liquids only when they are in approved containers.

8. Never ride on the truck, unless it is specifically designed for the driver to ride.

9. Never permit others to ride on the truck.

10. Do not indulge in horseplay.

11. Report damage or any unsafe conditions to your supervisor immediately.

## POWERED INDUSTRIAL TRUCKS

To comply with OSHA 29 CFR 1910.178, only trained and authorized operators are permitted to operate powered industrial trucks. Training programs must include safe operating and materials-handling practices as outlined in the standard. The supervisor must emphasize workers' safety awareness. Trained and authorized drivers should have badges or other clear authorization to drive, which they should display at all times.

Powered trucks have either a battery-powered motor or an internal combustion engine. Trucks should be maintained according to the manufacturer's recommendations. All trucks must meet the design and construction requirements established in ANSI/ASME B56.1–2004, Safety Standard for *Low Lift and High Lift Trucks*. Modifications and additions that affect capacity and safe operation should not be made by the customer or user without the manufacturer's prior written approval. All nameplates and markings must be accurate, in place, and legible.

Eleven different designations of trucks or tractors are authorized for use in various locations (see Fire Safety Standard for *Powered Industrial Trucks, Including Type Designations, Areas of Use, Conversions, Maintenance, and Operation*, ANSI/NFPA 505).

**Battery-Charging Installations.** This equipment must be located in areas designated for this specific purpose. The company must provide facilities for flushing and neutralizing spilled electrolyte, fire protection, and barriers to prevent trucks from damaging the charging apparatus. There must also be adequate ventilation for dispersal of gases from gassing batteries. Eyewash stations are necessary if workers are adding water to cells. Racks used to support batteries must be made of spark-resistant materials or be coated or covered to prevent spark generation.

A conveyor, overhead hoist, or equivalent equipment must be used for handling batteries. Reinstalled batteries must be properly positioned and secured in the truck. Use a carboy tilter or siphon for handling electrolyte. Teach workers that acid must always be poured into water; water must never be poured into acid—it overheats and splatters. During charging operations, vent caps must be kept in place to avoid electrolyte spray. Make sure vent caps are functioning. Battery or compartment covers must be open to dissipate heat.

Take all precautions to prevent open flames, sparks, or electric arcs in battery-charging areas. Tools and other metallic objects must be kept away from uncovered battery terminals. Employees charging or changing batteries must be authorized to do this work, trained in proper handling techniques, and required to wear protective clothing, including face shields, long sleeves, rubber boots, aprons, and gloves. Smoking is prohibited in the charging area.

**Refueling.** All internal combustion engines must be turned off before refueling. Refueling should be in the open or in specifically designated areas where ventilation is adequate to carry away fuel vapors. Liquid fuels, such as gasoline and diesel, must be handled and stored in accordance with the National Fire Protection Association's Flammable and Combustible Liquids Code, NFPA 30. Liquefied petroleum gas fuel (or LP-gas) must be stored in accordance with Storage and Handling of Liquefied Petroleum Gases, NFPA 58. LP-gas tanks must be secured with both straps while on the truck so as not to shake loose and possibly catch fire. Install no-smoking signs in prominent places. Smoking must be strictly prohibited in the service areas.

Train employees to follow general rules for driving powered industrial trucks in your operations. For each job, you should draw up a specific set of rules (referenced from OSHA 29 CFR 1910.178):

1. All traffic regulations must be observed, especially facility speed limits.
2. Safe distances must be maintained approximately three truck lengths from the truck in front. Trucks must be kept under the operator's control at all times.
3. The right-of-way must be yielded to ambulances, fire trucks, and other vehicles in emergency situations.

4. Other trucks traveling in the same direction must not be passed at intersections, blind spots, or other dangerous locations.

5. Drivers are required to slow down and proceed with caution, and sound horn at crosswalks, aisles, and other locations where vision is obstructed or when approaching pedestrians. If the forward view is obstructed because of the load being carried, the load should be adjusted to provide clear view of travel.

6. Railroad tracks must be crossed diagonally and drivers must never park within 8 ft (2.5 m) of the center of the nearest railroad track bed.

7. Drivers are required to keep their eyes on the direction of travel and to have a clear view of the path ahead at all times. They should never back up without looking. Be especially careful on loading docks (Figure 15–4).

8. Grades shall be ascended or descended slowly. For grades in excess of 10%, loaded trucks will be driven with the load upgrade. On all grades, load and load-engaging means are to be tilted back if applicable and raised only as far as necessary to clear the floor or road surface.

9. Trucks must always be operated at a speed that will permit them to stop safely. Drivers are required to slow down on wet and slippery floors. Stunt driving and horseplay are not tolerated.

10. Dockboards or bridge plates are to be driven over carefully and slowly and only after they have been properly secured. Their rated weight capacity must never be exceeded.

11. Elevators shall be approached slowly and then entered squarely after the elevator car is properly leveled. Once on the elevator, the controls should be neutralized, power shut off, and the brakes set.

12. Never run over loose objects on the roadway surface.

13. While negotiating turns, reduce speed to a safe level and turn the hand steering wheel in a smooth, sweeping motion. When maneuvering at a low speed, the hand steering wheel must be turned at a moderate and even rate.

14. Only stable or safely balanced loads shall be handled, and extra caution must be exercised when handling off-center loads.

15. Only loads within the rated load capacity of the truck shall be handled, and long or

**Figure 15–4.** Operators must keep alert when backing a truck and face in the direction of travel. *(Courtesy of Clark Equipment Co.)*

high (including multiple-tiered) loads that may affect capacity must be adjusted.

16. Load-engaging means must be placed under the load as far as possible; the mast is then carefully tilted backward to stabilize the load.

17. Extreme care must be used when tilting the load forward or backward, particularly when placing items on high tiers. Tilting forward with the load-engaging means elevated is not permitted except to pick up a load. Elevated loads must not be tilted forward, except when the load is in a deposit position over a rack or stack. When stacking or tiering, use only enough backward tilt to stabilize the load.

18. When operating in close quarters, keep hands out of the way so they cannot be pinched between steering controls and projecting stationary objects. Keep legs and feet inside the guard or the operating stations of the truck.

19. Do not use the reverse control on electric trucks for braking.

20. Park trucks only in designated areas—never in an aisle or doorway or where they will obstruct equipment or material. Fully lower the load-engaging means flat on the floor surface, neutralize the controls, shut off the power, and set the brake and wheel block when parking on an incline.

OSHA does not have a specific standard that requires the use of seat belts; however, recognition of powered industrial truck tip-over hazards and the need for an operator restraint system by certain powered industrial trucks is evidenced at ASME B56.1-2004. Therefore, OSHA would enforce this standard under Section 5(a) (1) of the Occupational Safety and Health Act of 1970 (known as the General Duty Clause).

Depending on its use, a power truck should have a dead-man control and be equipped with guards, such as a hand-enclosure guard or overhead guard. A lift truck should also have upper and lower limit switches to prevent overtravel. No powered industrial truck should be used for any purpose other than the one for which it was designed.

A forklift should not be used to elevate employees (for example, in servicing light fixtures or when stacking materials) unless a safety platform meeting OSHA requirements, with standard railing and toeboards, is fastened securely to the forks. Failure to use safety platforms has resulted in many serious incidents.

**Combustion By-Products.** When internal-combustion engine trucks are operated in enclosures, the concentration of carbon monoxide should not exceed limits specified by local or state laws. In no case should the time-weighted average concentration exceed 35 parts per million (ppm) for an eight-hour exposure. The atmosphere must contain a minimum of 19% oxygen, by volume—air usually contains 20.8% oxygen, by volume. Many companies equip their fuel-powered trucks with catalytic exhaust purifiers to burn the carbon monoxide before it reaches the atmosphere. Some power trucks are approved and marked only for use in specific locations. These include areas where flammable gases or vapors are present in quantities sufficient to produce explosive or ignitable mixtures, or areas where combustible dust or easily ignitable fibers or flyings are present. (Refer to ANSI/NFPA 505–2002, Fire Safety Standards of *Powered Industrial Trucks…Operation*, for classification of types.)

**Loads.** Operators should not carry unstable loads. If material of irregular shape must be transported, make sure it is securely placed and blocked, tied, or otherwise strapped down. If possible, crosstie the load after neatly stacking it. Never place an extra counterbalance on a forklift to help handle overloads because it puts added strain on the truck and endangers the operator and nearby employees. It is difficult to safely operate an overloaded truck.

Many stand-up-type electric trucks are being used in industry. They vary in design and operation, and workers must be trained in the operating procedures for each one they use.

Operators of powered industrial trucks should not drive up to anyone standing in front of a wall or other fixed object, or allow anyone to stand or pass under the elevated portion of any truck, whether loaded or empty. It is the operator's responsibility to protect pedestrians. Operators should not expect pedestrians to see them. Personnel are not permitted to ride on trucks as passengers; even if they are authorized operators. They are permitted only when they are in the space provided for a passenger. No one should be allowed to place arms or legs between uprights of the mast or outside the running lanes of the truck.

Raise or lower loads at the point of loading or unloading, not during travel. Operators should be trained to check the area around the truck before raising a load so they will not strike structural members of the building, electric wiring or cables, or piping (especially ones carrying gases or flammable fluids).

Operators must make sure there is sufficient headroom under overhead installations, such as lights, pipes, and sprinkler systems, before driving underneath them. Overhead guards must be used on lift trucks as protection against falling objects. An overhead guard is intended to offer protection from the impact of small packages, boxes, bagged materials, and so forth, but not to withstand the impact of a falling capacity load. Load backrest extensions must be used whenever necessary to minimize the possibility of the load or part of the load falling toward the rear.

Operators must make sure they check decking of semi trucks before entering with a lift truck. Floors on which power trucks travel should be kept in peak condition. They should have a load capacity great enough to support both truck and load. The combined weight of the heaviest truck and load should not exceed one-fifth of the design strength of the floor. For example, a floor originally designed to bear 1,000 pounds per square foot—or 5,000 kg/m$^2$—should not be allowed to bear more than 200 pounds per square foot. Maximum floor loadings should be prominently displayed (ANSI standard A58.1–1982, *Minimum Design Loads for Buildings and Other Structures*).

The 5-to-1 ratio, although not excessively conservative, is important in establishing a margin of safety as protection against possible miscalculations.

## DOCK PLATES

Dock plates (bridge plates) that are not built-in should be secured to prevent them from "walking" or sliding when used. Secure by drilling holes and dropping pins or bolts through them.

Plates should be large and strong enough to hold the heaviest load, including the equipment. A safety factor of five or six should be used. Plates can have the edges turned up, or have angle-iron welded to the edges, to keep trucks from running over their edges. Keep plates in good condition. Make sure they do not have curled corners or bends. They should be kept clean and dry—clear of snow and ice, and free of oil and grease. Canopies can be used to protect them from the weather.

Dock plates should be positioned with a forklift or other mechanical equipment to avoid back injuries and pinched fingers. The dock plate must be kept under control and not allowed to bounce. Handholds or other effective means must be provided on portable dock plates to permit safe handling.

Permanent, adjustable dock levelers are the safest and most efficient method of traversing between the dock and a vehicle. They should be of sufficient capacity to handle present and future loads, and be long enough to reduce operating grades. Dock levelers should be inspected and maintained on a regular basis (Figure 15–5).

Trucks or trailers backed up to the dock must have their brakes set and rear wheels chocked to prevent movement when power trucks are loading or unloading. This would also apply to railroad cars. This should be the lift truck operator's responsibility. Fixed jacks must be used when the trailer is not coupled to a tractor. When not in use, dock plates should be safely stored to avoid acting as tripping hazards. They should not be placed on edge so they can fall over. For details of loading highway trucks and cars, see later sections in this chapter.

## CONVEYORS

The primary rule regarding conveyors is simple: no one is allowed to ride or climb on or over them. Post prominent warning signs making this rule clear. If people must cross over conveyors, the company should provide bridges or stiles so that they do not have to crawl across or under the conveyor.

**Figure 15–5.** This hydraulic dock leveler prevents forklifts and other vehicles from rolling or being driven off the dock when the truck or trailer is not present. *(Courtesy of Rite Hite Corp.)*

In the case of certain roller conveyors, it is possible to provide a hinged section that opens and permits people to walk through without having to climb over. The hinged section should be returned to its original position as soon as the individual has reached the other side. Supervisors should train workers to make sure that no merchandise is on the roller conveyor when they use the hinged section.

Gravity conveyors, usually the chute or roller type, should be equipped with warning or protective devices to prevent workers' hands from being caught in descending material or from being jammed between material and a receiving table. If the conveyor jams up, first try to free it from the top side. If someone must enter the chute to free a jam, confined space entry and lockout procedures should be implemented. Adhere to OSHA's confined space requirements (29 CFR 1910.146). Inform your employees of these hazards and train them to handle such hazards safely.

Pinch points on conveyors—gears, chain drives, and revolving shafts—should be guarded. Emergency stop systems, such as tripwire cables, can be installed. Train workers not to remove these guards. If the conveyor needs servicing, all energy sources should be isolated and locked out.

Pneumatic or blower conveyors must be shut off and locked out before inspections or repairs are made. Inspection ports, equipped with transparent coverings made of durable, nonflammable, shatter-resistant material, permit inspections without stopping the operation.

Screw conveyors should be completely enclosed unless otherwise effectively guarded. Their principal danger is that workers may try to dislodge material or free a jam with their hands or feet, and be caught in the conveyor. Screw conveyors should be repaired only when all energy sources have been isolated and locked out and OSHA's 29 CFR 1910.147 lockout/tagout requirements are met. Any exposed sections should be covered with a metal grate with openings no larger than ½ in (13 mm), or with a solid cover—ANSI/ASME standard B20.1–2003, *Safety Standards for Conveyors and Related Equipment*. The cover should be strong enough to withstand abuse. Before removing any cover, grating, or guard, implement lockout procedures. Do not rely on interlocks. Never step or walk on covers, gratings, or guards.

## CRANES

All overhead and gantry cranes must meet the design specifications of the ANSI/ASME B30.17–1985, *Overhead and Gantry Cranes*, and OSHA 29 CFR 1910.129; all mobile, locomotive, and truck cranes must meet ANSI/ASME B30.5–1989, *Mobile and Locomotive Truck Cranes* (OSHA 29 CFR 1910.180).

Although many types of cranes are used in industry, safe operating procedures are mostly the same. Standard crane hand signals (Figures 15–6 and 15–7) should be thoroughly understood by both the operator and signaler. Only one person is assigned the duty to signal movements of the crane to the operator. The only exception is the emergency stop signal that anyone can use to stop operation of the crane. People working with or near a crane should keep out from under the load, be alert at all times, and watch and listen to warning signals, or combination of flashing lights, closely. When a warning signal sounds, they must move to a safe place immediately.

No crane should be loaded beyond its rated load capacity. Depending on the type of crane, the rated load capacity must be clearly marked on the beam and on the traveling block of cranes equipped with more than one means of hoisting. The weight of all auxiliary handling devices, such as hoist blocks, hooks, and slings, or other attachments must be considered as part of the load.

Substantial and durable rating charts with clearly legible letters and figures

**Figure 15–6.** Standard hand signals for controlling operation of overhead and gantry cranes. *(Reprinted with permission from ANSI/ASME B30.5-1989.)*

are to be fixed to the crane cab in a location clearly visible to the operator while seated at the control station, or nearby for cranes equipped with hand controls operated at floor level.

**Figure 15–7.** Standard hand signals suitable for crawler, locomotive, wheel, and truck-mounted cranes. *(Reprinted with permission from ANSI/ASME B30 series.)*

The hoist chain or hoist rope must be free of kinks and twists and must not be wrapped around the load. Loads must be attached to the load line hook by means of a sling or other approved device. Care must be taken to make certain that the sling clears all obstacles. The load must be well secured and properly balanced in the sling or lifting device before it is lifted more than a few inches. Before starting to hoist, make sure that multiple-part lines are not twisted around each other.

Hooks should be brought over the load slowly to prevent swinging. During hoisting operations, there must be no sudden acceleration or deceleration of the moving load. The load must not come in contact with any obstruction.

Cranes must not be used for side pulls, unless specifically authorized by a responsible person who can determine that the stability and integrity of the crane is not endangered and that the various parts of the crane will not be over-stressed. There must be no hoisting, lowering, or traveling while any employee is on the load or hook. The operator must not carry loads over people's heads. On overhead cranes, a warning signal must be given when starting the bridge and when the load or hook comes close to, or is about to pass over, any person's head so that workers can get out of the way.

Brakes must be tested each time a load approaching the rated load is handled. Loads must not be lowered below the point where less than two full wraps of rope remain on the hoisting drum. When two or more cranes are used to lift a load, only one qualified, responsible person must be in charge of the operation. He or she must analyze the operation and instruct all personnel involved about the proper positioning and rigging of the load, and all other movements. An operator must not leave the controls while the load is suspended. All necessary clothing and personal belongings that are stored in an operator's cab must not interfere with access or operation. Tools, oil cans, waste, extra fuses, and other necessary articles must be kept in a toolbox and not carelessly left in or around the cab. A portable fire extinguisher must be kept in the cab. Access to the cab must be by fixed ladder, stairs, or platform, but no steps over gaps exceeding 12 in. (30 cm) are allowed (OSHA CFR 1910.179, (c) Cabs, (2) Access to crane).

The upper load limit switch of each hoist must be tested under no-load conditions at the beginning of each operator's shift. If the switch does not operate properly, a qualified person must be notified immediately. The hoist limit switch that controls the upper limit of travel of the load block must never be used as an operating control.

**Inspections.** Make sure that cranes in regular service are inspected periodically. The intervals depend on the nature of the crane's critical components and the degree of its exposure to deterioration or malfunction. Daily inspections must be made of all control mechanisms for maladjustments, including deterioration or leakage in air or hydraulic systems.

Frequent inspections at daily to monthly intervals, or as specifically recommended by the manufacturer, must be conducted of the following items:
- all control mechanisms for excessive wear of components or contamination by lubricants or other foreign matter
- all safety devices for malfunction
- crane hooks for deformations, cracks, or more than 15% in excess of normal throat opening or more than a 10-degree twist from the plane of the unbent hook, and spring-loaded safety latch
- rope reeving not in compliance with the manufacturer's recommendations
- electrical apparatus for malfunctioning and excessive deterioration, dirt, and moisture accumulation

Periodic inspections, from one- to 12-month intervals, or as recommended by the manufacturer, cover the complete crane. These inspections include the items listed under frequent inspections, as well as:
- deformed, cracked, or corroded members in crane structure and hook, and spring-loaded safety latch
- loose bolts or rivets
- cracked or worn sheaves and drums, OSHA 1910.179J, Inspections, (2) Frequent inspections (iii)
- worn, cracked, or distorted parts, such as pins, bearings, shafts, gears, rollers, and locking devices
- excessive wear on brake and clutch system parts linings, pawls, and ratchets
- load, boom, angle, and other indicators over their full range to detect significant inaccuracies
- gasoline, diesel, electric, or other power plants for improper performance or noncompliance with safety requirements, excessive wear of chain-drive sprockets, and excessive chain stretch
- travel steering, braking, and locking devices for malfunction and excessively worn or damaged tires
- determine if the above items constitute a hazard and, if so, correct them; Accurate records must be kept; see OSHA requirements

Prior to initial use, all new and altered cranes must be thoroughly inspected to ensure the crane meets all compliance requirements of the applicable provisions of the OSHA standard. Cranes not in regular service and idle for one month but less than six months must meet standards of items outlined for frequent inspections. Cranes idle for six months or more must be inspected in accordance with OSHA requirements (29 CFR 1910.179 and 180). Standby cranes must be inspected at least semiannually. Cranes exposed to adverse environmental conditions must be inspected more often.

All wire ropes must be inspected thoroughly once a month. A full written, dated, and signed report of wire rope condition must be kept on file where it is readily available to appointed personnel. Any deterioration, resulting in appreciable loss of original strength, must be carefully noted. The rope may constitute a hazard if there is a reduction of rope diameter below normal dimensions caused by loss of core support, internal or external corrosion, or wear or breakage of outside wires; corroded or broken wires at end connections; or severe kinking, crushing, cutting, or unstranding.

All ropes not used for a month or more, due to shutdown or storage of a crane on which it is installed, must be given a thorough inspection before being placed in service. This inspection should identify all types of deterioration. It should be performed by a responsible person whose approval is required before the rope is returned to service. A written, dated report of the rope condition must be available for inspection.

On limited-travel ropes, heavy wear and/or broken wires may occur in sections that come in contact with equalizer sheaves or other sheaves where rope travel is limited, or saddles. Particular care must be taken to inspect all ropes at these locations, with special attention paid to all nonrotating ropes.

**Safety Rules.** These safety rules for electric overhead cranes are suggested by one crane manufacturer:

1. Only regularly authorized operators should use cranes.
2. When on duty, operators should remain in the crane cabs ready for prompt service.
3. Before traveling, make certain the hook is high enough to clear all obstacles.
4. Under no circumstances should one crane bump another.
5. Examine the crane at every shift for loose or defective gears, keys, runways, railings, warning bells, signs, switches, sweeps, ropes, cables, air or

hydraulic systems, deformed or cracked hooks, and hoist chains. Report defects. Keep the crane clean and well lubricated.

6. No one should go to the top of a crane without first opening the main switch, placing a warning sign on it, and locking it out. Operators should unlock the switch and remove the sign promptly when they come down.

7. After completion of a repair job, make sure that bolts, tools, and other materials have been removed to prevent damaging the machinery when the crane is restarted and to be sure nothing falls off the crane. Keep tools, oil cans, and other loose objects in a box provided for that purpose.

8. Do not carry a load over workers' heads on the floor. Sound a gong or siren when necessary. No one should ride the load or the crane hook.

9. If the power goes off, move controller to OFF position until power is again available.

10. Make sure the fire extinguisher on the crane is in good condition; and if used, that it is refilled immediately.

11. Never let an operator who is ill or not physically fit run a crane.

12. Do not drag slings, chains, or ropes.

13. If the operator feels it is unsafe to move a load, he or she should check with the supervisor.

14. Be sure that when an operator leaves the cage, he or she leaves the main switch open. A magnetic crane must be empty and its controller turned off.

15. When an outside crane is parked at the end of a shift, the brake should be set or the crane should be chained to the track.

16. Operators should know that if their crane fails to respond correctly, they should call the supervisor. Attempting to get out of difficulty by repeated operation may make the condition worse, not better.

## RAILROAD CARS

Spotting of railroad cars is usually done by switch-engine crews. When employees must move railroad cars, the supervisor or crew leader should consider the following factors:

- facilities or equipment available to move the cars
- distance the cars must be moved
- number of cars to be moved
- whether cars are full or empty—if full, the type of load carried
- whether the track is level or on a grade

After weighing these factors, you can decide whether to get power equipment, move the car by hand, or take material to (or from) the car's location. If possible, use power equipment to move the car.

If a car must be spotted, brakes must be set and wheels blocked when loading and unloading or moved without power equipment, and a car mover (operated by only one person) can be used. If it takes two workers to move a car, each one should have a car mover and they should work on opposite wheels. All hand-actuated car movers should be equipped with knuckle guards. A stationary car puller, if available, should be used.

When opening doors of boxcars, all workers should be alert to the danger of fingers being caught or of freight falling out. A bar or ratchet hoist or door puller should be used. Do not use lift trucks to open or close car doors.

Metal transfer plates must be securely anchored with bolts. Gangplanks and skids must be securely placed before using. Operators of mechanized equipment should avoid bumping doors or door posts. Metal bands should be removed or cut to avoid dangling ends. Stack special equipment, such as DF bars, at the ends of a car when empty, and close and latch doors before moving the car. Specially equipped cars with movable bulkheads have instructions in the cars that must be followed. When bulkhead partitions are difficult to move, examine rollers to determine the reason for binding to avoid having the bulkhead fall on those working in the car.

## HIGHWAY TRUCKS

When loading or unloading highway trucks, set the brakes and place wheel chocks in front of the rear wheels or use dock-mounted vehicle restraints to prevent the trucks from rolling away from the dock (Figure 15–8). With vehicle restraints, trailers are held in place by their underride protection bar, or ICC bar. Unlike wheel chocks, vehicle restraints hold the truck safely to the dock if the landing gear or jack collapses, or if the driver tries to pull away from the dock before loading or unloading is completed. They are also effective in icy and wet environments.

If a powered industrial truck is used inside a trailer that is not coupled to its tractor, place fixed jacks as needed to support the trailer. Make sure jacks are snug against the trailer bottom. Where automated systems are used, visual inspection should be conducted to make sure the trailer is secured. Check decking before entering a lift truck.

**Figure 15–8.** Two types of vehicle restraints to secure a truck. *(Courtesy of Rite Hite Corp.)*

## MOTORIZED OFF-ROAD EQUIPMENT

Heavy-duty trucks, mobile cranes, tractors, bulldozers, and other motorized equipment used in the production of stone, ore, and similar materials and also in construction work, have been involved in frequent, and often serious, incidents. In general, prevention of these incidents requires:

- safe equipment
- systematic maintenance and repairs
- safety training for operators
- safety training for repair personnel
- strict observance of safety procedures, including on-site traffic
- control of visitors, vendors, contractors, and their training

**Operation.** Manufacturers' manuals contain detailed information on the operation of equipment. Recommended driving practices are similar to those necessary for the safe operation of highway vehicles. Off-road driving involves certain hazards that require special training and safety measures. The modern heavy-duty truck or off-road vehicle is a carefully engineered and expensive piece of equipment and warrants operation only by drivers who are qualified physically and mentally. For safety, driving standards should be especially high.

The time required for prospective drivers or operators to become thoroughly familiar with the mechanical/hydraulic/electrical features of the equipment, safety rules, driver reports, and emergency conditions varies. In no case should even an experienced driver be permitted to operate equipment until the instructor or supervisor is satisfied with his or her abilities. After workers and supervisors have been trained, supervisors have the important responsibility of seeing that drivers continue to operate vehicles as they were instructed.

**Equipment.** Safe operation of heavy equipment begins with the purchase specifications. A good policy is to specify safeguards over power transfer components, safe oiling devices, handholds, guardrails, and other devices when the order is placed with the manufacturer. Before equipment is put into operation, it should be thoroughly inspected and necessary safety devices installed and checked.

Both operators and repair personnel, whether experienced or not, should know and follow the manufacturers' recommendations pertaining to lubrication, adjustments, repairs, and operating practices for equipment. A preventive maintenance program is essential to ensure worker safety and efficiency. Frequent and regular inspections and prompt repairs provide effective preventive maintenance.

In general, the operator is responsible for inspecting such mechanical conditions as hold-down bolts, brakes, clutches, clamps, hooks, and similar vital parts. Wire ropes should be lubricated (see manufacturers' instructions). For inspection guidelines, see the section that follows.

Many of the basic safety measures recommended for trucks also apply to motor graders and other earth-moving equipment. All machines should be inspected regularly by the operator, who should report any defects promptly. Equipment safety and efficiency of equipment are increased by adherence to periodic maintenance.

# ROPES, CHAINS, AND SLINGS

Special safety precautions apply to using ropes, rope slings, wire rope, chains and chain slings, and storing chains. You should thoroughly know the properties of these various types and the precautions pertaining to their use and maintenance. For additional help regarding ropes, chains, or slings, consult the supplier.

## FIBER ROPES

Fiber rope is used extensively in handling and moving materials. The rope is generally made from manila (abaca), sisal, or nylon. Manila or nylon ropes give the best uniform strength and service. Other types of rope include those made from polyester or polypropylene or a composite of both that is adaptable to special uses.

Sisal is not as satisfactory as manila because the strength varies in different grades. Sisal rope is about 80 % as strong as manila. Manila rope is yellowish with a somewhat silvery or pearly luster. Sisal rope is also yellowish, but often has a green tinge and lacks the luster of manila. Its fibers tend to splinter, and it is rather stiff. The safe working loads and breaking strengths of the rope that you are using should be known, and the limits must be observed.

**Maintenance Tips.** Precautions should be taken to keep ropes in good condition. Kinking, for example, strains the rope and may overstress the fibers.

It may be difficult to detect a weak spot made by a kink. To prevent a new rope from kinking while it is being uncoiled, first lay the rope coil on the floor with the bottom end down. Then pull the bottom end up through the coil and unwind the rope counterclockwise. If it uncoils in the other direction, turn the coil of rope over and pull the end out on the other side.

Rope should not be stored unless it has first been cleaned. Dirty rope can be hung in loops over a bar or beam and then sprayed with water to remove the dirt. The spray should not be so powerful that it forces the dirt into the fibers. After washing, the ropes should be allowed to dry and then shaken to remove any remaining dirt. Rope must be thoroughly dried out immediately after it becomes wet; otherwise it will deteriorate quickly. A wet rope should be hung up or laid in a loose coil until thoroughly dried. Rope will deteriorate more rapidly if it is frequently alternating between wet and dry, than if it remains wet. Rope should not be allowed to freeze.

Store rope in a dry place where air circulates freely. Small ropes can be hung up. Larger ropes can be laid on gratings so air can get underneath and around them. Rope should not be stored or used in an area where the atmosphere contains acid or acid fumes because it will quickly deteriorate. Signs of deterioration are dark brown or black spots.

Sharp bends should be avoided whenever possible because they create extreme tension in the fibers. When cinching or tying a rope, be sure the object has a large-enough diameter to prevent the rope from bending sharply. A pad can be placed around sharp corners.

If at all possible, ropes should not be dragged. Dragging abrades the outer fibers. If ropes pick up dirt and sand, abrasion within the lay of the rope will rapidly wear them out. Lengths of rope should be joined by splicing. A properly made short splice will retain approximately 80% of the rope's strength. A knot will retain only about 50%.

Ropes should be inspected at least every 30 days; every day if they are used to support scaffolding on which people work. The following procedure will assist in inspection of natural fiber rope:

1. Check for broken fibers and abrasions on the outside.
2. Inspect fibers by untwisting the rope in several places. If the inner yarns are bright, clear, and unspotted, the strength of the rope has been preserved to a large degree.
3. Unwind a piece of yarn 8 in (20 cm) long and ¼ in (0.6 cm) diameter from the rope. Try to break this with your hands. If the yarn breaks with little effort, the rope is not safe.
4. Inspect rope used around acid or caustic materials daily. If black or rusty brown spots are noted, test the fibers as described in steps 1, 2, and 3, and discard all rope that fails the tests.
5. As a general rule, rope that has lost its pliability or stretch, or in which the fibers have lost their luster and appear dry and brittle, should be viewed with suspicion and replaced. This is particularly important if the rope is used on scaffolding or for hoisting where workers could be seriously injured, or property could be damaged, by a break.

## ROPE SLINGS

Because of high-tensile strength needed for a sling, fiber rope slings should be made of manila only. The following are some precautions to be observed in using manila rope slings:

1. Make sure that the sling is in good condition and of sufficient strength. Take into account the factors of the splice and leg angles of the sling. Hooks, rings, and other fittings must be properly spliced.
2. Reduce the load by one-half after the sling has been in use for six months, even though the sling shows no sign of wear.
3. If the sling shows evidence of cuts, excessive wear, or other damage, destroy it.

Whether fiber or wire rope slings are used, certain general precautions should be taken. (These were given earlier under Fiber Ropes—Maintenance Tips.) In addition, it is best to keep the legs of slings as nearly vertical as possible (Figure 15–9).

**Figure 15–9.** Increasing the sling angle (the angle of the sling leg to the vertical) increases the stress on each leg of the sling, even though the load remains constant. The stress on each leg can be calculated by dividing the load weight by the cosine of the angle.

## WIRE ROPE SLINGS

Wire rope is used widely instead of fiber because it has greater strength and durability for the same diameter and weight. Its strength is constant either wet or dry and it maintains a constant length under varying weather conditions. Wire rope is made according to its intended use. A large number of wires give flexibility, while fewer strands with fewer wires are less flexible.

A wire rope used for general hoisting should not be subjected to a working load greater than those recommended in tables N-184-3 through N184-14 of the OSHA standard.. Wire ropes should be thoroughly inspected each day by a competent person, before being used, for wear or damage; damaged slings shall be immediately removed from service. In addition, a periodic inspection shall be conducted on all slings with the interval of the inspection to be determined by the use of the sling OSHA regulation (29 CFR 1910.184).

Proper storage of wire rope slings will add to their serviceable life as well as to protection for the worker. Slings should be maintained per manufacturers' directions, keeping them dry and free from chemicals, dirt, rust, and corrosion. Periodic use of all slings will add to their overall life. Slings should not be kept on the floor nor allowed to collect in a pile; rather they should be stored on pegs in an area away from damaging elements such as dust, water, and chemicals.

All sling hooks and connecting points should be checked daily for distortion and cracks prior to use. Hooks should be checked periodically for distortion in the hook of no greater than 15% of the original hook opening nor more than 10 degrees off of vertical for the tongue of the hook.

## CHAINS AND CHAIN SLINGS

Alloy steel chain, approximately twice as strong—size for size—as wrought-iron chain, has become standard material for chain slings. One advantage is that it is suitable for high-temperature operations. Continuous operation at temperatures of 800 F (425 C) (the highest temperature for which continuous operation is recommended) requires a reduction of 30 % in the regular working load limit. For intermittent service, these chains can be used at temperatures as high as 1,000 F (540 C), but at only 50 % of the regular working load limit.

An alloy steel chain should never be annealed when the chain is being serviced or repaired. The process reduces the hardness of the steel; and this, in turn, reduces the strength of the chain. Wrought-iron chain, on the other hand, should be annealed periodically by the manufacturer (or persons specially trained in the operation) if it is used where failure might endanger human life or property.

Impact loads, caused by faulty hitches, bumpy crane tracks, and slipping hookups, can add significant stress in the chain. The impact resistance of heat-treated alloy steel chain does not increase in proportion to the strength of the chain. Under a full working load, it will fail before a fully loaded wrought-iron or heat-treated carbon steel chain.

Chain slings preferably should be purchased complete from the manufacturer. When they need repair, they should be returned to the manufacturer. Query suppliers and manufacturers about details of strength and specifications of the slings.

**Load and Wear.** Chains should not be overloaded. The working load limit and break test limit for alloy steel chains are given in Table 15–A. Some of the principal causes of chain failure are:

1. brittleness caused by cold working of the metal surface
2. failure of the weld
3. repeated and severe bending or deformation of the links
4. metal fatigue
5. brittleness caused by defects in the metal
6. tensile failure—full elongation of the links

How much wear a chain can stand is determined by its use, the original factor of safety, and the loads the chain is expected to carry. A regular schedule of checking and calipering the links should be set up for each chain, depending on usage (Table 15–B).

### Table 15-A. Working Load Limits and Break Test Limits for Alloy Steel Chain*

| Nominal Size of Chain Bar in Inches | Working Load Limits in Pounds | Minimum Break Test in Pounds |
|:---:|:---:|:---:|
| ¼ | 3,250 | 10,000 |
| ⅜ | 6,600 | 19,000 |
| ½ | 11,250 | 32,500 |
| ⅝ | 16,500 | 50,000 |
| ¾ | 23,000 | 69,500 |
| ⅞ | 28,750 | 93,500 |
| 1 | 38,750 | 122,000 |
| 1⅛ | 44,500 | 143,000 |
| 1¼ | 57,500 | 180,000 |
| 1⅜ | 67,000 | 207,000 |
| 1½ | 80,000 | 244,000 |
| 1¾ | 100,000 | 325,000 |
| 1 in = 2.54 cm; 1 kg = 2.5 lb | | |

*The strengths for wrought-iron chain are slightly less than half these figures.

Specification for Alloy Chain, American Society for Testing and Materials, A391-1975.

To check elongation, chain should be calipered in gage lengths of 1 to 3 ft (preferably 5 ft or 1.5 m) when new, and a record of that test should be kept. The chain (or sling) should be discarded if it shows a stretch (includes link wear) of more than 5%. Some states specify that a chain must be destroyed or discarded when any 3-ft (0.9-m) length is found to have been stretched one-third of a link length.

A broken chain should never be spliced by inserting a bolt between two links. Nor should one link be passed through another and a nail or bolt inserted to hold them.

**Use.** Before making a lift, workers should check the chain to be sure it has no kinks, knots, or twists. Workers should lift the load gradually and uniformly and make certain it is not merely attached to the top of the hook. They should never hammer a link over the hook, as this will stretch the hook and the links of the chain. When lowering a crane load, brakes should be applied gradually to bring the load to a smooth stop.

| Table 15–B. Correction Table for the Reduction of Working Load Limits of Chain Due to Wear | | | |
|---|---|---|---|
| Reduce working load limits of chain by the following percentage when diameter of stock at worn section is as follows: | | | |
| **Nominal Original Chain Size (inches)** | **5 %** | **10 %** | **Remove from Service** |
| ¼ | 0.250…0.244 | 0.237 | 0.233 |
| ⅜ | 0.375…0.366 | 0.356 | 0.335 |
| ½ | 0.500…0.487 | 0.474 | 0.448 |
| ⅝ | 0.625…0.609 | 0.593 | 0.559 |
| ¾ | 0.750…0.731 | 0.711 | 0.671 |
| ⅞ | 0.875…0.853 | 0.830 | 0.783 |
| 1 | 1.000…0.975 | 0.949 | 0.895 |
| 1⅛ | 1.125…1.100 | 1.070 | 1.010 |
| 1¼ | 1.250…1.220 | 1.190 | 1.120 |
| 1⅜ | 1.375…1.340 | 1.310 | 1.230 |
| 1½ | 1.500…1.460 | 1.430 | 1.340 |
| 1⅝ | 1.625…1.590 | 1.540 | 1.450 |
| 1¾ | 1.750…1.710 | 1.660 | 1.570 |
| 1⅞ | 1.875…1.830 | 1.780 | 1.680 |
| 2 | 2.000…1.950 | 1.900 | 1.790 |

1 in = 2.54 cm

ANSI/ASME B30.17-1985, Overhead and Gantry Cranes (Top Running Bridge, Single Girder, Underhung Hoist).

**Tagging.** Each chain should be tagged to indicate its load capacity, date of last inspection, date of purchase, and type of material. This information must not be stamped on the lifting links, however, as stamping can cause stress points that weaken the chain (Figure 15–10). Chain attachments (rings, shackles, couplings, and end links) should be made of the same material to which they are fastened. Hooks should be made of forged or laminated steel and should be equipped with safety latches. Replace hooks that have been overloaded, loaded on the tips, or have a permanent set greater than 15 % of the normal throat opening.

**Storing Chains.** Proper chain storage serves two purposes: It preserves the chain and promotes safe housekeeping. Chains are best stored on racks inside dry build-

**Figure 15–10.** In a typical double-chain sling, all components, such as the oblong master link, the body chain, and the hook, are carefully matched for compatibility. Note the permanent identification tag. *(Courtesy of Columbus McKinnon Corp.)*

ings that have a fairly constant temperature. Chains should be stored where they will not be run over by trucks or other mobile equipment, nor exposed to corrosive chemicals.

Storage racks and bins should be constructed in such a way that air circulates freely around the chain and keeps it dry. Use power equipment to lift large and heavy chains. Attempting to lift a chain manually can cause an injury.

The supplier or manufacturer can arrange a time to give you on-location instructions on the care, use, inspection, and storage of chains. By instructing maintenance workers on chains, you will be in a better position to evaluate your needs.

## MATERIALS STORAGE

Both temporary and permanent storage of all materials must be neat and orderly. Materials piled haphazardly, or strewn about, increase the possibility of accidental injuries. The warehouse supervisor must direct the storage of raw materials (and sometimes processed stock) kept in quantity lots for a length of time. The production supervisor is usually responsible for storing limited amounts of materials and stock, for short periods, close to the processing operations. Planned materials storage minimizes handling required to bring materials into production and removing finished products from production to shipping.

When employees plan materials storage, they must make sure that materials do not obstruct fire-alarm boxes, sprinkler system controls, fire extinguishers, first-aid equipment, lights, electric switches, and fuse boxes. Exits and aisles must be kept clear at all times.

There should be at least 18 in (0.5 m) clearance below sprinkler heads to reduce interference with water distribution. This clearance should be increased to 36 in (0.9 m), if the material being stored is flammable.

Aisles designed for one-way traffic should be at least 3 ft (0.9 m) wider than the widest vehicle, when loaded. If materials are to be handled from the aisles, the turning radius of the power truck must be considered. Employees should keep materials out of the aisles and loading and unloading areas, which should be marked with white lines.

Storage is facilitated and hazards are reduced by using bins and racks. Material stored on racks, pallets, or skids can be moved easily and quickly from one workstation to another. When possible, crosstie materials piled on skids or pallets.

In an area where the same type of material is stored continuously, it is a good idea to paint a horizontal line on the wall to indicate the maximum height to which material may be piled. This will help keep the floor load within proper limits and sprinkler heads in the clear.

## CONTAINERS AND OTHER PACKING MATERIALS

Because packing containers vary considerably in size and shape, height limitations for stacking these materials vary. The weight of the container itself must also be figured when keeping within the proper floor-loading limit.

Sheets of heavy wrapping paper placed between layers of cartons will help to prevent the pile from shifting. Better still, cross-tying will prevent shifting and sagging, and will permit higher stacking. Wire-bound containers should be placed so that sharp ends do not stick into the passageway. When stacking corrugated paper cartons on top of each other, remember that increased humidity may cause them to slump.

Bagged materials can be crosstied with the mouths of the bags toward the inside of the pile so contents will not spill if the closure breaks. Piles over 5 ft (1.5 m) high should be stepped back one row. Step back an additional row for each additional 3 ft (0.9 m) of height. Sacks should be removed from the top of the pile, not halfway up or from a corner. Remove sacks from an entire layer before the next layer is started. This keeps the pile from collapsing and possibly injuring people and damaging material.

Small-diameter bar stock and pipe are usually stored in special racks, located so that when stock is removed, passersby are not endangered. The front of the racks should not face the main aisle, and material should not protrude into the aisles. When the storage area is set up, the proper floor load should be considered.

Large-diameter pipe and bar stock should be piled in layers separated by strips of wood or by iron bars. If wood is used, it should have blocks at both ends; if iron bars are used, the ends should be turned up. Again, floor load must be considered.

Lumber and pipe have a tendency to roll or slide. When making a pyramid pile, operators should set the pieces down rather than dropping them, and not use either hands or feet to stop rolling or sliding material; this can result in serious injury.

Sheet metal usually has sharp edges and should be handled with care, using leathers, leather gloves, or gloves with metal inserts. Large quantities of sheet metal should be handled in bundles by power equipment. If sheet-metal piles are separated by strips of wood, they will be easier to handle when the material is needed, and will be less likely to shift or slide.

Tinplate strip stock is heavy and razor-sharp. Should a load or partial load fall, it could badly injure anyone in its way. Two measures can be taken to prevent spillage and injuries: band the stock after shearing, and use wooden or metal stakes around the stock tables and pallets that hold the loads. It is the responsibility of everyone who handles the bundles to band loads properly and make sure the stakes are in place when the load is on the table.

Packing materials, such as straw, excelsior, or shredded paper, should be kept in a fire-resistant room equipped with sprinklers and dustproof electric equipment. Materials received in bales should be kept baled until used. Remove only enough material for immediate use or for one day's supply. Packing materials should be taken to the packing room and placed in metal (or metal-lined wood) bins. Bin covers should have fusible links that close automatically in case of fire. To prevent injury in case a counterweight rope breaks, the weight should be boxed in.

Steel and plastic strapping, which may be flat or round, requires that the worker be trained in both application and removal. In all cases, the worker should wear eye or face protection, or a combination of both, and leather-palm gloves. Heavy straps may require the wearing of steel-studded gloves.

Barrels or kegs stored on their sides should be placed in pyramids, with the bottom rows securely blocked to prevent rolling. If piled on end, the layers should be separated by planks. If pallets are used, they should be large enough to prevent overhang.

## HAZARDOUS LIQUIDS

**Acid Carboys.** These are best handled with special equipment such as carboy trucks. Boxed carboys should generally be stacked no higher than two tiers and never higher than three. No more than two tiers should be used for carboys of strong oxidizing agents, such as concentrated nitric acid or concentrated hydrogen peroxide. The best storage method is specially designed racks.

Before handling carboy boxes, inspect them thoroughly to make sure that nails have not rusted or that the wood has not been weakened by acid. Empty carboys should be completely drained and their caps replaced.

The safest way to draw off liquid from a carboy is to use suction from a vacuum pump or a siphon started either by a rubber bulb or an ejector. Another method is to use a carboy inclinator that holds the carboy by the top, sides, and bottom, and returns the carboy to upright position automatically when released. Pouring by hand or starting pipettes or siphons by mouth suction to draw off the contents of a carboy should never be permitted.

**Portable Containers.** When liquids are handled in portable containers, contents should be clearly labeled. Substances should be stored separately in safety storage cabinets that meet pertinent OSHA 29 CFR 1910.106 and NFPA 30 requirements. Purchasing agents should specify to suppliers that all chemicals be properly labeled showing contents, hazards, and precautions. Labels are discussed in Chapter 8, Industrial Hygiene.

**Drums.** Storage drums containing hazardous liquids should be placed on a rack, not stacked. It is best to have a separate rack for each type of material. Racks permit good housekeeping, easy access to the drums for inspection, and easy handling. Racks allow drums or barrels to be emptied through the bung. Self-closing spigots should be used, particularly where individuals are allowed to draw off their own supplies. If there is a chance that spigots might be hit by material or equipment, drums should be stored on end and a pump used to withdraw the contents. Drums containing flammable liquid, and the racks that hold them, should be grounded (Figure 15–11). In addition, the smaller container into which the drum is being emptied should be bonded to the edge of the drum by a flexible wire with a C-clamp at each end.

Storage areas for liquid chemicals should be naturally well ventilated. Natural ventilation is better than mechanical because the mechanical system may fail. Nevertheless, some rooms are required to have mechanical ventilation, especially near the floor, with an alarm if ventilation shuts off (NFPA 30). The floor should be made either of concrete or of some other material treated to reduce absorption of the liquids. The floor should be pitched toward one or more corrosion-resistant drains that can be cleaned easily.

Where hazardous chemicals are stored, handled, or used, emergency flood showers or eyewash fountains should be available. Operators should wear chemical goggles, rubber aprons, boots, gloves, and other protective equipment necessary to handle the particular liquid (Chapter 9, Personal Protective Equipment).

**Figure 15–11.** A storage tank reduces hazards, aids identification of the drum contents, and permits easy grounding.

Each type of hazardous material requires either special handling techniques or special protective clothing, or both. In all cases, the supervisor or the purchasing agent should consult the safety professional or the safety department for handling precautions.

**Additional Measures.** When handling hazardous liquids, take these extra precautions:

1. Wear proper protective equipment.
2. Keep floors clean; do not allow them to become slippery.
3. Drain all siphons, ejectors, and other emptying devices completely before removing them from carboys or drums.
4. To dilute acid, always add acid to water—never add water to acid. Add the acid slowly and stir constantly with a glass implement.
5. In case of an injury, give first aid. Flush burned areas with lots of water for 15 minutes. Call a doctor immediately.
6. Do not force hazardous chemicals from containers by injecting compressed air. The container may burst or the contents may ignite.
7. Do not try to wash or clean a container unless its label specifically requires that it be cleaned before it is returned.
8. Do not store hazardous chemicals near heat or steam pipes, or where

strong sunlight will strike them. The contents may expand and may cause a fire or explosion. Bottles may focus sunlight to ignite a fire in nearby combustibles.

9. Do not stir acids with metal implements.
10. Do not pour corrosives from a carboy by hand.
11. Do not move a carboy unless it is securely capped.

Pipelines to outside storage tanks are safer for large quantities of liquid chemicals or solvents, because they reduce and localize any spillage. When hazardous liquids are used in small quantities, only enough for one shift should be kept on the job and then in a closed/approved container. In many cases, local ordinances limit the amount of hazardous materials that can be stored or processed inside a building. The main supply should be stored in tanks located in an isolated place.

**Piping.** Pipes should be color-coded and labeled to identify their contents. Make sure workers understand the company's use of identifying colors and labels. If outside contractors work in the area, they should be adequately informed of piping system contents and any hazards.

When a valve is to be worked on, it should be closed. The section of pipe in which it is located should be thoroughly drained before the flanges are loosened. Appropriate personal protective equipment must be worn by workers.

Workers should loosen the flange bolts farthest away first so that any remaining pipe contents will be directed away from them. A blind (or blank piece) should be fastened over the flanges on the open pipe ends as soon as they are parted. If a line is opened often, a two-ended metal blank can be permanently installed on a flange bolt so that the blank can be pivoted on the bolt when the flanges are parted. One end of the blank is shaped in the form of a gasket, to permit flow of liquid. The other end is blind and is swung to cover the pipe opening when the flanges are unbolted and parted.

**Tank Cars.** These storage cars should be protected on sidings by derailers and by blue stop flags or blue lights before they are loaded or unloaded. Hand brakes should be set and wheels chocked. Before the car is opened, it should be bonded to the loading line. The track and the loading or unloading rack should be grounded, and all connections checked regularly. Chemical tank cars should be unloaded through the dome rather than through the bottom connection.

## GAS CYLINDERS

There are many regulations regarding the storage and handling of gas cylinders. Several government agencies and private organizations and their standards should be consulted—OSHA, the National Fire Protection Association, the Compressed Gas Association, and others. Care should be taken to comply with all local government codes as well.

Compressed gas cylinders should be transported and stored in the upright position. All cylinders should have protective caps in place and should be chained or otherwise secured firmly in place against a wall, post, or other solid object to prevent them from being knocked over. Different kinds of gases should either be separated by aisles or stored in separate sections of the building or storage yard. Fuel gas cylinders should be stored separately from oxygen cylinders. Empty cylinders must be stored apart from full cylinders. Oxygen must be separated from flammable gases by 20 ft (6.1 m) or a half-hour rated wall.

Set up storage areas away from heavy traffic. Containers should be stored where there is minimal exposure to excessive temperature, physical damage, or tampering. Never store cylinders of flammable gases near flammable or combustible substances.

Before cylinders are moved, check to make certain all valves are closed and that protective valve caps are in place. Always close the valves on empty cylinders. Never let anyone use a hammer or wrench to open valves. As stated before, cylinders should never be used as rollers to move heavy equipment. Handle all cylinders with care—a cylinder marked "empty" is never completely empty.

To transport cylinders, use a carrier that does not allow excessive movement, sudden or violent contacts, and upsets. When a two-wheeled truck with rounded back is used, chain the cylinder upright.

Never use a magnet to lift a cylinder. For short-distance moving, a cylinder may be rolled on its bottom edge, but never dragged. Cylinders should never be dropped or permitted to strike one another. Protective caps must be kept on cylinder valves when cylinders are not being used. Oxy-acetylene welding assemblies should be fitted with check valves to avoid inadvertent mixing of gases in the system. When in doubt about how to handle a compressed gas cylinder, or how to control a particular type of gas once it is released, ask the safety professional or the safety department for information and instructions.

## COMBUSTIBLE SOLIDS

Bulk storage of grains and granular or powdered organic chemicals and certain other materials presents fire and explosion hazards. Many materials that are not considered hazardous in solid form often become combustible when finely divided. Some of these are carbon, fertilizers, food products and by-products, metal powders, resins, waxes and soaps, spices, drugs and insecticides, wood, paper, tanning materials, chemical products, hard rubber, sulfur, starch, and tobacco. This list is by no means complete, mentioning only the general categories of materials that may generate explosive dusts.

**Dust.** To avoid dust explosions, prevent formation of an explosive mixture or eliminate all sources of ignition. Either the dust must be kept down, or enough air supplied to keep the mixture below the combustible limit. Good housekeeping and dust-collection equipment will go far toward preventing disaster. Make sure dust does not accumulate on floors and other structural parts of a building. A dust explosion is a series or progression of explosions. The initial one may be small, but it shakes up dust that has collected on rafters and equipment so that a second explosion is created. This chain reaction usually continues until the result is the same as if a single large explosion had taken place.

Wearing protective equipment is important when dusts are toxic. Such equipment may range from a respirator to a complete set of protective clothing. The National Safety Council, or a state or local safety agency, can recommend the proper type of protective clothing for specific jobs. Recommendations are detailed in Chapter 9, Personal Protective Equipment.

**Bins.** Bins in which solids are stored should have sloping bottoms to allow material to run out freely and to help prevent arching. For some materials, especially if they are in process, bins should have vibrators or agitators in the bottoms to keep material flowing.

If large rectangular bins, open at one side, are entered with power shovels or other power equipment, workers should take care not to undercut the material and endanger themselves if arched material should suddenly give way. Where practical, tops of bins should be covered with a 2-in (5-cm) mesh screen, or at least by a 6-in (15-cm) grating (or parallel bars on 6-in centers) to keep people from falling into the bins.

Tests for oxygen content and for presence of toxic materials should be made before anyone enters a bin or tank for any purpose. If such tests indicate the need for protection, workers should use air-supplied masks, in addition to a safety belt and lifeline, before entering. Filling, agitating, and emptying equipment should be deenergized, tagged, and locked out so it cannot be started again, except by the worker after leaving the bin or by the immediate supervisor after checking the worker out of the bin. A person working in a bin should have a companion employee, equipped to act in case of emergency, stationed on top of the bin. Bins should be entered from the top only. All aspects of OSHA's confined space programs also need to be met.

If flammable vapors, dusts, or gases are found, workers should use only spark-resistant equipment and electrical equipment approved for use in such atmospheres by a recognized testing laboratory.

## SUMMARY OF KEY POINTS

Key points covered in this chapter include:
- Manual handling of materials accounts for an estimated 25% of all occupational injuries. These injuries are caused by improper lifting, carrying too heavy a load, incorrect gripping, failing to observe proper foot or hand clearances, failing to use or wear proper safety equipment, and poor job design. Employees should assess their operations to see if manual handling can be eliminated, or powered equipment can be used to minimize the hazards, or workstations and areas can be redesigned using ergonomic principles.
- The largest number of injuries occurs to fingers and hands. Workers must be taught how to pick up and put down heavy, bulky, or long objects. Protective gloves, eyewear, and safety shoes should be worn to prevent injuries. If manual lifting cannot be eliminated, workers must be taught correct lifting practices. Researchers have found that no single lifting method is best for everyone. Workers should choose the method that is most comfortable for them. Workers must be taught how to handle specific shapes—box, carton, crate, drum, or barrel—to minimize risk of injury to themselves and damage to containers. Special equipment or tools can make the job easier and safer.
- Materials-handling equipment include hand tools, nonpowered and powered hand trucks, powered industrial trucks, dock plates, conveyors, cranes, railroad cars, highway trucks, and motorized equipment. Hand tools, such as chisels, crowbars, rollers, hooks, and shovels, should be in good condition and used safely to prevent injuries.
- Nonpowered hand trucks include two-wheel and four-wheel trucks. Four major hazards associated with these trucks are (1) running wheels off bridge

plates or platforms, (2) colliding with other trucks or obstructions, (3) jamming hands between the truck and other objects, and (4) running wheels over feet. Only qualified personnel should be allowed to handle these trucks. They should be used only for the specific purpose that they were designed for. Loads should be carefully placed so they do not shift, slide, or obstruct the handler's view of the path ahead.

- Major hazards associated with powered hand trucks are (1) catching the operator between the truck and another object, and (2) collisions with objects or people. The truck should be equipped with a dead-man control, wheel guards, and an ignition key that can be removed when the operator leaves the truck. Only qualified workers should be allowed to use these vehicles.

- Powered industrial trucks are either battery operated or run by internal combustion engines. These trucks must meet stringent standards and be operated only by qualified drivers. Battery-charging installations must be in special areas, with adequate precautions for handling spilled electrolyte and dealing with fire hazards. Refueling areas must be adequately ventilated and secured against fire. Loads must be carefully stacked and secured to avoid spilling the load or overturning the truck. All employees must be trained in use and maintenance of this equipment.

- Dock plates not built in must be fastened down to prevent them from moving or sliding when used. Plates should be large and strong enough to hold the heaviest load—a safety factor of five or six should be adequate. They should be inspected frequently and kept in good condition.

- The primary rule in using conveyors is simple: Never allow anyone to ride on them. Workers should know the specific safety rules for working on or around gravity, pneumatic or blower, or screw conveyors. Supervisors should train employees in how to maintain guards and other safety equipment, free conveyor jams, inspect conveyors for hazards, and cross safely from one side of a conveyor to the other.

- Although many types of cranes are used in industry, safe operating procedures are much the same. Supervisors should assign only one signal person for each crane and tell crane operators to obey only this person's instructions. Loads must be properly balanced and carefully lifted before being moved from one location to another. Cranes and their chains and ropes must be inspected regularly for signs of wear, corrosion, or deterioration. Supervisors should pay particular attention to all control mechanisms, safety devices, crane hooks, ropes, and electrical apparatus.

- Moving any railroad car should be done after considering the facilities or equipment available to move the car, number of cars and distance they are to be moved, and whether the track is level or on a grade. The supervisor may need power equipment, or may have enough workers to move the car by hand. Workers should be alert to the dangers involved in opening and closing doors, transferring material in and out of cars, and securing car doors.

- When loading or unloading highway trucks, workers should set brakes and place wheel chocks under the rear wheels or use a vehicle restraint. They should use a fixed jack to support a semi trailer not attached to a tractor.

- All motorized equipment requires special handling to prevent accidents. Workers must be taught to operate this equipment according to detailed instructions in manufacturers' manuals. Equipment must be thoroughly inspected and given safety clearance before it is put into service. A preventive maintenance program is necessary to ensure worker safety and efficiency.

- Ropes, chains, and slings are common materials handling equipment that require careful inspection and maintenance to remain in good working order. Supervisors should become familiar with the properties of various types of ropes, chains, and slings, and the precautions for their use and maintenance. This equipment should never be used to lift a load if it has a kink, knot, or twist in the rope or chain. Fiber rope must be kept clean and dry, and stored in a dry, warm place. If the rope becomes frayed or worn, it should be replaced immediately. Wire ropes have greater durability and strength, and are used to handle heavier loads. They must be carefully lubricated to prevent rust and excessive wear.

- Chains and chain slings must be inspected frequently for wear, failure of a weld, metal fatigue, deformation, or stress in the links. They should not be overloaded or forced onto a hook as this weakens the links. Each chain should be tagged to indicate its load capacity. Chains should be stored in dry buildings with fairly constant temperatures.

- Both temporary and permanent storage of all materials must be neat and orderly. Supervisors should make sure materials do not obstruct fire-alarm boxes, sprinkler system controls, fire extinguishers, first-aid equipment, lights, electric switches, fuse boxes, and exits and aisles. Use of bins and racks, pallets, and skids can reduce hazards associated with materials storage.

- Materials storage includes careful stacking and placement of containers and other packing materials, hazardous liquids, gas cylinders, and combustible solids. Supervisors must be familiar with hazards and precautions associated with each of these materials. Containers and other packing materials can be stacked safely by taking into consideration height and weight limitations to keep within safe floor-loading limits. Materials should be stacked to prevent loads from sliding or shifting.

- Hazardous liquids are best handled with special equipment. Acid carboys, portable containers, and drums must be moved only by trained personnel who are equipped with the proper tools and protective gear. Each type of hazardous material requires special clothing or handling techniques.

- Pipelines, tank cars, and gas cylinders also require special handling when used for storage. Pipelines should be color-coded and labeled to identify contents. Tank cars should be protected on sidings by derails and by blue stop flags or blue lights before they are loaded or unloaded. Gas cylinders must be carefully chained or fixed to a stationary object. Empty cylinders should be stored apart from full cylinders and kept away from flammable or spark-generating materials. Workers must be trained thoroughly in safe work practices to deal with these materials.

- Bulk storage of grains and granular or powdered chemicals or materials presents fire and explosion hazards. To avoid dust explosions, dust must be kept down or enough air supplied to keep the concentration below the combustible limit. Good housekeeping and dust-collection equipment will go far toward preventing disaster. Workers should wear protective equipment when working near toxic dust.

- If bins are used for storage, they should have a sloping bottom and, in some cases, agitators or vibrators to keep material moving. Supervisors must always check for oxygen content and for presence of toxic materials before allowing workers to enter a confined space such as a bin or tank for any reason. Procedures for permit-required entry into confined spaces should be followed. Workers should wear protective clothing and have a companion worker stationed outside in case of emergency.

# 16

# Electrical Safety

**After reading this chapter, you will be able to:**

- Explain the action of electricity on the human body and how to use grounding to prevent shock hazards
- Describe how to use plug- and cord-connected equipment and extension cords safely
- Test branch circuits and electrical equipment
- Explain the uses and limitations of ground-fault circuit interrupters
- List the classes and divisions of hazardous locations and the proper electrical safeguards to use in each classified location
- Instruct workers in the proper procedures for handling electrical equipment safely on or off the job

Electrical safety is one of an employee's most important responsibilities in ensuring safe production. Although everyone knows high-voltage electricity is dangerous, low-voltage alternating-current (AC) levels can also cause fatalities by affecting the heart. An estimated 31% of the known electrocution fatalities occur in the home, 24% in general industry, and the remaining in the electrical power generation and distribution industries. This means that approximately 55% of electrocutions are due to contact with so-called low-voltage circuits (up to 600 volts AC or 750 volts DC).

Electricity usually does not look hazardous. Many people have been electrocuted when they thought they knew what they were doing. Unless you are trained and experienced in the electrical work to be done, call an electrical expert. A key concept to remember with any electrical exposure is: Do not become a part of any electrical circuit.

Electricity can also claim lives from fire-related incidents. The U.S. Consumer Product Safety Commission estimates that there are 169,000 house fires of electrical origin each year, claiming 1,100 lives and injuring 5,600. Property losses are estimated at $1.1 billion per year.

## MYTHS AND MISCONCEPTIONS ABOUT ELECTRICITY

Misunderstandings about electricity can lead to serious injuries and property damage. Supervisors must make sure their workers know the facts. Some of the more common myths and misconceptions are as follows:

1. Myth: "Electricity takes the path of least resistance." Actually, current will take any conductive path, high or low resistance, in order to return to the source that provides it power. Small amounts of current will flow through paths of high resistance, but even small amounts can kill.

2. "Electricity tends to go to ground." A person is led to believe that electricity wants to go to ground and simply disappear. In reality, ground serves as just one of the electrical loops that misdirected current can use to get back to the grounded power source.

3. "If an electrical appliance or tool falls into a sink or tub of water, the item will short out and trip the circuit breaker." This is not reliable information because the sink or tub may be nonconductive and therefore not part of the loop to ground.

    Workers should be warned not to reach into water to retrieve any appliance or tool. The water serves as an electrical conducting path for the electricity to flow out of the electrical equipment. Any person reaching into water with one hand when some other part of the body is touching a grounded object could receive a severe or even fatal shock.

4. "AC reverse polarity is not hazardous." The National Electrical Code (NEC) requires that power tools, attachment plugs, and receptacles be properly wired so as not to reverse the designated polarity. Many tools have switches in only one of the two conductors serving the item. The switch is supposed to be on the "hot" conductor supplying the power. If the tool (e.g., a drill) is plugged into a receptacle with reverse polarity and accidentally dropped into water, all of the internal wiring will be energized.

   The increased energized surface area in the water will allow more fault current to flow and increase the hazard of serious or fatal shock. If the tool and the receptacle are properly polarized, this would minimize the shock hazard. Only the terminal on the switch would be energized, not the entire tool, as in the case of reverse polarity.

5. Myth: "It takes high voltage to kill; 120 volts AC is not dangerous." Current is the culprit that kills. Voltage is a factor in determining how much current will flow. Five milliamps is human level of sensation; 100 milliamps can kill. Under the right conditions, AC voltage as low as 60 volts can kill. Respect all voltages as having the potential to kill.

6. Myth: "Double-insulated power tools are doubly safe and will always provide safety." Read the manufacturer's operating instructions carefully, but never place your trust completely in any electrical safety device. Double-insulated power tools can be hazardous if dropped into water. Electrical current can flow out of the power tool into the water.

Every piece of equipment is a potential source of electric shock. Even an electric shock too small to cause injury can trigger an involuntary reaction that results in physical injury; for example, after touching a live wire, a person may lose his or her balance and fall off a ladder. Design, layout, installation, and preventive maintenance programs can minimize exposure to live or "hot" electrical conductors. An effective electrical safety policy, employee training, and hazard awareness program can further reduce incidents of electric shock. It is also recommended that an electrical inspection program be implemented and inspections conducted periodically, as conditions warrant.

This chapter will help you to prevent electric shock injuries in your facility.

Make sure your employees do not believe in, or cling to, any of the myths and misconceptions about electricity.

## ELECTRICAL FUNDAMENTALS REVIEW

### OHM'S LAW

Before discussing electrical hazards, a brief discussion of Ohm's law is necessary. Important terms are voltage (v), current, amperes (amp), milliamperes (ma), watts (w), ohms, resistance (R), impedance (A), and power.

Ohm's law simply states that 1 volt will cause a current of 1 ampere to flow through a conductor having the resistance of 1 ohm. As a formula, the relationship is represented by I (amps) = E/R (E = volts, R = resistance). An easy way to remember this formula, and the different ways it can be expressed, is to put the symbols in a circle (Figure 16–1).

1. Put your finger on I; I = E/R.

2. Put your finger on R; R = E/I.

3. Put your finger on E; E = I × R.

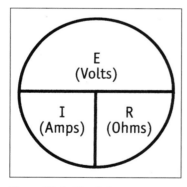

**Figure 16–1.** Ohm's law.

With this basic formula, you can better understand and explain the amount of electric current through a conductive body.

The above applies to direct current (DC); however, in alternating current (AC) usage, Ohm's law is E = I × Z, where Z is called the impedance.

To demonstrate an understanding of Ohm's law, try to solve this problem: Assume a person is working and perspiring, and has a hand-to-hand resistance of 1,000 ohms. This person contacts 100 volts with one hand and touches a ground surface with the other, completing a loop to the voltage source. What would be the current going through the body?

$$I = E/R = 100/1,000 = 0.1 \text{ amp } (100 \text{ ma})$$

Another important formula is the relationship of voltage, current, and power utilization (watts). Power is measured in watts and is equal to E (volts) × I (current). In other words, 1 watt would be equal to 1 ampere of current flowing through a resistor with 1 volt of potential difference. Another way to put this relationship to practical use would be to consider a 6-watt electric bulb—one requiring 6 watts of power to operate (about the size of a night light or panel-board light). Determine the current flowing through the filament of a 6-watt bulb that is being used in a 120-volt light socket.

$$P/E = 6/120 = 0.05 \text{ amp} = 50 \text{ ma}$$

This current is enough to affect the human heart and induce a fatal irregular heart beat. These examples of current levels from 50 ma to 20,000 ma are used to illustrate and clarify misconceptions about electricity and its effect on the human body.

## BODY RESISTANCE MODEL

The human body impedance, or resistance to electric current, has three distinct parts: internal body resistance and the two skin impedances associated with contact points on surfaces of different voltage potential. If the voltage applied is high enough, the skin offers virtually no impedance to the electric current (approximately 600 volts AC).

A body resistance model (Figure 16–2) would indicate approximately 1,000 ohms from hand-to-hand, or about 1,100 ohms from hand-to-foot. It is believed that the greatest number of injurious shocks involve a current pathway that is either hand-to-hand or hand-to-foot. Figure 16–2 illustrates that wet contact with an energy source (such as 120 volts AC) in one hand and contact with a grounded object with the other produces current through the chest area equal to:

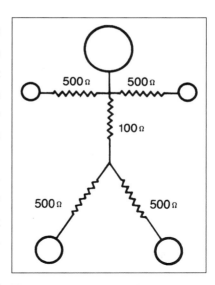

**Figure 16–2.** Human body resistance model.

$$I = E/R = 120/1,000 = 0.12 \text{ amp or } 120 \text{ ma}$$

This current level can be fatal (Table 16–A).

It is easy to analyze electrical shock hazards today. A small amount of current (50 ma) can cause the heart to start beating rapidly and irregularly—the potentially fatal condition known as ventricular fibrillation. The example given

## Table 16–A. Effects of Electrical Contact

Current in Amperes

| | |
|---|---|
| 4 | |
| 3 | Major burns of increasing extent and severity |
| 2 | |

Current in Milliamperes

| | |
|---|---|
| 1,000 | |
| 800 | |
| 700 | Increasing risk of burns at exit and entry |
| 600 | points |
| 500 | |
| 400 | |
| 300 | |
| 200 | |
| 100 | 70 to 200 milliamperes: risk of death from ventricular fibrillation if current pathway goes through heart |
| 80 | |
| 70 | |
| 60 | Risk of severe breathing difficulties |
| 50 | |
| 40 | |
| 30 | Severe shock |
| 20 | Risk of breathing difficulties due to muscular contractions |
| | Cannot release hand grip on conductor due to muscular contractions |
| 10 | |
| 8 | |
| 7 | Painful sensation (electric shock) |
| 6 | |
| 5 | |
| 4 | Increasingly unpleasant sensation or shock |
| 3 | Mild sensation |
| 2 | |
| | Threshold of sensation varies from person to person and points of skin contact |
| 1 | |
| 0 | |

Ventricular fibrillation means loss of heart's pumping action, loss of pulse, and risk of death from ineffective rapid beating of heart's ventricles.

Typical current pathways for heart risk:

a) Head to foot
b) Hand to opposite foot

(a)

(b)

**Current levels from domestic voltages carry biggest risk of ventricular fibrillation**

Source: *Canadian Occupational Safety*, January-February, 1988.

previously, that a 6-watt bulb used 50 ma of current when energized by 120 volts, means that the small amount of current required to light a 6-watt bulb is capable of causing death.

For instance, once ventricular fibrillation occurs, death will ensue in a few minutes. Resuscitation techniques, if applied immediately, can save the victim. Thus, contrary to many other beliefs, even low AC current is dangerous; and it does not take high voltage to kill, as shown in Table 16–A.

## ELECTRICAL CURRENT DANGER LEVELS

The danger threshold for ventricular fibrillation was established for typical adult workers as a result of research into the effects of electricity on the human body.

Figure 16–3 plots the range of electrical values that affect humans. Also shown is the requirement of Underwriters Laboratories for a Class A ground-fault circuit interrupter (GFCI) intended for protection. Note the time-current relationship at much lower currents and shorter time periods. The electrical current range below that, which causes muscles to freeze (called the "let go" range), is also illustrated.

From Figure 16–3, one can determine that a 100 ma current flowing for two seconds through a human adult body will cause death by electrocution. This doesn't seem like much current when you consider that a small, light-duty portable electric drill draws 30 times that much. But, because a current as small as 3 ma (3/1,000 amp) can cause painful shock, it is imperative that all electrical plugs, cords, and extension cords be kept in good condition and connected to a properly wired grounded circuit.

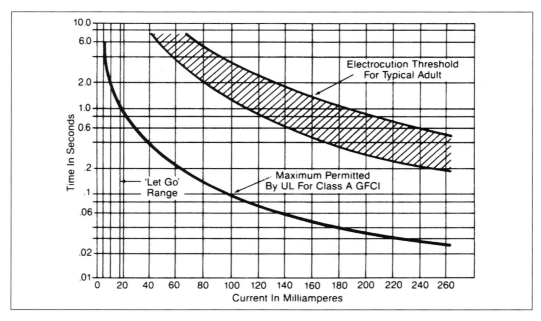

**Figure 16–3.** This chart shows the effects of electrical current intensity versus the time it flows through the human body.

## FACTORS ENHANCING ELECTRICAL SHOCK

The route that electrical current takes through the human body also affects the degree of injury. Voltage determines how much current flows. In cases where individuals come in contact with distribution lines, high voltage and high current can cause the moisture in the body to heat so rapidly that body parts literally explode. This extreme expansion is the result of body fluids changing to steam. This could result in a person being severely injured, but not electrocuted.

Any report of a shock hazard should be investigated. When investigating shock incidents—whether near misses or those that have caused injury—the factors of voltage, current, and body resistance can be used as investigative tools. The combination of one or more of these factors can be analyzed, and action can be taken to eliminate or control the hazard. Other factors that determine the degree of electric shock are as follows:

- wet and/or damp locations
- ground/grounded objects
- current loop from power source back to power sourc
- path of current through body and duration of contact
- area of body contact and pressure of contact
- physical size, condition, age of person
- type and/or amount of voltage
- personal protective equipment, gloves, shoes
- metal objects such as watches, necklaces, and rings
- miscellaneous:
  - poor workplace illumination
  - color blindness
  - lack of safety training and knowledge of electricity
  - no safe work procedures
  - noninsulated tools

## BASIC RULES OF ELECTRICAL ACTION

There are four basic rules of electrical action that everyone should know. They are as follows:

1. Electricity is not "live" until current flows.
2. Electric current will not flow until there is a complete loop out from, and back to, the power source.
3. Electrical current always returns to its source (i.e., the transformer or other source that supplied it).

4. When current flows, work (measured in watts) can be accomplished.

**Rule 1.** This rule explains why faulty power tools with live metal cases (due to a defective equipment-grounding conductor) can be used without causing a shock. Only when a person comes in contact with a ground loop with low enough resistance will current flow through the body and produce a shock.

**Rule 2.** When a current loop is formed, the electric current now has a path to return to its source. If the loop is broken, the current will stop flowing. To minimize electric shock hazards, look for ways to prevent ground loops. Insulated mats, insulated footwear, insulated gloves, and insulated tools are some methods that can be used.

**Rule 3.** Current will find any available loop with sufficiently low resistance or path to return to its source. In a grounded system, the NEC requires that the power source be electrically attached to an earth-driven ground rod as well as to the water piping. All other noncurrent-carrying metal parts must also be grounded. This creates many potential ground loops for current to get back to its source. Water pipes, metal ventilation ducts, metal door frames, T-bars holding suspended ceiling panels, and metal ridge rolls on a roof with a grounded lightning system are but a few of the many ground loops available.

**Rule 4.** If current flows through the body, it can do harm depending on the amount and duration of current flow. As noted in Table 16–A, the effect on the human body can be anything from a mild shock to complete tissue destruction.

These rules should help supervisors analyze how and why a worker received an electrical shock. Use the loop concept and Ohm's law to investigate actual or potential electric shock incidents and to determine appropriate safeguards for electrical hazards.

## BRANCH CIRCUITS AND GROUNDING CONCEPTS

### SINGLE-PHASE SERVICE

A typical pole-mounted transformer (single-phase) is shown in Figure 16–4. Industrial plants may have their transformer vaults, feeder lines, and branch circuits distributed throughout the building in conduit systems. This system is

**Figure 16–4.** Note how the elements are identified in this single-phase, service-branch circuit wiring.

used in residential, commercial, and industrial operations to provide 120 volts for lighting and small appliances. The 240-volt, single-phase power is used for heavy appliances and small motor loads. It is important that the function of the simple, single-phase, three-element, 120-volt electrical system be understood, and that misconceptions about grounding, bonding, and reverse polarity be clarified for your workers, especially your electricians.

The NEC requires that electrical wiring design ensures continuity of the system. In Figure 16–4, the ungrounded conductor from the service entrance panel (SEP) is wired to the small receptacle slot. The terminal screw and/or the metal connecting to the small slot is required to be a brass color. Insulation on the ungrounded conductor is generally black or red. Remember the phrase "black to brass" or the initials "B & B." Insulation on the grounded conductor is either white or gray. This conductor fastens to the large slot side of the receptacle. It is defined by the code as the "grounded conductor." An easy way to remember this wiring method is to think "white to light."

It is important that the receptacle be wired in this way. If it is wired in reverse, that is, the white insulated conductor to the brass screw and the black insulated wire to the silver screw, it produces reverse polarity, mentioned previously in this chapter.

*CAUTION: Do not consider yourself an electrical expert. Wiring and electrical repair should be done only by qualified personnel.*

## REVERSE POLARITY

Some individuals think that reverse polarity is not hazardous. An example of one hazardous situation could be an electric hand lamp. Figure 16–5 illustrates a hand lamp improperly wired and powered.

When the switch is turned off, the lamp socket shell is still energized. If a person accidentally touched the shell with one hand and a ground loop back to the transformer, a shock could result. If the lamp had no switch and was plugged in as shown, the lamp shell would be energized when the plug was inserted into the receptacle. Many two-prong plugs have blades that are the same size, and the situation described is just a matter of chance. If the plug is reversed (Figure 16–6), the voltage is applied to the bulb center terminal and the shell is at ground potential. Contact with the shell and ground would not create a shock hazard in this situation.

A second example: A person who works on a fluorescent lamp fixture deenergizes by turning off the wall switch. If the circuit has been wired backward, there will be 120 volts at the fixture, even though the switch is off (single-pole, single-throw switch).

Another example is the case of electric hair dryers or other plastic-covered electrical appliances. Figure 16–7 illustrates a hair dryer properly plugged into a receptacle with the correct polarity. When the appliance is plugged in with the switch in the hot or 120-volt leg, the voltage stops at the switch when it is in the OFF position. If the hair dryer were accidentally dropped into water, current could flow out of the plastic housing using the water as the conducting medium.

*CAUTION: Workers should never reach into water to retrieve an energized electrical appliance.*

This fault current might be in the 100-ma range (sufficient to kill but not sufficient to trip a 20-amp circuit breaker). When the appliance is polarized or plugged in, as shown in Figure 16–7, the switch connection is the only surface area where the hot conductor is fastened. It is so small that high resistance allows only a small current to be available ($I = E/R$). In this configuration, the fault current would be extremely low (unless the switch is in the ON position).

If the appliance is plugged in so that its polarity is reversed, as shown in Figure 16–8, when the switch is off, voltage will be present throughout all the internal wiring of the appliance. Now if it is dropped into water, there would be increased current flowing through the water. The heating element has a large surface area, which decreases resistance and increases current. A person who accidentally tries to retrieve the dryer is now in a very hazardous position because the voltage in the water could cause current to flow through the body.

**Figure 16–5.** When this hand lamp is plugged in so that the polarity is reversed, the shell is still energized even when the switch is turned off. (SEP is the service entrance panel.)

**Figure 16–6.** When this hand lamp is plugged in with the plug in correct polarity, current flows to the bulb center terminal and the shell is at ground potential.

**Figure 16–7.** This hair dryer is plugged in correctly so that the switch is on the hot conductor.

**Figure 16–8.** If the hair dryer is plugged in so that the polarity is reversed, voltage will be present throughout the internal wiring of the appliance even when the switch is turned off.

This illustrates the concept that reverse polarity is a problem whenever appliances with plastic housings are used in areas near sinks, or where water can wet the appliance. Remember, most motorized appliances have air passages for cooling. Wherever air can go, so can moisture and water.

If the appliance had a double-pole, double-throw switch, then it would make no difference how the plug was positioned in the outlet. The hazard would be minimized because the energized contact surface would be extremely small, resulting in a high resistance contact in the water and a resulting low available fault current. Later, we will see how GFCIs can be used to protect against shock hazards when using appliances with nonconductive housing around water.

## GROUNDING CONCEPTS

The terms *grounded conductor*, *equipment-grounding conductor*, *neutral*, and *ground* are a cause of many difficulties in understanding basic electrical concepts. Grounding falls into two safety categories. One of these is system safety grounding, and the other is equipment safety grounding.

*System grounding* protects power company equipment during emergencies, such as lightning strikes. *Equipment safety grounding* provides a system where all noncurrent-carrying metal parts (such as in a metal drill press frame or an electric refrigerator) are bonded together and kept at the same potential.

**Grounded Conductor.** The NEC defines "grounded conductor" as "a system or circuit that is intentionally grounded." Figure 16–9 shows the conductor from the SEP to the outlet as the grounded conductor. As shown in the illustration, for 120-volt AC, three-conductor outlets, the grounded conductor (white or gray insulation) is fastened to the silver screw, which is the large parallel slot.

**Figure 16–9.** Note the wiring terminology, particularly the relationship of the neutral to the grounded conductor and the equipment-grounding conductor.

The grounded conductor is a current-carrying conductor and also a connected ground at the SEP. From the SEP back to the power source, the grounded conductor is called neutral. On 120-volt circuits, refer to the grounded conductor from the SEP to the receptacle as just that, the grounded conductor, not the neutral.

**Neutral.** The term *neutral* refers to a neutral conductor that carries only unbalanced current from the other conductors in the case of normally balanced circuits. From the SEP to the power source, the term "neutral" is used. In circuits from the SEP to large 240-volt AC appliance receptacles, the grounded conductor would also be referred to as the neutral.

**Equipment-Grounding Conductor.** Figure 16–9 shows the equipment-grounding conductor going from the SEP to the appliance outlet. This path is sometimes also called the "grounding wire." This grounding conductor, if insulated, is colored green, or green with a yellow stripe. This conductor may also be a bare wire like those used in plastic sheathed cable. The NEC allows the conduit system to be the grounding conductor path also.

The grounding conductor normally does not carry a current. Only when an equipment malfunction occurs—allowing current to flow from the motor or electrical device to the equipment case—does the grounding conductor carry the current back to the power source and keep the equipment case at ground potential. (More on this can be found in the discussion on grounded power tools and equipment.)

Portable equipment and portable power tools can be grounded (equipment grounding) by use of a separate grounding wire from the equipment housing to a known ground. This method is satisfactory for stationary equipment as long as conditions do not change. It is less than satisfactory for portable equipment where the integrity of the grounding wire could be questionable and the quality of the ground is unknown. However, it is better than no grounding provision at all. If used, the grounding wire should be at least 1 in longer than the power cord, so that the plug can be removed from the receptacle before the ground is disconnected.

**Ground.** The NEC defines this term as a "conducting connection, whether intentional or accidental, between an electrical circuit or equipment and earth, or to some conducting body that serves in place of earth." Ground, then, means an electrical connection to a water pipe or a driven ground rod, or, where pipe is not available, to the rod alone. To "ground" something means to connect it electrically to the ground. Because the power source and the SEP are both connected to ground, the earth or ground loop becomes an electrical loop hazard.

**Figure 16–10.** A shock is caused when a person contacts the hot wire while standing on the ground.

Figure 16–10 illustrates that accidental contact with the hot conductor, while a person is standing on the ground, could allow a current to return to the power source through the ground loop formed by the person and the earth. This can cause a shock.

The metal framing studs used in new construction are usually grounded from the electrical conduit system. The T-bar hangers for suspended ceilings can be a ground loop because the electric lighting fixtures are grounded and electrically may cause the T-bars to be ground potential. Ventilation duct work may be ground potential because electric switches, dampers, and other devices may be mounted on the duct work.

## PLUG- AND CORD-CONNECTED EQUIPMENT AND EXTENSION CORDS

### GROUNDED POWER TOOLS AND EQUIPMENT

The earlier discussion of two-conductor appliances with plastic housings revealed the hazard of dropping ungrounded electrical devices in water. Now it is important to explain how proper grounding of appliances with metal cases can provide safety if the appliances are accidentally immersed in water.

Virtually all portable power tools, except double-insulated tools, used in industry and on construction sites are equipped with a three-prong plug. Unfortunately, not all receptacles are equipped with provisions for a three-prong plug.

This fact can lead a worker who is unwilling to make the effort to use an adapter to break off the U-blade, or ground, and defeat the protection of an equipment-grounding circuit. This is a highly dangerous practice on construction sites and in other areas that are often damp or wet. Cutting off the equip-

**Figure 16–11.** Removing the grounding pin destroys the grounding path and can allow the plug to be inserted in the receptacle in reverse polarity.

ment-grounding conductor U-blade not only destroys the grounding path, but also may allow the plug to be inserted in the receptacle in a reverse polarity mode (Figure 16–11). As shown in this figure, the grounding pin is cut off. In this situation, if a fault developed when the power tool was plugged into an outlet, the case would be energized with leakage voltage. The power tool would operate and the person using the power tool would be safe—but only until a ground loop was encountered. Then, current could flow through the person, through the ground loop, and back to the power source. This would cause an electric shock to the person. The loop resistance would determine the seriousness of the shock.

## DOUBLE-INSULATED POWER TOOLS

Double-insulated tools and equipment generally do not have an equipment-grounding conductor. Protection from shock depends on the dielectric properties of internal protective insulation and the external housing to insulate users from the electrical parts. External metal parts, such as chucks and saw blades, are internally insulated from the electrical system.

Today, many electrical power tool manufacturers produce a line of double-insulated tools. As a result, many of their advertisements lead us to believe that double-insulated power tools are "doubly safe." However, you should take the following precautions when using double-insulated power tools:

- Double-insulated tools are designed so the inner electrical parts are isolated physically and electrically from the outer housing. The housing is nonconductive. Dirt particles and other foreign matter from drilling and grinding operations may enter the housing through cooling vents. It can become lodged between the two shells, possibly voiding the required insulation properties.
- Double insulation does not protect against defects in cords, plugs, and receptacles. Continuous inspection and maintenance are required.
- A product with a dielectric housing—for example, plastic—protects users from shock if the interior wiring contacts the housing. Immersion in water, however, can allow a leakage path that may be either high or low resistance. Handling the product with wet hands, in high humidity, or outdoors

after a rainstorm can be hazardous. The best indication of the safety of a double-insulated tool is the Underwriters Laboratories (UL) (or other recognized testing laboratory's) label attached to the housing. The UL listing is evidence that the tool design and construction technique meet minimum standards. A tool marked "double insulated," but without a UL label, may not be safe.

Double-insulated tools and equipment should be inspected and tested, like all other electrical equipment. Remember, do not use double-insulated power tools or equipment where water and a ground loop are present. They should not be used in any situation where dampness, steam, or potential wetness can occur, unless protected by a GFCI.

## ATTACHMENT PLUGS AND RECEPTACLES

Attachment plugs are devices fastened to the end of a cord so that electrical contact can be made between the conductors in the equipment cord and conductors in the receptacle. The plugs and receptacles are uniquely designed for different voltages and currents, so only matching plugs will fit into the correct receptacle. In this way, a piece of equipment rated for one voltage/current combination cannot be plugged into a power system that is of a different voltage or current capacity (Figure 16–12).

The polarized three-prong plug is also designed with the equipment-grounding blade slightly longer than the two parallel blades. This provides equipment grounding before the equipment is energized. Conversely, when the plug is removed from the receptacle, the equipment-grounding pin is the last to leave, ensuring a grounded case until after power is removed. The parallel line blades may be the same width on some appliances since the three-prong plug can only be inserted in one way.

Figure 16–12 illustrates the

**Figure 16–12.** Some NEMA plug and receptacle blade configurations. (From OSHA Publication 3073, *An Illustrated Guide to Electrical Safety.*)

National Electrical Manufacturers Association (NEMA) standard plug and receptacle connector blade configurations. Each has been developed to standardize the use of plugs and receptacles for different voltages, amperes, and phases from 115 through 600 volts; from 10 through 60 amps; and for single- and three-phase systems.

Supervisors should watch for workers who jury-rig adapters. For example, trying to match a 20-amp attachment plug to a 50-amp receptacle configuration is not a safe practice. The danger is getting the voltage and current ratings mixed up and causing fire and/or shock hazards. Equipment attachment plugs and receptacles should match in voltage and current ratings to provide safe power to meet the equipment ratings.

## ELECTRICAL EXTENSION CORD DEFICIENCIES

Electrical extension cords are used in manufacturing, maintenance, construction, and in office operations. Defective, worn, or damaged extension cords have been the cause of death, injury, fire, and electrical equipment damage. Often the rush to meet schedules on time or lack of funds result in jury-rigged cords being fabricated and used. Purchasing staff might buy the cheapest units available, and workers end up with inferior extension cord sets that fail the contact tension test after a few cycles of use. Many times, neither the receiving department nor the workers test the extension cords prior to using. This results in the possibility of brand-new, defective units being put into service.

Other problems resulting from extension cord use include overloading, cutting off the grounding blade, and incorrect maintenance and repair. The following safety criteria should be strictly observed and used for training and inspection purposes:

- Before new or repaired extension cords are put into use, they must be tested. In addition, always purchase UL-listed extension cords. One supplier purchased and distributed several thousand units that were incorrectly wired (reverse polarity). These defective units were not UL listed. If extension cords are repaired by your company's electrical shop personnel, be sure they test these units prior to returning them to service. Follow the Testing Extension Cords procedures described later in this chapter.
- A visual and electrical inspection should be made of the extension cord each time it is used. Cords with cracked, worn, or otherwise damaged insulation or damaged attachment plugs should be removed from service immediately. Cords with the grounding prong missing or cut off must also be removed from service.

If extension cords are fabricated, only qualified electricians should do the work. The use of jury-rigged receptacles fabricated from junction boxes should be prohibited. Only UL-listed attachment plugs, cords, and receptacle ends should be used when fabricating cords.

- Use of two-conductor extension cords should be prohibited. Only grounding-type extension cord sets should be used.
- When cords must cross passageways, whether vehicular or personnel, the cords should be protected against physical damage and moisture, and identified with appropriate warnings. Cords should be used in this manner only for temporary or emergency use. They must never be used where they pass through exit doorways, hazardous storage areas, smoke barriers, or fire barriers.
- Extension cords must be of continuous length without splices or taps. Personnel should be instructed to remove from service any cord sets that have exposed wires, cut insulation, loose connections at the plug or receptacle ends, and jury-rigged splices.
- Extension cords must not be connected or disconnected until all electrical load has been removed from the extension cord receptacle end.
- Extension cords used in hazardous locations must be approved for the applicable hazards (see the Hazardous Location section in this chapter.)
- When the extension cord is not in use, it should be disconnected and neatly stored. Cords should be stored in a dry, room-temperature area and not exposed to excessively cold or hot temperatures.
- Never overload an extension cord electrically. If it is warm or hot to the touch, have a qualified person check the cord and the appliance to determine the problem. The wire size of the extension cord may be too small for the amperage of the appliance being used.
- Extension cords must not be draped over, or touch, hot surfaces such as steam lines, space heaters, radiators, or other heat sources.
- Do not plug in or pull out extension cords if you have wet hands.
- Extension cords must not be run through standing water, across wet floors, or other wet areas. If the extension cord is to be used to provide electrical power to equipment used in damp or wet locations, use GFCI protection.
- Extension cords used outdoors should be listed and marked for outdoor use and should be protected by GFCIs. Do not allow damaged cord sets to be used outside.
- Extension cords cannot be used in place of permanent wiring.

# BRANCH CIRCUIT AND EQUIPMENT TESTING METHODS

## TESTING BRANCH CIRCUIT WIRING

Branch circuit receptacles should be tested periodically. The frequency of testing should be established based on receptacle usage. In shop areas, quarterly testing may be necessary, while office areas may need only annual testing. A preventive maintenance program should be established, particularly because it is not unusual to find receptacles as old as the facility. A popular belief exists that receptacles never wear out or break. This is obviously false. For example, receptacles take severe abuse from employees disconnecting the plug from the receptacle by yanking on the cord. This can put severe strain on the contacts inside the receptacle and on the plastic face.

The electrical receptacle is a critical electrical system component. It must act as a strong mechanical connection for the appliance plug and provide a continuous electrical circuit for each of the prongs. The receptacle must be electrically correct or serious incidents can result. For this reason, a three-step testing procedure is recommended.

The first two steps can easily be done by most collateral-duty safety inspectors. The third step is one that should be performed by electrical maintenance personnel prior to accepting contractor wiring jobs. Also, the third step should be done on any circuit where electrical maintenance has been performed. Some inspectors use the third step as a spot check of ground-loop quality.

**Step 1.** Plug in a three-prong receptacle circuit tester and note the combination of indicator lights (Figure 16–13). (The circuit tester has three lights; the flashing light indicates which of three problems is present in the wires.) The tester checks the receptacle for proper connection of ground wire, correct polarity, or fault in any of the three wires. If these are normal, proceed to the next test. If the tester indicates a wiring problem, have it corrected as soon as possible. Remove the receptacle from all use until the problem is corrected by disengaging the circuit, if possible, or covering and tagging the receptacle "out of service." Retest after the problem is corrected.

**Step 2.** Once you are sure the receptacle is wired correctly, test the receptacle contact tension. A typical tension tester is shown in Figure 16–14. The tension should be 8 oz or more. If it is less than 8 oz, have it replaced. The first receptacle

**Figure 16–13.** A receptacle circuit tester checks the receptacle for proper connection of ground wire, correct polarity, and faults in any of the three wires.

**Figure 16–14.** Use a receptacle tension tester to be sure receptable contact tension is 8 oz or more.

function that loses its contact tension is usually the grounding contact circuit. Because this is a critical circuit, it illustrates the importance of checking for receptacle contact tension.

The frequency of testing must be based on receptacle usage. All electrical maintenance personnel should be equipped with receptacle testers and tension testers. Other maintenance employees could also be equipped with testers and taught how to use them. In this manner, the receptacles can be tested before maintenance workers use them.

**Step 3.** The ground loop impedance tester (GLIT) should be used next. This step may be conducted by the electrical maintenance person whenever any electrical circuit is reworked or repaired. An electrical preventive maintenance program should be established to perform regular (at least annual) ground loop testing in high usage areas. The safety inspector can use the GLIT for spot-checking.

To perform this test, simply insert the three-prong plug of the GLIT into the receptacle. Press the grooved portion of the red bar and observe the meter reading. A reading of 1 ohm or less is required. Any reading of more than 1 ohm should be reported to electrical maintenance for analysis and corrective action.

*WARNING: The GLIT must never be used to test a branch circuit when patient monitoring devices or computer equipment of any kind is in use.*

The GLIT should not be used until Steps 1 and 2 have been completed successfully. The tester will indicate the wiring fault. Prior to making the circuit operational, all wiring faults must be corrected and the circuit retested.

## TESTING EXTENSION CORDS

Extension cord testing and maintenance is critical to worker safety. The extension cord conducts electrical energy from a fixed outlet or source to a remote location. It must be wired correctly, or the cord can become the critical fault path.

Many electrical incidents have been caused by faulty or incorrectly repaired extension cords. Many times, the male plug on a three-conductor extension cord becomes damaged and the repair person installs a two-prong plug (Figure 16–15). This obviously means the receptacle end has no grounding path. It also means the plug can be inserted with correct polarity or reverse polarity.

Sometimes the repair may be made in such a way that the hot and ground are interchanged, as shown in Figure 16–16. This could happen on repair of the plug or the receptacle. In this configuration, when a grounded power tool is plugged into the extension cord, voltage will be applied to the tool casing. This could cause a dangerous or fatal electrical shock.

The same dangerous condition is illustrated in Figure 16–17. A repair was made replacing the broken three-prong plug with a two-prong plug. The repair person connected the white- and green-insulated conductors on one prong, then connected the black-insulated conductor to the other prong. Depending on how the plug was inserted into the receptacle, in one mode, the case would be grounded; and in the other mode, the case would be energized.

With a "hot" case, the user would have to find only a ground loop to the source and a fatal shock could occur. Interestingly enough, the power tool would operate in either mode.

Figure 16–18 illustrates a repaired or a jury-rigged extension cord fabricated with a three-prong plug and receptacle, but the connecting cord is only a two-conductor. This configuration would obviously show an open ground when tested. Because at times the cord may look like a three-conductor cord, the tester should be used to make sure the extension cord is safe.

New extension cords must be tested before being put into service. Many inspectors have found that new extension cords have open ground or reverse polarity. Supervisors should not assume that a new extension cord is good—they need to test it!

**Figure 16–15.** A three-conductor extension cord repaired with a two-prong plug.

**Figure 16–16.** A three-conductor extension cord with hot and ground wires interchanged.

**Figure 16–17.** This three-conductor extension cord is repaired with a two-prong plug that has the white and green wires tied together.

**Figure 16–18.** This extension cord with a three-prong plug and receptacle is connected to a two-conductor cord.

Testing extension cords must follow the same three steps used for electrical receptacle testing. These three steps are:

**Step 1.** Plug the extension cord into an electrical receptacle that has successfully passed the three-step receptacle testing procedure. Plug any three-prong receptacle circuit tester into the extension receptacle. Note the combination of indicator lights. If the tester shows the extension cord is safe, proceed to the next step. If the tester indicates a faulty condition, return the extension cord to electrical maintenance for repair. Once the cord is correctly repaired and passes the three-prong circuit tester test, it must be tested according to Step 2.

**Step 2.** Plug a reliable tension tester into the receptacle end. The parallel receptacle contact tension and the grounding contact tension should check out at 8 oz or more. If the tension is less than 8 oz, the receptacle end of the extension cord must be repaired. As with fixed electrical receptacles, the receptacle end of an extension cord loses its grounding contact tension first. This path is the critical human protection path and must be both electrically and mechanically in good condition.

**Step 3.** Once an extension cord has passed Steps 1 and 2, the ground-loop impedance tester may be used. The extension cord must be plugged into a properly connected receptacle (as noted in Step 1). Plug the GLIT into the receptacle end of the extension cord. A reading of 1 ohm or less is required. If you obtain a reading of more than 1 ohm, return the cord to maintenance for analysis and repair or replacement.

If a GLIT is not available, use an ohmmeter to test the continuity from the U-shaped grounding pin to the grounding outlet hole in the receptacle end of the extension cord. This test is done with the extension cord deenergized. The extension cord should not be plugged into an electrical receptacle when using an ohmmeter. Damage to the ohmmeter could result.

## TESTING PLUG- AND CORD-CONNECTED EQUIPMENT

The last element of the electrical systems test is the cord- and plug-connected equipment. The electrical inspector should be alert for jury-rigged repairs made on electric power tools and appliances. Often a visual inspection will disclose three-prong plugs with the grounding prong broken or cut off. The grounding path to the equipment case has been destroyed. The plug can now be plugged into the receptacle in the reverse polarity configuration.

Double-insulated equipment generally has a nonconductive case and will not be tested using the procedures in this section. Some manufacturers who have UL-listed double-insulation ratings may also provide a three-prong plug to ground any exposed noncurrent-carrying metal using the grounding path conductor. In these cases, the grounding path continuity can be tested.

A common error from a maintenance standpoint is installation of a three-prong plug on a two-conductor cord to the appliance. Obviously, there will be no grounding path if there are only two conductors in the cord. Anything can be jury-rigged, but normally the maintenance error is simply to hook up the black-and-white-insulated conductors to the brass and silver screws, respectively, and leave the third screw unconnected. Some hospital-grade plugs have transparent bases that allow visual inspection of the electrical connection to each prong. Even in this situation, you should still perform an electrical continuity test.

**Field Testing with an Ohmmeter.** Many electrical power tool testers require availability of electric power. The ohmmeter can be used in the field and locations where electric power is not available or is not obtained easily. An additional safety feature is that the plug- and cord-connected equipment tests are made on deenergized equipment. Testing of deenergized equipment in wet and damp locations also can be done safely.

The plug- and cord-connected equipment test using a self-contained battery-powered ohmmeter is simple and straightforward. The two-step testing sequence can be performed on three-prong plug grounded equipment.

## VOLTAGE DETECTOR TESTING

When conducting an electrical inspection, a voltage detector should be used in conjunction with the circuit tester, ohmmeter, and tension tester. There are several types of these devices on the market. A lightweight and self-contained tester makes an ideal inspection tool.

The voltage detector works like a radio receiver. It can receive or detect the 60-hertz electromagnetic signal from the voltage waveform surrounding an ungrounded ("hot") conductor. Whenever the front of the detector is placed near an energized (hot or ungrounded) conductor, the tester will provide an audible and visual warning.

The detector can also be used to test for properly grounded equipment. Whenever the tester is positioned on a properly grounded power tool (e.g., the electric drill shown in Figure 16–19), the tester will not give a warning indication. The reason is that the electromagnetic field is shielded from the detector so that no signal is picked up. If the grounding prong had been removed and the drill was not grounded, the electromagnetic waveform would radiate from the drill. The detector would receive the signal and give an alarm. As illus-

Safely and quickly check for voltage at AC outlets

Easily check power tools for proper grounding

Quickly check circuit breakers

Check fixed equipment grounding

**Figure 16–19.** A voltage detector can be used to check receptacle voltage, equipment grounding, and circuit-breaker condition.

trated, the detector can test for many things during an inspection. Receptacles can be checked for proper AC polarity.

Circuit breakers can be checked to determine if they are on or off. All fixed equipment can be checked for proper grounding. If the detector gives a warning indication on any metal-enclosed equipment, have it tagged out and repaired quickly. Ungrounded equipment can be just that or, in addition, there may be current leakage to the enclosure, making it hot with respect to ground. An experienced inspector could use a voltmeter to determine if there is voltage on the enclosure. The voltage detector should be used as an indicating tester, while qualitative testing should be done using other testing devices such as a multimeter. The use of the detector can speed up the inspection process by allowing you to check equipment grounding quickly and safely. Another unique feature of the voltage detector is that it is voltage sensitive.

When using the voltage detector, you must understand how it operates. If you put the tester on an energized, double-insulated power tool, you will see the red warning light turn on. In this situation, the voltage waveform is detected because there is no metal enclosure to shield the waveform. This does not mean that the plastic- or non-metallic-enclosed equipment is unsafe, only that it is energized and not a grounding-type piece of equipment. By understanding the operational features of a voltage detector, you can see the benefits it offers and the many electrical hazards it can detect that other equipment may overlook.

## GROUND-FAULT CIRCUIT INTERRUPTERS

### OPERATIONAL THEORY

The ground-fault circuit interrupter (GFCI) is a fast-acting device that monitors current flow to a protected load. It can sense any leakage current returning to the power supply by any electrical loop other than through the white (grounded) and the black (hot) conductors.

When any leakage current 5 ma or over is sensed, the GFCI, in a fraction of a second, shuts off the current on both the hot and grounded conductors, thereby interrupting the fault current to the appliance and the fault loop. This action is illustrated in Figure 16–20.

As long as $I_1$ is equal to $I_2$ (normal appliance operation with no ground-fault leakage), the GFCI switching system remains closed (N.C.). If a fault occurs between the metal case of an appliance and the hot conductor, fault current $I_3$ will cause an imbalance (5 ma or greater for personnel protection), allowing the GFCI switching system to open (as illustrated) and remove power from both the white and hot conductors.

Another type of ground fault can occur when a person contacts a hot conductor directly or touches an appliance with no (or faulty) equipment-grounding conductor. In this case, $I_4$ represents the fault current loop back to the transformer. This ground fault is generally the type to which workers are exposed.

The GFCI is intended to provide worker protection by deenergizing a circuit, or portion of a circuit, in approximately $\frac{1}{5140}$ of a second when the ground-fault current exceeds 5 ma. The GFCI should not be confused with ground-fault protection (GFP) devices that protect equipment from damaging line-to-line fault currents. Protection provided by GFCIs is independent of the condition of the equipment-grounding conductor.

**Figure 16–20.** A circuit diagram for a GFCI.

**Figure 16–21.** Three types of GFCIs: *(top)* a circuit breaker with a GFCI, *(middle)* a receptacle GFCI, *(bottom)* a portable GFCI.

The GFCI can provide protection to workers even when the equipment-grounding conductor is accidentally damaged and made inoperative. When replacing receptacles in older homes or business buildings (equipped with nongrounding-type receptacles), an exception in the NEC allows GFCIs to be used when a grounding means does not exist for installations of a grounding-type receptacle. The existing non-grounding receptacles are permitted to be replaced with grounding-type receptacles when power is provided to them through a GFCI-style receptacle or circuit breaker.

It is important to remember that a fuse or circuit breaker cannot provide hot-to-ground loop protection at the 5-ma level. The fuse or circuit breaker is designed to trip or open the circuit if a line-to-line or line-to-ground fault occurs that exceeds the circuit protection device rating. For a 15-amp circuit breaker, a short in excess of 15 amps or 15,000 ma would be required. The GFCI will trip if 0.005 amp or 5 ma starts to flow through a ground fault in a circuit it is protecting. This small amount of current (5 ma) flowing for the extremely short time required to trip the GFCI will not electrocute a person but will shock the person in the magnitude noted in Table 16–A.

GFCIs are available in several different types. Figure 16–21 illustrates three of the types available. Circuit breakers can be purchased with built-in GFCI protection. This combination unit can be secured in the circuit-breaker panel to prevent unauthorized persons from having access to the GFCI. The receptacle GFCI is the most convenient type. It can be tested and/or reset at the location where it is being used. It can also be installed so it is closest to the circuit-breaker panel and will provide GFCI protection to all receptacles on the load side of the GFCI. The portable type can be carried in a maintenance toolbox for instant use at a location where the available AC power is not GFCI-protected.

Other versions of GFCIs are available, including multiple outlet boxes for use by several power tools and for outdoor applications. The obvious difference between a regular electrical receptacle and a GFCI receptacle is the presence of the TEST and RESET buttons. The user should follow the manufacturer's instructions regarding the testing of the GFCI. Some recommend pushing the TEST button and resetting monthly. Defective units should be replaced immediately.

Be aware that a GFCI will not protect the user from line-to-line electrical contact. Suppose a person is standing on a surface insulated from ground (e.g., a dry, insulated floor mat), while holding a faulty appliance with a hot case in one hand. The worker then reaches with the other hand to unplug the appliance and contacts an exposed grounded (white) conductor. The worker has made a line-to-line contact. However, the GFCI will not protect her or him because there is no ground loop. To prevent this type of incident, it is essential to have an equipment-grounding conductor inspection program in addition to a GFCI program. All workers should be trained and encouraged to follow the electrical safety policy described later in this chapter.

GFCIs do not replace an ongoing electrical equipment inspection program. Rather, GFCIs should be considered as an additional safety factor for protection against the most common form of electrical shock and electrocution, the line-to-ground fault. It is recommended that GFCI protection be provided in any work area where damp or wet conditions can or do exist.

**Off-the-Job Electrical Safety.** The NEC requires GFCIs on 15-amp, 20-amp, 120-volt outside receptacles, garage circuits, and bathrooms. The NEC also requires that GFCIs be installed in crawl spaces at or below grade level and in unfinished basements. Receptacles installed within 6 ft (1.8 m) of a kitchen sink to serve countertop surfaces must also be GFCI protected. Other locations where the NEC requires GFCI protection include pools, fountains, commercial garages, pipeline heating, marinas and boat yards, mobile homes, and recreational vehicles. It is recommended that you consult the NEC for the specific requirements applicable to any of these occupancies.

## NUISANCE GFCI TRIPPING

When GFCIs are used in construction activities, the GFCI should be located as close as possible to the electrical equipment it protects. Excessive lengths of temporary wiring or long extension cords can cause ground-fault leakage current to flow by capacitive and inductive coupling. The combined leakage current can exceed 5 ma, causing the GFCI to trip.

Other nuisance tripping may be caused by one or several of the following items:

- wet electrical extension cord-to-tool connections
- wet power tools
- outdoor GFCIs not protected from rain or water sprays
- bad electrical equipment with case-to-hot-conductor fault
- too many power tools on one GFCI branch
- resistive heaters
- coiled extension cords (long lengths)
- poorly installed GFCI
- defective or damaged GFCI
- electromagnetic-induced current near high-voltage lines
- portable GFCI plugged into a GFCI-protected branch circuit

Remember that a GFCI does not prevent shock. It simply limits the duration so the heart is not affected. The shock lasts about $\frac{1}{5140}$ of a second (0.025 second) and can be intense enough to knock a person off a ladder or otherwise cause an accidental injury. Emphasize to your workers that they should not consider a GFCI adequate or general protection against electrical hazards, and that they must follow all safety steps rigorously.

## HAZARDOUS LOCATIONS

### OVERVIEW OF CLASSES AND DIVISIONS

This section provides only a summary of electrical equipment requirements as a guide for inspection or system analysis of various hazardous locations. For more in-depth design and engineering requirements, refer to subpart S, Electrical, in the OSHA general industry standards (29 CFR 1910) and Chapter 5 of the NEC. Hazardous locations are areas where flammable liquids, gases, or vapors, combustible dusts, or other easily ignitable materials exist (or can exist accidentally) in sufficient quantities to produce an explosion or fire. In hazardous locations, specially designed equipment and special installation techniques must be used to protect against the explosive and flammable potential of these substances.

Hazardous locations are classified as Class I, Class II, or Class III, depending on what type of hazardous substance is or may be present. In general, Class I locations are those in which flammable vapors and gases may be present. Class II locations are those in which combustible dusts may be found. Class III locations are those in which there are ignitable fibers and flyings.

Each of these classes is divided into two hazard categories, Division 1 and Division 2, depending on the likelihood of a flammable or ignitable concentration of a substance. Division 1 locations are designated as such because a flammable gas, vapor, dust, or easily ignitable material is normally present in hazardous quantities. In Division 2 locations, the existence of hazardous quantities of these materials is not normal, but they occasionally exist either accidentally or when material in storage is handled. In general, the installation requirements for Division 1 locations are more stringent than for Division 2 locations.

Additionally, Class I and Class II locations are also subdivided into groups of gases, vapors, and dusts having similar properties. Table 16–B summarizes the various hazardous (classified) locations.

## Table 16–B. Summary of Class I, II, and III Hazardous Locations

| Classes | Groups | Divisions | |
|---------|--------|-----------|---|
| | | 1 | 2 |
| Gases, Vapors, and Liquids (Art. 501) | A: Acetylene<br>B: Hydrogen, etc.<br>C: Ether, etc.<br>D: Hydrocarbons, Fuels, Solvents, etc. | Normally explosive and hazardous | Not normally present in an explosive concentration (but may accidentally exist) |
| III Dusts (Art. 502) | E: Metal dusts (conductive* and explosive)<br>F: Carbon dusts some are conductive,* and all are explosive)<br>G: Flour, Starch, Grain, Combustible Plastic or Chemical Dust (explosive) | Ignitable quantities of dust normally is or may be in suspension, or conductive dust may be present | Dust not normally suspended in an ignitable concentration (but may accidentally exist). Dust layers are present. |
| III Fibers and Flyings (Art. 503) | H: Textiles, wood-workings, etc. (easily ignitable, but not likely to be explosive) | Handled or used in manufacturing | Stored or handled in storage (exclusive of manufacturing) |

* Note: Electrically conducive dusts are dusts with a resitivity less than 106 ohm-centimeters

Source: *An Illustrated Guide to Electrical Safety,* OSHA Pub. No. 3073.

## EQUIPMENT REQUIREMENTS

General-purpose electrical equipment can cause explosions and fires in areas where flammable vapors, liquids and gases, and combustible dusts or fibers are present. Hazardous areas require special electrical equipment designed for the specific hazards involved. This includes explosion-proof equipment for flammable vapor, liquid, and gas hazards, and dust-ignition-proof equipment for combustible dust. Other kinds of equipment are nonsparking equipment, intrinsically safe equipment, and purged and pressurized equipment.

Many pieces of electrical equipment have parts—such as circuit controls, switches, or motors—that arc, spark, or produce heat under normal operating conditions. These energy sources can produce temperatures high enough to cause ignition. When electrical equipment must be installed in hazardous areas, the sparking, arcing, and heating nature of the equipment must be controlled.

Installations in hazardous locations must be (1) intrinsically safe, (2) approved for the hazardous location, or (3) of a type and design that provides protection from the hazards arising from the combustibility and flammability of the vapors, liquids, gases, dusts, or fibers present. Installations can have one or more of these options. Each option is described in the following paragraphs.

**Intrinsically Safe.** Equipment and wiring approved as intrinsically safe are acceptable in any hazardous (classified) location for which they are designed. Intrinsically safe equipment cannot release sufficient electrical or thermal energy under normal or abnormal conditions to ignite specific flammable or combustible materials present in the location.

Make sure that flammable gases and vapors do not come into contact with or pass through the equipment. Additionally, evaluate all interconnections between circuits to be sure that an unexpected source of ignition is not introduced through other nonintrinsically safe equipment. Separation of intrinsically safe and nonintrinsically safe wiring may be necessary to make sure that the circuits in hazardous (classified) locations remain safe.

**Approved for the Hazardous (Classified) Location.** Under this option, equipment must be approved for the class, division, and group of locations. Two types of equipment are designed specifically for hazardous (classified) locations: explosion-proof and dust-ignition-proof. Explosion-proof apparatus is intended for Class I locations, while dust-ignition-proof equipment is primarily intended for Class II and Class III locations. Equipment listed specifically for hazardous locations should have a recognized testing laboratory's label or mark, indicating the class, division, and group of locations where it may be installed. Equipment approved for use in a Division 1 location may be installed in a Division 2 location of the same class and group.

**Explosion-Proof.** Generally, equipment installed in Class I locations must be approved as explosion-proof. Because it is impractical to keep flammable gases outside of enclosures, arcing equipment must be installed in enclosures that are designed to withstand an explosion. This minimizes risk of an external explosion that occurs when flammable gas enters the enclosure and is ignited by the arcs.

Not only must the equipment be strong enough to withstand an internal explosion, but the enclosures must be designed to vent the resulting explosive gases. This venting must ensure that the gases are cooled to a temperature below that of ignition temperature of the hazardous substance involved, before being released into the hazardous atmosphere.

There are two common enclosure designs: threaded-joint enclosures (Figure 16–22) and ground-joint enclosures (Figure 16–23). When hot gases travel through small openings in either of these joints, they are cooled before reaching the surrounding hazardous atmosphere.

Other design requirements (e.g., sealing), prevent the gases, vapors, or fumes from passing into one portion of an electrical system from another. Motors, which typically contain sparking brushes or commutators and tend to heat up, must also be designed to prevent internal explosions.

**Dust-Ignition-Proof.** In Class II locations, equipment must generally be dust-ignition-proof. Section 502–1 of the NEC defines dust-ignition-proof as equipment enclosed in a manner that will exclude ignitable amounts of dust or amounts that might affect performance or rating. Where installed and protected in accordance with the NEC, the equipment will not permit arcs, sparks, or heat otherwise generated or liberated inside the enclosure to cause ignition of exterior accumulations or atmospheric suspensions of a specified dust on or in the vicinity of the enclosure.

Dust-ignition-proof equipment is designed to keep ignitable amounts of dust from entering the enclosure. In addition, dust can accumulate on electrical equipment, causing overheating as well as dehydration or gradual carbonization of organic dust deposits. Overheated equipment may malfunction and cause a fire. Dust that has carbonized is susceptible to spontaneous ignition or smolder-

Hot gases escape through openings designed into threaded joint.

**Figure 16–22.** This threaded-joint, explosion-proof enclosure is made of cast metal strong enough to withstand the maximum explosion pressure of a specific group of hazardous gases or vapors. Small openings designed into the threaded joint cool the hot gases as they escape. (From OSHA Publication 3073, *An Illustrated Guide to Electrical Safety.*)

Explosion-proof Enclosure

Accurately Machined Ground Joint of Flange. Maximum Gap: 0.0015 in.

Cool Gases

Threaded Boss

Minimum Flame Path is ⅜ in.

Hot Gases

**Figure 16–23.** This ground-joint enclosure cools gases as they escape through the ground joint of the flanges. (From OSHA Publication 3073, *An Illustrated Guide to Electrical Safety.*)

ing. Equipment must be designed to operate below the ignition temperature of the specific dust involved, even when blanketed. The shape of the enclosure must be designed to minimize dust accumulation when fixtures are out of reach of normal housekeeping activities (e.g., lighting fixture canopies).

Class II hazardous locations are composed of three groups—E, F, and G. (See NFPA 497M for an explanation of hazardous atmosphere groups A–G.) Special equipment designs are required for these locations to prevent dust from entering the electrical equipment enclosure. For example, assembly joints and motor shaft openings must be tight enough to prevent dust from entering the enclosure. The design must take into account the insulating effects of dust layers on equipment and make sure that the equipment will operate below the ignition temperature of the dust involved. If conductive combustible dusts are present, the design of equipment must take the special nature of these dusts into account.

In general, explosion-proof equipment is not designed for, and is not acceptable for, use in Class II locations, unless specifically approved. For example, because grain dust has a lower ignition temperature than that of many flammable vapors, equipment approved for Class I locations may operate at a temperature that is too high for Class II locations. In contrast, equipment that is dust-ignition-proof is generally acceptable for use in Class III locations because the same design considerations are involved.

**Safe Design for Hazardous Locations.** Under this option, equipment installed in hazardous locations must be of a type and design that provides protection from the hazards arising from combustibility and flammability of vapors, liquids, gases, dusts, or fibers. The employer has the responsibility of demonstrating that the installation meets this requirement. Guidelines for installing equipment under this option are contained in the NEC in effect at the time of installation of the equipment. Compliance with these guidelines is not the only means of complying with this option; however, the employer must demonstrate that installation is safe for the hazardous (classified) location.

## EQUIPMENT MARKING

Approved equipment must be marked to indicate the class, group, and operating temperature range (based on 40 C ambient temperature) for which it is designed to be used. Furthermore, the temperature marked on the equipment must not be greater than the ignition temperature of the specific gases or vapors in the area. There are, however, four exceptions to this marking requirement:

- First, equipment that produces heat but that has a maximum surface temperature of less than 100 C (212 F) is not required to be marked with operating temperature range. The heat normally released from this equipment cannot ignite gases, liquids, vapors, or dust.

- Second, any permanent lighting fixtures that are approved and marked for use in Class I, Division 2 locations do not need to be marked to show a specific group. This is because these fixtures are acceptable for use with all the chemical groups for Class I, that is, for Groups A, B, C, and D.

- Third, general-purpose equipment, other than lighting fixtures, considered acceptable for use in Division 2 locations, does not have to be labeled according to class, group, division, or operating temperature. This type of equipment does not contain any devices that might produce arcs or sparks. It is therefore not a potential ignition source. For example, squirrel-cage induction motors without brushes, switching mechanisms, or similar arc-producing devices are permitted in Class I, Division 2 locations (See NEC, Section 501–8(b)). Therefore, they need no marking.

- Fourth, for Class II, Division 2, and Class III locations, dust-tight equipment (other than lighting fixtures) is not required to be marked. In these locations, dust-tight equipment does not present a hazard so it need not be identified.

## INSPECTION GUIDELINES AND CHECKLIST

Before you conduct an electrical inspection, check your test equipment to be sure it works properly. As a minimum, you should have a circuit tester, a GFCI tester (combination circuit/GFCI testers are available), a contact tension tester, and a volt-ohmmeter. Avoid wearing metal jewelry or loose hanging necklaces. Wear protective footwear that is electrical hazard-rated per ANSI 241, 1991. Your footwear should have synthetic soles (do not wear leather soles). A clipboard, writing material, and the Table 16–C checklist should also be included. A camera is optional; if used, remember that before-and-after photos or slides make valuable visual aids for training use.

Table 16–C provides general guidelines that will assist you in checking for electrical hazards from the service entrance panel(s) to the equipment using the power. Further explanation of these guidelines is as follows:

| Table 16–C. Inspection Guidelines |
| --- |
| 1. **Service Entrance Panel**—Circuit I.D., Secure Mounting, Knockouts, Connectors, Clearances, Live Parts, Ratings |
| 2. **System Grounding**—Secure Connections, Corrosion, Access, Protection, Proper Size |
| 3. **Wiring (General)**—Temporary, Splices, Protected, J Box Covers, Insulation, Knockouts, Fittings, Workmanship |
| 4. **Electrical Equipment/Machinery**—Grounding, Wiring Size, Overcurrent/Disconnect Devices, Installation, Protection |
| 5. **Small Power Tools**—Attachment Plugs, Cords, Cord Clamps, Leakage, Grounding |
| 6. **Receptacles**—Proper Polarity, Adequate Number, Mounting, Covers, Connections, Protection, Adapters |
| 7. **Lighting**—Grounding, Connections, Attachment Plugs and Cords, Cord Clamps, Live Parts |
| 8. **GFCI Protection**—Bathrooms, Wet Locations, Fountains, Outdoor Circuits, Testing |

1. Service Entrance Panel—Check the branch circuit identification. It should be up to date and posted on the panel door. Be sure the panel and cable or conduit connectors are secure. There should be no storage within 3 ft (1 m) of the panel. No flammable materials of any kind should be stored in the same area or room. Look for corrosion and water in or around the area. Missing knockouts, covers, or openings must be covered properly to eliminate workers' exposure to live parts.

2. System Grounding—Check connection of the grounding electrode conductor to the metal cold-water pipe and to any driven ground rod. Also check any bonding jumper connections and any supplemental grounding elec-

trode fittings. These items should not be exposed to corrosion but should be accessible for maintenance and visual inspection.

3. Wiring (General)—Temporary wiring used for an extended time should be replaced with fixed wiring. Conduit and/or cable systems must be protected from damage by vehicles or other mobile equipment. All fittings and connections to junction boxes and other equipment must be secure. No exposed wiring can be allowed. Check for missing knockouts and cover plates. Jury-rigged splices on flexible cords and cables must be repaired correctly. Electrical equipment should be installed in a neat and professional manner. Check for damaged insulation on flexible cords and pendant drops.

4. Electrical Equipment/Machinery—The most important item is to test for proper grounding. All electrical equipment and machinery must be grounded effectively so that no potential difference exists between the metal enclosures. Use the voltage detector to find discrepancies and use other test equipment to determine the corrective action required. Disconnects should be identified clearly regarding the specific machinery they shut off, and should be placed near the machinery for emergency use. The disconnects should be exercised periodically to be sure they are operable. All electrical connections to equipment must be secure so that no wiring tension is transmitted to the electrical terminals within the equipment. The wiring installation should be protected from damage at all times while machinery is in use.

5. Small Power Tools—Attachment plugs should be checked for defective cord clamps and broken or missing blades. Connection of the cord to the power tool should be secure. Use your ohmmeter to check for leakage and an effective equipment-grounding conductor.

6. Receptacles—The receptacles should be tested for proper wiring configuration. Be sure to install enough receptacles to eliminate use of extension cords as much as possible. Covers should be in place and not broken. Multiple outlet adapters on a single outlet should be discouraged to prevent overloading. Surface-mounted receptacle boxes should be protected from damage by housekeeping or other mobile equipment.

7. Lighting—Cord- and plug-connected metal lamps and fixtures should be tested for grounding. Use the ohmmeter to do this. Check all cord clamps for secure connections. Frayed or old cords should be replaced.

8. GFCI Protection—Generally, GFCI protection is not required by the NEC on a retroactive basis. However, where employees are exposed to poten-

tial line-to-ground shock hazards, GFCI protection is recommended. This is especially important in workplaces where portable electrical equipment is being used in wet or damp areas in contact with earth or grounded conductive surfaces. Typically, areas where GFCIs should be installed include bathrooms, outdoor fountains, tubs, countertop sinks, crawl spaces under buildings, outdoor circuits, swimming pools, and other wet locations. Use your GFCI tester to be sure that the GFCI is operable. After years of use, they can become defective and will need to be replaced. Check to make sure that receptacles being provided with GFCI protection are labeled as such.

## SAFEGUARDS FOR PORTABLE HOME ELECTRICAL APPLIANCES

This section is primarily provided for use in off-the-job safety training. There are instances where home appliances are used on the job, and the following general precautions would be applicable to those situations. It is recommended that any electrical training include discussions of safeguards for small, portable, electrical home appliances. You should recommend installation of GFCIs in bathrooms and other locations required by the NEC for new dwelling construction.

When using electrical appliances, especially when children are present, basic safety precautions should always be followed. Appliance manufacturers provide written procedures for safe use of their appliances. Their instructions should be read by adults and children alike. Safety precautions should be understood. Strong emphasis should be given on the danger of using small electrical appliances in or near water. The following list established a commonsense baseline for the safe handling, use, and storage of electrical appliances in the home. To reduce risk of electrocution:

- Install GFCI-protected receptacles in bathrooms, near kitchen sinks, outdoors, and in other areas near water or where wet or damp conditions may exist.
- Always unplug and store appliances after using.
- Do not use any electrical appliance while bathing.
- Do not place or store appliances where they can fall or be pulled into a sink or tub (such as on towel bar on tub door).
- Do not place or drop any appliance into water or other liquid.
- Do not reach for an appliance that has fallen into water. Unplug the electrical cord immediately.

To reduce the risk of burns, electrocution, fire, or injury:

- Never leave an appliance unattended when it is plugged in.
- Closely supervise an appliance used by, on, or near children or invalids.
- Use the appliance only for its intended use as described in the manufacturer's manual. Do not use attachments not recommended by the manufacturer.
- Never operate an appliance if it has a damaged cord or plug, is not working properly, has been dropped or damaged, or has been immersed in water. Return the appliance to the dealer or repair center for examination and electrical or mechanical adjustment and repair.
- Keep the cord away from heated surfaces.
- Never block the air openings of an appliance. Keep the air openings free of lint, hair, paper, or other materials.
- Never use appliances when you are drowsy.
- Never drop or insert any object into any appliance opening.
- Do not use an appliance where flammable liquid vapors or aerosol (spray) products are being used or where oxygen is being administered.
- Do not attempt to repair appliances yourself. Most small electrical appliances do not lend themselves to normal repair procedures.
- Keep your appliance in a cool, dry place and out of the reach of children. When handling or storing your unit, do not put any stress on the line cord and the point where it enters the unit. Excessive stress could cause the cord to fray and to break.
- Never plug an appliance into a damaged receptacle or one that has a missing face plate.
- Keep all electrical appliances away from hot tubs, bathtubs, circulating water tubs, sinks, and swimming pools.

## SAFETY PROGRAM POLICY AND PROCEDURES

### SAFETY POLICY FOR ELECTRICAL EQUIPMENT

Each facility should have an electrical safety program policy. The policy should cover the responsibilities of all employees, including supervisors, workers, and the specialists who inspect, install, and maintain the electrical systems and equipment. The policy should stress management's concern for and support of safety issues. Individuals responsible for applying and enforcing the electrical policy should have standards of performance that include periodic assessment of their electrical safety performance.

In addition to policy and implementation procedures, an electrical safety program should also include four basic areas of concern: training and education, hazardous condition identification and reporting, work practices, and house-keeping. In addition, the professional electrician responsible for installing, repairing, and maintaining electrical equipment and systems should be familiar with NFPA 70E, *Electrical Safety Requirements of Employee Workplaces.*

Management should support an effective, preventive maintenance program. Use NFPA 70B, *Electrical Equipment Maintenance*, as a guide to implement or fine-tune this type of program. All employees should be made responsible for detecting and reporting unsafe electrical equipment. The following discussion provides suggestions for ensuring that these responsibilities are carried out effectively.

## SUPERVISOR RESPONSIBILITIES

**Training and Education.** Supervisors should have training courses to help them carry out the electrical safety program responsibilities for their specific areas. All employees exposed to electrical hazards should receive electrical safety training. In addition to electrical workers, this may include janitors, machine operators, and others. You must also monitor employees and assess performance against established facility safety program procedures.

**Hazardous Condition Reporting.** A written procedure promoting observation and reporting of electrical hazards should be implemented. A supervisor might also institute an employee recognition program in conjunction with hazard report-ing. This will reward employees who identify and locate electrical hazards so they can be controlled.

**Work Practices.** A sample of suggested work practices is included in this section under Employee Responsibilities. Employees should be rated on performance in following safe work practices.

You should also be familiar with OSHA standards as they apply to your responsibility.

**Housekeeping.** Housekeeping must be monitored closely by the supervisor. Floor-area problems always present challenges. Areas around electrical equipment, such as circuit-breaker panels, disconnects, and fixed power tools, should be kept free from stored items, debris, and any liquids or material that could create slippery floors or obstruct access to the equipment for maintenance or emergencies. When hazards in these areas are reported, they should be logged in, and work orders issued to correct them

## EMPLOYEE RESPONSIBILITIES

All employees have the responsibility of following safety procedures that they have been taught. They are also responsible for reasonable alertness and cooperation in maintaining safety procedures on the job.

**Training and Education.** Employees must receive training in electrical safety work practices, equipment operation, and required personal protective equipment. New job duties also require new safety training. Many injuries are caused by employees' lack of knowledge about equipment or its operation. Sometimes blame for incidents is placed on employees when, in reality, specific training was not provided. OSHA requires it under 29 CFR 1910.330–335.

**Hazardous Condition Reporting.** Employees should always report unsafe equipment, conditions, or procedures. Team effort is required by employees and supervisors. Getting equipment repaired should receive priority, even if it requires rescheduling a process or project. Under no condition should defective electrical equipment causing electrical shocks to employees be used to get a job done. The Electrical Safety Policy should be followed and deviations should be reported immediately.

**Work Practices.** Employees are responsible for following safe work practices, procedures, and safety policy as established by the employer. Workers should also be familiar with OSHA regulations as they apply to workplace safety.

**Housekeeping.** In the process of routine housekeeping, employees should be observant and report conditions that could cause electrical shock hazards. For example, the use of improperly grounded electrical equipment in areas that have water on the floor can expose workers to risk of electrical shock. Storing tools or other materials around electrical panels or equipment disconnects can create hazards as well as prevent immediate disconnection in an emergency. Cleaning electrical equipment with solvents can create health and physical safety problems.

## ELECTRICAL SAFETY POLICY

Supervisors must know all facets of their employers' electrical safety procedures and ensure that workers understand what is expected of them. You should stress that, as a minimum, the following items must be observed:

- Plug power equipment into wall receptacles with power switches in the OFF position.
- Unplug electrical equipment by grasping the plug and pulling. Do not pull or jerk the cord to unplug the equipment.
- Do not drape power cords over hot pipes, radiators, or sharp objects.
- Check the receptacle for missing or damaged parts. Do not plug equipment into defective receptacles.
- Check for frayed, cracked, or exposed wiring on equipment cords. Check for defective cord clamps at locations where power cords enter equipment or attachment plugs.
- Generally speaking, extension cords should not be used in office areas. Extension cords should be limited to use by maintenance personnel.
- "Cheater plugs," extension cords with junction box receptacle ends, or other jury-rigged equipment should not be used.
- Consumer electrical equipment or appliances should not be used if not properly grounded.
- Workers should know the location of electrical circuit-breaker panels that control equipment and lighting in their areas. Circuits and equipment disconnects must be identified.
- Temporary or permanent storage of materials must not be allowed within 3 ft (1 m) of any electrical panel or electrical equipment.
- When defective electrical equipment is identified by a worker, it should be tagged immediately and removed from service for repair or replacement.
- Any electrical equipment causing shocks or with high leakage potential must be tagged with a DANGER—DO NOT USE label or equivalent.

## ELECTRICAL DISTRIBUTION SYSTEM REVIEW

The electrical distribution system should be designed, installed, operated, and maintained to safely and reliably provide electrical power for all required operations. The facility should be analyzed to determine that the following requirements are documented and observed regarding overall electrical system operations and maintenance. If these requirements are not met, obtain professional electrical services to correct the situation and eliminate hazards.

- There is a program of preventive maintenance and periodic inspection to ensure that the electrical distribution system operates safely and reliably. Inspections and corrective actions are documented.
- The facility identifies components of the electrical distribution system to be included in the program. Consideration is given to reliability of receptacles, wiring, fuse and/or breaker panels, transformers, switch gear, and service equipment.
- There is a current set of documents that indicate the distribution of and controls for partial or complete shutdown of each electrical system.
- Electrical maintenance and operating personnel are provided with appropriate job training, authorization to perform specific work, and written records of this training.
- The capacity of electrical feeds and transformers is adequate for the electrical demands of the facility.
- Any special-purpose electrical subsystems or installed devices are maintained as required.
- The electrical distribution system is designed with enough receptacles and circuits to power devices used in each area without the use of extension cords.

## SUMMARY OF KEY POINTS

This chapter covered the following key points:
- Electrical safety is one of everyone's most important responsibilities. Supervisors must know the facts about electricity and be able to counter myths and misconceptions that workers may hold.
- The four basic rules of electrical action are (1) electricity isn't "live" until current flows, (2) current won't flow until there is a complete loop out from and back to the power source, (3) current always returns to its source, and (4) when current flows, work measured in watts is performed. These four rules help supervisors to analyze electrical shock hazards or investigate how and why electrical injuries occurred in their departments.
- According to Ohm's law, it is current that can injure or kill a worker, while voltage determines how—either by burns or electrocution. The human body presents resistance to current. If a worker establishes an electrical loop

between herself or himself and a ground, the current will pass through the body on the way to the ground. Even a small amount of current can cause the heart to go into ventricular fibrillation, which can produce sudden death if the heart is not resuscitated within a few minutes.

- Factors that can enhance shock potential include wet or damp locations, nearby grounded objects that the worker or electrical tool may touch, lack of worker training, and type and amount of voltage and current.

- The keys to preventing electrical shock hazards are (1) to train workers thoroughly in safe work practices; (2) to wire, insulate, and ground all electrical equipment properly so there are no ground loops between an electrical tool or machine and the worker; and (3) to routinely inspect, maintain, and replace such equipment, along with all cords, outlets, plugs, attachments, and power sources. Employees should understand the single-phase, three-element, 120-volt electrical system that powers industrial and commercial operations in the United States. The 240-volt, single-phase power is used for heavy-duty appliances and small motor loads. If the system is not wired properly, or if a tool or appliance is not plugged in properly, it can create reverse polarity. In reverse polarity, the entire casing of a tool or machine is energized and creates a shock hazard, particularly if accidentally immersed in water.

- A ground is a conducting connection, whether intentional or accidental, between an electrical circuit or equipment and earth, or to some conducting body that serves in the place of earth. A ground loop can become an electrical hazard if workers accidentally form part of the loop. Grounding falls into two safety categories—system grounding and equipment safety grounding.

- System grounding provides protection of the power company equipment and the consumer. Equipment safety grounding provides a system where all non-current-carrying metal parts are bonded together and kept at the same potential. Grounding concepts include the grounded conductor, neutral conductor, and equipment-grounding conductor.

- Power tools and equipment can be designed to protect workers against shock hazards by being double insulated or by being equipped with three-pronged plugs in which one prong acts as a ground. However, these safeguards are not foolproof, and workers must remember to inspect and test their equipment regularly. Double-insulated equipment should not be used in damp or wet conditions or immersed in water.

- Attachment plugs and receptacles are designed for specific voltage/current combinations and should not be used for any other combination. To do so can cause fires or shock hazards to workers. Electrical extension cords must be the proper guage to carry the current load required. These cords should be UL-listed, tested before being put into use, and inspected regularly.

- Employees can use several testing procedures for branch circuits and electrical equipment. Branch circuit receptacles should be tested periodically to ensure that electrical outlets are safe and in good working order. Inspectors often use a three-step process using a three-prong receptacle circuit tester, a receptacle contact tension tester, and a ground-loop impedance tester. Extension cords should be tested using the same three-step procedure to make sure cords are in good repair and can carry the required current safely.

- Cord- and plug-connected equipment should be tested using a two-step procedure of ground-to-pin testing and an appliance leakage test. Inspectors should be alert for jury-rigged repairs made on electric power tools and appliances, and for signs of wear or deterioration. An ohmmeter can be used where electric power is not available for testing equipment. A voltage detector can be used in conjunction with the circuit tester, ohmmeter, and tension tester to identify electrical hazards. It can detect voltage waveforms surrounding an ungrounded, unshielded, or hot conductor. It is also voltage-sensitive and can detect voltage from a distance.

- The ground-fault circuit interrupter (GFCI) is a fast-acting device that monitors current flow to a protected load. If it senses a current leakage, it will shut off the current on grounded and hot conductors. The GFCI protects workers against electrical hazards caused by ground faults, but it will not protect workers from line-to-line electrical contact. GFCIs should be used where workers are exposed to wet or humid conditions and where they may come in contact with ground or grounded equipment. GFCIs do not prevent shock but limit the duration so the heart is not affected. Hazardous locations are classified as Class I (flammable vapors and gases present), Class II (combustible dusts present), and Class III (ignitable fibers and flyings present). Each class is divided into two hazard categories: Division 1—flammable gas, vapor, dust, or easily ignitable material is present in hazardous quantities, and Division 2—hazardous quantities of these materials in hazardous locations must consist of equipment and wiring that is (1) intrinsically safe, (2) approved for hazardous locations, or (3) of a type or design that provides protection from the hazards present. Generally, equipment in Class I locations must be approved as explosion proof, while Class II equipment must be dust-ignition proof. With some exceptions, equipment must be clearly marked to indicate the class, group, and operating temperature range for which it is designed.

- Inspectors must be familiar with often-violated electrical standards. This knowledge will help them identify and correct electrical hazards frequently occurring in the workplace. Guidelines established for inspecting electrical equipment and wiring include the following items: service entrance panel, system grounding, electrical equipment machinery, small power tools, receptacles, lighting, and GFCI protection.

- Workers should also instruct workers in off-the-job safety training to guard against electrical hazards in the home. These precautions can prevent burns, electrocution, fire, or injury to workers and their family members.

- An electrical safety program should include training and education, hazardous condition reporting, safe work practices, and housekeeping. Management personnel and employees should know clearly what their responsibilities are and what is expected of them. Part of the safety program involves setting up an electrical distribution system review to provide safe, reliable electric power for all required operations.

**17**

**Fire Safety**

**After reading this chapter, you will be able to:**

- Explain the basic principles of fire safety, including the chemistry of fires
- Identify fire hazards, causes of fires, and safeguards required to prevent fires
- Conduct regular, periodic inspections to ensure that work areas remain in a fire-safe condition
- Describe the use and operation of fire protection equipment and systems
- Teach employees procedures for reporting fires and evacuating work areas
- Develop an effective fire protection education program for both on and off the job

An effective fire protection program requires careful planning with a focus on prevention. Although the overall program may be developed and administered by the director of safety, fire protection, security, engineering, or maintenance, each supervisor has a direct responsibility in the program's success.

Fire protection is a science in itself. This manual cannot cover all aspects of fire prevention and protection but this chapter presents condensed, basic information to help organizations conduct fire-safe operations. More specialized information is available from the sources cited throughout this chapter.

## BASIC PRINCIPLES

Fire is an oxidation process that emits light, heat, and other products of combustion. In order to explain or understand fire development, experts have devised various fire models. One of the more common models is shown as a tetrahedron in Figure 17–1. To sustain most fires, four elements must be available at the same time: elevated temperature, oxygen, fuel, and an uninhibited chain reaction. Fire extinguishing agents act by removing one or more of the tetrahedron's sides. As Figure 17–1 shows, an uninhibited chain reaction is not necessary for a surface-glowing type of fire to continue. An example of this type of fire would be a hot coal in a barbeque. Under the proper conditions, many materials will burn. A good example is steel, which burns quite readily, in the form of steel wool.

Fire begins with an ignition source in the presence of a fuel source and can spread to other fuel sources by conduction, convection, and radiation. Conduction transfers heat through contact with solid material. Convection transfers heat through heated air. Radiation heat transfers through electromagnetic waves given off by flames.

Experienced employees are an excellent source of information to determine necessary fire prevention measures. Employees' knowledge of day-to-day operations may reveal potential ignition or fuel sources not easily observed. Employees may also be helpful in developing housekeeping standards for an area or operation. Routine departmental housekeeping can be very effective in removing waste that may serve as a fuel source. Preventive maintenance of equipment can help prevent ignition sources that arise from stressed motors or overloaded processes. Make sure your employees follow safe housekeeping practices and maintain equipment as required to prevent fires.

Organizations should evaluate the need for specific fire protection equip-

**Figure 17–1.** In the presence of oxygen, fuel, and elevated temperature, an uninhibited chain reaction can produce a flaming fire, but without a chain reaction you can still produce a deep-seated, surface-glowing fire.

ment and see to it that such equipment is provided. Supervisors should be educated and trained on the fire equipment provided for their particular operations or department. Supervisors and employees should be routinely trained in their role in using such equipment.

Continuous training in fire-safe work procedures, regular inspections of work areas, and close supervision of employee job performance are primary requirements of a successful fire prevention program. Although fire protection equipment may be maintained by others, the supervisor has a significant responsibility for the safety of his or her employees, materials, and equipment.

## Understanding Fire Chemistry

Fire prevention and control programs should begin with an understanding of basic fire chemistry. Ordinary fire results when a substance (fuel) in the presence of air (oxygen) is heated to a critical temperature, called its *ignition temperature*.

This important concept is best illustrated by the fire tetrahedron (Figure

17–1). By looking at this model, you can understand how fires may be prevented or extinguished. The methods used include:

**Oxygen Removal.** Removing or lowering oxygen levels to prevent fires is not common because fire occurs in the same percentage of oxygen in the air that humans need for breathing. In some cases, oxygen levels can be reduced below the minimum percentage needed for combustion by purging and rendering the atmosphere inert in closed containers or processing systems. Fire-fighting foam extinguishes fires by smothering (and cooling) action.

**Fuel Removal.** To prevent fires, try to keep the quantity of stored flammable and combustible materials at a minimum. Routine housekeeping is an excellent method of preventing the accumulation of waste that may help spread a fire by serving as a fuel source.

**Heat Source Control.** Eliminating and controlling heat sources also are elementary steps in fire prevention. Conscientious workers can control use of welding and cutting equipment, torches, heating equipment, spark-producing equipment, electricity, and smoking materials. Preventive maintenance programs can keep motors and other equipment from overheating. Keep heat and ignition sources away from fuel.

Table 17–A lists common sources of ignition that cause industrial fires, gives examples in each case, and suggests preventive measures. The section on Causes of Fire discusses each source in detail.

For many years, the principle of extinguishing fires focused on the fire triangle and the removal of any of its three sides, representative of three components. The fire triangle was modified into a fire tetrahedron (Figure 17–1) to take into consideration the chemical chain reaction needed to sustain combustion. The tetrahedron has four sides or faces. Each side represents a component necessary to sustain a fire. Removing one or more of the faces will put out the fire. Because each face is directly adjacent to and connected to the others, a tetrahedron represents their interdependency.

To extinguish a fire, the following steps can be taken:
1. Reduce or remove the oxygen by smothering (for example, by shutting the lid over a tank of burning solvent, or by covering it with foam), or by dilution (replacing the air with an inert gas such as carbon dioxide).
2. Remove or seal off the fuel by mechanical means, or divert or shut off the flow of liquids or gases fueling the fire.

## Table 17–A. How to Control Sources of Ignition in Industrial Fires

| Sources of Ignition | Examples | Preventive Measures |
| --- | --- | --- |
| Electrical equipment | Electrical defects, generally due to poor maintenance, mostly in wiring, motors, switches, lamps, and hot elements | Use only approved equipment. Follow the National Electrical Code. Establish regular maintenance. |
| Friction | Hot bearings, misaligned or broken machine parts, choking or jamming of materials, poor adjustment | Follow a regular schedule of inspection, maintenance, and lubrication. |
| Foreign substances | Tramp metal, which produces sparks when struck by rapidly revolving machinery (a common cause in textile industry) | Keep foreign material from stock. Use magnetic or other separators to remove tramp metal. |
| Open flames | Cutting and welding torches, gas and oil burners, misuse of gasoline torches | Follow established welding precautions. Keep burners clean and properly adjusted. Use "hot work" permits that specify acceptable work conditions. |
| Smoking and matches | Dangerous near flammable liquids and in areas where combustibles are used or stored | Smoke only in permitted areas. Use prescribed receptacles. Make sure matches are out. |
| Spontaneous ignition | Deposits in ducts and flues, low-grade storage, industrial wastes, oily wastes and rubbish | Clean ducts and flues frequently. Remove waste daily. Isolate stored materials likely to heat spontaneously. |
| Hot surfaces | Exposure of combustibles to furnaces, hot ducts or flues, electric lamps or irons, hot metal being processed | Provide ample clearances, insulation, air circulation. Check heating apparatus before leaving it unattended. |
| Combustion sparks | Rubbish-burning, foundry cupolas, furnaces and fireboxes, and process equipment | Use incinerators of approved design. Provide spark arresters on stacks. Operate equipment carefully. |
| Overheated materials | abnormal process temperatures, materials in driers, overheating of flammable liquids | Have careful supervision and competent operators, supplemented by well-maintained automatic temperature controls. |
| Static electricity | Dangerous in presence of flammable vapors, occurs at spreading and coating rolls or where liquid flows from pipes | Ground equipment. Use static eliminators. Humidify the atmosphere. |

Adapted from *Factory Mutual Record.*

3. Cool the burning material below its ignition point with a suitable cooling agent (hose streams or water extinguishers).

4. Interrupt the chemical chain reaction of the fire (using dry chemical agents).

When employees understand the fire tetrahedron and its practical application, they will be educated to participate in fire prevention and control methods.

## Determining Fire Hazards

You can contribute to a fire protection program in two ways: first, identify existing fire hazards and, second, take action to resolve them. To do this, seek the best technical advice available from fire protection experts.

**Self-Inspections.** Develop a comprehensive inspection checklist (in conjunction with hourly employees who understand the operations and fire protection specialists). The comprehensive list should specify any places, materials procedures, classes of equipment, conditions, and circumstances where fire hazards may exist. For each potential fire hazard, identify the specific controls that are in place, such as precise fire-safe practices that must be followed or fire abatement equipment. Specify which personnel—for example, machine operator, electrician, maintenance crew, or porter—have specific responsibilities in maintaining controls or emergency response (see discussion of inspections in Chapter 6, Safety and Health Inspections).

After developing and reviewing a comprehensive fire inspection checklist, you will have a detailed picture of how many fire protection measures are actually in effect. You will be able to remedy any neglected items. The sample fire prevention checklist shown in Figure 17–2 can serve as a guide. This sample checklist highlights some general categories that may be used by a supervisor or safety team to develop a more detailed site-specific inspection checklist. You should make your own list and include more specific information than can be listed in a general example.

Additional sources that may be helpful in developing inspection lists or inspection protocol may be the insurance company that insures your building, your local fire department or the National Safety Council. *The NFPA Inspection Manual*, a pocket-sized book, is a valuable reference tool for the beginner as well as the experienced inspector. It covers common and special fire hazards, eliminating or safeguarding a fire, and human safety in all types of properties.

## FIRE PREVENTION CHECKLIST

### ELECTRICAL EQUIPMENT

☐ No makeshift wiring
☐ Extension cords serviceable
☐ Motors and tools free of dirt and grease
☐ Lights clear of combustible materials
☐ Safest cleaning solvents used

☐ Fuse and control boxes clean and closed
☐ Circuits properly fused or otherwise protected
☐ Equipment approved for use in hazardous areas (if required)
☐ Ground connections clean and tight and have electrical continuity

### FRICTION

☐ Machinery properly lubricated

☐ Machinery properly adjusted and/or aligned

### SPECIAL FIRE-HAZARD MATERIALS

☐ Storage of special flammables isolated

☐ Nonmetal stock free of tramp metal

### WELDING AND CUTTING

☐ Area surveyed for fire safety

☐ Combustibles removed or covered
☐ Permit issued

### OPEN FLAMES

☐ Kept away from spray rooms and booths

☐ Portable torches clear of flammable surfaces
☐ No gas leaks

### PORTABLE HEATERS

☐ Set up with ample horizontal and overhead clearances
☐ Secured against tipping or upset
☐ Combustibles removed or covered

☐ Safely mounted on noncombustible surface
☐ Not used as rubbish burners
☐ Use of steel drums prohibited

### HOT SURFACES

☐ Hot pipes clear of combustible materials
☐ Ample clearance around boilers and furnaces

☐ Soldering irons kept off combustible surfaces
☐ Ashes in metal containers

### SMOKING AND MATCHES

☐ "No smoking" and "smoking" areas clearly marked
☐ Butt containers available and serviceable

☐ No discarded smoking materials in prohibited areas

### SPONTANEOUS IGNITION

☐ Flammable waste material in closed, metal containers
☐ Flammable waste material containers emptied frequently

☐ Piled material, cool, dry, and well ventilated
☐ Trash receptacles emptied daily

### STATIC ELECTRICITY

☐ Flammable liquid dispensing vessels grounded or bonded
☐ Moving machinery grounded

☐ Proper humidity maintained

### HOUSEKEEPING

☐ No accumulations of rubbish
☐ Safe storage of flammables
☐ Passageways clear of obstacles
☐ Automatic sprinklers unobstructed

☐ Premises free of unnecessary combustible materials
☐ No leaks or dripping of flammables and floor free of spills
☐ Fire doors unblocked and operating freely with fusible links intact

### EXTINGUISHING EQUIPMENT

☐ Proper type
☐ In proper location
☐ Access unobstructed
☐ Clearly marked

☐ In working order
☐ Service date current
☐ Personnel trained in use of equipment

**Figure 17–2.** This sample checklist serves as a guide for the supervisor in drawing up an inspection list. It should be reviewed regularly to keep it up to date.

Fire inspections should be made with a critical eye. Every shortcoming should be listed. You should not hedge in listing certain hazards for fear that they might reflect poor practice. Omission of a pertinent detail might result in a fire later; such an outcome would reflect poor supervision.

### Informing the Workforce

Periodic inspections are an important part of any fire protection program. Under a complete program, your responsibilities extend further. As you become acquainted with actual or potential fire hazards, educate the personnel in your department about each hazard and explain how it relates to them individually. If an employee has specific tasks to perform to control a fire hazard, ensure that employee has been properly trained on his or her responsibilities.

Educate employees on all the physical safeguards that have been provided to prevent injury and destruction by fire. You should stress to individuals the precautions necessary for their jobs—the safe practices that complement mechanical protection. Employees who understand the reason for precautions and the possible consequences if fire-safety rules are not followed are more likely to comply. Patient explanation and persistent enforcement, in every case, are important fire prevention duties of the supervisor.

## CAUSES OF FIRES

All employees should be alert for potential causes of fires. A discussion of some common sources of fire and other fire prevention concerns are also noted in the following sections.

### Electrical Equipment

Special fire hazards are presented by electrical motors, switches, lights, and other electrical equipment exposed to flammable vapors, dusts, gases, or fibers. The NFPA's code and standards pamphlets designate the standards governing a particular hazard and indicate the special protective equipment needed. The National Electrical Code (NEC), ANSI/NFPA 70, gives the specifications for the equipment required. Substandard substitutions or replacements must not be made. The NEC updates every three years. (See Chapter 16, Electrical Safety, for more details.)

Haphazard wiring, poor connects, and temporary repairs must be brought up to standard. Fuses should be of the proper type and size. Circuit breakers should be checked to see that they have not been locked in the closed position (which results in overloading), and to see that moving parts do not stick.

Cleaning electrical equipment with solvents can be hazardous because many solvents are flammable and toxic. Keep in mind that a solvent may not present a fire hazard, but still be a health hazard. For example, carbon tetrachloride is nonflammable, but its vapors are extremely toxic. Before a solvent is used, therefore, determine both its toxic and flammable properties.

Many of the cleaning solvents encountered in industry are mixtures of different chemicals. They are usually marketed under trade names or code numbers. Most industrial cleaning solvents have some health or physical risks associated with their usage. Before any commercially available solvent is used, you must carefully review the information provided on the Material Safety Data Sheet (MSDS). Without such knowledge, the hazards cannot be evaluated, nor can the required safety controls be used (see Chapter 8, Industrial Hygiene.)

### Friction

Friction generated by overhead transmission bearings and shafting are frequent sources of ignition where dust and lint accumulate. Examples of such locations are grain elevators and woodworking plants. Bearings can overheat and ignite dust in the surrounding area. Keep bearings lubricated so they do not run hot. Pressure lubrication fittings should be kept in place, and oil holes of bearings should be kept covered to prevent combustible dust and grit from entering the bearings and causing overheating. Remove accumulations of combustible dust or other combustible material as part of a rigid housekeeping routine.

### Flammable Liquids

Liquids are defined as flammable or combustible depending on their flashpoint. *Flashpoint* means the minimum temperature at which a liquid gives off vapor within a test vessel in sufficient concentration to form an ignitable mixture with air near the surface of the liquid. The test is designed to determine if a liquid being used in normal workplace temperatures may be a likely ignition source. A spark or other source of ignition, which might otherwise be harmless, can start a fire or touch off explosive forces in the presence of flammable vapors that have evaporated from liquids and then mixed with air.

A liquid is termed flammable if the liquid emits enough vapors to burn at temperatures less than 100 F. Liquids that give off enough vapors to support combustion when heated over 100 F are termed a combustible liquid.

Combustible and flammable liquids are subdivided into classes depending on the boiling point and flashpoint. For example, a liquid whose flashpoint is below 100 F and whose boiling point is below 73 F would be classified as a Flammable Liquid Class 1A. These liquids would easily give off enough flammable vapors to support ignition in normal work place temperatures. MSDSs provide additional information about any chemical and the potential fire prevention procedures required for safe usage.

Nearly all industrial plants use flammable liquids. It is the responsibility of the managers and supervisors to ensure that safe practices are followed in storage, handling, and dispensing of such liquids. Workers must be trained in safe operating practices, and the supervisor must reinforce the training in the worksite.

Because all flammable liquids are volatile, they are continually giving off invisible vapors. Use the following guidelines when training workers on safe handling of these materials:

- Flammable liquids should be stored in and dispensed from approved safety containers equipped with vapor-tight, self-closing caps or covers.
- Flammable liquids should be used only in areas or under conditions where ventilation is adequate to prevent explosive air-vapor mixtures from forming.
- When volatile or flammable liquids are being used, warning signs should be prominent, notifying personnel that all open flames are hazardous and must be kept away.
- Vapors of flammable liquids are normally heavier than air and accumulate near the floor or other lowest possible level where they may not be easily detected. The vapors may then be ignited when they reach an ignition source such as a pilot light on a water heater or they encounter a spark.
- Wherever flammable liquids are used, it is essential that ignition sources—open flames and sparks—be eliminated or alternative preventive measures be taken.

Depending on volumes stored and handling methods, some of the following safeguards may be required:

- an approved flammable liquids storage vault or room
- special explosion-proof fixtures and equipment
- automatic suppression system
- explosive-relief devices and panels
- self-closing faucets and safety vents for drums
- flammable-liquids storage cabinets

- safety cans
- special ventilating equipment

Specific guidelines to help determine which procedures and equipment need to be implemented at a particular site can be found in the NFPA guidelines or OSHA 29 CFR 1910.106. Fire department inspectors or companies that insure your building can also be of invaluable assistance in determining the minimal requirements of the local code.

## Flammable Gases

Many flammable gases are stored in high-pressure cylinders or in bulk tanks. The gases are then piped to the user's location. Cylinder storage and gas usage areas should be well planned for fire protection. The storage and usage areas should normally include fire-resistant separations, automatic sprinklers, special ventilation, explosion-relieving vents, separation of gases from other materials, and separation of incompatible gases.

## Explosive Dusts

Finely divided materials not only tend to burn more readily than do bulk materials, but also can create explosive atmospheres if suspended in air. Industries that generate explosive dusts normally have planned programs to limit the amount of dust accumulations and the number of ignition sources and have equipment to limit the effects of the destructive forces developed in dust explosions.

The programs also include other measures, such as specific cleaning procedures. Ignition sources are limited through special electrical equipment, no-smoking policies, welding and cutting policies, and open-flame limitations. Areas expected to have dust often have explosion-relief panels and, potentially, explosion-suppressive systems.

## Plastics

Over the past two decades, general plastic usage has increased dramatically. Plastic containers are used for combustible, flammable, and nonflammable materials. At times, plastic containers are stored empty or within corrugated cartons. Noncontainer use of plastics has also increased.

Regardless of storage configurations, plastics tend to burn hotter and faster and create more smoke than cellulosic materials such as wood or paper. Standard sprinkler systems may not control the high-challenge fires that plastics create. Supervisors of industrial plastic processes should be aware of the increased hazard this material represents.

## Ordinary Combustibles

Paper and wood products are often referred to as ordinary combustibles. Rack storage and solid pile storage of these materials provide conditions that promote fire growth. Materials that are stacked above one another provide good flue spaces while blocking sprinkler water patterns, thus acting to spread a fire. Sprinkler systems may be designed to overcome such fire growth.

A second category of ordinary combustibles that pose rapid fire-growth hazards is wooden pallets. Most sprinkler systems will not control fires in wooden pallets stacked over 6 ft tall (1.8 m). If the sprinkler density has been increased, a maximum 8 ft (2.4 m) tall stack may be possible. In either case, the supervisor is usually faced with the problem of limiting pallet heights, while meeting production and space limitations. Extra pallets are often stored outside or in separate buildings to reduce exposure to serious fires.

NFPA 30, *Flammable and Combustible Liquids*, published by the National Fire Protection Association, is an excellent source of detailed information on this subject.

## Other Hazardous Materials

There are many other types of materials that must be kept isolated to prevent fire. For example, some chemicals, such as sodium and potassium, decompose violently in the presence of water, evolve hydrogen, and ignite spontaneously. Yellow phosphorus may also ignite spontaneously on exposure to air. Numerous other chemical combinations may react with the evolution of heat and produce fire or explosion—in some cases, without air or oxygen being present. These materials must be handled in a special manner. NFPA 49, *Hazardous Chemicals Data*, lists about 100 such items and provides information on unusual shipping containers, fire hazards, life hazards, storage, fire-fighting phases, and additional data, where applicable. Check the NFPA for other publications related to hazardous materials.

Some materials, such as ethers or organic peroxides, may become unstable during long periods of storage and may eventually explode. In such cases, using the oldest stock first is both good fire safety and sound housekeeping. Part of your job, as supervisor, is seeing that this principle is followed.

When materials are used or stored in your department, you should find out their hazards—whether they explode when heated, react with water, heat spontaneously, yield hazardous decomposition products, or otherwise react in combination with other materials. Check the MSDS for recommended extinguishing methods and reactivity data.

Obtain detailed information on all potentially hazardous materials and devise appropriate fire safety measures. Your company's fire insurance carrier can be asked to help.

Each type of industry tends to have unique fire protection requirements. The guidelines for protecting such items as computer facilities, ovens, furnaces, boilers, and other specific equipment are covered by various NFPA design standards.

## Welding and Cutting

Ideally, welding and cutting operations should be conducted in a separate, well-ventilated room with a fire-resistant floor. Of course, this safety measure is not always practical.

If welding and cutting must be carried on in other locations, these operations must not be performed until (1) the areas have been surveyed for fire safety by experts who know the hazards, (2) the necessary precautions to prevent fires have been taken, and (3) a permit has been issued. Permits should be issued for a specific time period, usually one shift or less if the task may be completed within the shift. At the beginning of each shift, the permit should be re-issued by the incoming supervisors. This ensures that the supervisor is aware of the work, is familiar with the hazards, and has had the opportunity to verify that no new hazards were introduced during the previous shift. (For more information on hot work permit programs, see National Safety Council's *Accident Prevention Manual for Business & Industry, Engineering & Technology* volume.)

If welding must be done over wood floors, they should be swept clean, wetted down, and then covered with fire-retardant blankets, metal, or other non-combustible covering. Pieces of hot metal and sparks must be kept from falling through floor openings onto combustible materials.

Sheet metal or flame-resistant canvas or curtains should be used around welding operations to keep sparks from reaching combustible materials. Welding curtains also can be used to prevent casual observers from watching the weld and unknowingly burning their eyes (welder flash). Supervisors should ensure that the reflected light from the welds are also shielded so no employees suffer eye burns or discomfort from the welding task.

Welding or cutting should not be permitted in or near rooms containing flammable or combustible liquids, vapors, or dusts. Nor should these operations be done in or near closed tanks that contain—or have contained—flammable liquids, until the tanks have been thoroughly drained and purged, and tested free from flammable gases or vapors.

No welding or cutting should be done on a surface until combustible coverings or deposits have been removed. It is important that combustible dusts or vapors not be created during welding. Be sure to provide fire-extinguishing equipment at each welding or cutting location. Where extra-hazardous conditions cannot be eliminated completely or protected by isolation, consider installing the backup protection of a fire hose.

Station a watcher at the welding operation who can prevent stray sparks or slag from starting fires, or who can immediately extinguish fires that do start despite all the precautions taken. The area should be under fire surveillance for at least one-half hour after welding or cutting has been completed. Many fires are not detected until a few minutes after they ignite.

## Open Flames

No open flames should be allowed in or near spray rooms or spray booths. Occasionally, workers may need to do indoor spray-painting or spray-cleaning outside of a standard spray room or booth. In such cases, provide adequate ventilation and eliminate all possible ignition sources, such as spark-producing devices and open flames.

Gasoline, kerosene, or alcohol torches should be placed so their flames are at least 18 in (46 cm) away from wood surfaces. They should not be used in the presence of dusts or vapors, near flammable or combustible liquids, paper, excelsior, or similar material. Torches should never be left unattended while they are burning.

## Portable Heaters

Gasoline furnaces, portable heaters, and salamanders always present a serious fire hazard. Discourage workers from using them. Upright models of fuel oil salamanders have been involved in numerous incidents. Replace these hazards with other types of low-profile heaters.

Fuel used in portable heaters should be restricted to liquefied petroleum gas, coal, coke, fuel oil, or kerosene. The area must be well ventilated because the products of combustion contain carbon monoxide.

All of these heating devices require special attention with respect to clearances and mounting. A clearance of 2 ft (0.6 m) horizontally and 6 ft (1.8 m) vertically should be maintained between a heater and any combustible material. If coal or coke is used, the heater should be supported on legs 6 in (15 cm) high or on 4 in (10 cm) of tile blocks and set on a noncombustible surface.

Combustible material overhead should be removed or shielded by non combustible insulating board or sheet material with an air space between it and the combustible material. A natural-draft hood and flue of noncombustible material should be installed.

As a fire precaution, each unit must be carefully watched. Heaters should be shut down and allowed to cool off before being refueled. Coal and coke salamanders should not be moved until the fire is out.

All portable heating devices must be equipped with suitable handles for safe and easy carrying. Make sure they are secured or protected against workers accidentally upsetting them.

Serious fire hazard occurs when heaters are improvised from old steel drums or empty paint containers, with scrap wood, tar paper, or other waste used as fuel. Residue or vapors can spread the fire. Let workers know that this practice is prohibited.

## Hot Surfaces

If possible, smoke pipes from heating appliances should not pass through ceilings or floors. If a smoke pipe must be run through a combustible wall, provide a galvanized double-thimble, with clearance equal to the diameter of the pipe and ventilated on both sides of the wall for protection.

Soldering irons must not be placed directly on wood benches or other combustibles. Special insulated rests, which will prevent dangerous heat transfer, can be used instead.

## Smoking and Matches

With increasing local, state, and federal regulation, and with increasing public awareness of the hazards of smoking, management usually has very specific policies about cigarette, cigar, or pipe smoking in the workplace. Smoking and nonsmoking areas must be clearly defined and marked with conspicuous signs. Reasons for these restrictions must be explained to employees, and rigid enforcement must be maintained at all times.

Fire-safe metal containers and safety ashtrays should be provided in any areas where smoking is permitted. Periodically check "no-smoking" areas, especially when they include stairways and other out-of-the-way places, for evidence of discarded smoking materials.

## Spontaneous Ignition

Spontaneous heating is a chemical action in which the oxidation (burning) of a fuel produces a slow generation of heat. When adjacent materials provide sufficient insulating properties, the heating process can continue until spontaneous ignition occurs. Conditions leading to spontaneous ignition exist where there is sufficient air for oxidation but not enough ventilation to carry away the heat generated by the oxidation. Any factor that accelerates oxidation while other conditions remain constant obviously increases the likelihood of ignition.

**Figure 17–3.** Check containers for oily waste to make sure the lid closes snugly against the top.

Gloves, rags, or waste saturated with oil, solvents, paint, or petroleum products often cause fires because no provision is made for the generated heat to escape. By keeping such refuse in airtight metal containers with self-closing covers (Figure 17–3), you limit the oxygen supply, and a fire will quickly extinguish itself. These containers should be emptied daily.

Materials like unslaked lime and sodium chlorate are susceptible to spontaneous ignition, especially when wet. Such chemicals should be kept cool and dry and away from combustible material. Other precautions against spontaneous ignition include total exclusion of air or good ventilation. The former is practical with small quantities of material through the use of airtight containers. Ventilation can best be ensured by storing material in small piles or by turning over a large pile at regular intervals.

Monitoring the temperature within the interior of a mass of material in various locations may determine spontaneous heating. Exterior temperatures are not likely to provide a good index.

## Static Electricity

Sparks due to static electricity may be a hazard wherever flammable vapors, gases, or combustible dusts exist. Precautions against static electricity are required in such areas. Static charges result from friction between small particles or from the contact and separation of two substances, one or both of which are nonconductive.

Static charges can be produced in many ways—for example, by the flow of liquids, such as gasoline through a nonconductive hose, by passing dry and powdered materials down a nonconductive chute, or through the friction of machine parts.

Although it is impossible to prevent generation of static electricity under all circumstances, the hazard of static sparks can be avoided by preventing buildup of static charges. One or more of the following preventive methods can be used:

- grounding
- bonding
- maintaining a specific level of humidity
- ionizing the atmosphere

A combination of these methods may be advisable in some instances where the accumulation of static charges presents a severe hazard. NFPA 77, *Recommended Practice on Static Electricity*, gives additional information and guidance on this subject.

Grounding is accomplished by mechanically connecting a conductive machine or vessel (in which the generation of static may be a hazard) to ground by means of a low-resistance conductor. Another method is to make the entire floor and structure of the building conductive so that all equipment in contact with it will be grounded. If you use the first method, be sure to check the continuity of the ground circuit. Connections must be clean and the conductor unbroken. With the second method, make certain that the floor is free of wax, oil, or other insulating films.

As people learn from walking across a rug and then touching a doorknob or other conductive object, the human body can also carry an electric charge. Workers can ground themselves by wearing conductive shoes with a floor made of conductive material. The conductive parts of such shoes should be made of nonferrous metal. As an added measure of safety, the conductive flooring may be made of spark-resistant metal. Ferrous contacts increase the risk of friction sparks.

When humidity is low, the hazard of static electricity is greatest. When humidity is high, the moisture content of the air serves as a conductor to drain off static charges as they are formed. When humidity is added to prevent accumulation of static charges, make sure that an effective relative humidity—usually 60 to 70 %—is maintained. However, the minimum humidity required for safety may vary over a considerable range under different conditions. In fact, under some conditions, the static charge cannot be controlled by humidifying the air. Consult with engineering authorities to determine the best approach.

When air is ionized, it has sufficient conductivity to prevent static accumulation. Ionization is produced by electrical discharges, radiation from radioactive substances, or gas flames. Only an electrostatic neutralizer designed for use in hazardous locations should be used. Otherwise, the neutralizer may itself be a source of ignition and touch off flammable vapor or dust. Neutralizers must be kept in good condition.

## EFFECTIVE HOUSEKEEPING FOR FIRE SAFETY

Good housekeeping is another important part of an effective fire protection program. As supervisor, you are responsible for ensuring that proper housekeeping standards are developed and adequate storage capacity is in place for your work area. Once the standard has been set, enforce housekeeping rules at all times. Each person should be held personally responsible for preventing the accumulation of unnecessary combustible materials in the work area. Workers should be held accountable for their areas at the end of their shift. Here are the precautions to take:

- Combustible materials should be present in work areas only in quantities required for the job. They should be removed to a designated, safe storage area at the end of each workday.
- Quick-burning and flammable materials should be stored only in designated locations. Such locations always should be away from ignition sources and have special fire-extinguishing provisions. Covered metal receptacles are best.
- Vessels or pipes containing flammable liquids or gases must be airtight. Any spills should be cleaned up immediately.
- Workers should be careful not to contaminate their clothing with flammable liquids. If contamination does occur, these individuals must be required to change their clothing before continuing to work.

- Passageways and fire doors should be unobstructed. Stairwell doors must never be propped open, and material should never be stored in stairwells. Material must never block automatic sprinklers (within 18 in (46 cm) of sprinkler head) or be piled around fire extinguisher locations or sprinkler and standpipe controls.

To obtain proper distribution of water, a minimum of 18 in (46 cm) of clear space is required below sprinkler deflectors. However, clearance of 24 to 36 in (60 to 90 cm) is recommended. If there are no sprinklers, clearance of 3 ft (0.9 m) between piled material and the ceiling is required to permit use of hose streams. Double these distances when stock is piled more than 15 ft (4.5 m) high.

Be sure to check applicable codes, especially *Life Safety Code*, ANSI/NFPA 101–2009. Local fire departments or state governments may have even more stringent codes.

## FIRE PREVENTION INSPECTIONS

Fire prevention inspections will vary from facility to facility and department to department. However, all fire prevention inspections have four basic goals:
1. Minimize the size of fires by ensuring that combustible and flammable material storage is controlled.
2. Control ignition sources to reduce the possibility of fire.
3. Make sure that fire protection equipment is operational.
4. Make sure that emergency evacuation procedures and routes are understood and are maintained.

Supervisors typically have some responsibility for fire prevention inspections as part of their daily, weekly, and monthly inspection activities. Fire prevention inspections generally are conducted at the supervisory level as part of routine safety inspection activities.

The best way to ensure that proper items are inspected is to prepare a department checklist. The checklist should be as specific as possible in defining what is, and what is not, permitted. This checklist can then be used to audit the department and can be given to employees to make sure that levels of housekeeping and other fire prevention items are clearly defined to all parties.

In order to prepare a checklist, the information in this chapter may be used, along with information from local or corporate safety coordinators, fire department inspectors, and insurance inspectors. Once the checklist has been prepared, the inspection frequency should be defined. Deficient items should be noted and the corrective action taken as soon as possible. Follow-up is also necessary at a predetermined interval.

## Fire Protection Equipment

One of the most important audit items will be to determine that the fixed fire protection equipment is operational. The checklist items should include the fire protection water valves in your department, which should be locked or supervised open. Other fire protection systems are normally controlled by electrical panels, which can be audited to ensure that the systems are on and that no trouble signals are showing. Automatic sprinkler patterns can be audited by making sure that there are at least 18 in (46 cm) between products and the sprinkler deflectors.

Fire extinguishers should be tagged and marked with the date of the last inspection. Gauges should show charge, and seals should be in place. The audit form should ensure that the extinguishers are hung in designated locations and that access is not blocked. The nozzles should be free and clear.

## Fire Protection Systems

Fires can be extinguished by exhausting the fuel source, by manual fire fighting, or by fixed fire-fighting systems. Fire extinguishment typically occurs by lowering the temperature through dousing the fire with water, reducing the oxygen available through carbon dioxide or foam systems, or by interrupting the chemical chain reaction by using a halon or dry chemical system.

All systems and system components should be listed by recognized testing agencies according to their fire-fighting purposes. In addition, all systems should be inspected and/or tested annually or as required by applicable standards or manufacturers' instructions. Automatic sprinklers are available in wet-pipe systems, dry-pipe systems for piping subject to freezing, deluge systems for high-challenge fires, or pre-action systems for locations subject to water damage.

Halon, carbon dioxide, and foam systems typically are actuated by fire detection systems. These systems generally are heat, smoke, or flame detectors. Detection systems also are often used to reduce the threat to life and property by providing early warning to initiate manual fire fighting or evacuation of the workplace.

# ALARMS, EQUIPMENT, AND EVACUATION

A fully effective fire protection program includes fire extinguishment. Whenever a fire breaks out, take immediate action:

1. Sound the fire alarm right away—regardless of the size of the fire.
2. If trained, attempt to extinguish or control the fire with appropriate fire-extinguishing equipment.
3. Fulfill your responsibilities as specified in your employer's emergency plan. Typical duties that may be assigned to supervisors including conducting a walk-though of the area if possible to see that no employees are unaware of the fire, or taking a head count to ensure that all employees are out of danger.

## Fire Alarms

Facilities in areas where municipal fire departments are available usually have an alarm box close to the facility entrance or located in one of the buildings. Others may have auxiliary alarm boxes, connected to the municipal fire alarm system at various points in the facility. Another system is a direct connection to the nearest fire station that may register water flow alarms in the sprinkler system, be activated by fire detectors, or be set off manually. In some cases, the telephone may be the only means for signaling a fire alarm.

Whatever the alarm system used, all employees should be carefully instructed in how, when, and where to report a fire. These three steps are extremely important. Many fires get out of control simply because someone did not know how, when, or where to signal the alarm.

## Extinguishers

As the supervisor, you need to know there are four classes of fires that might break out in your area. Before employees can combat beginning fires effectively, they must be familiar with and understand these classes of fires. Briefly, they are:

- Class A—Fires in ordinary combustible materials, such as wood, paper, cloth, rubber, and many plastics. These fires may be fought using the quenching and cooling effects of water or of solutions containing large percentages of water.
- Class B—Fires involving flammable liquids or flammable gases, greases, oils, tars, oil-based paints, lacquers, and similar materials. Class B fire extinguishing is most effective where smothering or exclusion of air and interrupting the chemical chain reaction through use of foams or dry chemicals is used.

- Class C—Fires in or near live electrical equipment, so that the use of a non-conductive extinguishing agent is of first importance. The actual material that is burning is either Class A or Class B in nature.
- Class D—Fires that occur in combustible metals, such as magnesium, lithium, and sodium. Special extinguishing agents and techniques are needed for fires of this type.

Each of the fire extinguishers in the department or at a job location should have a plate identifying the class of fire for which it is intended (see Figure 17–4), its operating instructions, and its servicing instructions. The data plate should have the identifying symbol/name of a recognized testing facility to indicate that the unit has been listed or approved. Equipment that does not bear an approval label should be brought to the attention of management. Only listed or approved equipment should be furnished. See "Fire Protection Equipment List" (Underwriters Laboratories Inc., Northbrook, Ill. 60062). Also, "Approved Equipment for Industrial Fire Protection" (Factory Mutual Research Corp., Norwood, Mass. 02062).

Any employee expected to use a fire extinguisher must be trained. See to it that every worker knows the important details about each of the fire-extinguishing agents provided in the particular job area or shop (Table 17–B). Fire extinguisher manufacturers and distributors can be contacted for additional training materials, and local fire departments may be willing to provide training assistance. Information on selection, installation, use, inspection, and maintenance of extinguishers can be obtained from the NFPA or OSHA.

**Figure 17–4.** These fire extinguisher symbols indicate not only the class(es) of fire for which the fire extinguisher is rated for use, but also the class(es) of fire for which the fire extinguisher should not be used (NFPA 10 Portable Fire Extinguishers). (*Reprinted with permission from the National Fire Protection Association.*)

# Table 17–B. Fire Extinguisher Selection Chart

| | CLASS A | | | | CLASS A/B | | CLASS B/C | | | |
| | WATER TYPES | | MULTIPURPOSE DRY CHEMICAL | | AFFF FOAM | HALON 1211 | CARBON DIOXIDE | DRY CHEMICAL TYPES | | HALON 1211 |
| | Stored Pressure* | Pump Tank* | Stored Pressure | Cartridge Operated | Stored Pressure* | Stored Pressure | Self Expelling | Stored Pressure | Cartridge Operated | Stored Pressure |
|---|---|---|---|---|---|---|---|---|---|---|
| Sizes Available | 2 ½ gal | 2 ½ and 5 gal | 2 ½ - 30 lb ALSO (Wheeled 150-350 lb) | 5-30 lb ALSO (Wheeled 50-350 lb) | 2 ½ gal | 9 to 22 lb | 5-20 lb ALSO Wheeled 50-100 lb | 2 ½-30 lb ALSO Wheeled 150-350 lb | 4-30 lb ALSO Wheeled 50-350 lb | 2 to 22 lb |
| Horizontal Range (Approx) | 30 to 40 ft | 30 to 40 ft | 10-15 ft (Wheeled 15-45 ft) | 10-20 ft (Wheeled 15-45 ft) | 20-25 ft | 14 to 16 ft | 3-8 ft (Wheeled 10 ft) | 10-15 ft (Wheeled 15-45 ft) | 10-20 ft (Wheeled 15-45 ft) | 10 to 16 ft |
| Discharge Time (Approx) | 1 min | 1 to 2 min | 8-25 sec (Wheeled 30-60 sec) | 8-25 sec (Wheeled 20-60 sec) | 50 sec | 10 to 18 sec | 8-15 sec (Wheeled 8-30 sec) | 8-25 sec (Wheeled 30-60 sec) | 8-25 sec (Wheeled 20-60 sec) | 8 to 18 sec |

| | CLASS/A/B/C | | CLASS D | |
| | MULTIPURPOSE DRY CHEMICAL HALON 1211 | | DRY POWDER | |
| | Stored Pressure | Cartridge Operated | Stored Pressure | Cartridge Operated |
|---|---|---|---|---|
| Sizes Available | 2 ½ - 30 lb ALSO (wheeled 150-350) | 5-30 lb ALSO Wheeled 50-350 lb | 9 to 22 lb | 30 lb ALSO Wheeled 150-350 lb |
| Horizontal Range (Approx) | 10-15 ft (Wheeled 15-45 ft) | 10-20 ft (Wheeled 15-45 ft) | 14-16 ft | 5 ft (Wheeled 15 ft) |
| Discharge Time (Approx) | 8-25 sec (Wheeled 30-60 sec) | 8-25 sec (Wheeled 20-60 sec) | 10 to 18 sec | 20 sec (Wheeled 150 lb to 70 sec 350 lb, 1 ¾ min) |

*Note: Must be protected from freezing

547

**Figure 17–5.** Fire equipment should be conspicuously located, appropriately marked, and regularly inspected.

Normally, the location and installation of fire safety equipment at strategic places are the responsibility of higher management. However, you should know the location of units in your area, and recommend relocating them or obtaining additional units if such changes will afford more adequate protection.

Check to see that extinguishers are not blocked and that signs indicating their location are conspicuous (Figure 17–5). Marking the area of the floor directly under a fire extinguisher is an excellent way to keep employees from placing obstructions in front of the equipment.

The location of extinguishers can be identified by painting the support of the extinguisher with standard fire-protection red. Fire-protection equipment itself, such as sprinkler system piping, is sometimes painted or marked in red.

Make sure that each worker knows the location of the nearest unit and knows the importance of keeping areas around extinguisher units clear.

It is a good idea to schedule fire drills for employees trained and expected to use fire extinguishers and evacuation drills for all employees not trained or expected to respond to a fire. The company's safety professional or fire chief, the insurance company's safety engineer, or the local fire department representative will assist in conducting such drills.

Every organization should have a specific program for periodic inspection and servicing of portable fire-extinguishing equipment (see CFR 1910.157(e)(2)). Fire extinguishers must be visually inspected monthly. This routine work is usually outside the scope of the department supervisor and/or maintenance personnel. However, you can do some things to assist in this program. For example, during regular department inspections, you should double-check each data card to determine when each extinguisher was last inspected and serviced. Management should be notified when inspections have been missed or are outdated. If the extinguisher seal is broken, its pressure gauge reads below normal, or any other unsatisfactory condition is observed, report these conditions immediately. Defective extinguishers may not work.

Fire extinguishers must meet the following requirements:

- Be kept fully charged and in their designated places.
- Be located along normal paths of travel where practical.
- Not be obstructed or obscured from view.
- Not be mounted higher than 5 ft (1.5 m) (to the top of the extinguisher) if they weigh 40 lb (18 kg) or less. If heavier than 40 lb, extinguishers must not be mounted higher than 3 ½ ft (1 m). There shall be a clearance of at least 4 in (10 cm) between the bottom of the extinguisher and the floor.
- Be inspected by management or a designated employee, at least monthly, to make sure that they are in their designated places, have not been tampered with or actuated, and do not have corrosion or other impairment.
- Be examined at least yearly and/or recharged or repaired to ensure operability and safety. A tag must be attached to show the maintenance or recharge date, and the signature or initials of the person performing the service.
- Be hydrostatically tested. Extinguisher servicing agencies should be contacted to perform this service at appropriate intervals.
- Be selected on the basis of type of hazard, degree of hazard, and area to be protected.
- Be placed so that the maximum travel (walking) distances between extinguishers, unless there are extremely hazardous conditions, do not exceed 75 ft (23 m) for Class A extinguishers or 50 ft (15 m) where Class B extinguishers are used for hazardous-area protection. The travel distance requirement does not apply when Class B extinguishers are used for spot-hazard protection.

## Fire Brigades

Many plants and construction job sites have organized fire brigades. There are two basic types of fire brigades:

1. *Incipient stage fire* brigades are organized to fight an incipient stage fire. Incipient stage fire refers to a fire in the initial or beginning stage and can be controlled or extinguished by portable fire extinguishers, Class II standpipe, or small hose systems without the need for protective clothing or breathing apparatus.
2. Interior structure fire brigades are expected to conduct interior structural fire fighting. *Interior structural fire fighting* refers to the physical activity of fire suppression, rescue or both, inside of buildings or enclosed structures, that are involved in a fire situation beyond the incipient stage.

Each type of brigade requires that all members undergo training commensurate with their responsibilities. No employee should attempt to respond to an emergency, other than sounding an alarm, unless that employee has been properly trained to do so.

Regardless of your role in the fire-brigade organization, you should be familiar with the location and operation of the following items in and adjacent to your department—standpipes and valves, sprinkler system valves, electric switches for fans and lights, fire-alarm boxes and telephones, fire doors, and emergency power for equipment and/or lighting. In particular, you must make sure that enclosed stairway doors are kept closed. Such doors are provided to keep smoke, combustion gases, and heat out of exit passageways and stairs, as well as to help prevent the rapid spread of fire. Doors should be equipped with self-closing devices and should not be propped open. In cases where the exit way is used frequently, doors may be held open with fire-actuated detectors that will automatically close the door if fire breaks out.

To become familiar with the items listed above, accompany the fire inspector on inspection tours to learn how the items operate. It can be extremely dangerous to let anyone without proper knowledge operate the equipment. Activating it at the wrong time or using it incorrectly might seriously endanger people's lives in the event of fire.

### Follow-up for Fire Safety

Frequent inspections of the area, correction of hazardous conditions, and instruction of the workers in fire prevention and extinguishment measures are still not enough to ensure a fire-safe work area. Make sure you instill fire-safe attitudes in your personnel.

Make fire prevention and extinguishment the subject of frequent safety talks between workers. Also, discuss these issues in safety meetings so that workers will be fully aware that fire prevention is a vital part of the overall departmental safety program.

### Special Fire Protection Problems

**Construction.** When construction is going on at an existing facility, fire protection problems are compounded. You and the construction supervisor should work together in determining the hazards in each field of operations and in identifying the limits of your fire-fighting capabilities. On the basis of this knowledge, you should take steps to upgrade fire protection equipment and practices.

Construction activities may require measures such as using temporary wiring or portable heaters and gasoline engines that must be refueled on the spot. As a result, manufacturing processes, ordinarily not hazardous, may become so under these conditions.

Ensure that employees are aware of additional ignition sources created by the construction operations. The construction supervisor should inform workers of additional fire hazard exposures present from plant activity.

When the construction work involves plumbing, the water supply available for fire extinguishment may be decreased or may need to be shut off. To offset this possible shortage of water, obtain additional auxiliary fire-extinguishing equipment such as temporary hose lines or water tank trucks.

**Sprinkler System Shutdown.** Whenever a sprinkler system must be shut down, special precautions must be taken to ensure maximum fire safety. Check to see that all preliminary work is completed. Set out extra fire-fighting equipment, and notify fire-fighting units which areas will be without sprinkler protection. One procedure requires that a bright red tag be attached to the sprinkler valve when it is closed and that the insurance carrier be notified. The work is completed as quickly as possible, at which time the tag is removed and the insurance company notified that service is restored. To make certain that the valve is left wide open, it is tested and then locked in the open position. All sprinkler valves should be checked regularly to make sure they have not been tampered with, locked, or shut.

Whenever the sprinkler system must be shut down for alterations or repairs, you and your maintenance people should do the work before or after normal working hours. If work must be done during the most hazardous times, special precautions may have to be taken, such as having hose lines laid and furnishing extra fire patrols.

Some plant and construction supervisors make a practice of attending each other's safety meetings. They benefit from exchange of their respective points of view and learn about each other's problems. This joint approach to common exposures is recommended.

## Evacuation

Prevention of fire is the primary objective of any fire protection program. Nevertheless, each program must include provisions to ensure the safety of employees in the event of a fire. All site personnel must be included in the overall plan.

Provisions should be made for personnel on site who are not employees. Examples would be over-the-road truck drivers, venders, visitors, janitors, or security personnel.

Most facilities will have a plan to respond to emergencies. The plan may be as simple as an informal evacuation plan, or as complicated as a fully written emergency response and disaster recovery plan for various natural and man-made emergencies. Normally, your responsibilities during an emergency are for the people and department under your authority.

You should be trained on a written emergency plan at periodic intervals and instruct your workforce on their individual roles. You should develop a clear personal plan for complying with the facility's emergency plan and responding to evacuation, spills, fires, fire fighting, natural disasters, and utility failures. Make sure you know proper procedures for reporting emergencies and the meaning of alarm signals. You should actively participate in routine emergency drills.

In some cases, you or a member of your workforce will serve on the emergency response team and will have specific duties outside the department during an emergency. You should plan how to respond to an emergency with those members missing. Workers should be trained to respond in specific assigned duties, such as shutdown of equipment, accounting for all employees, assembling in assigned places, taking charge of specific responsibilities, sounding the all-clear, etc.

From time to time, the local fire department will tour your facility. The company can benefit by your active participation in these tours. The more the fire fighters know about flammable and combustible materials in the facility, the quicker the fire department can respond to minimize fire losses.

You can do much to prepare for the safety of fellow employees in the event of a serious fire. As a matter of routine, you should make sure that every employee knows the evacuation alarms, both primary and alternate exit and escape routes, what to do, and where to be during and after an evacuation.

You should make sure that each person knows that he or she, upon being alerted, must proceed at a fast walking pace—not a run—to an assigned exit or, if it is blocked, to the nearest clear one. Day in and day out, you should emphasize that exit routes and fire doors must be kept unobstructed.

Such essential steps in emergency procedures will best be instilled by periodic fire drills. Plan and conduct practice evacuations at regular intervals. You should also do everything possible to integrate fire emergency training with facility-wide drills and the company's overall evacuation plan. When it is impossible to hold drills, give oral instructions regarding evacuation procedures, and distribute printed information describing these policies to all employees.

## FIRE PROTECTION EDUCATION

Fire protection education for the work force typically involves the following:
- emergency response
- fire fighting
- home fire safety

Emergency response training usually will be necessary one or more times per year. The work force will need to know how to turn in and how to respond to alarms. The specific procedures should be rehearsed.

Workers also should practice use of portable fire extinguishers and hoses at least yearly if they are expected to use the equipment. A yearly review of the fixed fire protection systems and manual operation of those systems is usually necessary. If any of the employees travels, hotel fire safety is an appropriate topic to address. Fire Prevention Week in October is normally a convenient time to review these items. Safety meetings also provide a convenient forum. Brochures and written materials are available from the NFPA. OSHA's Emergency Action Plan (29 CFR 1910.38) can provide additional guidelines.

Fire Prevention Week is also a good time to address home fire safety. The proper use of smoking materials, smoke detector usage, and home fire drills are appropriate topics. The beginning of the heating season is a good time to introduce concepts of proper use of wood-burning equipment. In December, review the hazards and protective measures regarding holiday decorations and natural Christmas trees.

## PROTECTIVE INSURANCE REQUIREMENTS

The property (fire) insurance carrier generally conducts inspection tours of the facility on a routine basis. These tours are intended to identify items that can help to reduce loss in a fire or minimize ignition sources. The supervisors of an area will often accompany the insurance representative on the tour to understand the insurance concerns. The representative also can be used as a resource to help solve fire protection problems.

The insurance carrier will require specific conditions that you must implement and maintain. No-smoking requirements are one such condition and the use of a cutting and welding permit system may be another. The locking or supervision of water supply valves, and testing and maintenance of fire protection equipment are other typical required programs.

Another common insurance requirement involves those times when fire protection systems are removed from service. The insurance carrier often will require a call and a specific tagging system. The carrier then tracks the time when the system is out of service to ensure that the system is returned to service. The carrier may require alternative fire protection such as standby charged hoses during those periods. The notification of the fire department is also a good procedure.

## SUMMARY OF KEY POINTS

This chapter covered the following key points:

- The primary requirements for a successful fire prevention program include continuous training in fire-safe work practices, regular inspections of work areas, and close supervision of employee job performance. Ultimate responsibility for fire safety rests with the supervisor.
- Fire chemistry can be illustrated by the tetrahedron model in which elevated temperature, oxygen, fuel, and chain reaction must occur simultaneously for a fire to continue burning. Fires can be extinguished by one or more of the following: removing or reducing oxygen supply, removing fuel sources, reducing or cooling heat sources, and interrupting the chain reaction.
- Common causes of fires include faulty electrical equipment, friction, flammable liquids, flammable gases, explosive dusts, plastics, and ordinary combustibles such as paper or wood. Supervisors should ensure that electrical equipment is properly grounded, insulated, and maintained, and that all flammable or combustible materials are properly stored, ventilated, and contained.
- Other causes of fires include operations such as welding and cutting, open flames, portable heaters, hot surfaces, smoking and lighted matches, spontaneous combustion, and static electricity. Stringent safe work practices and

housekeeping measures can reduce the possibility of fire from these causes.

- Supervisors must establish good housekeeping standards and hold each employee accountable for following these procedures. These rules include preventing the accumulation of combustible materials, changing clothing contaminated with flammable liquids, keeping passageways and fire doors unobstructed, and making sure sprinkler systems are not blocked by stacked material.

- Supervisors should make regular inspections to identify fire hazards in their areas and take action to resolve them. Workers should be continuously informed of these hazards and the safeguards required to prevent fires from occurring. A fire prevention inspection has four goals: (1) minimize the size of fires by controlling combustible and flammable materials, (2) controlling ignition sources, (3) making sure fire protection equipment operates, and (4) making sure that emergency evacuation procedures and routes are understood and are maintained.

- Supervisors should make sure that employees are trained to operate fire alarms and fire-fighting equipment, and understand evacuation procedures. Employees should be trained how, when, and where to report a fire by activating fire alarms. They should also understand the hazards of Class A, B, C, and D fires, and explain company specific procedures for dealing with each type of fire. Fire fighting and extinguishing equipment should have this information clearly printed on the data plate.

- Such equipment must be inspected periodically and repaired or replaced if found defective. Supervisors must follow up on inspections and instruction sessions to make sure employees are complying with safety procedures.

- Evacuation procedures should be communicated to every employee in the event of a serious fire or other emergency. Workers should know evacuation alarm signals, where to exit the facility, and how to secure their work areas before leaving. Supervisors should be trained on their exact responsibilities in an emergency.

- In some facilities, companies organize fire brigades. Supervisors must know how the fire brigade operates and the specific responsibilities of each brigade member. Supervisors should also know the location and operation of all fire safety equipment and exit facilities. Periodic drills are desirable to practice evacuation procedures.

- Employees expected to respond to an emergency should receive detailed training and should practice using fire-extinguishing equipment.

- In some cases, construction on site, sprinkler system shutdowns, or the presence of hazardous materials present special fire safety problems. Supervisors must make sure special fire prevention measures are taken.
- Insurance companies generally inspect facilities on a routine basis. Supervisors can accompany these inspectors on their tours and use them as a resource to help solve fire protection problems.

# Case Studies

The following case studies are provided for classroom instructors and for supervisors conducting on-the-job training. The instructor should present only the case and allow the students to solve the problem through discussion and questioning. The background information, possible solutions, and summary are provided for the instructor's guidance.

# CASE I

A maintenance man was working on a metal stair platform and wanted to use a ¼-in electric drill. The drill had a three-wire power cord. An extension cord running from the source of power also was a three-wire cord, but was not long enough. In order to connect the drill to the extension cord, the man obtained another short extension cord from the tool crib.

As the man started to drill, he was electrocuted.

What caused the incident? What could have been done to prevent this incident and similar incidents in the future?

## Guide and Background Information for Case I

Explain to the group, if asked, that the drill and cord were in proper working order, with no shorts. Also, the extension from the power source was okay. However, the extension cord from the tool crib had been improperly wired. The grounding lead had been connected to the "hot" terminal so that the frame of the drill was energized. When the man grasped the grounded metal stair platform, he was electrocuted.

Once this fact is brought out, have the group pursue the question of preventing another such incident. In the state in which this incident occurred, three workers had been electrocuted because of improperly wired electrical connections. (This report was from a State Industrial Commission.)

Following are other pertinent facts:

- The maintenance man was not standing on damp or wet ground. The area was dry.
- The extension cord was made up by a tool crib attendant.
- There were no rules in effect concerning procedure for maintenance and repair of equipment.
- The cord had not been tested.

## Possible Solutions for Case I

1. All extension cords and electrical equipment should be immediately tested by a qualified electrician before allowing further use of such equipment.
2. A strict rule should be put into effect that only qualified persons be allowed to repair equipment.
3. Electric tools and equipment should be inspected on a regular schedule to make certain they are properly wired, grounded, and in proper working

order. The equipment should also be marked to indicate to the user that it has been inspected. The date of inspection should be shown.

4. All electric tools must be grounded regardless of use or location.

5. Make certain all workers understand that they are not to make repairs to tools or equipment.

6. If tool crib attendants are to maintain equipment, they should be properly trained.

## Summary

Summarize the discussion by pointing out that the unsafe condition must be eliminated—in this case, the improperly wired extension cord. The only way this can be done is to test all extensions and make certain there are no other defective ones. Once all the electrical equipment has been checked out, some procedures must be set up to make certain the equipment is kept in good condition.

Because there are also unsafe procedures involved, provisions must be made to help prevent a recurrence. In this case the tool crib attendant did not follow safe procedures and the maintenance man did not make certain that the extension was okay. Safe procedures should be enacted and enforced. Stress the importance of eliminating all possible causes.

# CASE II

A freight handler attempted to sharpen the point of a bale hook on a grinding wheel. The foreman of the department saw him, but assumed that "anybody can use a little bench grinder." The freight handler caught the point of his hook between the tool rest and the wheel. The wheel broke and a large piece of it struck him in the face. He was permanently disabled by the injury.

What could have been done to prevent this incident or similar future incidents?

## Guide and Background Information for Case II

Explain to the group, if asked, that this bench grinder was properly guarded and mounted. Also, the wheel was properly tested before mounting.

Following are other pertinent facts:

- The tool rest was properly set ¼-in from the wheel.
- Eye protection (goggles) was available, but obviously not used.

- The eye shield was evidently not in place.
- There was no regulation regarding unauthorized use of equipment.
- The freight handler was not from this foreman's department.
- A grinding wheel of this type is proper for sharpening bale hooks.

### Possible Solutions for Case II

1. It should be obvious to the group that the primary cause of this incident was one or more unsafe procedures. The foreman made a mistake by assuming the freight handler knew how to use a bench grinder. The freight handler committed several unsafe procedures. First, he did not use the eye protection that was available. Second, he must have put the hook in a position that let the point slip between the tool rest and the wheel.
2. While you may get such comments from the group as "fire the foreman," the solution should focus on a rigidly enforced rule, "no unauthorized use of equipment."
3. The following procedures can also be set up:
   a. a policy regarding the sharpening of tools, such as returning tools to the tool crib for sharpening or replacement
   b. a training program if the freight handlers are to do the sharpening
   c. a lockout device interlocked with the starting switch to prevent the wheel from operating when the eye shield is not in the proper position

### Summary

Point out to the group that the two cases, Case I and Case II, bring out the importance of unsafe conditions and unsafe procedures as the causes of incidents. Stress the importance of searching for all possible causes.

Point out that in Case II there were a number of places where the incident prevention program needed tightening in order to prevent similar incidents in the future.

Emphasize the importance of checking procedures and conditions in advance to prevent incidents of this type. Mention the importance of a job safety analysis as a tool for preventing incidents.

## CASE III

A tool truck driver, making his routine crib stops, picked up a crib attendant.

The attendant had requested a ride to another part of the facility because he knew the driver would be going in that direction.

The truck driver deviated from his aisle route and angled through a cleared but darkened area. This area was being prepared for new machinery installation and at that time of night was not fully lighted.

The attendant, who was sitting on the right side of the cab, suddenly noticed that the truck was headed for a steel building column. Before he could warn the driver, the left front corner of the truck struck and glanced off the column.

The impact threw the driver against the column and about 15 ft (4.6 m) away from the truck. The truck continued for approximately 50 ft (15.3 m) before the rider could get behind the wheel to apply the brakes and bring the truck to a stop.

The driver suffered a skull fracture, concussion, and severe injuries to the left arm and chest. He was taken immediately to the local hospital, where he died from a blood clot about three weeks later.

What could have been done to prevent this incident or similar incidents in the future?

## Guide and Background Information for Case III

Explain to the group, if asked, that these trucks are not designed to carry passengers. In order for a rider to sit on the seat, the driver must move over, which puts him or her in an awkward position.

Following are other pertinent facts:

- The aisles were not marked in this particular area.
- The area was not "roped off" and there were no signs to indicate equipment was being installed.
- The truck was not equipped with a seat belt.
- The incident happened on the second shift (about 10 p.m.).
- The machinery installation had been going on for an extended period of time.
- There were no rules concerning riders.
- The driver was experienced.

## Possible Solutions for Case III

1. The driver's failure to stay within the main aisle was an important factor in the incident. Properly marked aisles might have prevented the driver from taking a shortcut.

2. When there are properly marked aisles, truck drivers should be instructed and trained in proper procedures.

3. Better illumination might also have prevented the incident. Because the installation had been going on for some time, the area should have been properly illuminated.

4. The area could have been roped off or marked.

5. A rigidly enforced rule against "no riders" should have been instituted.

6. Installation of seat belts in equipment of this type is a possibility, but because of the nature of the work, not generally done.

## Summary

Stress to the group that the incident was caused by a combination of factors:

- The unsafe conditions were poor lighting, unmarked aisles, and lack of signs.
- The unsafe procedures were the driver's "shortcutting" and picking up a rider. The lack of a rule against riders did not exonerate the driver because he had to make room for the rider and had to be aware that he was not in the best position to control the truck. The rider also must have been aware of the situation when he moved into the seat.

## CASE IV

A truck driver and a millwright were in the process of moving a pump, weighing approximately 2,000 lbs, that was mounted on a skid. They were using a mobile crane, and as the operator was backing up with the crane boom elevated at about a 30-degree angle, the rear wheels of the crane rose off the ground. To compensate for this overbalanced condition, the operator, while proceeding backwards, raised the crane boom to an approximate 60-degree angle. His co-worker at the time was observing the load on the crane boom and did not notice an overhead 440-volt line.

The crane boom became entangled in the overhead 440-volt wires, located approximately 18 ft (5.4 m) above ground. When the operator observed this, he jumped off the operating platform and was not injured. The insulation on the 440-volt line was not cut, so the crane boom did not become energized.

What could have been done to prevent this incident or similar incidents in the future?

## Guide and Background Information for Case IV

Explain to the group, if the question comes up, that this man was not injured. However, if the insulation had been cut, this man could have been electrocuted.

Following are other pertinent facts:

- The crane was of proper size to handle the load.
- It was not possible to have the electric current turned off.
- The crane was designed to operate both forward the backward.
- Both employees were experienced and had worked together before. Both worked for the same company.
- The normal procedure in this company was for a man to walk at the head of the mast in order to guide the movement of the crane. In this case, the millwright should have taken over the job.

## Possible Solutions for Case IV

1. The millwright or another person should have walked at the head of the mast and made certain all was clear.
2. The job was poorly planned. When the rear wheels of the crane were raised off the ground at the 30-degree angle, this should have been the signal to stop the operation and do some checking. The pump should have been placed on dollies and rolled under the wires since it was quite obvious that with only 18 ft (5.4 m) of clearance and the boom raised to 60 degrees, there was going to be contact.

## Summary

Better job procedures should be instituted and definite procedures established and followed in all operations of this type. A thorough inspection should be made of the area and the route to be followed before any attempt is made to move loads of this nature.

# CASE V

A warehouse employee was in the act of removing a part from the top of a storage bin. To reach it, she climbed on a wooden container instead of getting a ladder available for this purpose.

While standing on the edge of the container, it began to tip. The employee, trying to check her fall, grasped for the top of the bin with her left hand. As she did so, the ring she wore on her ring finger became caught on a bolt end, damaging the finger so badly that it had to be amputated.

What could have been done to prevent this incident or similar incidents?

### Guide and Background Information for Case V

If the question is raised, there were no rules in this company prohibiting the wearing of jewelry at work.

Following are other pertinent facts:

- The bins contained parts that were not in great demand.
- There was only one ladder in the storage area. The storage area was large and contained three aisles.
- The ladder, at the time of the incident, was not in the aisle in which the employee was injured.

### Possible Solutions for Case V

1. A rule prohibiting the wearing of jewelry at work could be instituted. Better training on the hazards of wearing rings would help.
2. Availability of a ladder in each aisle would make one always handy and thus more likely to be used.
3. Better housekeeping would help because the wooden container the employee stood on was not in its proper place.
4. The use of posters also would help with the problem.

### Summary

Point out the hazards of wearing rings at work. Stress the need for education on the subject because it may be difficult to enforce a rule against wearing rings because of certain religious beliefs.

## CASE VI

An employee asked another employee to exchange a trim knife for him at the tool crib. The man placed the knife in the side pocket of his coveralls and started for the crib. On the way, he stopped to remove some stock from a shelf 8 ft (2.4 m) from the floor. He was unable to reach the stock from the floor, so he used the bottom shelf on the rack as a platform to stand on. As he bent his right leg to step up on the shelf, the sharp point of the knife was forced up and penetrated his body approximately 1 in. A deep laceration was the only injury.

What could have been done to prevent his injury or similar incidents in the future?

### Guide and Background Information for Case VI

If asked, explain that there were no holders or tool bags provided for carrying tools of this type, such as punches, awls, screwdrivers, etc. Also, it was common practice for workers to exchange tools for each other.

Following are other important facts that may help:

- The trim knife was not one of the tools used in the injured employee's own work. He did, however, use other similar tools.
- The injured employee started out to obtain stock, but was asked to exchange the trim knife on his way.
- There was a small stepladder nearby that the injured employee could have used.
- There were no rules prohibiting any employee from exchanging tools, even if the attendant at the tool crib knew it was for another person.

### Possible Solutions for Case VI

1. Proper belt holders should be provided to carry tools because these employees used sharp-pointed tools frequently.
2. A rule prohibiting employees from exchanging tools for each other could be put in effect.
3. Better training in the handling of tools would help.
4. More emphasis on the hazards of using shelves as a stepladder also would help prevent such incidents.

### Summary

Stress the number of incidents resulting from improperly handled tools. Emphasize the importance of a sound training program on the proper use and handling of sharp and pointed tools. Also stress the need for rigid rules for carrying tools of this type.

## CASE VII

A worker was assigned to clean out sludge in the large degreaser he had been operating. His supervisor advised him to use gloves and respiratory protection before entering the degreasing tank and told him to make sure he had a co-worker watch him while he worked. An hour later, the supervisor noted the co-worker eating lunch. He asked the co-worker where the degreaser operator was. The co-worker responded that the operator had told him to go to lunch, and that he would be finished in the degreaser in five minutes. The supervisor and co-worker went immediately to check on the operator, whom they saw face down in the degreasing tank. The supervisor hurriedly descended into the tank to rescue the worker, but had to leave without retrieving him when he immediately began to feel light-headed. The co-worker called emergency responders who, equipped with self-contained breathing apparatus, retrieved the unconscious operator. The operator survived, but spent several days in the hospital recuperating and has experienced some memory loss and problems with coordination.

What could have been done to prevent this incident and similar situations in the future?

## Guide and Background Information for Case VII

Explain that this is a typical confined-space incident, in which the rescuer is as likely (or more likely) to be injured as the initial victim. In more than half the recorded cases, the unprepared rescuer has also been a fatality.

Other pertinent facts include:

- The degreaser operator had not received training in proper use of respiratory protection or confined-space entry procedures.
- The operator was able to work in the pit for at least an hour before being overcome because:
  - His organic vapor respirator did provide protection against the initial concentration of solvent vapors in the tank.
  - The tank was not an oxygen-deficient atmosphere initially (although under other circumstances, it might have been one), but became one later.
  - The worker stirred and shoveled up the sludge. Solvent vapors were released from air pockets in the solid material, so the concentration of vapors increased in the tank, and began to displace oxygen. The respirator no longer protected him because the absorbent material had become saturated with the solvent, and more importantly, the respirator could not supply oxygen (which had been depleted by the end of the hour). When trying to rescue the worker, the supervisor began to experience the effects of a lack of oxygen.
- The cloth gloves the worker wore may have added to his solvent exposure because they may have allowed the solvent to be absorbed through the skin.

## Possible Solutions for Case VII

1. The job indicated poor or no planning. The operator, co-worker, and supervisor should have received training on the hazards and proper procedures for entry into confined spaces. These include:
    a. use of a permit entry system
    b. pre-entry sampling of the atmosphere for flammable or toxic gases and vapors
    c. use of air-supplied or self-contained breathing apparatus where dangerous chemical exposures could occur, as well as impervious clothing
    d. use of lifelines and belts to aid in retrieval, if necessary
    e. presence of an attendant throughout the duration of the work
2. The supervisor should have completed a confined-space entry permit and communicated to the operator the equipment and precautions that were necessary, and ensured that these were used.
3. The operator should have received training in the proper use of respiratory protection and protective clothing, and the distinction between air-purifying and air-supplying respirators.
4. Hazard communication training should have been provided to these workers to acquaint them with the toxic properties of the chemicals they worked with.

## Summary

Better training should be provided and procedures instituted to prevent future incidents of this type. The hazards of the confined space should be assessed and understood before anyone attempts an entry.

## CASE VIII

A worker in your department has taken an excessive number of sick days. You ask the company nurse what the problem might be. She says the worker has come in to ask for aspirin almost every day she has been at work for the last two months.

You go to observe the worker at her operation. Her job, which is in the rework area, involves using pliers to remove wires from a part before it is repaired. You ask the worker to demonstrate how she does her job. She shows you that she grasps the part in her left hand, and twists it as she uses the pliers to grasp and pull the wire out with her right hand.

You note that she does not use the stool provided. The worker says that it is uncomfortable; because she is short, her legs dangle. She explains that she has been absent due to pain in her back and numbness and tingling in her hands.

## Guide and Background Information for Case VIII

Explain to the group, if asked, that there are at least two obvious ergonomic problems at the workstation: dangling feet and repetitive, forceful grasping combined with twisting of the wrist. The latter is probably responsible for the worker's back pain.

Other pertinent facts include:

- Her doctor told her that her hand problem was due to hormonal changes caused by menopause and that her back pain was likely caused by stress. Her doctor had not seen her work area; however, the doctor told the worker to take aspirin to relieve the symptoms.
- She told you she did not want to complain about her discomfort because she was afraid of losing her job.

## Possible Solutions for Case VIII

1. Institute an ergonomics program that could include:
   a. observation of work operations with employee involvement and input
   b. medical surveillance to determine if and where workers are being injured
   c. management commitment to making positive ergonomic changes where necessary
   d. development of an ergonomically trained team of managers, engineers, medical personnel, supervisors, workers, and others
2. For this particular injured worker, the supervisor should observe which actions require forceful repetitive movement of the wrists. It is apparent that the worker had to twist both her left and right wrists while applying force to the pliers with her right hand. These repetitive motions have probably contributed to what appears to be carpal tunnel syndrome. Some solutions might be:
   a. automating the operation
   b. using a grip and clamp to hold the part, and using ergonomically designed pliers that prevent the wrist from bending
   c. reducing the work rate

  d. rotating the worker to other jobs that require use of different muscle
     groups

## Summary

Institute an ergonomics program. Evaluate this and all similar operations to
determine if other workers are experiencing similar problems. Investigate and
establish specific solutions for these operations and obtain employee feedback
on how well the solutions are working.

# Glossary

**Abrasive blasting.** A process for cleaning surfaces by means of such materials as sand, alumina, or steel grit in a stream of high-pressure air.

**Accident.** An occurrence in a sequence of unplanned events that may produce unintended injury, death, or property damage, interrupting the activity in process.

**Accident causes.** Hazards and those factors (management oversight, system limitation/failure, etc.) that, individually or in combination, directly cause accidents.

**Acclimatization.** The process of becoming accustomed to new conditions (i.e., heat, cold).

**Accuracy (instrument).** Often used incorrectly as precision (see Precision). Accuracy refers to the agreement between the true value and a reading or observation obtained from an instrument or a technique.

**ACGIH.** American Conference of Governmental Industrial Hygienists. ACGIH develops and publishes recommended occupational exposure limits for hundreds of chemical substances and physical agents. (See TLV.)

**Acoustic, Acoustical.** Containing, producing, arising from, actuated by, related to, or associated with sound.

**Acoustic trauma.** Hearing loss caused by a sudden loud noise in one ear, or by a sudden blow to the head. In most cases, hearing loss is temporary, although there may be some permanent loss.

**Acrylic.** A family of synthetic resins made by polymerizing esters of acrylic acids.

**Action level.** Term used by OSHA and NIOSH to express the level of toxicant that requires medical surveillance, usually one half the PEL.

**Acuity.** The sense that pertains to the sensitivity of receptors used in hearing or vision.

**Acute effect.** An adverse effect on a human or animal body, with several symptoms developing rapidly and coming quickly to a crisis.

**Acute exposure.** Symptoms usually appear within 24 hours of an exposure; the exposure is usually brief and highly concentrated.

**Acute toxicity.** The adverse (acute) effects resulting from a single dose of or exposure to a substance.

**Adhesion.** The ability of one substance to stick to another. There are two types of adhesion: mechanical, which depends on the penetration of the surface; and molecular or polar adhesion, in which adhesion to a smooth surface is obtained because of polar groups such as carboxyl groups.

**Administrative controls.** Methods of controlling employee exposures by job rotation, work assignment, or time periods away from the hazard.

**Adsorption.** The condensation of gases, liquids, or dissolved substances on the surfaces of solids.

**Aerosol.** Liquid droplets or solid particles dispersed in air; are of fine enough particle size (0.01 to 100 micrometers) to remain so dispersed for a period of time.

**Agency or agent.** The principal object, such as a tool, machine, or equipment, involved in an accident; it is usually the object inflicting injury or property damage.

**Air.** The mixture of gases that surrounds the earth; its major components are as follows: 78.08% nitrogen, 20.95% oxygen, 0.03% carbon dioxide, and 0.93% argon. Water vapor (humidity) varies. (See Standard air.)

**Air cleaner.** A device designed to remove atmospheric airborne impurities, such as dusts, gases, vapors, fumes, and smoke.

**Air conditioning.** The process of treating air so as to control simultaneously its temperature, humidity, cleanliness, and distribution to meet requirements of the conditioned space.

**Air filter.** An air-cleaning device to remove light particulate matter from normal atmospheric air.

**Air hammer.** A percussion-type pneumatic tool, fitted with a handle at one end of the shank and a tool chuck at the other, into which a variety of tools may be inserted.

**Air-line respirator.** A respirator that is connected to a compressed breathing air source by a hose.

**Air monitoring.** The sampling for and measuring of pollutants in the atmosphere.

**Air-purifying respirator.** Respirator that uses filters or sorbents to remove harmful substances from the air.

**Air-regulating valve.** An adjustable valve used to regulate airflow to the facepiece, helmet, or hood of an air-line respirator.

**Air-supplied respirator.** Respirator that provides a supply of breathable air from a clean source outside of the contaminated work area.

**Algorithm.** A precisely stated procedure or set of instructions that can be applied stepwise to solve a problem.

**Aliphatic.** (Derived from the Greek word for oil.) Pertaining to an open-chain carbon compound. Usually applied to petroleum products derived from a paraffin base and having a straight or branched chain and saturated or unsaturated molecular structure. Substances such as methane and ethane are typical aliphatic hydrocarbons.

**Alkali.** Any chemical substance that forms soluble soaps with fatty acids. Alkalis are also referred to as bases. They may cause severe burns to the skin. Alkalis turn litmus paper blue and have pH values from 8 to 14.

**Allergic reaction.** An abnormal physiological response to chemical or physical stimuli by a sensitive person.

**Allergy.** An abnormal response of a hypersensitive person to chemical and physical stimuli. Allergic manifestations of major importance occur in about 10% of the population.

**Alloy.** A mixture of metals.

**Alpha particle (alpha-ray, alpha-radiation).** A small, electrically charged particle of very high velocity thrown off by many radioactive materials, including uranium and radium. It is made up of two neutrons and two protons. Its electric charge is positive.

**Ambient noise.** The all-encompassing noise associated with a given environment, being usually a composite of sounds from many sources.

**Ampere.** The standard unit for measuring the strength of an electrical current.

**Anemometer.** A device to measure air velocity.

**Anesthetic.** A chemical that causes a total or partial loss of sensation. Overexposure to anesthetics can cause impaired judgment, dizziness, drowsiness, headache, unconsciousness, and even death. Examples of such chemicals include alcohol, paint remover, and degreasers.

**Anneal.** To treat by heat with subsequent cooling for drawing the temper of metals, that is, to soften and render them less brittle. (See Temper.)

**ANSI.** The American National Standards Institute is a voluntary membership organization (run with private funding) that develops consensus standards nationally for a wide variety of devices and procedures.

**Antidote.** A remedy to relieve, prevent, or counteract the effects of a poison.

**Antioxidant.** A compound that retards deterioration by oxidation. Antioxidants for human food and animal feeds, sometimes referred to as freshness preservers, retard rancidity of fats and lessen loss of fat-soluble vitamins (A, D, E, K). Antioxidants also are added to rubber, motor lubricants, and other materials to inhibit deterioration.

**Appearance.** A description of a substance at normal room temperature and normal atmospheric conditions. Appearance includes color, size, and consistency of a material.

**Approved.** Tested (and listed) as satisfactory meeting predetermined requirements by an authority having jurisdiction.

**Arc welding.** One form of electrical welding using either uncoated or coated rods.

**Arc-welding electrode.** A component of the welding circuit through which current is conducted between the electrode holder and the arc.

**Article.** A manufactured item formed to a specific shape or design which has an end-use function dependent in whole, or in part, upon its shape or use during end use and which does not release or otherwise result in exposure to a hazardous chemical under normal conditions of use.

**Asbestos.** A hydrated magnesium silicate in fibrous form.

**Asphyxia.** Suffocation from lack of oxygen. Chemical asphyxia is produced by a substance such as carbon monoxide that combines with hemoglobin to reduce the blood's capacity to transport oxygen. Simple asphyxia is the result of exposure to a substance such as methane, which displaces oxygen.

**Asphyxiant.** A vapor or gas that can cause unconsciousness or death by suffocation (lack of oxygen).

**ASTM.** American Society for Testing and Materials is a voluntary membership organization with members from a broad spectrum of individuals, agencies, and industries concerned with materials. As the world's largest source of voluntary consensus standards for materials, products, systems, and services, ASTM is a resource for sampling and testing methods, health and safety aspects of materials, safe performance guidelines, and effects of physical and biological agents and chemicals.

**Atmosphere-supply respirator.** A respirator that provides breathing air from a source independent of the surrounding atmosphere. There are two types: air-line and self-contained breathing apparatus.

**Atmospheric pressure.** The pressure exerted in all directions by the atmosphere. At sea level, mean atmospheric pressure is 29.92 inches Hg, 14.7 psi, or 407 inches w.g.

**Atmospheric tank.** A storage tank that has been designed to operate at pressures from atmospheric through 0.5 psig (3.5 kPa).

**Atomic energy.** Energy released in nuclear reactions. Of particular interest is the energy released when a neutron splits an atom's nucleus into smaller pieces (fission) or when two nuclei are joined together under millions of degrees of heat (fusion). "Atomic energy" is really a popular misnomer. It is more accurately called "nuclear energy."

**Atomic power.** See Atomic energy.

**Atomic waste.** The radioactive ash produced by the splitting of uranium (nuclear) fuel, as in a nuclear reactor. It may include products made radioactive in such a device.

**Audible range.** The frequency range over which normal ears hear approximately 20 Hz through 20,000 Hz. Above the range of 20,000 Hz, the term "ultrasonic" is used. Below 20 Hz, the term "subsonic" is used.

**Audiogram.** A record of loss of hearing level measured at several different frequencies—usually 500 to 6,000 Hz. The audiogram may be presented graphically or numerically. Hearing level is shown as a function of frequency.

**Audiometer.** A signal generator or instrument for measuring objectively the sensitivity of hearing in decibels.

**Autoignition temperature.** The temperature to which a closed, or nearly closed, container must be heated in order that a flammable liquid, when introduced into the container, will ignite spontaneously or burn.

**Babbitt.** An alloy of tin, antimony, copper, and lead used as a bearing metal.

**Background noise.** Noise coming from sources other than the particular noise source being monitored. (See Background radiation.)

**Background radiation.** The radiation coming from sources other than the radioactive material to be measured. This "background" is primarily due to cosmic rays, which constantly bombard the earth from space.

**Bacteria.** Microscopic organisms living in soil, water, organic matter, or the bodies of plants and animals characterized by lack of a distinct nucleus and lack of ability to photosynthesize. Singular: Bacterium.

**Bag house.** A term commonly used for the housing containing bag filters for recovery of fumes of arsenic, lead, sulfur, etc., from flues of smelters.

**Balancing by dampers.** Method of designing local exhaust system ducts using adjustable dampers to distribute airflow after installation.

**Balancing by static pressure.** Method of designing local exhaust system ducts by selecting the duct diameters that generate the static pressure to distribute airflow without dampers.

**Ball mill.** A grinding device using balls usually made of steel or stone in a rotating container.

**Base.** A compound that reacts with an acid to form a salt. It is another term for alkali. It turns litmus paper blue.

**Bauxite.** Impure mixture of aluminum oxides and hydroxides; the principal source of aluminum.

**Beat elbow.** Bursitis of the elbow; occurs from use of heavy vibrating tools.

**Beat knee.** Bursitis of the knee joints due to friction or vibration; common in mining.

**Beehive kiln.** A kiln shaped like a large beehive; used usually for calcining ceramics.

**Benign.** Not malignant. A benign tumor is one that does not metastasize or invade tissue. Benign tumors may still be lethal, due to pressure on vital organs.

**Beta particle (beta-radiation).** A small, electrically charged particle thrown off by many radioactive materials. It is identical with the electron. Beta particles emerge from radioactive material at high speeds.

**Billet.** A piece of semifinished iron or steel, nearly square in section, made by rolling and cutting an ingot.

**Binder.** The nonvolatile portion of a coating

vehicle, which is the film-forming ingredient used to bind the paint pigment particles together.

**Biodegradable.** Capable of being broken down into innocuous products by the action of living things.

**Biohazard.** A combination of the words "biological hazard." Organisms or products of organisms that present a risk to humans.

**Biohazard area.** Any area (a complete operating complex, a single facility, a room within a facility, etc.) in which work has been or is being performed with biohazardous agents or materials.

**Biohazard control.** Any set of equipment and procedures utilized to prevent or minimize the exposure of humans and their environment to biohazardous agents or materials.

**Biomechanics.** The study of the human body as a system operating under two sets of law: the laws of Newtonian mechanics and the biological laws of life.

**Black light.** Ultraviolet (UV) light radiation between 3,000 and 4,000 angstroms (0.3 and 0.4 micrometers).

**Black liquor.** A liquor composed of alkaline and organic matter resulting from digestion of wood pulp and cooking acid during the manufacture of paper.

**Bleaching bath.** Chemical solution used to bleach colors from a garment in preparation for dyeing it; a solution of chlorine or sodium hypochlorite is commonly used.

**Blind spot.** Normal defect in visual field due to position at which optic nerve enters the eye.

**Boiling point.** The temperature at which a liquid changes to a vapor state, expressed in degrees.

**Bonding.** The interconnecting of two objects by means of a clamp and bare wire. Its purpose is to equalize the electrical potential between objects. (See Grounding.)

**Brake horsepower.** The horsepower required to drive a unit; it includes the energy losses in the unit and can be determined only by actual test. It does not include driver losses between motor and unit.

**Brass.** An alloy of copper and zinc; may also contain a small portion of lead.

**Brattice.** A partition constructed in underground passageways to control ventilation in mines.

**Braze.** To solder with any alloy that is relatively infusible.

**Brazing furnace.** Used for heating metals to be joined by brazing. Requires a high temperature.

**Breathing tube.** A tube through which air or oxygen flows to the facepiece, helmet, or hood.

**Breathing zone.** Imaginary globe of two-foot radius surrounding the head.

**Broach.** A cutting tool for cutting nonround holes.

**Bubble tube.** A device used to calibrate air-sampling pumps.

**Buffer.** Any substance in a fluid that tends to resist the change in pH when acid or alkali is added.

**Bulk density.** Mass of powdered or granulated solid material per unit of volume.

**Bulk plant.** That portion of a property where flammable or combustible liquids are received by tank vessel, pipelines, tank car, or tank vehicle and are sorted or blended in bulk for the purpose of distributing such liquids by tank vessel, pipeline, tank car, tank vehicle, or container.

**Burns.** Result from the application of too much heat to the skin. First-degree burns show redness of the unbroken skin; second-degree, skin blisters and some breaking of the skin; third-degree, skin blisters and destruction of the skin and underlying tissues, which can include charring and blackening.

**Calcination.** The heat treatment of solid material to bring about thermal decomposition, to lose moisture or other volatile material, or to be oxidized or reduced.

**Calender.** An assembly of rollers for producing a desired finish on paper, rubber, artificial leather, plastics, or other sheet material.

**Calking.** The process or material used to fill seams of boats, cracks in tile, etc.

**Capture velocity.** Air velocity at any point in front of the hood necessary to overcome opposing air currents and to capture contaminated air by causing it to flow into the exhaust hood.

**Carbon black.** Essentially a pure carbon, best known as common soot. Commercial carbon black is produced by making soot under controlled conditions. It is sometimes called channel black, furnace black, acetylene black, or thermal black.

**Carbon monoxide.** A colorless, odorless, toxic gas produced by any process that involves the incomplete combustion of carbon-containing substances.

**Carcinogen.** A substance or agent that can cause a growth of abnormal tissue or tumors in humans or animals.

**Carcinogenic.** Cancer-producing.

**Carding.** The process of combing or untangling wool, cotton, etc.

**Carpal tunnel.** A passage in the wrist through which the median nerve and many tendons pass to the hand from the forearm.

**Carpal tunnel syndrome.** A common affliction caused by compression of the median nerve in the carpal tunnel. Often associated with tingling, pain, or numbness in the thumb and first three fingers.

**CAS number.** Identifies a particular chemical by the Chemical Abstract Service, a service of the American Chemical Society that indexes and compiles abstracts of worldwide chemical literature called "Chemical Abstracts."

**Case-hardening.** A process of surface-hardening metals by raising the carbon or nitrogen content of the outer surface.

**Cask (or coffin).** A thick-walled container (usually lead) used for transporting radioactive materials.

**Casting.** The pouring of a molten material into a mold and permitting it to solidify to the desired shape.

**Catalyst.** A substance that changes the speed of a chemical reaction but undergoes no permanent change itself. In respirator use, a substance that converts a toxic gas (or vapor) into a less toxic gas (or vapor). Usually catalysts greatly increase the reaction rate, as in conversion of petroleum to gasoline by cracking. In paint manufacture, catalysts, which hasten the film-forming, generally become part of the final product. In most uses, however, they do not, and can often be used over again.

**Cathode.** The negative electrode.

**Catwalk.** A narrow suspended footway usually for inspection or maintenance purposes.

**Caustic.** Something that strongly irritates, burns, corrodes, or destroys living tissue. (See Alkali.)

**cc.** Cubic centimeter; a volume measurement in the metric system, equal in capacity to one milliliter (ml)—approximately 20 drops. There are 16.4 cc in one cubic inch.

**Ceiling limit (C).** A maximum airborne concentration of a toxic substance in the work environment. ACGIH terminology. (See TLV.)

**Cellulose($C_6H_{10}OH_5$).** A carbohydrate that makes up the structural material of vegetable tissues and fibers.

**Celsius.** The Celsius temperature scale is a designation of the scale previously known as the centigrade scale.

**Cement, Portland.** Portland cement commonly consists of hydraulic calcium silicates to which the addition of certain materials in limited amounts is permitted. Ordinarily, the mixture consists of calcareous materials such as limestone, chalk, shells, marl, clay, shale, blast furnace slag, etc. In some specifications, iron ore and limestone are added. The mixture is fused by calcining at temperatures usually up to 1,500 C.

**Central nervous system.** The brain and spinal cord. These organs supervise and coordinate the activity of the entire nervous system. Sensory impulses are transmitted into the central nervous system, and motor impulses are transmitted out.

**Centrifuge.** An apparatus that uses centrifugal force to separate or remove particulate matter suspended in a liquid.

**Ceramic.** A term applied to pottery, brick, and the tile products molded from clay and subsequently calcined.

**CFR.** Code of Federal Regulations. A collection of the regulations that have been promulgated under U.S. law.

**Chemical.** Any element, chemical compound, or mixture of elements and/or compounds.

**Chemical burns.** Generally similar to those caused by heat. After emergency first aid, their treatment is the same as that for thermal burns.

**Chemical cartridge.** A chemical cartridge is used with a respirator for removal of low concentrations of specific vapors and gases.

**Chemical cartridge respirator.** A respirator that uses various chemical substances to purify inhaled air of certain gases and vapors.

**Chemical engineering.** That branch of engineering concerned with the development and application of manufacturing processes in which chemical or certain physical changes of materials are involved.

**Chemical manufacturer.** An employer with a workplace where chemicals are produced for use or distribution.

**Chemical name.** The scientific designation of a chemical in accordance with the nomenclature system developed by the International Union of Pure and Applied Chemistry (IUPAC) or the Chemical Abstracts Service (CAS), or a name that will clearly identify the chemical for the purpose of conducting a hazard evaluation.

**Chemical reaction.** A change in the arrangement of atoms or molecules to yield substances of different composition and properties. Common types of reaction are combination, decomposition, double decomposition, replacement, and double replacement.

**CHEMTREC.** Chemical Transportation Emergency Center.

**Chronic.** Persistent, prolonged, repeated.

**Chronic exposure.** Symptoms are usually delayed and cumulative and result from repeated exposure to low-level concentrations of hazardous chemicals.

**Chronic toxicity.** Adverse (chronic) effects resulting from repeated doses of or exposures to a substance over a relatively prolonged period of time.

**Circuit.** A complete path over which electrical current may flow.

**Circuit breaker.** A device that automatically interrupts the flow of an electrical current when the current becomes excessive.

**Clays.** A great variety of aluminum-silicate-bearing rocks, plastic when wet and hard when dry. Used in pottery, stoneware, tile, bricks, cements, fillers, and abrasives. Kaolin is one type of clay. Some clay deposits may include appreciable quartz. Commercial grades of clays may contain up to 20% quartz.

**Clean Air Act.** Federal law enacted to regulate/reduce air pollution. Administered by the EPA.

**Clean Water Act.** Federal law enacted to regulate/reduce water pollution. Administered by the EPA.

**Coagulation.** Formation of a clot or gelatinous mass.

**Coated welding rods.** The coatings of welding rods vary. For the welding of iron and most steel, the rods contain manganese, titanium, and a silicate.

*Code of Federal Regulations (CFR).* The rules promulgated under U.S. law and published in the Federal Register and actually in force at the end of a calendar year are incorporated in this code.

**Coffin.** A thick-walled container (usually lead) used for transporting radioactive materials; also called cask.

**Cohesion.** Molecular forces of attraction between particles of like composition.

**Colloid mill.** A machine that grinds materials into very fine state of suspension, often simultaneously placing this in suspension in a liquid.

**Combustible.** Able to catch fire and burn.

**Combustible liquids.** Combustible liquids are those having a flash point at or above 37.8 C (100 F).

**Comfort ventilation.** Airflow intended to maintain comfort of room occupants (heat, humidity, and odor).

**Comfort zone.** The average range of effective temperatures over which the majority (50% or more) of adults feel comfortable.

**Common name.** Any designation or identification such as code name, code number, trade name, brand name, or generic name used to identify a chemical other than by its chemical name.

**Communicable.** Refers to a disease whose causative agent is readily transferred from one person to another.

**Compaction.** The consolidation of solid particles between rolls, or by tamp, piston, screw, or other means of applying mechanical pressure.

**Compound.** A substance composed of two or more elements joined according to the laws of chemical combination. Each compound has its own characteristic properties different from those of its constituent elements.

**Compressed gas.** (1) Any gas or mixture of gases having, in a container, an absolute pressure exceeding 40 psi at 70 F (21.2 C); or (2) a gas or mixture of gases having, in a container, an absolute pressure exceeding 104 psi at 130 F (54.4 C) regardless of the pressure at 70 F (21.2 C); or (3) a liquid having a vapor pressure exceeding 40 psi at 100 F (37.8 C) as determined by ASTM D-323-72.

**Concentration.** The amount of a given substance in a stated unit of measure. Common methods of stating concentration are percent by weight or by volume, weight per unit volume, normality, etc.

**Condensate.** The liquid resulting from the process of condensation.

**Condensation.** Act or process of reducing from one form to another denser form, such as steam to water.

**Conditions to avoid.** Conditions encountered during handling or storage that could cause a substance to become unstable.

**Conductive hearing loss.** Type of hearing loss; not caused by noise exposure, but due to any disorder in the middle or external ear that prevents sound from reaching the inner ear.

**Confined space.** Any area that has limited openings for entry and exit that would make escape difficult in an emergency, has a lack of ventilation, contains known and potential hazards, and is not intended nor designated for continuous human occupancy. (See Permit Required Confined Space standard 29 CFR 1910.146(b)(23))

**Contact dermatitis.** Dermatitis caused by contact with a substance—gaseous, liquid, or solid. May be due to primary irritation or an allergy.

**Container.** Any bag, barrel, drum, bottle, box, can, cylinder, reaction vessel, storage tank, or the like that contains a hazardous chemical. For the purposes of the HCS, engines, fuel tanks, and other operating systems in a vehicle are not considered to be containers.

**Controlled area.** A specified area in which exposure of personnel to radiation or radioactive material is controlled and which is under the supervision of a person who has knowledge of the appropriate radiation protection practices, including pertinent regulations, and who has responsibility for applying them.

**Convection.** The motions in fluids resulting from differences in density and the action of gravity.

**Coolants.** Coolants are transfer agents used in a flow system to convey heat from its source.

**Corrective lens.** A lens ground to the wearer's individual prescription.

**Corrosion.** Physical change, usually deterioration or destruction, brought about through chemical or electrochemical action as contrasted with erosion caused by mechanical action.

**Corrosive (COR).** As defined by DOT, a corrosive material is a liquid or solid that causes visible destruction or irreversible alterations in human skin tissue at the site of contact or—in the case of leakage from its packaging—a liquid that has a severe corrosion rate on steel. Two common examples are caustic soda and sulfuric acid.

**Cosmic rays.** High-energy rays of great penetrating power that bombard the earth from space.

**Cottrell precipitator.** A device for dust collection using high-voltage electrodes.

**Counter.** A device for counting. (See Geiger counter and Scintillation counter.)

**Covered electrode.** A composite filler metal electrode consisting of a core of bare electrode or metal-cored electrode to which a covering (sufficient to provide a slag layer on the weld metal) has been applied; the covering may contain materials providing such functions as shielding from the atmosphere, deoxidation, and arc stabilization and can serve as a source of metallic additions to the weld.

**CPR.** Cardiopulmonary resuscitation.

**CPSC.** Consumer Products Safety Commission is a federal agency with responsibility for regulating hazardous materials when they appear in consumer goods. For CPSC purposes, hazards are defined in the Hazardous Substances Act and the Poison Prevention Packaging Act of 1970.

**Cristobalite.** A crystalline form of free silica, extremely hard and inert chemically; very resistant to heat. Quartz in refractory bricks and amorphous silica in diatomaceous earth are altered to cristobalite when exposed to high temperatures (calcined).

**Critical pressure.** The pressure under which a substance may exist as a gas in equilibrium with a liquid at the critical temperature.

**Critical temperature.** The temperature above which a gas cannot be liquefied by pressure alone.

**Crucible.** A heat-resistant, barrel-shaped pot used to hold metal during melting in a furnace.

**Crude petroleum.** Hydrocarbon mixtures that have a flash point below 65.6 C (150 F) and that have not been processed in a refinery.

**Cry-, cryo-** (prefix). Very cold.

**Cryogenics.** The field of science dealing with the behavior of matter at very low temperatures.

**Cubic centimeter (cc).** A volumetric measurement that is also equal to one milliliter (mL.)

**Cubic meter (m³).** A measure of volume in the metric system.

**Cutting fluids (oils).** The cutting fluids used in industry today are usually oil or oil-water emulsions used to cool and lubricate a cutting tool. Cutting oils are usually light or heavy petroleum fractions.

**CW laser.** Continuous wave laser. (See Laser.)

**Cyanide (as CN).** Cyanides inhibit tissue oxidation upon inhalation or ingestion and cause death.

**Cyanosis.** Blue appearance of the skin, especially on the face and extremities, indicating a lack of sufficient oxygen in the arterial blood.

**Cyclone separator.** A dust-collecting device that has the ability to separate particles by size. Typically used to collect respirable dust samples.

**Damage risk criterion.** The suggested base line of noise tolerance which, if not exceeded, should result in no hearing loss due to noise. A damage risk criterion may include in its statement a specification of such factors as time of exposure, noise level and frequency, amount of hearing loss that is considered significant, percentage of the population to be protected, and method of measuring the noise.

**Damp.** A harmful gas or mixture of gases occurring in coal mining.

**Dampers.** Adjustable sources of airflow resistance used to distribute airflow in a ventilation system.

**Dangerous to life or health, immediately (IDLH).** Used to describe very hazardous atmospheres where employee exposure can cause serious injury or death within a short time or serious delayed effects.

**dBA.** Sound level in decibels read on the A-scale of a sound level meter. The A scale discriminates against very low frequencies (as does the human ear) and is, therefore, better for measuring general sound levels. (See also Decibel.)

**Decibel (dB).** A unit used to express sound power level (Lw). Sound power is the total acoustic output of a sound source in watts (W). By definition, sound power level, in decibels, is: Lw = 10 log W divided by W o, where W is the sound power of the source and W o is the reference sound power.

**Decomposition.** Breakdown of a material or substance (by heat, chemical reaction, electrolysis, decay, or other process) into parts or elements or simpler compounds.

**Decontaminate.** To make safe by eliminating poisonous or otherwise harmful substances, such as noxious chemicals or radioactive material.

**Density.** The mass (weight) per unit volume of a substance. For example, lead is much more dense than aluminum.

**Depressant.** A substance that reduces a functional activity or an instinctive desire of the body, such as appetite.

**Dermal toxicity.** Adverse effects resulting from skin exposure to a substance. Ordinarily used to denote effects in experimental animals.

**Dermatitis.** Inflammation of the skin.

**Dermatosis.** A broader term than dermatitis; it includes any cutaneous abnormality, encompassing folliculitis, acne, pigmentary changes, and nodules and tumors.

**Designated representative.** Any individual or organization to whom an employee gives written authorization to exercise said employee's rights under the standard. A recognized or certified collective bargaining agent is treated automatically as a designated representative without regard to written employee authorization.

**Diaphragm.** (1) The musculomembranous partition separating the abdominal and thoracic cavities; (2) any separating membrane or structure; (3) a disk with one or more openings in it, or with an adjustable opening, mounted in relation to a lens, by which part of a light may be excluded from an area.

**Diatomaceous earth.** A soft, gritty, amorphous silica composed of minute siliceous skeletons of small aquatic plants. Used in filtration and decolorization of liquids, insulation, filler in dynamite, wax, textiles, plastics, paint, and rubber. Calcined and flux-calcined diatomaceous earth contains appreciable amounts of cristobalite, and dust levels should be controlled the same as for cristobalite.

**Die.** A hard metal or plastic form used to shape material to a particular contour or section.

**Differential pressure.** The difference in static pressure between two locations.

**Diffusion rate.** A measure of the tendency of one gas or vapor to disperse into or mix with another gas or vapor.

**Dike.** A barrier constructed to control or confine substances and prevent their movement.

**Diluent.** A liquid that is blended with a mixture to reduce concentration of the active agents.

**Dilution.** The process of increasing the proportion of solvent or diluent (liquid) to solute or particulate matter (solid).

**Dilution ventilation.** Airflow designed to dilute contaminants to acceptable levels. Also referred to as general ventilation or exhaust. (See General exhaust.)

**Direct-reading instrumentation.** Those instruments that give an immediate indication of the concentration of matter.

**Disinfectant.** An agent that frees from infection by killing the vegetative cells of microorganisms.

**Dispersion.** The general term describing systems consisting of particulate matter suspended in air or other fluid; also, the mixing and dilution of contaminant in the ambient environment.

**Distal.** Away from the central axis of the body.

**Distillery.** A plant, or that portion of a plant, where flammable or combustible liquids produced by fermentation are concentrated, and where the concentrated products may also be mixed, stored, or packaged.

**Distributor.** A business, other than a chemical manufacturer or importer, which supplies hazardous chemicals to other distributors or employees.

**DOL.** U.S. Department of Labor; includes the Occupational Safety and Health Administration (OSHA) and Mine Safety and Health Administration (MSHA).

**Dose.** A term used (1) to express the amount of a chemical or of ionizing radiation energy absorbed in a unit volume or an organ or individual. Dose rate is the dose delivered per unit of time. (See also Roentgen, Rad, Rem.) (2) Used to express amount of exposure to a chemical substance.

**Dose equivalent, maximum permissible (MPD).** The largest equivalent received within a specified period that is permitted by a regulatory agency or other authoritative group on the assumption that receipt of such dose equivalent creates no appreciable somatic or genetic injury. Different levels of MPD may be set for different groups within a population. (By popular usage, "dose, maximum permissible," is an accepted synonym.)

**Dosimeter (dose meter).** An instrument used to determine the full-shift exposure a person has received to a physical hazard.

**DOT hazard class.** DOT requires that hazardous materials offered for shipment be labeled with the proper DOT hazard class. These classes include corrosive, flammable liquid, organic peroxide, ORM-E, poison B, etc. The DOT hazard class may not adequately describe all the hazard properties of the material.

**Drier.** Any catalytic material that, when added to a drying oil, accelerates drying or hardening of the film.

**Drop forge.** To forge between dies by a drop hammer or drop press.

**Droplet.** A liquid particle suspended in a gas. The liquid particle is generally of such size and density that it settles rapidly and remains airborne for an appreciable length of time only in a turbulent atmosphere.

**Dross.** The scum that forms on the surface of molten metals, largely oxides and impurities.

**Dry chemical.** A powdered fire-extinguishing agent usually composed of sodium bicarbonate, monoammonium phosphate, potassium bicarbonate, etc.

**Duct.** A conduit used for conveying air at low pressures.

**Dust collector.** An air-cleaning device to remove heavy particulate loadings from exhaust systems before discharge to outdoors; usual range is loadings of 0.003 gr/cu ft (0.007 mg/m$^3$) and higher.

**Dusts.** Solid particles generated by handling, crushing, grinding, rapid impact, detonation, and decrepitation of organic or inorganic materials, such as rock, ore, metal, coal, wood, and grain. Dusts do not tend to flocculate except under electrostatic forces; they do not diffuse in air but settle under the influence of gravity.

**Dynometer.** Apparatus for measuring force or work output external to a subject. Often used to compare external output with associated physiological phenomena to assess physiological work efficiency.

**EAP.** Employee Assistance Program.

**Ear.** The entire hearing apparatus, consisting of three parts: external ear, the middle ear or tympanic cavity, and the inner ear or labyrinth. Sometimes the pinna is called the ear.

**Effective temperature.** An arbitrary index that combines into a single value the effect of temperature, humidity, and air movement on the sensation of warmth and cold on the human body.

**Effective temperature index.** An empirically determined index of the degree of warmth perceived on exposure to different combinations of temperature, humidity, and air movement. The determination of effective temperature requires simultaneous determinations of dry bulb and wet bulb temperatures.

**Effluent.** Generally something that flows out or forth, like an outflow of a sewer, storage tank, canal, or other channel.

**Electrical current.** The flow of electricity measured in amperes or milliamperes.

**Electrical precipitator.** A device that removes particles from an airstream by charging the particles and collecting the charged particles on a suitable surface.

**Electrolysis.** The process of conduction of an electric current by means of a chemical solution.

**Electromagnetic radiation.** The propagation of varying electric and magnetic fields through space at the speed of light, exhibiting the characteristics of wave motion.

**Electroplate.** To cover with a metal coating (plate) by means of electrolysis.

**Element.** Solid, liquid, or gaseous matter that cannot be further decomposed into simpler substances by chemical means.

**Elutriator.** A device used to separate particles according to mass and aerodynamic size by maintaining a laminar flow system at a rate that permits the particles of greatest mass to settle rapidly while the smaller particles are kept airborne by the resistance force of the flowing air for longer times and distances. The various times and distances of deposit may be used to determine representative fractions of particle mass and size.

**Emery.** Aluminum oxide, natural and synthetic abrasive.

**Emission factor.** Statistical average of the amount of a specific pollutant emitted from each type of polluting source in relation to a unit quantity of material handled, processed, or burned.

**Emission inventory.** A list of primary air pollutants emitted into a given community's atmosphere in amounts per day, by type of source.

**Emission standards.** The maximum amount of pollutant permitted to be discharged from a single polluting source.

**Employee.** A worker who may be exposed to hazardous chemicals under normal operating conditions or in foreseeable emergencies. However, not included in this definition are workers such as office workers or bank tellers who encounter hazardous chemicals only in nonroutine, isolated cases.

**Employer.** A person engaged in a business where chemicals are used, distributed, or produced for use or distribution, including a contractor or subcontractor.

**Emulsifier or emulsifying agent.** A chemical that holds one insoluble liquid in suspension in another. Casein, for example, is a natural emulsifier in milk, keeping butterfat droplets dispersed.

**Emulsion.** A suspension, each in the other, of two or more unlike liquids that usually will not dissolve in each other.

**Enamel.** A paintlike oily substance that produces a glossy finish to a surface to which it is applied. Often contains various synthetic resins. It is lead free. In contrast is the ceramic enamel, that is, porcelain enamel, which contains lead.

**Engineering controls.** Methods of controlling employee exposures by modifying the source or reducing the quantity of hazards.

**Entrance loss.** The loss in static pressure of a fluid that flows from an area into and through a hood or duct opening. The loss in static pressure is due to friction and turbulence resulting from the increased gas velocity and configuration of the entrance area.

**Entry loss.** Loss in pressure caused by air flowing into a duct or hood.

**Environmental toxicity.** Information obtained as a result of conducting environmental testing designed to study the effects of various substances on aquatic and plant life.

**Enzymes.** Delicate chemical substances, mostly proteins, which enter into and bring about chemical reactions in living organisms.

**EPA.** U. S. Environmental Protection Agency.

**EPA number.** The number assigned to chemicals regulated by the Environmental Protection Agency.

**Ergonomics.** A multidisciplinary activity dealing with interactions between people and their total working environment plus stresses related to such environmental elements as atmosphere, light, and sound as well as all tools and equipment of the workplace.

**Esters.** Organic compounds that may be made by interaction between an alcohol and an acid, or by other means. Esters are nonionic compounds, including solvents and natural fats.

**Etch.** To cut or eat away material with acid or other corrosive substance.

**Evaporation.** The process by which a liquid is changed into the vapor state.

**Evaporation rate.** The ratio of time required to evaporate a measured volume of a liquid to the time required to evaporate the same volume of a reference liquid (ethyl ether) under ideal test conditions. The higher the ratio, the slower the evaporation rate. FAST evaporation rate greater than 3.0. Examples: Methyl Ethyl Ketone (MEK) = 3.8, Acetone = 5.6, Hexane = 8.3.MEDIUM evaporation rate 0.8 to 3.0. Examples: 190 proof (95%) Ethyl Alcohol = 1.4, VM&P Naphtha = 1.4, MIBK = 1.6.SLOW evaporation rate less than 0.8. Examples: Xylene = 0.6, Isobutyl Alcohol = 0.6, Normal Butyl Alcohol = 0.4, Water = 0.3, Mineral Spirits = 0.1.

**Exhalation valve.** A device that allows exhaled air to leave a respirator and prevents outside air from entering through the valve.

**Exhaust ventilation.** The removal of air (usually by mechanical means) from any space. The flow of air between two points is due to the occurrence of a pressure difference between the two points. This pressure difference will cause air to flow from the high pressure to the low pressure zone.

**Explosive limit.** See Flammable limit.

**Explosive.** A reaction that causes a sudden, almost instantaneous release of pressure, gas, and heat.

**Exposure.** Contact with a chemical, biological, or physical hazard.

**Extinguishing medium.** The fire-fighting substance used to control a material in the event of a fire. It is usually named by its generic name, such as fog, foam, water, etc.

**Extrusion.** The forcing of raw material through a die or a form in either a heated or cold state, and in a solid state or in partial solution.

**Eyepiece.** Gastight, transparent window(s) in a full facepiece through which the wearer may see.

**Eye protection.** Recommended safety glasses, chemical splash goggles, face shields, etc. to be utilized when handling a hazardous material.

**Face velocity.** Average air velocity into the exhaust system measured at the opening into the hood or booth.

**Facepiece.** That portion of a respirator that covers the wearer's nose and mouth in a half-mask facepiece, or nose, mouth, and eyes in a full facepiece.

**Facing.** In foundry work, the final touch-up work of the mold surface to come in contact with metal is called the facing operation, and the fine-powdered material used is called the facing.

**Fainting.** Technically called syncope, a temporary loss of consciousness as a result of a diminished supply of blood to the brain.

**Fallout.** Dust particles that contain radioactive fission products resulting from a nuclear explosion. The wind can carry fallout particles many miles.

**Fan static pressure.** The pressure added to a system by a fan. It equals the sum of pressure losses in the system minus the velocity pressure in the air at the fan inlet.

**Farmer's lung.** Fungus infection and ensuing hypersensitivity from grain dust.

**Fatal accident.** An accident that results in one or more deaths within one year.

**FDA.** The U.S. Food and Drug Administration; under the provisions of the federal Food, Drug and Cosmetic Act, the FDA establishes requirements for the labeling of foods and drugs to protect consumers from misbranded, unwholesome, ineffective, and hazardous products. FDA also regulates materials used in contact with food and the conditions under which such materials are approved.

***Federal Register.*** Publication of U.S. government documents officially promulgated under the law, documents whose validity depends on such publication. (See Code of Federal Regulations.)

**Fertilizer.** Plant food usually sold in mixed formula containing basic plant nutrients: compounds of nitrogen, potassium, phosphorus, sulfur, and sometimes other minerals.

**Fever.** A condition in which the body temperature is above its regular or normal level.

**FIFRA.** Federal Insecticide, Fungicide, and Rodenticide Act; regulations administered by EPA under this Act require that certain useful poisons, such as chemical pesticides sold to the public, contain labels that carry health hazard warnings to protect users.

**Film badge.** A piece of masked photographic film worn by nuclear workers. Because it is darkened by nuclear radiation, radiation exposure can be checked by inspection of the film.

**Filter.** (1) A device for separating components of a signal on the basis of its frequency. It allows components in one or more frequency bands to pass relatively unattenuated, and it attenuates greatly components in other frequency bands. (2) A fibrous medium used in respirators to remove solid or liquid particles from the airstream entering the respirator. (3) A sheet of material that is interposed between patient and the source of x rays to absorb a selective part of the x rays. (4) A fibrous or membrane medium used to collect air samples of dust, fume, or mist.

**Filter, HEPA.** High-efficiency particulate air filter that is at least 99.97 % efficient in removing thermally generated monodisperse dioctylphthalate smoke particles.

**Firebrick.** A special clay that is capable of resisting high temperatures without melting or crumbling.

**Fire damp.** In mining, the accumulation of an explosive gas, chiefly methane gas. Miners refer to all dangerous underground gases as "damps."

**Fire point.** The lowest temperature at which a material can evolve vapors to support continuous combustion.

**First aid.** Emergency measures to be taken before regular medical help can be obtained.

**Fission.** The splitting of an atomic nucleus into two parts accompanied by the release of a large amount of radioactivity and heat.

**Flame propagation.** See Propagation of flame.

**Flammable.** A chemical that falls into one of the following categories: Aerosol flammable; gas flammable (includes forming a flammable mixture with air at a 13 % concentration of the chemical, or less, and forming a flammable range, when mixed with air, that is wider than 12 % by volume, regardless of the lower limit); liquid flammable; and solid flammable.

**Flammable limits.** Flammables have a minimum concentration below which propagation of flame does not occur on contact with a source of ignition. This is known as the lower flammable explosive limit (LEL). There is also a maximum concentration of vapor or gas in air above which propagation of flame does not occur. This is known as the upper flammable explosive limit (UEL). These units are expressed in percent of gas or vapor in air by volume.

**Flammable liquid.** Any liquid having a flash point below 37.8 C (100 F).

**Flammable range.** The difference between the lower and upper flammable limits, expressed in terms of percentage of vapor or gas in air by volume; is also often referred to as the "explosive range."

**Flange.** In an exhaust system, a rim or edge added to a hood to reduce the quantity of air entering the hood from behind the hood.

**Flashback.** Occurs when flame from a torch burns back into the tip, the torch, or the hose.

**Flash blindness.** Temporary visual disturbance resulting from viewing an intense light source.

**Flash ignition.** See Flash point.

**Flash point.** The lowest temperature at which a liquid gives off enough vapor to form an ignitable mixture with air and produce a flame when a source of ignition is present.

**Flask.** In foundry work, the assembly of the cope and the drag constitutes the flask. It is the wooden or iron frame containing sand into which molten metal is poured. Some flasks may have three or four parts.

**Flocculation.** The process of forming a very fluffy mass of material held together by weak forces of adhesion.

**Flotation.** A method of ore concentration in which the mineral is caused to float due to chemical frothing agents while the impurities sink.

**Flow coefficient.** A correction factor used for figuring volume flow rate of a fluid through an orifice. This factor includes the effects of contraction and turbulence loss (covered by the coefficient of discharge), plus the compressibility effect, and the effect of an upstream velocity other than zero. Because the latter two effects are negligible in many instances, the flow coefficient is often equal to the coefficient of discharge.

**Flow meter.** An instrument for measuring the rate of flow of a fluid.

**Fluid.** A substance tending to flow or conform to the outline of its container. It may be liquid, vapor, gas, or solid (like raw rubber).

**Fluorescence.** Emission of light from a crystal after the absorption of energy.

**Fluorescent screen.** A screen coated with a fluorescent substance so that it emits light when irradiated with x rays.

**Flux.** Usually refers to a substance used to clean surfaces and promote fusion in soldering. However, fluxes of various chemical nature are used in the smelting of ores, in the ceramic industry, in assaying silver and gold ores, and in other endeavors. The most common fluxes are silica, various silicates, lime, sodium, and potassium carbonate, and litharge and red lead in the ceramic industry. (See Galvanizing.)

**Fly ash.** Finely divided particles of ash entrained in flue gases arising from the combustion of fuel.

**Foot-candle.** A unit of illumination. The illumination at a point on a surface that is one foot from, and perpendicular to, a uniform point source of one candle.

**Foot-pound.** A unit of work equal to the amount of energy needed to raise one pound one foot upward.

**Foot-pounds of torque.** A measurement of the physiological stress exerted upon any joint during the performance of a task. The product of the force exerted and the distance from the point of application to the point of stress. Physiologically, torque that does not produce motion but causes work stress, the severity of which depends on the duration and magnitude of the torque. In lifting an object or holding it elevated, torque is exerted and applied to the lumbar vertebrae.

**Force.** That which changes the state of rest or motion in matter.

**Foreseeable emergency.** Any potential occurrence such as, but not limited to, equipment failure, rupture of containers, or failure of control equipment that could result in an uncontrolled release of a hazardous chemical into the workplace.

**Frequency (in hertz or Hz).** Rate at which pressure oscillations are produced. One hertz is equivalent to one cycle per second.

**Friable.** Particles or compounds in a crumbled or pulverized state.

**Friction factor.** A factor used in calculating loss of pressure due to friction of a liquid, solid, or gas flowing through a pipe or duct.

**Friction loss.** The pressure loss due to friction.

**Fume.** Airborne particulate formed by the evaporation of solid materials, e.g., metal fume emitted during welding. Usually less than one micron in diameter.

**Fuse.** A wire or strip of easily melted metal, usually set in a plug, placed in an electrical circuit as a safeguard. If the current becomes too strong, the metal melts, thus breaking the circuit.

**Fusion.** The joining of atomic nuclei to form a heavier nucleus, accomplished under conditions of extreme heat (millions of degrees). If two nuclei of light atoms fuse, the fusion is accompanied by the release of a great deal of energy. The energy of the sun is believed to be derived from the fusion of hydrogen atoms to form helium. In welding, the melting together of filler metal and base metal (substrate), or of base metal only, which results in coalescence.

**Gage pressure.** Pressure measured with respect to atmospheric pressure.

**Galvanizing.** An old but still used method of providing a protective coating for metals by dipping them in a bath of molten zinc.

**Gamma-rays (gamma radiation).** The most penetrating of all radiation. Gamma-rays are very high-energy x rays.

**Gangue.** In mining or quarrying, useless chipped rock.

**Gas.** A state of matter in which the material has very low density and viscosity; can expand and contract greatly in response to changes in temperature and pressure; easily diffuses; neither a solid nor a liquid.

**Gas metal arc welding (GMAW).** An arc-welding process that produces coalescence of metals by heating them with an arc between a continuous filler metal (consumable) electrode and the work; shielding is obtained entirely from an externally supplied gas or gas mixture; some methods of this process are called MIG or $CO_2$ welding.

**Gas tungsten arc welding (GTAW).** An arc-welding process that produces coalescence of metals by heating them with an arc between a tungsten (nonconsumable) electrode and the work; shielding is obtained from a gas or gas mixture. Pressure may or may not be used and filler metal may or may not be used. (This process has sometimes been called TIG welding.)

**Gate.** A groove in a mold to act as a passage for molten metal.

**Geiger counter.** A gas-filled electrical device that counts the presence of atomic particles or rays by detecting the ions produced. (Sometimes called a "Geiger-Mueller" counter.)

**General exhaust.** A system for exhausting air containing contaminants from a general work area; usually accomplished via dilution.

**General ventilation.** System of ventilation consisting of either natural or mechanically induced fresh air movements to mix with the dilute contaminants in the workroom air.

**Generic name.** A nonproprietary name for a material.

**Genetic effects.** Mutations or other changes that are produced by irradiation of the germ plasma.

**Genetically significant dose (GSD).** The dose which, if received by every member of the population, would be expected to produce the same total genetic injury to the population as do the actual doses received by various individuals.

**Glove box.** A sealed enclosure in which all handling of items inside the box is carried out through long impervious gloves sealed to ports in the walls of the enclosure.

**Gob.** "Gob pile" is waste mineral material, such as from coal mines, sometimes containing sufficient coal that gob fires may arise from spontaneous combustion.

**Grab sample.** A sample that is taken within a very short time period.

**Gram (g).** A metric unit of weight; one ounce equals 28.4 grams.

**Gravity, specific.** The ratio of the mass of a unit

volume of a substance to the mass of the same volume of a standard substance at a standard temperature. Water at 4 C (39.2 F) is the standard substance usually referred to. For gases, dry air, at the same temperature and pressure as the gas, is often taken as the standard substance.

**Gravity, standard.** A gravitational force that will produce an acceleration equal to 9.8 m/sec$^2$ or 32.17 ft/sec$^2$. The actual force of gravity varies slightly with altitude and latitude. The standard was arbitrarily established as that at sea level and 45-degree latitude.

**Gray iron.** The same as cast iron and, in general, any iron containing high carbon.

**Ground-fault circuit interrupter (GFCI).** A device that measures the amount of current flowing to and from an electrical source. When a difference is sensed, indicating a leakage of current that could cause an injury, the device very quickly breaks the circuit.

**Grounding.** The procedure used to carry an electrical charge to ground through a conductive path. (See Bonding.)

**Grooving.** Designing a tool with grooves on the handle to accommodate the fingers of the user. A bad practice because of the great variation in the size of workers' hands. Grooving interferes with sensory feedback. Intense pain may be caused by the grooves to an arthritic hand.

**Half-life, radioactive.** For a single radioactive decay process, the time required for the activity to decrease to half its value by that process.

**Halogenated hydrocarbon.** A chemical material that has carbon plus one or more of the halogen elements: chlorine, fluorine, bromine, or iodine.

**Hammer mill.** A machine for reducing the size of stone or other bulk material by means of hammers usually placed on a rotating axle inside a steel cylinder.

**Hand protection.** Specific type of gloves or other hand protection required to prevent harmful exposure to hazardous materials.

**Hardness.** A relative term to describe the penetrating quality of radiation. The higher the energy of the radiation, the more penetrating (harder) is the radiation.

**Hazard.** An unsafe condition which, if left uncontrolled, may contribute to an incident.

**Hazardous chemical.** Any chemical that is a physical or health hazard.

**Hazard warning.** Any words, pictures, symbols, or combination thereof appearing on a label or other appropriate form of warning that conveys the hazard(s) of the chemical(s) in the container(s).

**Hazardous decomposition products.** Any hazardous materials that may be produced in dangerous amounts if the material reacts with other agents, burns, or is exposed to other processes, such as welding. Examples of hazardous decomposition products formed when certain materials are heated include carbon monoxide and carbon dioxide.

**Hazardous material.** Any substance or compound that has the capability of producing adverse effects on the health and safety of humans.

**Heading.** In mining, a horizontal passage or drift of a tunnel, also the end of a drift or gallery. In tanning, a layer of ground bark over the tanning liquor.

**Health hazard.** A chemical for which there is statistically significant evidence based on at least one study conducted in accordance with established scientific principles that acute or chronic health effects may occur in exposed employees.

**Hearing conservation.** The prevention or minimizing of noise-induced deafness through the use of hearing protection devices; also, the control of noise through engineering methods, annual audiometric tests, and employee training.

**Hearing level.** The deviation in decibels of an individual's hearing threshold from the zero reference of the audiometer.

**Heat stress.** Relative amount of thermal strain from the environment.

**Heat stress index.** Index that combines the environmental heat and metabolic heat of the body into an expression of stress in terms of requirement for evaporation of sweat.

**Heatstroke.** A serious disorder resulting from exposure to excess heat. It results from sweat suppression and increased storage of body heat. Symptoms include hot, dry skin, high temperature, mental confusion, convulsions, and coma. Heatstroke is fatal if not treated promptly.

**Heat treatment.** Any of several processes of metal modification such as annealing.

**Helmet.** A device that shields the eyes, face, neck, and other parts of the head.

**HEPA filter.** See Filter, HEPA.

**Hertz.** The frequency measured in cycles per second. 1 cps = 1 Hz.

**High-frequency loss.** Refers to a hearing deficit starting with 2,000 Hz and higher.

**Homogenizer.** A machine that forces liquids under high pressure through a perforated shield against a hard surface to blend or emulsify the mixture.

**Hood.** (1) Enclosure, part of a local exhaust system; (2) a device that completely covers the head, neck, and portions of the shoulders.

**Horsepower.** A unit of power, equivalent to 33,000 foot-pounds per minute (746 W). (See Brake horsepower.)

**Hot.** In addition to meaning "having a relatively high temperature," this is a colloquial term meaning "highly radioactive."

**Human-equipment interface.** Area of physical or perceptual contact between man and equipment. The design characteristics of the human-equipment interface determine the quality of information. Poorly designed interfaces may lead to excessive fatigue or localized trauma, e.g., calluses.

**Humidify.** To add water vapor to the atmosphere; to add water vapor or moisture to any material.

**Humidity.** (1) Absolute humidity is the weight of water vapor per unit volume, pounds per cubic foot or grams per cubic centimeter. (2) Relative humidity is the ratio of the actual partial vapor pressure of the water vapor in a space to the saturation pressure of pure water at the same temperature.

**Hydration.** The process of converting raw material into pulp by prolonged beating in water; to combine with water or the elements of water.

**Hydrocarbons.** Organic compounds composed solely of carbon and hydrogen.

**ICC.** Interstate Commerce Commission.

**Identity.** Any chemical or common name indicated on the MSDS for the chemical. The identity used shall permit cross-references to be made among the required list of hazardous chemicals, the label, and the MSDS.

**IDLH.** Immediately dangerous to life or health.

**Ignitable.** Capable of being set afire.

**Immediate use.** The hazardous chemical will be used by, or under the control of, only the person who transfers it from a labeled container and only within the work shift in which it was transferred.

**Immiscible.** Not miscible. Any liquid that will not mix with another liquid, in which case it forms two separate layers or exhibits cloudiness or turbidity.

**Impaction.** The forcible contact of particles of matter; a term often used synonymously with impingement, but generally reserved for the case where particles are contacting a dry surface.

**Impervious.** A material that does not allow another substance to pass through or penetrate it.

**Impingement.** As used in air sampling, impingement refers to a process for the collection of particulate matter in which particles containing gas are directed against a wetted glass plate and the particles are retained by the liquid.

**Importer.** The first business, with employees and within the Customs Territory of the United States, that receives hazardous chemicals produced in other countries for the purpose of supplying them to distributors or employees within the United States.

**Inches of mercury column.** A unit used in measuring pressure. One inch of mercury column equals a pressure of 1.66 kPa (0.491 lb per sq in.).

**Inches of water column.** A unit used in measuring pressure. One inch of water column equals a pressure of 0.25 kPa (0.036 lb per sq in.).

**Incidence rate (as defined by OSHA).** The number of injuries and/or illnesses or lost workdays per 100 full-time employees per year or 200,000 hours of exposure.

**Incident.** An injury, illness, near miss, and other situation that has the potential to cause harm.

**Incompatible.** Materials that could cause dangerous reactions from direct contact with one another.

**Incubation.** Holding cultures of microorganisms under conditions favorable to their growth.

**Induration.** Heat hardening that may involve little more than thermal dehydration.

**Industrial hygiene.** The science (or art) devoted to the recognition, evaluation, and control of those environmental factors or stresses (i.e., chemical, physical, biological, and ergonomic) that may cause sickness, impaired health, or significant discomfort to employees or residents of the community.

**Inert gas.** A gas that does not normally combine chemically with another substance.

**Inert gas, welding.** An electric welding operation utilizing an inert gas such as helium to flush away the air to prevent oxidation of the metal being welded.

**Infrared radiation.** Electromagnetic energy with wavelengths from 770 nm to 12,000 nm.

**Ingestion.** (1) The process of taking substances into the stomach, as food, drink, medicine, etc. (2) With regard to certain cells, the act of engulfing or taking up bacteria and other foreign matter.

**Ingot.** A block of iron or steel cast in a mold for ease in handling before processing.

**Inhalation.** The breathing in of a substance in the form of a gas, vapor, fume, mist, or dust.

**Inhalation valve.** A device that allows respirable air to enter the facepiece and prevents exhaled air from leaving the facepiece through the intake opening.

**Injury.** Damage or harm to the body as a result of violence, infection, or other source.

**Inorganic.** Term used to designate compounds that generally do not contain carbon. Source: matter, other than vegetable or animal. Examples: sulfuric acid and salt. Exceptions are carbon monoxide, carbon dioxide.

**Insoluble.** Incapable of being dissolved.

**Intermediate.** A chemical formed as a "middle-step" in a series of chemical reactions, especially in the manufacture of organic dyes and pigments. In many cases, it may be isolated and used to form a variety of desired products. In other cases, the intermediate may be unstable or used up at once.

**Intoxication.** Means either drunkenness or poisoning.

**Inverse square law.** The propagation of energy through space is inversely proportional to the square of the distance it must travel. An object 3 m (9.8 ft) away from an energy source receives $\frac{1}{9}$ as much energy as an object 1 m (3.3 ft) away.

**Ionizing radiation.** Refers to (1) electrically charged or neutral particles, or (2) electromagnetic radiation that will interact with gases, liquids, or solids to produce ions. There are five major types: alpha, beta, x, gamma, and neutrons.

**Irradiation.** The exposure of something to radiation.

**Irritant.** A substance that produces an irritating effect when it contacts skin, eyes, nose, or respiratory system.

**Isotope.** One of two or more atomic species of an element differing in atomic weight but having the same atomic number. Each contains the same number of protons but a different number of neutrons.

**Jigs and fixtures.** Often used interchangeably; precisely, a jig holds work in position and guides the tools acting on the work, while a fixture holds but does not guide.

**Kaolin.** A type of clay composed of mixed silicates and used for refractories, ceramics, tile, and stoneware. In some deposits, free silica may be present as an impurity.

**kg.** Kilogram; a metric unit of weight, about 2.2 U.S. pounds. (Also see mg.)

**Kilogram (kg).** A unit of weight in the metric system equal to 2.2 lb.

**Kinetic energy.** Energy due to motion. (See Work.)

**L.** Liter; a metric unit of capacity. A liter is about the same as one quart (0.946 L).

**Label.** Any written, printed, or graphic material that is displayed on or affixed to containers of hazardous chemicals.

**Lacquer.** A colloidal dispersion or solution of nitrocellulose, or similar film-forming compounds, resins, and plasticizers in solvents and diluents used as a protective and decorative coating for various surfaces.

**Lapping.** The operation of polishing or sanding surfaces such as metal or glass to a precise dimension.

**Laser.** The acronym for Light Amplification by Stimulated Emission of Radiation. The gas laser is a type in which laser action takes place in a gas medium, usually a continuous wave (CW) laser, which has a continuous output—with an off time of less than 1% of the pulse duration time.

**Laser light region.** A portion of the electromagnetic spectrum that includes ultraviolet, visible, and infrared light.

**Latex.** Original meaning: milky extract from rubber tree, containing about 35% rubber hydrocarbon; the remainder is water, proteins, and sugars. This word also is applied to water emulsions of synthetic rubbers or resins. In emulsion paints, the film-forming resin is in the form of latex.

**Lathe.** A machine tool used to perform cutting operations on wood or metal by the rotation of the workpiece.

**LC.** Lethal concentration; a concentration of a substance being tested that will kill a test animal.

**LD.** Lethal dose; a dosage of a substance being tested that will kill a test animal.

**Lead poisoning.** Poisoning produced when lead compounds are swallowed or inhaled. Inorganic lead compounds commonly cause symptoms of lead colic and lead anemia. Organic lead compounds can attack the nervous system.

**Leakage radiation.** Radiation emanating from diagnostic equipment in addition to the useful beam; also, radiation produced when the exposure switch or timer is not activated.

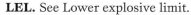

**LEL.** See Lower explosive limit.

**Lethal.** Capable of causing death.

**LFL.** Lower flammable limit. See LEL.

**Liquefied petroleum gas.** A compressed or liquefied gas usually composed of propane, some butane, and lesser quantities of other light hydrocarbons and impurities; obtained as a by-product in petroleum refining. Used chiefly as a fuel and in chemical synthesis.

**Liquid.** A state of matter in which the substance is a formless fluid that flows in accordance with a law of gravity.

**Liter (L).** A metric measure of capacity—one quart equals 0.9 L.

**Local exhaust.** A system for capturing and exhausting contaminants from the air at the point where the contaminants are produced.

**Local exhaust ventilation.** A ventilation system that captures and removes contaminants at the point they are being produced before they escape into the workroom air.

**Loudness.** The intensive attribute of an auditory sensation, in terms of which sounds may be ordered on a scale extending from soft to loud. Loudness depends primarily on the sound pressure of the stimulus, but it also depends on the frequency and wave form of the stimulus.

**Low-pressure tank.** A storage tank designed to operate at pressures at more than 0.5 psig but not more than 15 psig (3.5 to 103 kPa).

**Lower explosive limit (LEL).** The lower limit of flammability of a gas or vapor at ordinary ambient temperatures expressed in percent of the gas or vapor in air by volume.

**LP-gas.** See Liquefied petroleum gas.

**Lumen.** The flux on one square foot of a sphere, one foot in radius, with a light source of one candle at the center that radiates uniformly in all directions.

**M.** Meter; a unit of length in the metric system.

One meter is about 39 inches.

**M³.** Cubic meter; a metric measure of volume, about 35.3 cubic feet or 1.3 cubic yards.

**Makeup air.** Clean, tempered outdoor air supplied to a workspace to replace air removed by exhaust ventilation or some industrial process.

**Manometer.** Instrument for measuring pressure; essentially a U-tube partially filled with a liquid (usually water, mercury, or a light oil), so constructed that the amount of displacement of the liquid indicates the pressure being exerted on the instrument.

**Maser.** Microwave Amplification by Stimulated Emission of Radiation. When used in the term "optical maser," it is often interpreted as molecular amplification by stimulated emission of radiation.

**Material Safety Data Sheet (MSDS).** Written or printed material concerning a hazardous chemical that is prepared in accordance with paragraph (g) of the Hazard Communication Standard (HCS).

**Maximum permissible concentration (MPC).** The concentrations set by the National Committee on Radiation Protection (NCRP). They are recommended maximum average concentrations of radionuclides to which a worker may be exposed, assuming that he or she works eight hours a day, five days a week, 50 weeks a year.

**Maximum permissible dose (MPD).** Currently, a permissible dose is defined as the dose of ionizing radiation that, in light of present knowledge, is not expected to cause appreciable bodily injury to a person at any time during his or her lifetime.

**Maximum use concentration (MUC).** The product of the protection factor of the respiratory protection equipment and the permissible exposure limit (PEL).

**Mechanical filter respirator.** A respirator used to protect against airborne particulate matter like dusts, mists, metal fume, and smoke. Mechanical filter respirators do not provide protection against gases, vapors, or oxygen-deficient atmospheres.

**Mechanical ventilation.** A powered device, such as a motor-driven fan or vacuum hose attachment, for exhausting contaminants from a workplace, vessel, or enclosure.

**Mega.** One million—for example, 1 megacurie = 1 million curies.

**Melt.** In glass industry, the total batch of ingredients that may be introduced into a pot or furnace.

**Melting point.** The temperature at which a solid substance changes to a liquid state.

**Membrane filter.** A filter medium made from various polymeric materials, such as cellulose, polyethylene, or tetrapolyethylene. Membrane filters usually exhibit narrow ranges of effective pore diameters and, therefore, are very useful in collecting and sizing microscopic and submicroscopic particles and in sterilizing liquids.

**Meter.** See M.

**Mev.** Million electron volts.

**mg.** Milligram; a metric unit of weight. There are 1,000 milligrams in one gram (g) of a substance. One gram is equivalent to almost $^4/_{100}$ of an ounce.

**mg/kg.** Milligrams per kilogram; an expression of toxicological dose.

**mg/m³.** Milligrams per cubic meter; a unit used to express concentrations of dusts, gases, fumes, or mists in air.

**Mica.** A large group of silicates of varying composition, but similar in physical properties. All have excellent cleavage and can be split into very thin sheets. Used in electrical insulation.

**Microphone.** An electroacoustic transducer that responds to sound waves and delivers essentially equivalent electric waves.

**Milliampere.** $^1/_{1000}$ of an ampere.

**Milligram (mg).** A unit of weight in the metric system. One thousand milligrams equal one gram. (See mg.)

**Milligrams per cubic meter (mg/m³).** Unit used to measure air concentrations of dusts, gases, mists, and fumes. (See mg/m³.)

**Milliliter (mL).** A metric unit used to measure volume. One milliliter equals one cubic centimeter.

**Millimeters of mercury (mmHg).** The unit of pressure equal to the pressure exerted by a column of liquid mercury one millimeter high at a standard temperature. (See mmHg.)

**Millwright.** A mechanic engaged in the erection and maintenance of machinery.

**Mineral pitch.** Tar from petroleum or coal in distinction to wood tar.

**Mineral spirits.** A petroleum fraction with boiling range between 149 and 230 C (300 and 400 F).

**Mist.** Suspended liquid droplets generated by condensation from the gaseous to the liquid state or by breaking up a liquid into a dispersed state, such as by splashing, foaming, or atomizing.

**Mixture.** A combination of two or more substances that may be separated by mechanical means. The components may not be uniformly dispersed. (See also Solution.)

**mL.** Milliliter; a metric unit of capacity, equal in volume to one cubic centimeter (cc), or about $^1/_{16}$ of a cubic inch. There are 1,000 milliliters in one liter (L).

**mmHg.** Millimeters (mm) of mercury (Hg); a unit of measurement for low pressures or partial vacuums.

**Mold.** (1) A growth of fungi forming a furry patch, as on stale bread or cheese. (2) A hollow form or matrix into which molten material is poured to produce a cast.

**Monaural hearing.** Refers to hearing with one ear only.

**MPE.** Maximum permissible exposure.

**MPL.** May be either maximum permissible level, or limit, or dose. It refers to the tolerable dose rate of humans exposed to nuclear radiation.

**MSDS.** Material Safety Data Sheet.

**MSHA.** The Mine Safety and Health Administration of the U.S. Department of Labor; federal agency with safety and health regulatory and enforcement authority for the mining industry established by the Mine Safety and Health Act.

**Mutation.** A transformation of the gene that may alter characteristics of the offspring.

**Naphthas.** Hydrocarbons of the petroleum type that contain substantial portions of paraffins and naphthalenes.

**Narcosis.** A state of stupor, unconsciousness, or arrested activity produced by the influence of narcotics or other chemicals.

**Natural gas.** A combustible gas composed largely of methane and other hydrocarbons with variable amounts of nitrogen and noncombustible gases.

**Natural radioactivity.** The radioactive background or, more properly, the radioactivity that is associated with the heavy naturally occurring elements.

**Natural ventilation.** Air movement caused by wind, temperature difference, or other nonmechanical factors.

**Nausea.** An unpleasant sensation that often precedes vomiting.

**Neurotoxin.** Chemicals that produce their primary effect on the nervous system.

**Neutral wire.** Wire carrying electrical current back to the source, thus completing a circuit.

**Neutralize.** To eliminate potential hazards by inactivating strong acids, caustics, and oxidizers. For example, acids can be neutralized by adding an appropriate amount of caustic substance.

**NFPA.** The National Fire Protection Association; a voluntary organization whose aim is to promote and improve fire protection and prevention. The NFPA publishes 16 volumes of codes known as the National Fire Codes.

**NIOSH.** The National Institute for Occupational Safety and Health; a federal agency that conducts research on health and safety concerns, tests and certifies respirators, and trains occupational health and safety professionals.

**Noise.** Any unwanted sound.

**Noise-induced hearing loss.** The terminology used to refer to the slowly progressive inner ear hearing loss that results from exposure to continuous noise over a long period of time, as opposed to hearing loss from acoustic trauma or physical injury to the ear.

**Nonferrous metal.** Metal (such as nickel, brass, or bronze) that does not include any appreciable amount of iron.

**Nonflammable.** Not easily ignited, or if ignited, not burning rapidly.

**Nonionizing radiation.** Electromagnetic radiation that does not cause ionization. Includes ultraviolet, laser, infrared, microwave, and radiofrequency radiation.

**Nonsparking tools.** Tools made from beryllium—or copper or aluminum—bronze that greatly reduce the possibility of igniting dusts, gases, or flammable vapors. Although these tools may emit some sparks when striking metal, the sparks have a low heat content and are not likely to ignite most flammable liquids.

**Nonvolatile matter.** The portion of a material that does not evaporate at ordinary temperatures.

**NRC.** (1) National Response Center; a notification center of the Coast Guard that must be contacted when significant oil or chemical spills or other environmentally related incidents occur; (2) Nuclear Regulatory

Commission of the U.S. Department of Energy.

**Nuclear energy.** The energy released in a nuclear reaction, such as fission or fusion; mistakenly called "atomic energy."

**Nuclear reaction.** Result of the bombardment of a nucleus with atomic or subatomic particles or very high energy radiation.

**Nuclear reactor.** A machine for producing a controlled chain reaction in fissionable material. It is the heart of nuclear power plants where it serves as a source of heat.

**Nuisance dust.** Has long history of little adverse effect on the lungs and does not produce significant organic disease or toxic effect when exposures are kept under reasonable control.

**Odor.** That property of a substance that affects the sense of smell.

**Ohm.** The unit of electrical resistance.

**Ohm's Law.** The steady current through certain electrical circuits is directly proportional to the applied electromotive force (voltage).

**Oil dermatitis.** Blackheads and acne caused by oils and waxes that plug the hair follicles and sweat ducts.

**Olefins.** A class of unsaturated hydrocarbons characterized by relatively great chemical activity. Obtained from petroleum and natural gas. Examples: butene, ethylene, and propylene.

**Olfactory.** Relating to the sense of smell. The olfactory organ in the nasal cavity is the sensing element that detects odors and transmits information to the brain through the olfactory nerves.

**Ophthalmologist.** A physician who specializes in the structure, function, and diseases of the eye.

**Oral.** Used in or taken into the body through the mouth.

**Oral toxicity.** Adverse effects resulting from taking a substance into the body via the mouth. Typically used to denote effects in experimental animals.

**Organic.** Term used to designate chemicals that contain carbon. To date nearly one million organic compounds have been synthesized or isolated. Many occur in nature; others are produced by chemical synthesis. (See also Inorganic.)

**Organic peroxide.** An organic compound that contains the bivalent -O-O- structure and which may be considered to be a structural derivative of hydrogen peroxide, where one or both of the hydrogen atoms have been replaced by an organic radical.

**Orifice.** The opening that serves as an entrance and/or outlet of a body cavity or organ, especially the opening of a canal or a passage.

**Orifice meter.** A flow meter, employing as the measure of flow rate the difference between the pressures measured on the upstream and downstream sides of a restriction within a pipe or duct.

**Oscillation.** The variation, usually with time, of the magnitude of a quantity with respect to a specified reference when the magnitude is alternately greater and smaller than the reference.

**OSHA.** Occupational Safety and Health Administration; federal agency in the U.S. Department of Labor with safety and health regulatory and enforcement authority for most U.S. industry and business.

**Osmosis.** The passage of fluid through a semipermeable membrane as a result of osmotic pressure.

**Overexposure.** Exposure to a hazardous material beyond the allowable exposure levels.

**Oxidation.** Process of combining oxygen with some other substance; technically, a chemical change in which an atom loses one or more electrons whether or not oxygen is involved. Opposite of reduction.

**Oxidizer.** A chemical, other then a blasting agent

or explosive, that initiates or promotes combustion in other materials, thereby causing a fire either by itself or through the release of oxygen or other gases.

**Oxygen deficiency.** An atmosphere having less than the percentage of oxygen found in normal air.

**Paraffins, Paraffin series.** Those straight- or branched-chain hydrocarbon components of crude oil and natural gas whose molecules are saturated (i.e., carbon atoms attached to each other by single bonds) and therefore very stable. Examples: methane and ethane.

**Particle.** A small discrete mass of solid or liquid matter.

**Particulate.** A particle of solid or liquid matter.

**Particulate matter.** A suspension of fine solid or liquid particles in air, such as dust, fog, fume, mist, smoke, or sprays. Particulate matter suspended in air is commonly known as an aerosol.

**PEL.** Permissible exposure limit; the legally enforced exposure limit for a substance established by OSHA regulatory authority. The PEL indicates the permissible concentration of air contaminants to which nearly all workers may be repeatedly exposed eight hours a day, 40 hours a week, over a working lifetime (30 years) without adverse health effects.

**Pelleting.** In various industries, a material in powder form that may be made into pellets or briquettes for convenience. The pellet is a distinctly small briquette. (See Pelletizing.)

**Pelletizing.** Refers primarily to extrusion by pellet mills. The word "pellet," however, carries its lay meaning in that it is also applied to other small extrusions and to some balled products. Pellets are generally regarded as being larger than grains but smaller than briquettes.

**Percent impairment of hearing.** Percent hearing loss; an estimate of a person's ability to hear correctly. It is usually based, by means of an arbitrary rule, on the pure tone audiogram. The specific rule for calculating this quantity from the audiogram varies from state to state according to rule or law.

**Percent volatile.** The percentage of a liquid or solid (by volume) that will evaporate at an ambient temperature of 70 F (unless some other temperature is stated). Examples: butane, gasoline, and paint thinner (mineral spirits) are 100 % volatile; their individual evaporation rates vary, but over a period of time each will evaporate completely.

**Permanent disability** (or permanent impairment). Any degree of permanent, nonfatal injury. It includes any injury that results in the partial loss, to complete loss of use, of any part of the body, or any permanent impairment of functions of the body or a part thereof.

**Permissible exposure limit (PEL).** An exposure limit that is published and enforced by U.S. OSHA as a legal standard.

**Personal protective equipment.** Devices worn by workers to protect against hazards in the environment.

**Pesticides.** General term for that group of chemicals used to control or kill such pests as rats, insects, fungi, bacteria, weeds, etc., that prey on man or agricultural products. Among these are insecticides, herbicides, fungicides, rodenticides, miticides, fumigants, and repellents.

**Petrochemical.** A term applied to chemical substances produced from petroleum products and natural gas.

**pH.** Means used to express the degree of acidity or alkalinity of a solution with neutrality indicated at seven.

**Phosphors.** Materials capable of absorbing energy from suitable sources, such as visible light, cathode rays, or ultraviolet radiation, and then emitting a portion of the energy in the ultraviolet, visible, or infrared region of the electromagnetic spectrum. In short, they are fluorescent or luminescent.

**Physical hazard.** A chemical for which there is scientifically valid evidence that it is a combustible liquid, a compressed gas, explosive, flammable, an organic peroxide, an oxidizer, pyrophoric, unstable, or water reactive.

**Pig.** (1) A container (usually lead) used to ship or store radioactive materials. The thick walls protect the person handling the container from radiation. (2) In metal refining, a small ingot from the casting of blast furnace metal.

**Pigment.** A finely divided, insoluble substance that imparts color to the material to which it is added.

**Pilot plant.** Small-scale operation preliminary to a major enterprise. Common in the chemical industry.

**Pink noise.** Noise that has been weighted, especially at the low end of the spectrum, so that the energy per band (usually octave band) is about constant over the spectrum.

**Pitot tube.** A device consisting of two concentric tubes, one serving to measure the total or impact pressure existing in an airstream, the other to measure the static pressure only. When the annular space between the tubes and the interior of the center tube are connected across a pressure-measuring device, the pressure difference automatically nullifies the static pressure, and the velocity pressure alone is registered.

**Plasma arc welding (PAW).** An arc-welding process that produces coalescence of metals by heating them with a constricted arc between an electrode and the workpiece (transferred arc) or the electrode and the constricting nozzle (nontransferred arc). Shielding is obtained by the hot, ionized gas issuing from the orifice, which may be supplemented by an auxiliary source of shielding gas. Shielding gas can be an inert gas or a mixture of gases. Pressure may or may not be used, and filler metal may or may not be supplied.

**Plastic.** Officially defined as any one of a large group of materials that contains, as an essential ingredient, an organic substance of large molecular weight. Two basic types: thermosetting (irreversibly rigid) and thermoplastic (reversibly rigid). Before compounding and processing, plastics often are referred to as (synthetic) resins. Final form may be as film, sheet, solid, or foam; flexible or rigid.

**Plenum chamber.** An air compartment connected to one or more ducts or connected to a slot in a hood used for air distribution.

**Plutonium.** A heavy element that undergoes fission under the impact of neutrons. It is a useful fuel in nuclear reactors. Plutonium cannot be found in nature, but can be produced and "burned" in reactors.

**Point source.** A source of radiation whose dimensions are small enough compared with distance between source and receptor for those dimensions to be neglected in calculations.

**Poison.** A material introduced into a reactor core to absorb neutrons. Any substance that, when taken into the body, is injurious to health.

**Poison, Class A.** A DOT hazard class for extremely dangerous poisons, that is, poisonous gases or liquids of such nature that a very small amount of the gas or vapor of the liquid, mixed with air is dangerous to life. Some examples: phosgene, cyanogen, hydrocyanic acid, nitrogen peroxide.

**Poison, Class B.** A DOT hazard class for liquid, solid, paste, or semisolid substances—other than Class A poisons or irritating materials—that are known (or presumed on the basis of animal tests) to be so toxic to man as to afford a hazard to health during transportation. Some examples: arsenic, beryllium chloride, cyanide, mercuric oxide.

**Pollution.** Contamination of soil, water, or atmosphere beyond that which is natural.

**Polymer.** A high molecular weight material formed by the joining together of many simple molecules (monomers). There may be hundreds or even thousands of the original molecules linked end to end and often crosslinked. Rubber and cellulose are naturally occurring polymers. Most resins are chemically produced polymers.

**Portal.** Place of entrance.

**Portland cement.** See Cement, Portland.

**Positive displacement pump.** Any type of air mover pump in which leakage is negligible, so that the pump delivers a constant volume of fluid, building up to any pressure necessary to deliver that volume (unless, of course, the motor stalls or the pump breaks).

**Potential energy.** Energy due to position of one body with respect to another or to the relative parts of the same body.

**Power.** Time rate at which work is done; units are the watt (one joule per second) and the horsepower (33,000 foot-pounds per minute). One horsepower equals 746 watts.

**PPE, personal protective equipment.** Includes items such as gloves, goggles, respirators, and protective clothing.

**ppm.** Parts per million; part of air by volume of vapor or gas or other contaminant.

**Precision.** The degree of agreement of repeated measurements of the same property, expressed in terms of dispersion of test results about the mean result obtained by repetitive testing of a homogeneous sample under specified conditions.

**Pressure.** Force applied to, or distributed over a surface; measured as force per unit area. (See Atmospheric pressure, Gage pressure, Standard temperature and pressure, Static pressure, and Velocity pressure.)

**Pressure vessel.** A storage tank or vessel designed to operate at pressures greater than 15 psig (103 kPa).

**PRF laser.** A pulsed recurrence frequency laser, which is a pulsed-typed laser with properties similar to a CW laser if the frequency is very high.

**Probe.** A tube used for sampling or for measuring pressures at a distance from the actual collection or measuring apparatus. It is commonly used for reaching inside stacks or ducts.

**Produce.** To manufacture, process, formulate, or repackage a chemical.

**Propagation of flame.** The spread of flame through the entire volume of a flammable vapor-air mixture from a single source of ignition.

**Protection factor (PF).** With respiratory protective equipment, the ratio of the ambient airborne concentration of the contaminant to the concentration inside the facepiece.

**Protective atmosphere.** A gas envelope surrounding the part to be brazed, welded, or thermal sprayed, with the gas composition controlled with respect to chemical composition, dew point, pressure, flow rate, etc.

**Protective coating.** A thin layer of metal or organic material, such as paint, applied to a surface primarily to protect it from oxidation, weathering, and corrosion.

**psi.** Pounds per square inch. For technical accuracy, pressure must be expressed as psig (pounds per square inch gauge) or psia (pounds per square inch absolute); that is, gauge pressure plus sea level atmospheric pressure, or psig plus about 14.7 pounds per square inch. (See also mmHg.)

**Pulsed laser.** A class of laser characterized by operation in a pulsed mode, i.e., emission occurs in one or more flashes of short duration (pulse length).

**Pumice.** A natural silicate, such as volcanic ash or lava. Used as an abrasive.

**Push-pull hood.** A hood consisting of an air supply system on one side of the contaminant source blowing across the source and into an exhaust hood on the other side.

**Pyrethrum.** A pesticide obtained from the dried, powdered flowers of the plant of the same name; mixed with petroleum distillates, it is used as an insecticide.

**Pyrophoric.** A chemical that will ignite spontaneously in air at a temperature of 130 F (54.4 C) or below.

**Q fever.** Disease caused by rickettsial organism that infects meat and livestock handlers; similar but not identical to tick fever.

**Q-switched laser** (also known as Q-spoiled). A pulsed laser capable of extremely high peak powers for very short durations (pulse length of several nanoseconds).

**Quality.** A term used to describe the penetrating power of x rays or gamma rays.

**Quartz.** Vitreous, hard, chemically resistant, free silica, the most common form in nature. The main constituent in sandstone, igneous rocks, and common sands.

**Quenching.** A heat-treating operation in which metal raised to the desired temperature is quickly cooled by immersion in an oil bath.

**rad.** Radiation absorbed dose.

**Radiation (nuclear).** The emission of atomic particles or electromagnetic radiation from the nucleus of an atom.

**Radiation (thermal).** The transmission of energy by means of electromagnetic waves longer than visible light. Radiant energy of any wavelength may, when absorbed, become thermal energy and result in the increase in the temperature of the absorbing body.

**Radiation protection guide (RPG).** The radiation dose that should not be exceeded without careful consideration of the reasons for doing so; every effort should be made to encourage the maintenance of radiation doses as far below this guide as is practical.

**Radiation source.** An apparatus or a material emitting or capable of emitting ionizing radiation.

**Radiator.** That which is capable of emitting energy in wave form.

**Radioactive.** The property of an isotope or element that is characterized by spontaneous decay to emit radiation.

**Random noise.** A sound or electrical wave whose instantaneous amplitudes occur as a function of time, according to a normal (Gaussian) distribution curve. Random noise is an oscillation whose instantaneous magnitude is not specified for any given instant of time.

**Rated line voltage.** The range of potentials in volts of an electrical supply line.

**Reaction.** A chemical transformation or change; the interaction of two or more substances to form new substances.

**Reactive.** See Unstable.

**Reactivity.** A description of the tendency of a substance to undergo chemical reaction with the release of energy. Undesirable effects—such as pressure buildup, temperature increase, formation of noxious, toxic, or corrosive by-products—may occur because of the reactivity of a substance to heating, burning, direct contact with other materials, or other conditions in use or in storage.

**Reagent.** Any substance used in a chemical reaction to produce, measure, examine, or detect another substance.

**Recoil energy.** The emitted energy that is shared by the reaction products when a nucleus undergoes a nuclear reaction, such as fission or radioactive decay.

**Reducing agent.** In a reduction reaction (which always occurs simultaneously with an oxidation reaction), the reducing agent is the chemical or substance that (1) combines with oxygen or (2) loses electrons to the reaction.

**Refractory.** A material that is especially resistant to the action of heat, such as fireclay, magnetite, graphite, and silica; used for lining furnaces, etc.

**Regimen.** A regulation of the mode of living, diet, sleep, exercise, etc., for a hygienic or therapeutic purpose; sometimes mistakenly called regime.

**Relative humidity.** See Humidity.

**Reliability.** The degree to which an instrument, component, or system retains its performance characteristics over a period of time.

**Rem.** Roentgen equivalent man, a dose unit that equals the dose in rads multiplied by the appropriate value of RBE for the particular radiation.

**Resistance.** (1) In electricity, any condition that retards current or the flow of electrons; it is measured in ohms. (2) Opposition to the flow of air, as through a canister, cartridge, particulate filter, or orifice. (3) A property of conductors, depending on their dimensions, material, and temperature, which determines the current produced by a given difference in electrical potential.

**Respirable size particulates.** Particles in the size range that permits them to penetrate deep into the lungs upon inhalation.

**Respirator.** A device to protect the wearer from inhalation of harmful contaminants.

**Respiratory protection.** Devices that will protect the wearer's respiratory system from overexposure by inhalation of airborne contaminants.

**Respiratory system.** Consists of (in descending order)—the nose, mouth, nasal passages, nasal pharynx, pharynx, larynx, trachea, bronchi, bronchioles, air sacs (alveoli) of the lungs, and muscles of respiration.

**Responsible party.** Someone who can provide additional information on the hazardous chemical and appropriate emergency procedures, if necessary.

**Reverbatory furnace.** A furnace in which heat is supplied by burning of fuel in a space between the charge and a low roof.

**Riser.** In metal casting, a channel in a mold to permit escape of gases.

**Roentgen (R).** A unit of radioactive dose or exposure (pronounced rentgen). (See rad.)

**Route of entry.** The path by which chemicals can enter the body; primarily inhalation, ingestion, and skin absorption.

**Rosin.** Specifically applies to the resin of the pine tree and chiefly derives from the manufacture of turpentine. Widely used in the manufacture of soap, flux.

**Rotameter.** A flow meter, consisting of a precision-bored, tapered, transparent tube with a solid float inside.

**Rotary kiln.** Any of several types of kilns used to heat material, such as in the Portland cement industry.

**Rouge.** A finely powdered form of iron oxide used as a polishing agent.

**Safety.** The control of recognized hazards to attain an acceptable level of risk.

**Safety can.** An approved container, of not more than 19 L (5 gal) capacity, having a spring-closing lid and spout cover, and so designed that it will safely relieve internal pressure when subjected to fire exposure.

**Safety program.** Activities designed to assist employees in the recognition, understanding, and control of hazards in the workplace.

**Salamander.** A small furnace, usually cylindrical in shape, without grates, used for heating.

**Salt.** A product of the reaction between an acid and a base.

**Sampling.** The withdrawal or isolation of a fractional part of a whole.

**Sandblasting.** A process for cleaning metal castings and other surfaces with sand by a high-pressure airstream.

**Sandhog.** Any worker doing tunneling work requiring atmospheric pressure control.

**Sanitize.** To reduce the microbial flora in or on articles, such as eating utensils, to levels judged safe by public health authorities.

**SCBA.** See Self-contained breathing apparatus.

**Sealed source.** A radioactive source sealed in a container or having a bonded cover, where the container or cover has sufficient mechanical strength to prevent contact with and dispersion of the radioactive material under the conditions of use and wear for which it was designed.

**Self-contained breathing apparatus (SCBA).** A respiratory protection device that consists of a supply or a means of obtaining respirable air, oxygen, or oxygen-generating material carried by the wearer.

**Self-ignition.** See Auto-ignition temperature.

**Sensible.** Capable of being perceived by the sense organs.

**Sensitivity.** The minimum amount of contaminant that can repeatedly be detected by an instrument.

**Sensitization.** The process of rendering an individual sensitive to the action of a chemical.

**Sensitizer.** A substance which, on first exposure, causes little or no reaction in man or test animals but which, on repeated exposure, may cause a marked response not necessarily limited to the contact site.

**Sensory feedback.** Use of external signals perceived by sense organs (e.g., eye, ear) to indicate quality or level of performance of an event triggered by voluntary action.

**Shakeout.** In the foundry industry, the separation of the solid, but still not cold, casting from its molding sand.

**Shale.** Many meanings in industry, but in geology a common fossil rock formed from clay, mud, or silt somewhat stratified but without characteristic cleavage.

**Shale oil.** Some shale is bituminous and on distillation yields a tarry oil.

**Shield, shielding.** Interposed material (like a wall) that protects workers from harmful radiations released by radioactive materials.

**Shielded metal arc welding (SMAW).** An arc-welding process that produces coalescence of metals by heating them with an arc between a covered metal electrode and the work. Shielding is obtained from decomposition of the electrode covering. Pressure is not used and filler metal is obtained from the electrode.

**Shock.** Primarily the rapid fall in blood pressure following an injury, operation, or the administration of anesthesia.

**Shootblasting.** A process for cleaning of metal castings or other surfaces by small steel shot in a high-pressure airstream. This process is a substitute for sandblasting to avoid silicosis.

**Short-term exposure limit (STEL).** ACGIH-recommended exposure limit. Maximum concentration to which workers can be exposed for a short period of time (15 minutes) for only four times throughout the day with at least one hour between exposures.

**Silica gel.** A regenerative absorbent consisting of the amorphous silica manufactured by the action of HCl on sodium silicate. Hard, glossy, quartzlike in appearance. Used in dehydrating and in drying and as a catalyst carrier.

**Silicates.** Compounds of silicon, oxygen, and one or more metals with or without hydrogen. Silicate dusts cause nonspecific dust reactions, but generally do not interfere with pulmonary function or result in disability.

**Silicon.** A nonmetallic element being, next to oxygen, the chief elementary constituent of the earth's crust.

**Silicones.** Unique group of compounds made

by molecular combination of the element silicon or certain of its compounds with organic chemicals. Produced in a variety of forms, including silicone fluids, resins, and rubber. Silicones have special properties, such as water repellency, wide temperature resistance, high durability, and great dielectric strength.

**Silver solder.** A solder of varying components but usually containing an appreciable amount of cadmium.

**Sintering.** Process of making coherent powder of earthy substances by heating, but without melting.

**Skin absorption.** Ability of some hazardous chemicals to pass directly through the skin and enter the bloodstream.

**Skin dose.** A special instance of tissue dose, referring to the dose immediately on the surface of the skin.

**Skin sensitizer.** See Sensitizer.

**Skin toxicity.** See Dermal toxicity.

**Slag.** The dross of flux and impurities that rise to the surface of molten metal during melting and refining.

**Sludge.** In general, any muddy or slushy mass.

**Slurry.** A thick, creamy liquid resulting from the mixing and grinding of limestone, clay, and other raw materials with water.

**Smelting.** One step in the procurement of metals from ore—hence, to reduce, to refine, to flux, or to scorify.

**Smog.** Irritating hazard resulting from the sun's effect on certain pollutants in the air.

**Smoke.** An air suspension (aerosol) of particles, originating from combustion or sublimation.

**Soaking pit.** A device in steel manufacturing in which ingots with still molten interiors stand in a heated, upright chamber until solidification is complete.

**Soap.** Ordinarily a metal salt of a fatty acid, usually sodium stearate, sodium oleate, sodium palmitate, or some combination of these.

**Soapstone.** Complex silicate of varied composition similar to some talcs with wide industrial application such as in rubber manufacture.

**Solder.** A material used for joining metal surfaces together by filling a joint or covering a junction.

**Solid-state laser.** A type of laser that utilizes a solid crystal such as ruby or glass. This type is most commonly used in impulse lasers.

**Solubility in water.** A term expressing the percentage of a material (by weight) that will dissolve in water at ambient temperature. Solubility information can be useful in determining spill cleanup methods and fire-extinguishing agents and methods for diluting a material. Terms used to express solubility are:

Negligible: Less than 0.1%

Slight: 0.1 to 1.0%

Moderate: 1 to 10%

Appreciable: More than 10% Complete

**Solution.** Mixture in which the components lose their identities and are uniformly dispersed. All solutions consist of a solvent (water or other fluid) and the substance dissolved, called the "solute." A true solution is homogeneous, such as salt in water.

**Solvent.** A substance that dissolves another substance.

**Soot.** Agglomerations of particles of carbon impregnated with "tar," formed in the incomplete combustion of carbonaceous material.

**Sorbent(s).** (1) A material that removes toxic gases and vapors from air inhaled through a canister or cartridge. (2) Material used to collect gases and vapors during air-sampling. (3) Nonreactive materials used to clean up chemical spills. Examples: clay and vermiculite.

**Sound.** An oscillation in pressure, stress, particle displacement, particle velocity, etc., which is propagated in an elastic material, in a medium with internal forces (e.g., elastic, viscous), or the superposition of such propagated oscillations.

**Sound absorption.** The change of sound energy into some other form, usually heat, as it passes through a medium or strikes a surface.

**Sound analyzer.** A device for measuring the band-pressure level or pressure-spectrum level of a sound as a function of frequency.

**Sound level.** A weighted sound pressure level, obtained by the use of metering characteristics and the weighting A, B, or C specified in ANSI S1.4.

**Sound-level meter and octave-band analyzer.** Instruments for measuring sound pressure levels in decibels referenced to 0.0002 microbar.

**Sound pressure level (SPL).** The level, in decibels, of a sound is 20 times the logarithm to the base 10 of the ratio of the pressure of the sound to the reference pressure. The reference pressure must be explicitly stated.

**Sour gas.** Slang for either natural gas or a gasoline contaminated with odor-causing sulfur compounds. In natural gas, the contaminant is usually hydrogen sulfide; in gasolines, it is usually mercaptans.

**Source.** Any substance that emits radiation. Usually refers to a piece of radioactive material conveniently packaged for scientific or industrial use.

**Spasm.** Tightening or contraction of any set(s) of muscles.

**Special fire-fighting procedures.** Special procedures and/or personal protective equipment that is necessary when a particular substance is involved in a fire.

**Specific chemical identity.** The chemical name, Chemical Abstracts Service (CAS) Registry Number, or any other information that reveals the precise chemical designation of a substance.

**Specific gravity.** The weight of a material compared to the weight of an equal volume of water; an expression of the density (or heaviness) of the material. (See Gravity, weight.)

**Specific weight.** The weight per unit volume of a substance, same as density.

**Spectrophotometer.** An instrument used for comparing the relative intensities of the corresponding colors produced by chemical reaction.

**Spectrum.** The distribution in frequency of the magnitudes (and sometimes phases) of the components of a wave. Spectrum also is used to signify a continuous range of frequencies, usually wide in extent, within which waves have some specified common characteristics. Also, the pattern of red-to-violet light observed when a beam of sunlight passes through a prism and then projects upon a surface.

**Speech interference level (SIL).** The speech interference level of a noise is the average, in decibels, of the sound pressure levels of the noise in the three octave bands of frequency 600-1,200, 1,200-2,400, and 2,400-4,800 Hz.

**Speech perception test.** A measurement of hearing acuity by the administration of a carefully controlled list of words. The identification of correct responses is evaluated in terms of norms established by the average performance of normal listeners.

**Spontaneously combustible.** A material that ignites as a result of retained heat from processing, or which will oxidize to generate heat and ignite, or which absorbs moisture to generate heat and ignite.

**Spot welding.** One form of electrical-resistance welding in which the current and pressure are restricted to the spots of metal surfaces directly in contact.

**Spray coating painting.** The result of the application of a spray in painting as a substitute for brush painting or dipping.

**Stability.** An expression of the ability of a material to remain unchanged. For MSDS purposes, a material is stable if it remains in the same form under expected and reasonable conditions of storage or use. Conditions that may cause instability (dangerous change) are stated—i.e., temperatures above 150 F, shock from dropping, etc.

**Stain.** A dye used to color microorganisms as an aid to visual inspection.

**Stamping.** Many different usages in industry, but commonly, the crushing of ores by pulverizing.

**Standard air.** Air at standard temperature and pressure. The most common values are 21.1 C (70 F) and 101.3 kPa (29.92 in. Hg).

**Standard conditions.** In industrial ventilation, 21.1 C (70 F), 50% relative humidity, and 101.3 kPa (29.92 in. of mercury) atmosphere pressure.

**Standard Industrial Classification (SIC) Code.** A classification system, by the U.S. Government, for places of employment according to major type of activity.

**Standard man.** A theoretical physically fit man of standard (average) height, weight dimensions, and other parameters (blood composition, percentage of water, mass of salivary glands, to name a few), used in studies of how heat or ionizing radiation affects humans.

**Standard temperature and pressure.** See Standard air.

**Static pressure.** The potential pressure exerted in all directions by a fluid at rest.

**STEL.** Short-term exposure limit; ACGIH terminology. See TLV.

**Sterile.** Free of living microorganisms.

**Sterility.** Inability to reproduce.

**Sterilization.** The process of making sterile; the killing of all forms of life.

**Stink damp.** In mining, hydrogen sulfide.

**Stress.** A physical, chemical, or emotional factor that causes bodily or mental tension and may be a factor in disease causation or fatigue.

**Stressor.** Any agent or thing causing a condition of stress.

**Strip mine.** A mine in which coal or ore is extracted from the earth's surface after removal of overlayers of soil, clay, and rock.

**Supplied-air respirators.** Air-line respirators or self-contained breathing apparatus.

**Supplied-air suit.** A one- or two-piece suit that is impermeable to most particulate and gaseous contaminants and is provided with an adequate supply of respirable air.

**Surface-active agent or surfactant.** Any of a group of compounds added to a liquid to modify surface of interfacial tension. In synthetic detergents, which is the best known use of surface-active agents, reduction of interfacial tension provides cleansing action.

**Surface coating.** Term used to include paint, lacquer, varnish, and other chemical compositions used for protecting and/or decorating surfaces. See Protective coating.

**Suspect carcinogen.** A material that is believed to be capable of causing cancer but for which there is limited scientific evidence.

**Sweating.** (1) Visible perspiration; (2) the process of uniting metal parts by heating solder so that it runs between the parts.

**Sweetening.** The process by which petroleum products are improved in odor by chemically changing certain sulfur compounds of objectionable odor into compounds having little or no odor.

**Swing grinder.** A large power-driven grinding wheel mounted on a counterbalanced swivel-supported arm guided by two handles.

**Synthesis.** The reaction or series of reactions by which a complex compound is obtained from simpler compounds or elements.

**Synthetic.** (From Greek work synthetikos—that which is put together.) "Man-made 'synthetic' should not be thought of as a substitute for the natural," states Encyclopedia of the Chemical Process Industries; it adds, "Synthetic chemicals are frequently more pure and uniform than those obtained naturally." Example: synthetic indigo.

**Synthetic detergents.** Chemically tailored cleaning agents soluble in water or other solvents. Originally developed as soap substitutes. Because they do not form insoluble precipitates, they are especially valuable in hard water. They may be composed of surface-active agents alone, but generally are combinations of surface-active agents and other substances, such as complex phosphates, to enhance detergency.

**Synthetic rubber.** Artificially made polymer with rubberlike properties. Various types have varying composition and properties. Major types designated as S-type, butyl, neoprene (chloroprense polymers), and N-type. Several synthetics duplicate the chemical structure of natural rubber.

**Systemic toxicity.** Adverse effects caused by a substance that affects the body in a general rather than local manner.

**Tailings.** In mining or metal recovery processes, the gangue rock residue from which all or most of the metal has been extracted.

**Talc.** A hydrous magnesium silicate used in ceramics, cosmetics, paint, and pharmaceuticals, and as filler in soap, putty, and plaster.

**Tall oil.** (Name derived from Swedish word tallolja; material first investigated in Sweden—not synonymous with U.S. pine oil).

Natural mixture of rosin acids, fatty acids, sterols, high-molecular weight alcohols, and other materials derived primarily from waste liquors of sulfate wood pulp manufacture. Dark brown, viscous, oily liquid often called liquid rosin.

**Tar.** A loose term embracing wood, coal, or petroleum exudations. In general, represents complex mixture of chemicals of top fractional distillation systems.

**Tar crude.** Organic raw material derived from distillation of coal tar and used for chemicals.

**Tare.** A deduction of weight, made in allowance for the weight of a container or medium. The initial weight of a filter, for example.

**Temper.** To relieve the internal stresses in metal or glass and to increase ductility by heating the material to a point below its critical temperature and cooling slowly.

**Tempering.** The process of heating or cooling makeup air to the proper temperature.

**Temporary threshold shift (TTS).** The hearing loss suffered as the result of noise exposure, all or part of which is recovered during an arbitrary period of time when one is removed from the noise.

**Temporary total disability.** An injury that does not result in death or permanent disability, but which renders the injured person unable to perform regular duties or activities on one or more calendar days after the day of injury. (This is a definition established by OSHA.)

**Tendon.** Fibrous component of a muscle. It frequently attaches at the area of application of tensile force. When its cross section is small, stresses in the tendon are high, particularly because the total force of many muscle fibers is applied at the single terminal tendon. (See Tenosynovitis.)

**Tenosynovitis.** Inflammation of the connective tissue sheath of a tendon.

**Teratogen.** A substance or agent to which exposure of a pregnant female can result in malformations in the fetus. An example is the drug thalidomide.

**Terminal velocity.** The terminal rate of fall of a particle through a fluid as induced by gravity or other external force; the rate at which frictional drag balances the accelerating force (or the external force).

**Thermal pollution.** Discharge of heat into bodies of water to the point that increased warmth activates all sewage, depletes the oxygen the water needs to cleanse itself, and eventually destroys some of the fish and other organisms in the water.

**Thermonuclear reaction.** A fusion reaction, that is, a reaction in which two light nuclei combine to form a heavier atom, releasing a large amount of energy. This is believed to be the sun's source of energy. It is called thermonuclear because it occurs only at a very high temperature.

**Thermoplastic.** Capable of being repeatedly softened by heat.

**Thermosetting.** Capable of undergoing a chemical change from a soft to a hardened substance when heated.

**Thermosetting plastics.** Those plastics that are heat-set in their final processing to a permanently hard state. Examples: phenolics, ureas, and melamines.

**Thinner.** A liquid used to increase the fluidity of paints, varnishes, and shellac.

**Threshold.** The level where the first effects occur; also the point at which a person just begins to notice a tone is becoming audible.

**Threshold limit value (TLV).** A guideline for exposure developed by ACGIH. Refers to airborne concentrations of substances and represents conditions under which it is believed that nearly all workers may be repeatedly exposed day after day without adverse effect. Some workers may be affected at, or below, the TLV limits.

**Threshold limit value ceiling (TLV-C).** The ceiling level of the exposure that should never be exceeded. This value has been established as the maximum level to be used in computing the TWA and STEL limits. The ceiling value is under the IDLH limit for a given substance.

**Threshold limit value short-term exposure limit (TLV-STEL).** An exposure level considered safe to work in for short periods of time (15 minutes), four times a day maximum, with at least 60 minutes between exposures. No irritation or other adverse effects should be experienced. This level is higher than the TLV-TWA for a given substance.

**Threshold limit value time-weighted average (TLV-TWA).** The time-weighted average concentration of a substance for a normal 8-hour workday and a 40-hour workweek to which nearly all workers may be repeatedly exposed, day after day, without adverse effect.

**Time-weighted average concentration** (Refers to concentrations of airborne toxic materials that have been weighted for a certain time duration, usually 8 hours.

**Tinning.** Any work with tin, such as tin roofing; but in particular in soldering, the primary coating with solder of the two surfaces to be united.

**TLV-UEL.** See Upper explosive limit. (Also see Lower explosive limit.)

**Tolerance.** (1) The ability of a living organism to resist the usually anticipated stress. (2) Also, the limits of permissible inaccuracy in the fabrication of an article above or below its design specifications.

**Tolerance dose.** See Maximum permissible concentration and MPL.

**Toxic substance.** Any substance that can cause acute or chronic injury to the human body, or that is suspected of being able to cause diseases or injury under some conditions.

**Toxicity.** The sum of adverse effects resulting from exposure to a material, generally by the mouth, skin, or respiratory tract.

**Toxin.** A poisonous substance that is derived from an organism.

**Trade name.** The commercial name or trademark by which something is known.

**Trade secret.** Any confidential formula pattern, process, device, information or compilation of information (including chemical name or other unique chemical identifier) that is used in an employer's business, and that gives the employer an opportunity to obtain an advantage over competitors who do not know or use it.

**Trauma.** An injury or wound brought about by an outside force.

**TSCA.** Toxic Substances Control Act; federal environmental legislation, administered by EPA, for regulating the manufacture, handling, and use of materials classified as "toxic substances."

**Tumbling.** An industrial process, such as in founding, in which small castings are cleaned by friction in a rotating drum (tumbling mill, tumbling barrel), which may contain sand, sawdust, stone, etc.

**Turbid.** Cloudy.

**Turning vanes.** Curved pieces added to elbows or fan inlet boxes to direct air and so reduce turbulence losses.

**Ultraviolet.** Those wavelengths of the electromagnetic spectrum that are shorter than those of visible light and longer than x rays, 10 cm to 100 cm wavelength.

**UN number.** A registry number assigned to dangerous commonly carried goods by the United Nations Committee of Experts on the Transport of Dangerous Goods. The UN number is required in shipping documentation and on packaging as part of the DOT regulations for shipping hazardous materials.

**Unsafe condition.** That part of the work environment that contributed or could have contributed to an incident.

**Unstable.** A chemical that, in the pure state, or as produced or transported, will vigorously polymerize, decompose, condense, or become self-reactive under conditions of shock, pressure, or temperature. Such chemicals are also referred to as reactive.

**Upper explosive limit (UEL).** The highest concentration (expressed in percent vapor or gas in the air by volume) of a substance that will burn or explode when an ignition source is present.

*USC.* United States Code, the official compilation of federal statutes.

**USDA.** U.S. Department of Agriculture.

**Use.** To package, handle, react, or transfer a substance.

**Valve.** A device that controls the direction of air or fluid flow or the rate and pressure at which air or fluid is delivered, or both.

**Vapor.** The gaseous form of substances that are normally in the solid or liquid state (at room temperature and pressure).

**Vat dyes.** Water insoluble, complex coal tar dyes that can be chemically reduced in a heated solution to a soluble form that will impregnate fibers. Subsequent oxidation then produces insoluble color dyestuffs that are remarkably fast to washing, light, and chemicals.

**Vector.** (1) Term applied to an insect or any living carrier that transports a pathogenic microorganism from the sick to the well, inoculating the latter; the organism may or may not pass through any developmental cycle. (2) Anything (e.g., velocity, mechanical force, electromotive force) having magnitude, direction, and sense that can be represented by a straight line of appropriate length and direction.

**Velocity.** A vector that specifies the time rate of change or displacement with respect to a reference.

**Velocity pressure.** The kinetic pressure in the direction of flow necessary to cause a fluid at rest to flow at a given velocity. When added to static pressure, it gives total pressure.

**Velometer.** A device for measuring air velocity.

**Ventilation.** Circulating fresh air to replace contaminated air.

**Ventilation, dilution.** Airflow designed to dilute contaminants to acceptable levels.

**Ventilation, mechanical.** Air movement caused by a fan or other air-moving device.

**Ventilation, natural.** Air movement caused by wind, temperature difference, or other nonmechanical factors.

**Vibration.** An oscillation motion about an equilibrium position produced by a disturbing force.

**Vinyl.** A general term applied to a class of resins such as polyvinyl chloride, acetate, butyral, etc.

**Viscose rayon.** The type of rayon produced from the reaction of carbon disulfide with cellulose and the hardening of the resulting viscous fluid by passing it through dilute sulfuric acid, causing the evolution of hydrogen sulfide gas.

**Viscosity.** The tendency of a fluid to resist internal flow without regard to its density.

**Visual acuity.** Ability of the eye to sharply perceive the shape of objects in the direct line of vision.

**Volatile.** The percentage of a liquid or solid (by volume) that will evaporate at an ambient temperature of 70 F (unless some other temperature is stated). Examples: butane, gasoline, and paint thinner (mineral spirits) are 100% volatile; their individual evaporation rates vary, but over a period of time each will evaporate completely.

**Volt.** The practical unit of electromotive force or difference in potential between two points in an electrical field. Just as pressure in a water pipe causes water to flow, electrical pressure—or voltage—pushes the current of electrons through wires to receptacles, such as light fixtures and outlets.

**Volume flow rate.** The quantity (measured in units of volume) of a fluid flowing per unit of time, as cubic feet per minute, gallons per hour, or cubic meters per second.

**Vulcanization.** Process of combining rubber (natural, synthetic, or latex) with sulfur and accelerators in presence of zinc oxide under heat and usually pressure in order to change the material permanently from a thermoplastic to a thermosetting composition, or from a plastic to an elastic condition. Strength, elasticity, and abrasion resistance also are improved.

**Water column.** A unit used in measuring pressure. (See also Inches of water column.)

**Water curtain or waterfall booth.** In spray painting, the water running down a wall into which excess paint spray is drawn or blown by fans; the water carries the paint downward to a collecting point.

**Waterproofing agents.** These usually are formulations of three distinct materials: (1) a coating material, (2) a solvent, and (3) a plasticizer. Among the materials used in waterproofing are cellulose esters and ether, polyvinyl chloride resins or acetates, and variations of vinyl chloride-vinylidine chloride polymers.

**Water-reactive.** A chemical that reacts with water to release a gas that is either flammable or presents a health hazard.

**Watt (w).** A unit of electrical power equal to one joule per second.

**Weight.** The force with which a body is attracted toward the earth.

**Weld.** A localized coalescence of metals or nonmetals produced either by heating the materials to suitable temperatures, with or without the application of pressure, or by the application of pressure alone, and with or without the use of filler material.

**Welding.** The several types of welding are electric arc welding, oxyacetylene welding, spot welding, and inert or shielded gas welding utilizing helium or argon. The hazards involved in welding stem from (1) the fumes from the weld metal, such as lead or cadmium metal, (2) the gases created by the process, or (3) the fumes or gases arising from the flux.

**Welding rod.** A rod or heavy wire that is melted and fused into metals in arc welding.

**White damp.** In mining, carbon monoxide.

**White noise.** A noise whose spectrum density (or spectrum level) is substantially independent of frequency over a specified range.

**Work.** When a force acts against a resistance to produce motion in a body, the force is said to do work. Work is measured by the product of the force acting and the distance moved against a resistance. The units of measurement are the erg (the joule is $1 \times 10^7$ ergs) and the footpound.

**Work area.** A room or defined space in a workplace where hazardous chemicals are produced or used and where employees are present.

**Workers.** All persons gainfully employed, including owners, managers, other paid employees, the self-employed, and unpaid family workers, but excluding private household workers.

**Workers' Compensation.** Insurance plan that compensates injured workers or their survivors.

**Work-hardening.** The property of a metal to become harder and more brittle on being "worked," that is, bent repeatedly or drawn.

**Work hours.** The total number of hours worked by all employees.

**Work injuries** (including occupational illnesses). Those that arise out of and in the course of gainful employment regardless of where the incident occurs. Excluded are work injuries to private household workers and injuries occurring in connection with farm chores, which are classified as home injuries.

**Workplace.** An establishment, job site, or project at one geographical location containing one or more work areas.

**Work strain.** The natural physiological response reaction of the body to the application of work stress. The locus of the reaction may often be remote from the point of application of work stress. Work strain is not necessarily traumatic, but may appear as trauma when excessive, either directly or cumulatively, and must be considered by the industrial engineer in equipment and task design. Thus, a moderate increase of heart rate is nontraumatic work strain resulting from physical work strain if caused by undue work stress on the wrists.

**Work stress.** Biomechanically, any external force acting on the body during the performance of a task. It always produces work strain. Application of work stress to the human body is the inevitable consequence of performance of any task, and is, therefore, only synonymous with "stressful work conditions" when excessive. Work stress analysis is an integral part of task design.

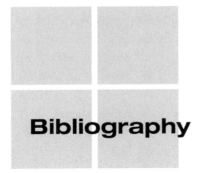

# Bibliography

American Conference of Governmental Industrial Hygienists, 1330 Kemper Meadow Drive, Cincinnati, OH 45240. *Threshold Limit Values for Chemical Substances and Physical Agents in the Workroom Environment* (Issued annually). www.acgih.org.

American Industrial Hygiene Association, 2700 Prosperity Avenue, Suite 250, Fairfax, VA 22031. *Industrial Ventilation—A Manual of Recommended Practice* (Latest edition). www.aiha.org.

American National Standards Institute, 25 West 43rd Street, 4th floor, New York, NY 10036. "Catalog of American National Standards" (Issued annually). www.ansi.org.

American Red Cross National Headquarters, 2025 E Street, N.W., Washington, DC 20006. *First Aid Textbook* (Latest edition). www.redcross.org.

American Society for Training and Development (ASTD), 1640 King Street, Box 1443, Alexandria, VA 22313. *Literature Index* (Quarterly). www.astd.org.

American Welding Society (AWS), 550 LeJeune Road, N.W., P.O. Box 351040, Miami, FL 33135. *Welding Handbook*. www.aws.org.

Bird, F.E., Jr. *Management Guide to Loss Control*. Loganville, GA: Institute Publications ILCI, 1974.

Bird, F.E., Jr., and R.G. Loftus. *Loss Control Management*. Loganville, GA: Institute Publications, ILCI, 1976.

Blanchard, K., et al. *The One Minute Manager Gets Fit*. New York: Nightingale-Conant, 1989.

"Care and Operating Instructions for Various Electric Tools." Milwaukee Electric Tool Corp., 13135 West Lisbon Road, Brookfield, WI 53005-2550. www.milwaukeetool.com.

CHEMTREC: Chemical Transportation Emergency Center. 1300 Wilson Blvd., Arlington, VA 22209. www.chemtrec.com.

Computer Databases:

CHRIS: Special Chemical Hazards Response Information Systems. www.chrismanual.com/Intro/othersys.htm.

CAMEO: Computer-Aided Management of Emergency Operations.

www.epa.gov/oem/content/cameo/index.htm.

Currance, P.L. and A.C. Bronstein. *Emergency Care for Hazardous Materials Exposure No. 2*. St. Louis, MO: Mosby, 1994.

DeCristoforo, R.J. *Complete Book of Stationary Power Tool Techniques*. New York: Sterling, 1988.

Employee Involvement Association (formerly National Association of Suggestion Systems (NASS)), P.O. Box 2307, Dayton OH 45401-2307. www.eianet.org.

"Engineering Anthropometry," G. Savendy, (editor). *Handbook of Human Factors/Ergonomics*. New York: John M. Wiley & Sons, 1987, pp. 154–68.

Factory Mutual Engineering Corporation, 1151 Boston-Providence Turnpike, Norwood, MA 02062. *Factory Mutual System Approval Guide* (Annual). *Handbook of Property Conservation. Loss Prevention Data Books* (Periodic). www.fmglobal.com.

Fallon, W.K. (editor). *AMA Management Handbook*. 2nd edition. New York: AMACOM, 1983.

Ferry, T.S. *Modern Accident Investigation and Analysis*. 2nd edition. New York: John M. Wiley & Sons, 1988.

Firenze, R.J. *The Process of Hazard Control*. Dubuque, IA: Kendall/Hunt Publishing Company, 1978.

Grandjean, E. *Fitting the Task to the Man: An Ergonomics Approach*. 4th edition. London: Taylor and Francis, 1988.

Hammer, W. *Occupational Safety Management and Engineering*. 4th edition. Englewood Cliffs, NJ: Prentice-Hall, Inc., 1989.

*Hawley's Condensed Chemical Dictionary,* 12th edition. New York: Reinhold Publishing Corporation, 1993.

Heinrich, H.W., D. Petersen, and N. Roos. *Industrial Accident Prevention: A Safety Management Approach*. 5th rev. edition. New York: McGraw-Hill Book Company, 1980.

Human Factors Section, Eastman Kodak Laboratory. *Ergonomics Design for People at Work,* vol. 1. New York: Van Nostrand Reinhold, 1983.

Illuminating Engineering Society, 120 Wall Street, 17th floor, New York, NY 10005-4001. *IES Lighting Handbook (The Standard Lighting Guide). Practice for Industrial Lighting* (ANSI/IES RP7). www.ies.org.

Johnson, W.G. *MORT Safety Assurance Systems*. New York: Marcel Dekker, 1980.

Kroemer, K.H.E. "Coupling the hand with the handle," *Human Factors* 28 (1986): 337–39.

Mosby, C.V. International Agency for Research on Cancer (IARC) Monographs. www.iarc.fr.

National Archives and Records Administration, Office of the Federal Register, 800 North Capitol Street, N.W., Suite 700, Washington DC 20001. *Code of Federal Regulations.* www.archives.gov/federal-register/.

National Fire Protection Association, 1 Batterymarch Park, Quincy, MA 02269. *Fire Protection Handbook. Fire Protection Guide on Hazardous Materials. Flammable and Combustible Liquids Code Handbook. Industrial Fire Hazards Handbook. Inspection Manual. Life Safety Code Handbook. National Electrical Code Handbook.* "Standards and Recommended Practices." (Catalog available). www.nfpa.org.

National Institute of Occupational Safety and Health (NIOSH), 1600 Clifton Road, Atlanta, GA 30333. *Certified Personal Protective Equipment.* "Criteria Documents." *The Industrial Environment: Its Evaluation and Control. Machine Guarding—Assessment of Need.* "Publications Catalog." *Safety Program Practices in High versus Low Accident Rate Companies. Toxic Substances List.* "Welding Safely." Public Health Service, Centers for Disease Control, Department of Health and Human Services. www.niosh.org.

National Safety Council, 1121 Spring Lake Drive, Itasca, IL 60143-3201. *Injury Facts* (Annual). *Accident Prevention Manual for Business & Industry* (Multiple volumes). *Electrical Inspection Illustrated. Family Safety and Health* (Quarterly). *Fundamentals of Industrial Hygiene. Safeguarding Concepts Illustrated. Occupational Safety & Health Data Sheets. Power Press Safety Manual. Safety and Health* (Monthly).

"Office Ergonomics: Work Station Dimensions," in D.C. Alexander, and B.M. Pulat, (editors). *Industrial Ergonomics.* Norcross, GA: Institute of Industrial Engineers, 1985, pp. 187–201.

Ottoboni, M.A. *The Dose Makes the Poison: A Plain Language Guide to Toxicology.* 2nd edition. New York: Van Nostrand Reinhold, 1991.

Petersen, D.C. *Techniques of Safety Management.* 3rd edition. Goshen, NY: Aloray, Inc., 1989.

Prevent Blindness America. 211 West Wacker Drive, Suite 1700, Chicago, IL 60606 www.preventblindness.org.

Sax, N.I. and R.J. Lewis. *Dangerous Properties of Industrial Materials.* 9th edition, 3 vols. New York: Van Nostrand Reinhold, 1993.

Simonds, R.H. and J.V. Grimaldi. *Safety Management.* 5th edition. Homewood, IL: Richard D. Irwin, Inc., 1989.

Steil, L., et al. *Listening: It Can Change Your Life.* New York: John M. Wiley & Sons, 1984.

"Testing individual capability to lift materials: Repeatability of a dynamic test compared with static testing," *Journal of Safety Research* 16 (1985): 1–7.

TOXNET: Toxicology Data Network. National Library of Medicine, 8600 Rockville Pike, Bethesda, MD 20894. http://toxnet.nlm.nih.gov.

Underwriters Laboratories Inc. 2600 N.W. Lake Rd., Camas, WA 98607-8542. www.ul.com/global/eng/pages.

U.S. Department of Commerce, www.commerce.gov.

U.S. Department of Labor, Occupational Safety and Health Administration (OSHA), 200 Constitution Avenue, N.W., Washington, DC 20210. *An Illustrated Guide to Electrical Safety. A Brief Guide to Recordkeeping Requirements for Occupational Injuries and Illnesses.* www.osha.gov.

"VDT workstation design," in Helander, M.G., (editor) *Handbook of Human Computer Interaction.* New York: Elsevier Publications, 1989.

Wolvin, A.D. and C. Coakley. *Listening.* 4th edition. Dubuque, IA: Brown and Benchmark, 1992.

# Index